Springer Series on Polymer and Composite Materials

Series editor

Susheel Kalia, Dehradun, India

More information about this series at http://www.springer.com/series/13173

Deepalekshmi Ponnamma
Kishor Kumar Sadasivuni · Chaoying Wan
Sabu Thomas · Mariam Al-Ali AlMa'adeed
Editors

Flexible and Stretchable Electronic Composites

Springer

Editors
Deepalekshmi Ponnamma
Center for Advanced Materials
Qatar University
Doha
Qatar

Sabu Thomas
School of Chemical Sciences
Mahatma Gandhi University
Kottayam
India

Kishor Kumar Sadasivuni
Center for Advanced Materials
Qatar University
Doha
Qatar

Mariam Al-Ali AlMa'adeed
Center for Advanced Materials
Qatar University
Doha
Qatar

Chaoying Wan
International Institute for
Nanocomposites Manufacturing
University of Warwick
Coventry
UK

ISSN 2364-1878 ISSN 2364-1886 (electronic)
Springer Series on Polymer and Composite Materials
ISBN 978-3-319-23662-9 ISBN 978-3-319-23663-6 (eBook)
DOI 10.1007/978-3-319-23663-6

Library of Congress Control Number: 2015950005

Springer Cham Heidelberg New York Dordrecht London

Printed on acid-free paper

Springer International Publishing AG Switzerland is part of Springer Science+Business Media (www.springer.com)

Contents

Natural Polyisoprene Composites and Their Electronic Applications

Deepalekshmi Ponnamma, Kishor Kumar Sadasivuni, K.T. Varughese, Sabu Thomas and Mariam Al-Ali AlMa'adeed

Abstract Rubber-based composites have been recognized as efficient materials for the fabrication of technologically important products. Various particles are successfully incorporated into cis-polyisoprene or natural rubber (NR) in recent years both in solution and in melt forms. Potential electronic applications of such composites specifically containing carbon nanotubes, graphene, graphene-like structures, fibers, metallic fillers, and inorganic fillers have been realized in this article. Advanced performances of NR composites obtained via different methods are compared with those of the neat polymer. Special attention is paid to the structural changes occurring in the matrix under the influence of fillers. Other issues regarding the technology limitations, research challenges, and future trends are also discussed. The main objective of this review is threefold: (1) to present the latest electronic applications of NR composite technology and development, (2) to describe the need for fundamental research in this field, and (3) to outline major challenges in rubber composite preparation. At first an overview of NR composites, then their preparation methods, and thereafter their applications are described. In short, other than summarizing different classes of particles filled NR composites and their applications, this review highlights different ways to create smaller, cheaper, lighter, and faster devices based on such materials. The developed materials are highly useful in the fields of electronics and diffusion as well as in the marine and transport industries.

Keywords Rubber · Actuator · Sensor · Electronics · Microwave absorbers · Supercapacitors

D. Ponnamma (✉) · K.K. Sadasivuni · M.A.-A. AlMa'adeed
Center for Advanced Materials, Qatar University, P.O. Box 2713, Doha, Qatar
e-mail: lekshmi_deepa@yahoo.com

K.T. Varughese
Polymer Laboratory, Dielectric Materials Division, Central Power Research Institute, Bangalore 560080, India

D. Ponnamma et al. (eds.), *Flexible and Stretchable Electronic Composites*, Springer Series on Polymer and Composite Materials,
DOI 10.1007/978-3-319-23663-6_1

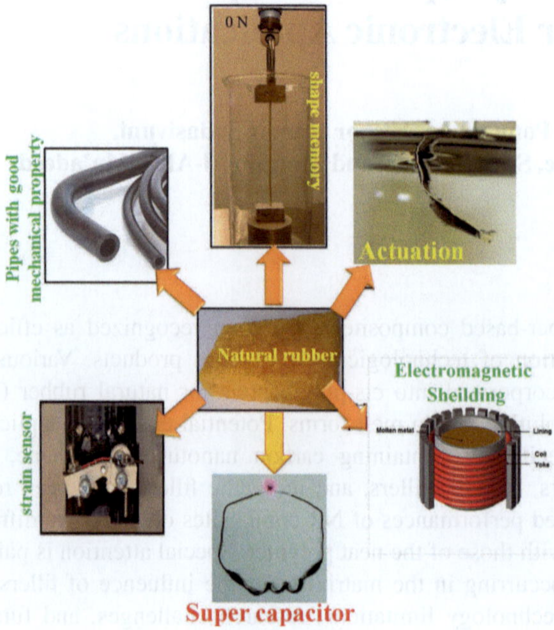

1 Introduction

Natural rubber (NR) is one of the most important biosynthesized polymers displaying excellent chemical and physical properties. Among other rubbers, it was the first industrially exploited [1, 2]. This elastomer has tremendous applications in various areas [3–5]. The source of this polymer is *Hevea Brasiliensis* tree and is present in the form of a chemical-free biomacromolecule, natural rubber latex (NRL). NRL is a stable colloidal dispersion of cis-1,4-polyisoprene (PI) of high molecular mass in aqueous medium and can be stabilized with ammonia. The highly stereo-regular microstructure of NR provides unique mechanical properties

S. Thomas
School of Chemical Sciences, Mahatma Gandhi University, Kottayam 686560, Kerala, India

S. Thomas
Center for Nanoscience and Nanotechnology, Mahatma Gandhi University, Priyadarshini Hills P.O, Kottayam 686 560, Kerala, India

M.A.-A. AlMa'adeed
Materials Science and Technology Program, Qatar University, P.O. Box 2713, Doha, Qatar

(high elasticity) and crystallizing nature [6–10] to it. NR molecules consist mainly of cis-PI units without any trans configuration, in contrast to the synthetic polyisoprenes [11]. The very flexible backbone leads to very low glass transition temperature (T_g) of about −64 to −70 °C. Raw NRL finds wide applications in manufacturing medical products such as gloves, condoms, and safety bags due to its excellent elasticity, flexibility, antivirus permeation, good formability, and biodegradability [12–14]. However, NRL products that have short shelf lives and life cycles [15, 16] especially for medical gloves and condoms, low tensile strength, poor tear resistance and low resistance to burning [17, 18] are observed. Even though these drawbacks sound no problem for the medical products, industrially important materials such as tyres require superior performance. Due to this, several attempts to incorporate graphene, carbon nanotubes (CNTs), carbon black (CB) [19], ultrafine calcium carbonate [20], modified montmorillonite (MMT) [21], silica [22], and starch [23] fillers in dry NR or NRL [24] have been reported [25, 26].

Based on the dimensions of filler materials, it can be macrosized, microsized, and nanosized. For nanosized fillers, still the classification depends on the structure or dimension. More specifically, layered clay minerals and graphene sheets come under one-dimensional fillers, nanotubes and nanofibers are two-dimensional, and spherical particles are three-dimensional [25, 27–36] based on the number of dimensions in nanorange. Nanocomposites will be formed if one of the reinforcing filler phases has its dimension in the nanometer range. For nanofillers, maximum improvement in combined characteristics is generally observed at low loading levels, without significantly increasing the density of the compound or reducing light transmission. For instance, when the microsized fillers having lower surface area at 10–30 wt% loading impart good properties to the NR matrix, the nanofillers at 2–3 wt% loading can give superior properties due to its high surface area [37, 38]. The high reinforcing efficiency of nanofilled rubber composites, even at low loading of filler, can also be attributed to the nanoscale dispersion and the very high aspect ratio (length-to-thickness ratio) of the filler [39–43]. So the use of nanofillers in regulating the performances of NR products by alleviating the aforementioned disadvantages of NR is very fascinating and has great interest. The effect of two-dimensional nanoclay with high aspect ratios on NR microstructure when stretched was studied by Carretero et al. [44, 45] and by Huang and coworkers [46, 47]. Also, the reinforcement mechanism of NR composites with nanoalumina [48], carbon nanofibers [49], cellulose nanofibers [3, 50], and kaolin and silica [51] was studied extensively. Renewable and biodegradable biomaterials such as cellulose [52–54] consist of crystalline nanofibers of 10–40 nm diameter and 1.5 g/cm^3 density with a length of few micrometers. Such nanofibers possess high value of tensile strength (10 GPa) and Young's modulus (134 GPa) [55].

Out of the several techniques of filler reinforcement of NR, melt intercalation [37, 56], latex compounding [38], and solution-mixing methods [57] are the most common. In the nanocomposites, the dispersed nanofillers make some structural reinforcement which is responsible for the enhanced chemical and physical properties. Nair et al. [58–60] have found a three-dimensional filler network within the NR matrix responsible for the high solvent resistance and mechanical properties, when nano crab chitin whiskers were used as reinforcement. However, direct methods of reinforcement have certain limitations to reinforce NR with fillers of higher specific surface such as spherical inorganic nanoparticles. This is probably because of the necessity to solve strong self-aggregation of nanoparticles prior to composite manufacture [61–63]. In order to achieve significant nanodispersion, self-assembly process is reported for NR/silica (SiO_2) nanocomposite preparation which could achieve better thermal properties [64].

Discovery of conducting fillers such as CNTs has opened up another way to fabricate conducting rubbers. CNTs can be regarded as rolled-up graphene sheets with essentially sp^2 hybridized carbon units in the form of cylinders with nano-diameters [65–70]. The extraordinary mechanical, thermal, and electrical properties of these nanomaterials make them potentially applicable in material science, optics, and electronics. Though CNTs prove to be ideal candidates to enhance NR properties [71], simple mixing methods were reported to be insufficient for great improvement. The general issues such as homogeneous CNT dispersion, orientations and aspect ratio, polymer morphology, CNT polymer, and CNT–CNT interfacial interactions have to be considered during composite fabrication [72]. Among the nanofillers, graphene layers also have remarkable importance in suturing the rubber properties, especially because these fillers are easier and cheaper to make compared to CNTs. The influence of graphene on the microwave properties of elastomeric composites is highly noteworthy [73–76]. In addition to CNTs and graphitic fillers, metal-coated fibers such as aluminum- or nickel-coated carbon or glass fiber and nickel-coated mica, polyaniline, short carbon fiber, and conductive CB impart conductivity to NR composites. The major electronic applications of such composites include various touch control switches, and strain and pressure sensors for applications such as robot hands or artificial limbs. Gaskets made from electrically conductive rubber composites can prevent electromagnetic induction (EMI) radiation leakage. Moreover, these materials are considered for use in different microelectronic devices and microwave applications, such as absorbing materials.

However, the above-mentioned problems associated with CNTs are often observed for all nanofillers. In order to fabricate high-performance composite products, the compatibility between the filler and the hydrophobic polymer matrix is really necessary and can be accomplished by physical or chemical modifications of the nanoparticle surfaces [10, 43, 77–83]. Using various techniques, a lot of technologically important materials have been fabricated out of NR-based composites. The objective of this critical review is to explore all electronic applications of NR composites (both micro and nano) and to highlight the high performance of nanocomposites.

2 NR-Based Composites

2.1 Synthesis Methods

The primary difficulty associated with the filler particles is to achieve a uniform dispersion in the matrix polymer, and due to this reason, processing remains a key challenge in fully utilizing the advantages of filler reinforcement. Nonuniform dispersion results in the formation of filler aggregates which can degrade the mechanical as well as physical performances of the composites through defect sites. Particle dispersions also depend on the viscosity of the medium, diffusion of particles, and particle–particle and particle–matrix interactions [84]. Better dispersion and superior properties can be achieved if the interparticle distance is comparable to the radius of gyration of the polymer. In the composite, the entire polymer matrix can be a part of an interphase and this influences processing as well. Anyway, common processing methods for polymer composites including solution processing [57], melt processing [56, 85], and reaction processing [84] are applicable for NR composites also. Latex compounding [38] method is employed especially for fabricating latex-based composites.

2.1.1 Solution Processing

Fillers (especially nano) are generally dispersed in a suitable solvent and then mixed with polymer solution by shear mixing, magnetic stirring, or sonication [86] in this technique. Sonication can be of bath (mild sonication) or probe (high-power sonication). Finally after mixing, the composite can be obtained by vaporizing the solvent. Figure 1 shows the flowchart for the solution processing. This method is considered as an effective way to prepare composite films with homogeneously distributed nanofillers [87]. However, since this method depends mostly on the efficient filler dispersion in the used solvent, the selection of solvent is rather significant. Often functionalization methods are employed in order to avoid reaggregation of fillers, and solvent selection is done on the basis of these functional groups and solubility of the polymer. Sometimes the latex can be directly used for solution mixing. This was done by diluting the NRL and adjusting the pH. The fillers in this case would be made in such a way that with addition, NRL composite will precipitate [88].

2.1.2 Melt Processing

The main principle behind this method is softening of rubber when heated. During this process, rubber transforms into viscous liquid, and at this state, additives can be well mixed into the melt by rheological shear mixing [87]. In addition to

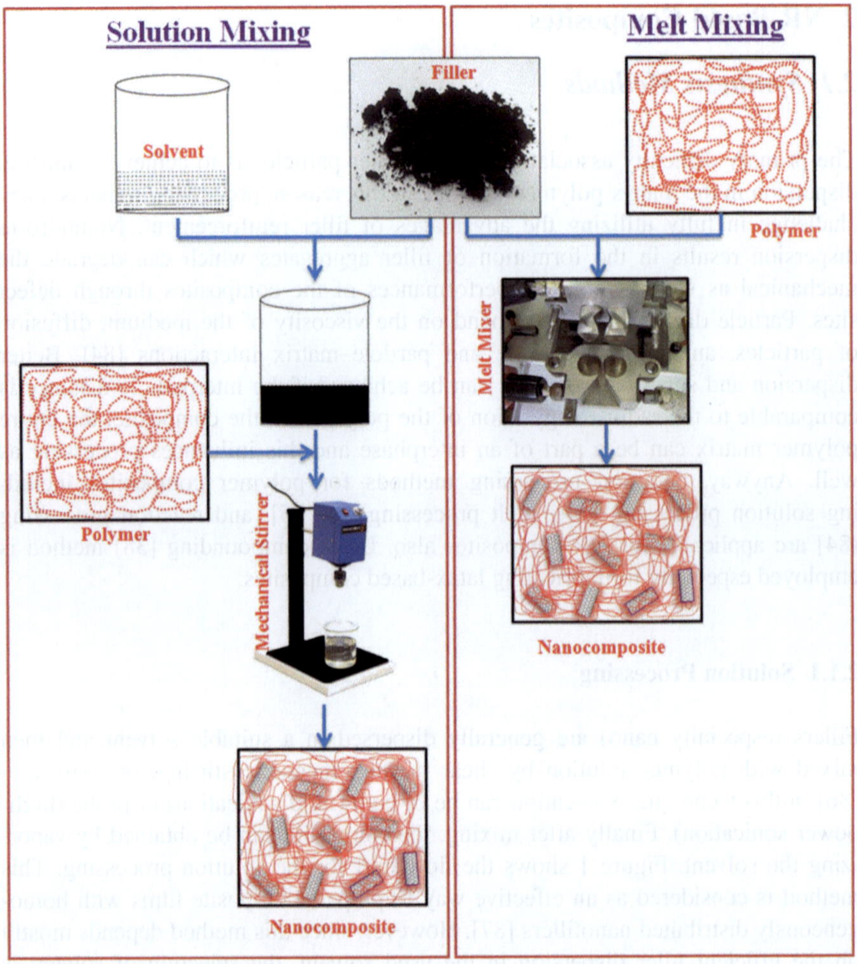

Fig. 1 Flowchart presenting the different steps of the solution and melt mixing processing of a polymer composite

the common melt mixers, large extruders are also used for bulk sample mixing. The process of melt mixing is also included in the flowchart (Fig. 1). Starch/NR/clay nanocomposites were prepared via melt extrusion technique in a high-speed mixer [10, 89]. Open two-roll mill mixing was also practiced at room temperature. The vulcanization ingredients, plasticizers if any, and fillers were added to the elastomer during mixing. NR composite containing clay or organoclay, and octadecylamine in the absence of clay was synthesized, and its effect on the cross-linking degree of NR is investigated [90]. In short, the melt mixing possesses the advantages of speed and simplicity, and compatibility with standard industrial techniques [91, 92].

2.1.3 Latex-Stage Compounding

Fillers along with curing agents such as sulfur in the presence of ZnO, stearic acid, TMTD, MBTS, etc. are compounded with NRL. At first, the curing agents are mixed (or compounded) with NRL followed by filler dispersion. This process improves cross-linking between the polymer chains in the nanocomposites, and thus, better properties can be achieved [38]. Even though in situ polymerization method is another efficient method to significantly improve the dispersion of fillers in polymers and to achieve stable filler–polymer interaction, for NR little reports exist on this. This is because of the availability of NR in its polymer form and the difficulty in controlling the polymerization of isoprene into cis-polyisoprene. Various methods to synthesize NR/CNT composites are discussed by Ponnamma et al. [42] in their review.

2.2 *NR Composite Applications*

2.2.1 Electromagnetic Interference Shielding

Metals are generally used to shield unwanted electromagnetic (EM) radiations; however, their heavy weight, high cost, and corrosive nature demand the need of cheaper polymeric materials. Though polymers are insulating, conducting nanoparticles produce their nanocomposites possessing good shielding effectiveness (SE). There is an increase in demand for the EMI shielding particularly at low-frequency range (30 kHz–1.5 GHz) due to a significant increase in the use of portable wireless devices such as cell phones and radios which interfere with increased sensitiveness [93]. Due to the structural and electrical advantages of nanocarbon materials, these are largely used to fabricate EMI shielding. Before going to the experimental details of the shielding materials, the theory behind EMI shielding is discussed here.

Figure 2 shows the experimental setup used by Zhang et al. to calculate the EMI SE of CNT-filled shape memory polymer. This setup is the same for NR samples as well. In general, all devices contain a network analyzer, transmitting horn antenna, and a receiving waveguide probe as shown in the figure. According to the

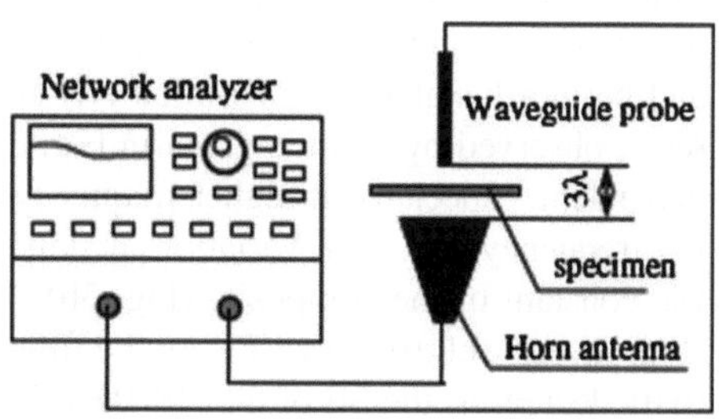

Fig. 2 The schematic of an EMI shielding effectiveness measurement system [94]. Copyright 2007. Reprinted with permission from Elsevier

frequency to be shielded, horn antenna and waveguide probe are changed. Here, the distance between the mutually perpendicular waveguide probe and the horn antenna is maintained as three times the wavelength for a given measurement frequency. The output from the network analyzer is put into the horn antenna, and the energy received after transmitting through the sample is measured by the waveguide probe. EMI SE is defined as the difference of the transmitted energy between the nanocomposite sample and a reference material. If the distance between the shield material and the source of radiation is higher than $\lambda/2\pi$, it is said to be in far field, and if the distance is less than $\lambda/2\pi$, it is in the near field [94].

The SE evaluates the reduction of EMI at a specific frequency (f) as illustrated by Eq. 1.

$$\mathrm{SE} = 10\log\frac{P_0}{P_t} = 20\log\frac{E_0}{E_t} = 20\log\frac{H_0}{H_t} \tag{1}$$

where P_0, E_0, and H_0 are the energy and electric and magnetic field intensities incident on the shield and P_t, E_t, and H_t are the counterparts transmitted through the shield.

During EMI shielding, reflection, absorption, and re-reflection of the radiation can occur. These reflections as well as absorption loss depend on the electrical conductivity (σ) and magnetic permeability (μ), and such multiple reflection loss can be minimized when the distance between the reflecting surfaces or interfaces is larger than the skin depth (δ).

$$\delta = \frac{1}{\sqrt{\pi f \mu \sigma}} \tag{2}$$

SE can also be explained as the sum of reflection loss and absorption loss as shown in Eq. 3

$$\mathrm{SE} = 20\ \log\frac{Z_0\delta\sigma}{2\sqrt{2}} + 8.68\frac{t}{\delta} \tag{3}$$

where t is the thickness of the material and $Z_0 = 120\ \pi$. If $t/\delta \leq 1.3$, the equation reduces to

$$\mathrm{SE} = 20\ \log\left(1 + \frac{Z_0 t\sigma}{2}\right) \tag{4}$$

Thus, in short, the SE depends on the frequency of the radiation, thickness of the material, and resistivity. The EMI characteristics of nanocarbon particles and barium hexaferrite (BF)-filled NR are examined by Yakuphanoglu et al. The result observed by them is given in Fig. 3. The SE of the composites is frequency dependent, especially at higher-frequency range, and it increases with the increasing frequency (Fig. 3a). Variation of skin depth was in accordance with the dielectric constant of the composite (Fig. 3b). The SE also increases with filler loading (Fig. 3c) as well (0–20 wt%). As the filler loading increases and the volume resistivity decreases, the SE of the NR/BF nanocomposites increases rapidly [95].

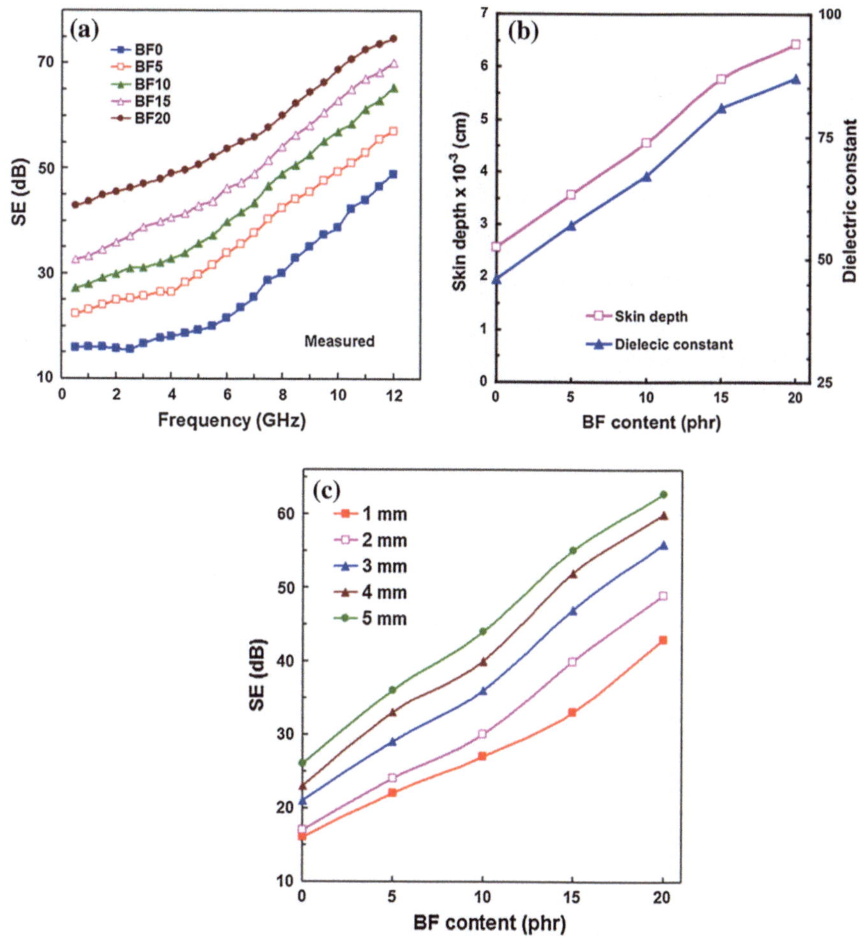

Fig. 3 **a** Shielding effectiveness as a function of frequency, **b** variation of skin depth and dielectric constant with BF content, and **c** SE with thickness of NR/BF nanocomposites [95]. Copyright 2014. Reprinted with permission from Springer

Since the EMI shielding studies are mainly based on the shielding studies of microwave radiations, this is deeply discussed in the following Sect. (2.2.2).

2.2.2 Microwave Absorption

Very recently, a large number of studies have come out on the microwave absorbing ability of NR nanocomposites. Microwave absorbing rubber composites of NR and epoxidized natural rubber (ENR), filled with conductive CB and aluminum powder, were prepared, and their shielding effectiveness was studied as shown in

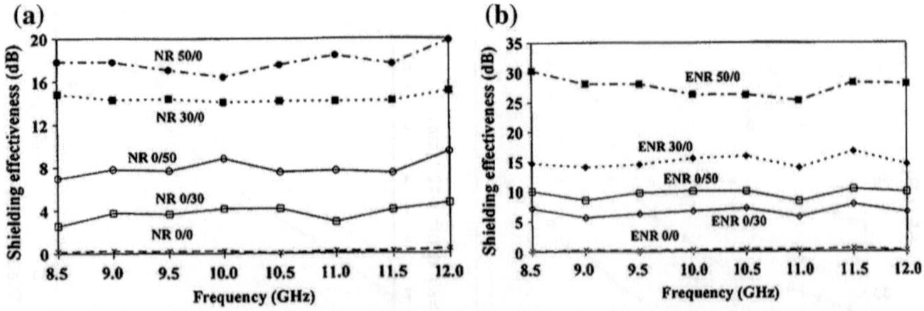

Fig. 4 Effect of single filler on SE: **a** NR composites; **b** ENR composites. The former number represents CB content, and the latter number represents aluminum powder content [98]. Copyright 2012. Reprinted with permission from Scientific Research Open Access

Fig. 4 [96]. For this, X-band microwave in the frequency range of 8–12 GHz is used. It is found that increase of filler loading-enhanced SE of the rubber composites and also the conductive CB was more effective in shielding than aluminum powder. Binary filler-filled rubber composites showed higher SE than that of single filler-filled rubber composites. SE of each sample was in the same range under the entire frequencies (8.5–12 GHz), and the effect of fillers on SE of two rubber systems was similar. Unfilled rubbers showed very low SE, <1 dB. SE of the rubber increased from <1 dB to 18–28 dB after adding 50 phr of CB or to 30–40 dB after adding 50 phr of CB and 50 phr of aluminum powder. The maximum SE of rubber composites containing 50 phr of aluminum powder (0/50 samples) was 10 dB. The SE and the mechanical properties were in the order of ENR > NR, and both these properties were correlated with electrical conductivity as well. Although the conductivity was higher, aluminum powder was less effective than conductive CB because of the lower volume fraction and larger grain size. The EMI SE of CB-filled rubber composites (50/0 samples) observed here (18–28 dB) was higher than that of other rubber composites containing 50–60 phr of the same filler CB, i.e., nitrile rubber (12 dB), ethylene vinyl acetate (EVA) (18 dB), ethylene propylene diene terpolymer (EPDM) (8–10 dB), and EVA/EPDM blend (22.5 dB).

The factors affecting the EMI SE include conductivity or polarity of rubbers and conductive fillers, filler content, particle size/shape, filler dispersion and distribution, specimen size and shape, and testing method [97]. Magnetic and microwave absorbing properties of thermoplastic natural rubber (TPNR)-filled 4–12 wt% magnetite (Fe_3O_4) nanocomposites prepared by melt blending were investigated. TPNR matrix was prepared from polypropylene (PP), NR, and liquid natural rubber (LNR) in the ratio of 70:20:10 with the LNR as the compatibilizer. The TPNR–Fe_3O_4 composites containing nanocubic spinel-structured Fe_3O_4 with 20–50 nm size showed good microwave absorption property. The absorption or minimum reflection loss (RL) continuously increases, and the dip shifts to a lower-frequency region with the increase in both filler content in nanocomposites

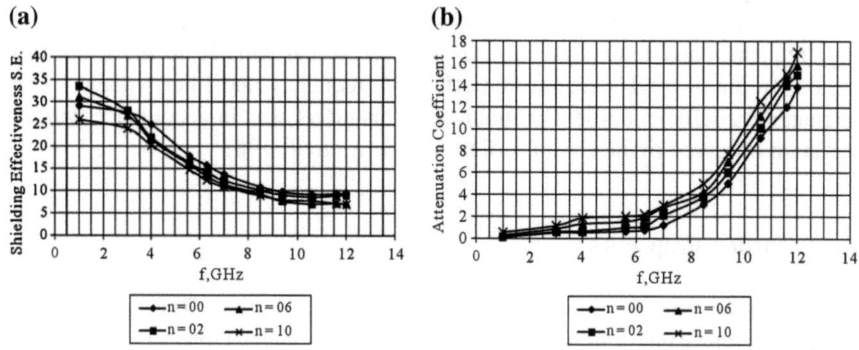

Fig. 5 Frequency dependence of **a** SE and **b** attenuation coefficient α of electromagnetic waves at various filler concentrations (n-phr of graphene) [98]. Copyright 2012. Reprinted with permission from Scientific Research Open Access

and the sample thickness. The RL is 25.51 dB at 12.65 GHz, and the absorbing bandwidth in which the RL is less than 10 dB is 2.7 GHz when the filler content is 12 wt% at 9 mm sample thickness [97]. At low frequencies, the dielectric loss in the samples is influenced by direct current conductivity, while at higher frequencies, the loss is due to alternating current conductivity. The microwave absorption properties analyzed by using peculiar absorber method reveal the existence of two matching conditions of minimum reflection due to the geometrical cancellation of the incidence and reflected waves at the surface of the absorber at low and high frequencies for the same thickness of the samples. The increment of filler content in the nanocomposites can shift the maximum attenuation to a lower frequency, increase maximum absorbing effect, and also enlarge the absorbing bandwidth. The minimum RL was found increasing and moving toward the lower-frequency region with increase in sample thickness. In short, the microwave absorbing properties of nanocomposites were determined by the thickness and the composition of filler, which can vary the dip frequency, microwave absorbing properties, and the absorbing capacity and width of the nanocomposites.

Microwave (reflection coefficient, attenuation coefficient, SE) properties of NR/GNP nanocomposites revealed interesting dependence of these parameters on frequency and filler amount (Fig. 5). The reflection coefficient increased with filler amount and frequency, whereas the attenuation coefficient was independent of frequency and filler amount in the 1–6 GHz range and increased slightly from 6 to 9 GHz and drastically in the 9–12 GHz range. The dependence was almost linear, and both the frequency and filler amount influence the SE values (10–34 dB range) as well (Fig. 5a). Since the attenuation was not great enough to compensate the increasing reflectance, the SE decreased gradually with the increasing frequency, especially up to 7 GHz. However, as a whole, the effect was more pronounced at 6–12 GHz frequency range and at 6 phr of GNP [98].

The lowering in the attenuation coefficient values (Fig. 5b) of composites is attributed to the presence of a foaming agent which makes the composite structure porous. As a whole, the attenuation coefficient slightly increases with the increasing frequency and filler amount. The increase is particularly in the 6–12 GHz region. In the 1–6 GHz region, the effect of frequency and filler amount on the attenuation coefficient is less pronounced. This effect was in accordance with the variation observed in the complex dielectric permittivity [98].

2.2.3 Sensing Properties

Sensors based on NR composites are widely reported [68]. The sensors come under different categories such as strain sensor to detect large strains, pressure sensors to detect pressure variations, and solvent and vapor sensors to detect organic solvents and volatile components. On NR substrate, thin layer of the conducting polymer, polypyrrole (PPy), is coated by vapor-phase polymerization technique in vacuum [99] using the oxidant $FeCl_3$. This material is successfully used as a strain sensor to detect large mechanical strains, the sensing performance of which can be regulated by various parameters. Prestraining the rubber substrate prior to deposition, high concentration of $FeCl_3$, and formation of monomer vapor from diluted pyrrole monomer resulted in maximum stability, consistency, and best sensing performance of the composite. Oxygen plasma exposure produced the highest reduction in water contact angle on the rubber's surface to improve the hydrophilicity without sacrificing the rubber's elasticity and flexibility. The developed strain sensor was successfully implemented on an air muscle in conjunction with a PID controller where the extension/contraction of the muscle can be accurately controlled. It also has a capability to measure large bending strains and hence be used as a rotary sensor. The performance of the sensor can be explained by examining at the strain distribution on the NR substrate during the bending motion at different angular positions (Fig. 6). The length of the NR substrate was broken down into four segments, and the physical change in length of each segment was measured. The disadvantage of this sensor is that it shows some hysteresis.

In order to understand the strain profile of the NR/PPy sensor, the electrical resistance versus time profile is taken as given in Fig. 7a. The upward shift of the electrical resistance over the time indicates that the absolute electrical resistance value cannot be used to represent the absolute strain. So normalized $\Delta R/R_0$ values

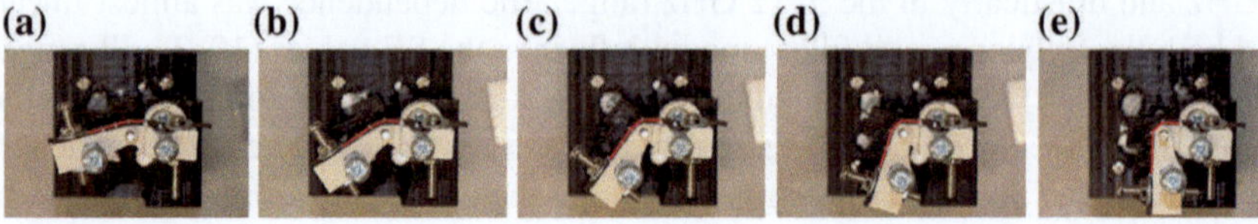

Fig. 6 The strain distribution along the length of the rubber at **a** 10°, **b** 30°, **c** 50°, **d** 70°, and **e** 90°. The different segments are represented by the *red lines* [99]. Copyright 2012. Reprinted with permission from Elsevier

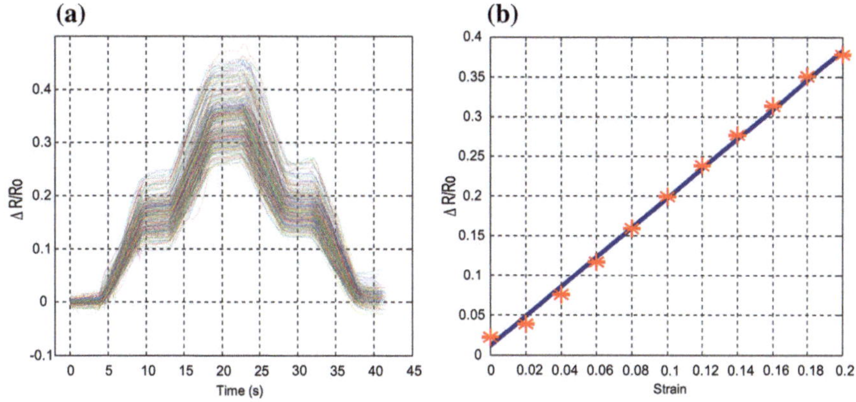

Fig. 7 **a** The change of resistance profile over 800 cycles and **b** the relationship between the strain and relative change in electrical resistance of the strain sensor from linear motion strain [99]. Copyright 2012. Reprinted with permission from Elsevier

were determined and plotted against strain (Fig. 7b). The sensitivity can be calculated by gauge factor (GF) which is defined by Eq. 5.

$$\mathrm{GF} = \left[{}^{(R - R_0)}\!/_{R_0}\right] \Big/ \left[\left({}^{L - L_0}\!/_{L_0}\right)\right] \tag{5}$$

where R is the final resistance and R_0 the initial resistance of the sample which varies its length from L_0 to L over the application of strain.

From Fig. 7b, the gauge factor for NR/PPy sensor is calculated to be approximately 1.86. The $\Delta R/R_0$ of each strain value was averaged to determine the mathematical relationship between the averaged $\Delta R/R_0$ and strain.

However, nanocomposites of NR extend stable and reversible piezoresistive strain sensing over 3000 tested cycles upon applying a small compressive strain (~16.7 %) under a relatively high frequency (0.5 Hz, 2 s/cycle). This is done by adding CNT and carbon nanofibers in NR composites pretreated with CB. The large reinforcement effect makes this composite suitable for heavy-duty pressure sensors, i.e., healthy motion monitoring of industrial machinery vibrations [100].

The strain profile of the CNT/graphite flake-sandwiched NR sensor is illustrated in Fig. 8a. The resistance increases with strain, and the CNT sensor (because of the smaller CNT size) showed higher measurable strain limit (620 %) compared with the graphite sensor (246 %). The sensitivity was calculated by GF as given in Eq. 5. The sensitivity values for the MWCNT sensors were 5.5, 10.2, 16.3, 23.1, 31.5, and 43.4 at 100, 200, 300, 400, 500, and 620 % strain, respectively, where as for the graphite sensors (Fig. 8b) the values were 23.7, 37.5, 148.1, and 346.6 at 50, 100, 200, and 246 % strain, respectively. The responses over time were checked by attaching one end of the MWCNT strain sensor to a disk, which was rotating with an angular velocity of 2 rad/s. The results obtained are given in Fig. 8c where the applied sinusoidal strain (with a period of ~3.04 s) ranged between 150 and 500 %, and the total number of cycles was more than 400 [101].

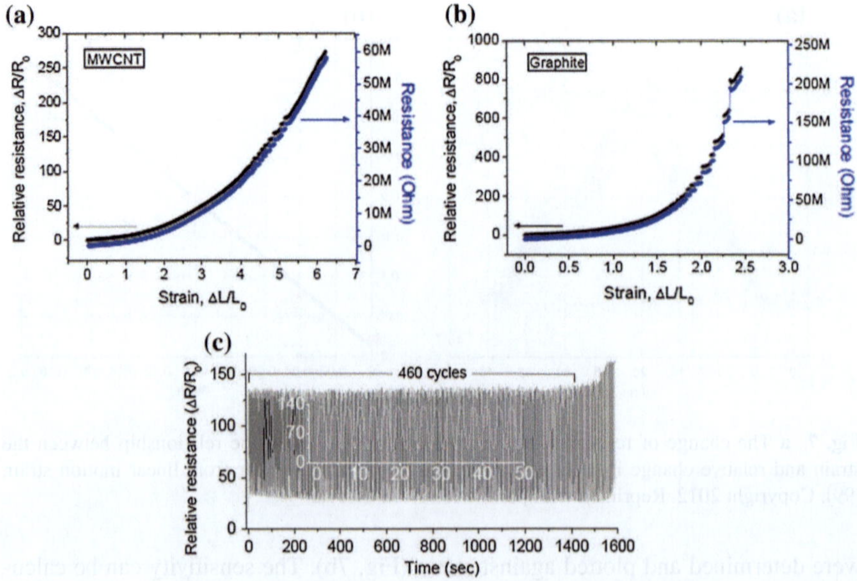

Fig. 8 Relative resistance versus strain response curves for the **a** MWCNT sensor, **b** graphite sensor, and **c** multiple cycle tests for the strain range of 150–500 % (*Inset image* the first 16 full cycles carried out over a time period of 486 s) [101]. Copyright 2014. Reprinted with permission from MDPI

The superior mechanical and electrical properties of CNTs became the basis for fabricating large number of sensors. Their nanoscale electrical properties are often related to the mechanical deformation as in the previous case [68, 102]. Macroscale strain sensor based on PI-MWCNT composites (PMCNTCs) was prepared. Better dispersion of the nanoparticles was ensured by dispersing CNTs in chloroform and thereafter mixing with the PI matrix solution. About four orders of magnitude of nonreversible change in the electric resistivity at large (40 %) stretch were observed due to the entangled structure of nanotube, thus making the PMCNTC an attractive tensile and compressive strain sensor. On the other hand, the high structure carbon nanoparticles have shown about six orders of magnitude of reversible change in the electric resistivity at large (40 %) stretch. This is due to the high porosity of carbon agglomerates which provides better adhesion to polymeric globules and mobility of nanoparticles. The maximum change at stretch for both types of composites was observed near the percolation threshold. PMCNTC can be used for small tensile strain sensing, but high structure CB–PI composite are preferable for large tensile strain sensing. The resistance of the composites was examined with regard to the force of stretch F and the absolute mechanical deformation in the direction of the force. No variation in the electric resistance was found for the samples with 7.7 mass parts of highly structured carbon black (HSCB) and 3.63 mass parts of MWCNT even at 100 % deformation of the composite. Similarly, there was a relatively weak dependence of R on the

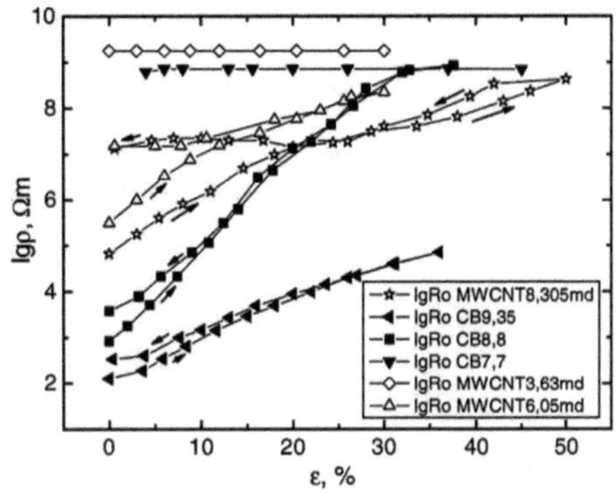

Fig. 9 The electric resistivity ρ of the PI–MWCNT composite and PI–HSCB composite versus small tensile strain ε [102]. Copyright 2007. Reprinted with permission from Elsevier

deformation at 9.35 mass parts of HSCB. Figure 9 clearly shows the variation in resistivity of different samples with respect to strain percentage.

Of all the composites examined, the best strain performance was obtained for the samples with 8.8 mass parts of HSCB and 8.3 mass parts of MWCNT, respectively, which belong to the region of percolation threshold. The resistivity of these samples changed more than six orders upon a 40 % stretch for 8.8 mass parts of HSCB composite and more than four orders for 8.3 mass parts of MWCNT composite. When the materials are released, the resistivity of HSCB composite returned to its previous value, while in the case of MWCNT composite, a relatively large hysteresis of $\rho(\varepsilon)$ was observed because of the entangled structure and high stiffness of nanotubes.

In another study, novel biocompatible NO sensor was made by spin trap (iron(II)-diethyldithiocarbamate complex, FeDETC) incorporated in a latex rubber matrix [103]. The sensor was optimized for the latex rubber matrix temperature of polymerization and FeDETC concentration inside the matrix. The NO trapped in the sensor can be detected by electron paramagnetic resonance (EPR). The EPR signals showed variation with respect to different conditions. The best sensor was made with a latex rubber matrix polymerized at 10 °C and with a FeDETC concentration of 14.8 mM. High stability of the EPR response was detected for 40 days at room temperature.

2.2.4 Dielectric Property

Neat NR is an insulator, but it becomes conductive when conducting nanoparticles are incorporated in it. In addition to the electrical conductivity, some of these composites are good dielectric materials as well. NRL composites with polyaniline (Pani) were prepared by the in situ polymerization of aniline in the presence of latex [104]. The dielectric properties of different compositions of the samples in microwave frequencies were determined (Fig. 10a).

From variation of the dielectric constant with frequency (S bands 2–4 GHz) of Pani–NR semi-interpenetrating networks, it is clear that the dielectric constant was

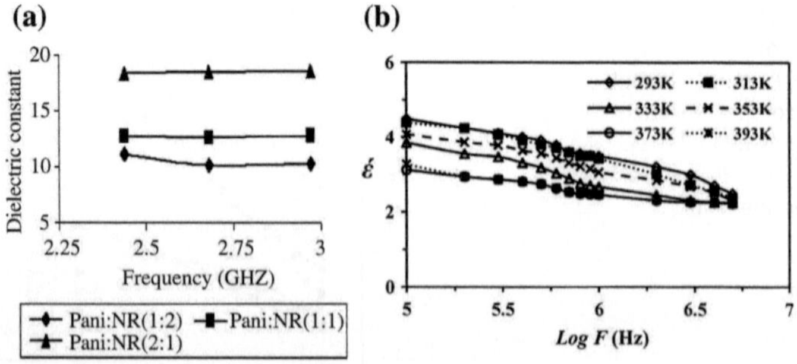

Fig. 10 **a** Variation of dielectric constant with frequency for the Pani–NR composites [104]. Copyright 2007. Reprinted with permission from John Wiley & Sons. **b** Dielectric constant for NR as a function of frequency at different temperatures [105]. Copyright 2012. Reprinted with permission from John Wiley & Sons

less affected by frequency. But the value changes from NR (i.e., 2) to Pani–NR, the maximum shown for the 2:1 composition. The absorption coefficient and penetration depth calculated for 2:1 composition were, respectively, the highest and the lowest. NR allowed the passage of electromagnetic radiation, and when the Pani content was increased, the electromagnetic radiations were restricted to penetration. The in situ technique also contributes toward high dielectric constant. The conducting NR could be used in various applications, including antistatic coatings, antistatic chappels, and anticorrosion applications.

Figure 10b shows the effect of both frequency and temperature on the dielectric constant for NR [105]. Both dielectric constant and tan δ of NR are less dependent on frequency and temperature due to the absence of polarization and ionic properties of the material. Since NR is a nonpolar molecule, its dielectric constant will be the square of the refractive index (1.59) [106, 107]; however, the experimental value obtained is higher than the theoretical one due to the presence of interfacial polarization which arises due to the presence of impurities such as the compounding ingredients. At high frequencies, the values become about 2.5 for all temperature ranges, which can be ascribed to the mobility of a polar group in the polymer chains, which is too slow to contribute to the dielectric constant at high frequencies. By following the experiment, the dielectric constant, dielectric loss, and tan δ of barium ferrite–NR composites were studied [105]. All these values decreased as the frequency increases, and at higher frequencies, the effect significantly weakens. ε' for barium ferrite was related to the preparation method, and at lower filler loading, the ferrite content strongly influences the dielectric properties. Up to 30 Phr ferrite loading, ε' sharply increases, and thereafter, the dependence of ε' on ferrite content becomes minimal as a result of ferrite particle interaction. Ferrite content also enhanced the dielectric loss and tan δ. They could observe a strong correlation between magnetic initial permeability and dielectric constant for hard magnetic material polymer composites as well.

Oil palm fibers were modified with silane, and the effect of modification on dielectric properties of NR is checked for the biocomposite. At low level of fiber loading (20 %), the dielectric properties show only a nominal change, and with fiber treatment, the loss factor increases [108]. Even though the effect of the oil palm fibers and their chemical treatment with silanes on the dielectric properties is evidenced well, other factors such as fiber orientation and quantity of chemical used were not addressed in this work. For the sisal/coir hybrid fiber-reinforced NR composite, the dielectric constant values were higher due to the polarization effect induced by fibers' incorporation. The lignocellulosic fiber addition increased the number of polar groups and decreased the volume resistivity of the composites. At 15.6 % volume, percolation threshold is reached. Upon increase in frequency, the interfacial and orientation polarization decreased and thus the dielectric constant [109].

Fiber treatment, such as acetylation, benzoylation and reaction with alkali, peroxide and permanganate were carried out to improve the adhesion between fibers and matrix. Modification also reduces the moisture absorption capacity of the fibers due to reduced interaction between polar –OH groups of lignocellulosic fibers and water molecules. For composites containing chemically treated fibers, the dielectric constant values were found to be lower for all frequencies due to the increased fiber hydrophobicity. Chemical treatment increased the volume resistivity of the composites as well. In the untreated state, the –OH groups of fibers form strong hydrogen bonds and are thus unreactive. Alkali treatment destroys the hydrogen bonding and makes them more reactive and leads to decreased orientation polarization. It also can lead to fibrillation and thus can enhance the surface area [110]. This is the reason why alkali-treated fibrous composites showed considerable decrease in dielectric constant. In composites, the fibers and NR matrix interlock each other. The addition of two-component dry bonding agent consisting of hexamethylenetetramine and resorcinol improved the interfacial adhesion between the matrix and fibers and reduced the dielectric constant. It also enhanced the electrical resistivity and dissipation factor of the composites. The added fibers and different chemical treatments for them increased the dielectric dissipation factor. A dielectric relaxation has been observed at a frequency of 5 MHz. The dielectric constant of treated fibers decreases in the order, alkali > acetylated > benzoylated > permanganate > peroxide > bonding agent [106]. The conductivity of the composites was also found to increase with increase in fiber loading as the number of conductive particles increased. Finally, the application of such low-cost composites in antistatic materials is explained. Depending on the concentration of alkali, the dielectric constant varies. Fibers treated with 4 % alkali showed the lowest value. Of the three types of silanes, fluorosilane and vinyl silane, respectively, exhibit the minimum and maximum dielectric constant. Mercerization prior to silane treatment is an effective means of attaining better property retention to composites exposed to moisture. Silane coupling agents have inherent natural attraction with both the natural fiber and resin matrix [111].

Effect of alumina particles on the properties NR revealed the ionic polarization of the microalumina particles and the induced dipole moments set up in the

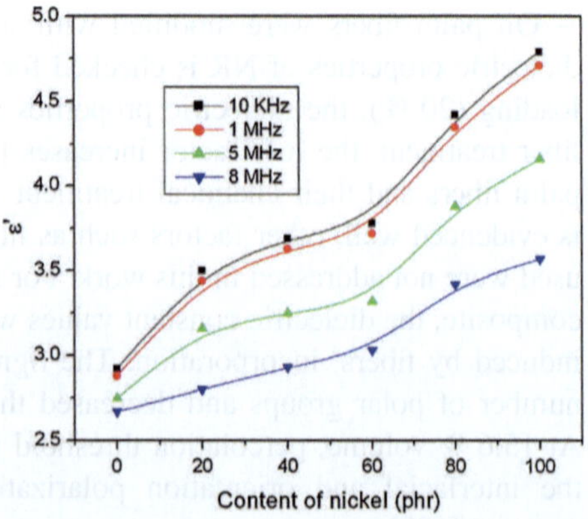

Fig. 11 Variation of permittivity with content of nickel in the composites per hundred weight ratio of rubber (phr) is plotted in the horizontal axis [119]. Copyright 2009. Reprinted with permission from Elsevier

matrix. At 0.00968 volume fraction, the composite showed the highest dielectric constant of 2.123 and the electrical conductivity of 2.683×10^{-8} $(\Omega.\text{m})^{-1}$. The soft and flexible composites containing small amount of alumina can store additional elastic energy within the matrix under electric field and are useful in fabricating biomimetic actuators and/or artificial muscles. ε' of magnetic NR composites containing uniformly distributed nickel nanoparticles decreased with increase in frequency and increased with increase in nickel loading (Fig. 11). Metallic inclusions in any insulating material enhance the dielectric permittivity of the material as reported [112, 113]. Interfacial polarization due to the heterogeneity of the samples is attributed as the reason for the variation of dielectric permittivity with frequency as well as filler concentration [114]. The dielectric dispersion observed in the frequency range of 100–8 MHz at 40 °C shows the linear increase in permittivity with the nickel particle loading. Here, the dielectric loss in the material is small and can be useful in making capacitors. The mechanism of ε' is understood by interfacial polarization, enhancement of electrical conductivity [115], and formation of internal barrier layer capacitors [116, 117]. Interfacial polarization arises at the interface of a composite with electrical heterogeneity due to the migration of charge carriers through different phases, and charges accumulate at the interface [118]. External electric field moves these charges, and motions prevented within the medium cause polarization.

Permittivity increases with filler loading almost linearly with all frequencies and at all temperatures. Dielectric loss of blank rubber, as well as the composites, was observed to increase with temperature [119]. The decrease in permittivity with temperature is attributed to the high volume expansivity of rubber at elevated temperatures. Dielectric loss of the composites was found to increase with temperature, and a relaxation peak was only observed in blank rubber. For filled samples, the dielectric relaxation peaks appear to exist below 100 kHz. A weak interfacial polarization present in the rubber matrix can be attributed to the observance of relaxation peaks in blank rubber.

In another work, the relaxation phenomena occurring in NR and polyurethane (PU)/NR blend-based nanocomposites, reinforced by layered silicates, were

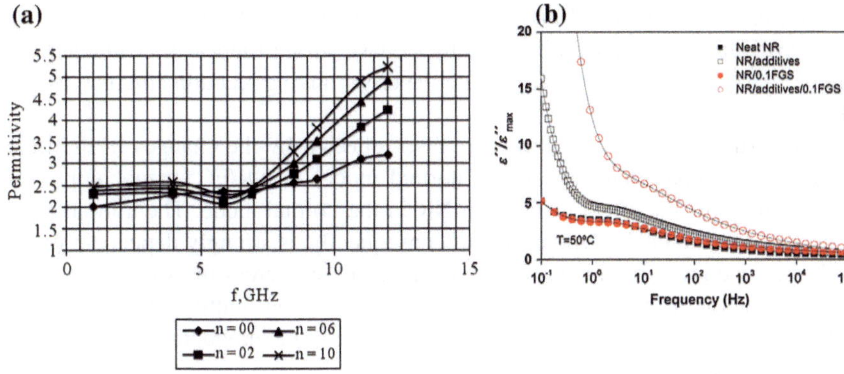

Fig. 12 Frequency dependence of **a** relative dielectric permittivity ε' at various filler content (n-phr of graphene) [98]. Copyright 2012. Reprinted with permission from Scientific Research Open Access. **b** Normalized dielectric loss ε'' of the nonvulcanized nanocomposites in the normal mode. *Solid lines* correspond to HN fittings and *dashed lines* to individual processes [123]. Copyright 2012. Reprinted with permission from Elsevier

examined during 10^{-1} to 10^6 Hz frequency. NR is nonpolar, and its performance is only slightly affected by the presence of layered silicates. It has low dielectric permittivity and loss, and its relaxation is visible in the low T_g range. But PU exhibits four relaxation processes—ascending relaxation rate, interfacial polarization (IP), glass/rubber transition, and local motions of polar side groups and small segments of the polymer chain. Both matrix and filler influence relaxation phenomena, and IP is present in all nanocomposites being the slowest recorded process [120].

The influence of graphene nanoparticles (GNPs) [98] at 2.0–10.0 phr on the dielectric properties (dielectric permittivity, dielectric loss angle tangent) of NR nanocomposites has been investigated in the wide frequency range (1–12 GHz). The dielectric permittivity increased slightly with the increasing frequency and filler amount, especially significant in the 9–12 GHz range. The tendency was the same in the case of dielectric loss angle tangent, although the impact of the filler amount is less marked, i.e., the imaginary part of the complex relative dielectric permittivity is less sensitive to alternations of the filler amount and frequency than the real part. Figure 12a shows the frequency dependence of the real part of the complex relative dielectric permittivity. The close dielectric permittivity value for various filler concentrations up to 7 GHz is clearly seen. As explained at >7 GHz, the values show significant difference between the nonfilled and filled composites. The values obtained for the nonfilled composites are close to the ones for NR at 1000 Hz reported, which are 2.40–2.70 [121]. At higher than 9 GHz, the high dielectric permittivity comes mainly due to the composite polarity [122].

The effect of functionalized graphene sheets (FGSs) on the dielectric properties of NR nanocomposites is illustrated in Fig. 12b. The shift toward higher frequencies is observed in the presence of the additives and becomes more significant with FGS. Molecular dynamics of NR is often affected by vulcanizing additives which can decrease the segmental motions [123]. FGSs have defects, free radicals, and other irregularities on its surface and thus cause strong rubber–filler interactions. These interfacial interactions accelerate the cross-linking reaction and increase the electrical conductivity.

Those rubber–filler interactions are also reflected by the changes in the time-scale of the global chain dynamics (normal mode) in NR/FGS nanocomposites. The figure gives the effect of FGS on NR in the normal mode. The normal mode is assigned to the chain dynamics and, consequently, is strongly dependent on the molecular weight and on the mobility of the chain-end cross-links in the naturally occurring network of the NR. The final material having multifunctional property enhancements coupled with low cost and large-scale production makes wider applications for it.

2.2.5 Thermal Conductivity

CNTs are reported to be excellent in improving the thermal conductivity of NR which is also significant in electronics. In situ measurements of thermal conductivity, heat capacity per unit volume, and T_g of pure PI and PI/single-walled carbon nanotube (SWCNT) composites cross-linked by high-pressure treatment (4 h at 513 K and 1 GPa) revealed the enhancement in thermal conductivity of both the elastomeric and liquid PI by ~120 % at 5 wt% SWCNT loading [124]. T_g of the cross-linked elastomeric state increased by 12 K and that of the liquid PI by ~7 K at the same filler concentration. CNTs enhanced broadening of T_g in the high-pressure-treated densified elastomeric states but not in the untreated samples. The increase observed for cross-linked state can be attributed to the filler-induced increase in the cross-link density. The heat capacity decreased by ~30 % in both the glassy and liquid/elastomeric states (above and below T_g) due to the large reduction in the vibrational and configurational heat capacity of PI in the presence of SWCNTs. It is also observed that the thermal conductivity (κ) of the composite linearly increased with SWCNT content. Also, the pressure-induced densification decreases the interparticle (CNT–CNT) distances without agglomeration and improves the thermal and electrical contact between the constituent materials of the composite. In the case of a semicrystalline polymer, high-pressure treatment can increase the degree of crystallinity and thus can strongly enhance the thermal conductivity of the matrix. Thus, thermal as well as electrical percolation networks in polymer CNT composites and the nature of thermal and electric transport in them are better studied by high-pressure treatments.

Variation of κ with temperature at 0.02 GPa is shown in Fig. 13a. All samples show similar amorphous behavior where κ is almost constant or slightly increasing with temperature. A small peak is also observed distinguishing two temperature ranges, which is most apparent in PI-SWCNT5. Below the peak, κ moderately increases with temperature, whereas above the peak it becomes temperature independent. More than three times larger change for PI-SWCNT5 is noted compared to PI. For pure PI, the change is less pronounced. The change occurring in κ with

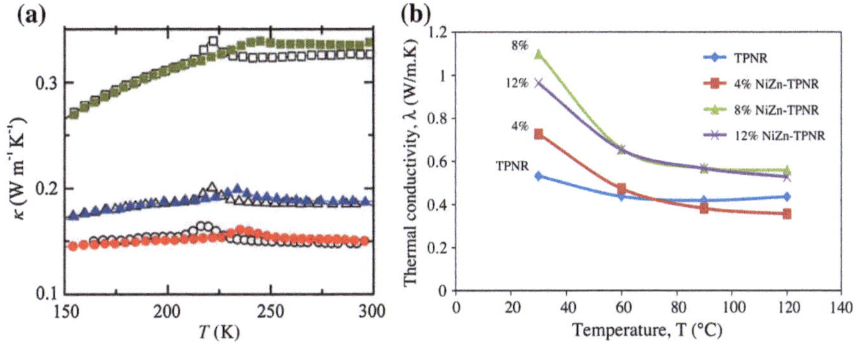

Fig. 13 **a** Simultaneously measured results for thermal conductivity [124]. Copyright 2009. Reprinted with permission from American Chemical Society. **b** Thermal conductivity dependence on temperature for TPNR and different loaded NiZn–TPNR nanocomposite samples at 30, 60, 90, and 120 °C [126]. Copyright 2013. Reprinted with permission from Elsevier

Table 1 Thermal conductivity of NR nanocomposites

Sl. No.	Type of filler and concentration (wt%)	Thermal conductivity W/(mK) (composite/neat)	References
1	RGO at 5	0219/0157 for solution treated 0188/0157 for mill mixed	[128]
2	Alumina and boron carbide at 12	09/02	[129]
3	Graphene at 2	01963/01741	[130]
4	Nanosilica at 20	0133/0097	[131]
5	Nanographene platelets at 2	2130/013	[132]

respect to temperature is attributed to the thermal expansivity increase at the T_g and the stronger decrease in density above T_g than below [125]. High-temperature and high-pressure treatment causes changes in the number of covalent bonds between the chains, density, and structural order, and affects κ values. During the treatment, both chain scission and cross-linking can occur, and the density as well as the order can decrease or increase due to decreased or increased degree of crystallinity. But here, the authors have ruled out the possibility of chain scission because of sample cross-linking. Comparison of samples of pure PI cross-linked at 1 GPa and 0.5 GPa by submerging in water–methanol mixture revealed that the samples become progressively denser with increasing pressure of the high-temperature treatment. The sulfidic and carbon–carbon cross-links further decrease the already weak tendency of cold crystallization in PI. Since PI is difficult to crystallize, the major part is inevitably amorphous and the difference is likely small. Moreover, a decreased crystallinity would also decrease κ. The increase in κ at room temperature and near atmospheric pressure is therefore attributable to the increased number of covalent bonds between the chains and increased density. The κ values for a few NR nanocomposites are given in Table 1.

Thermal conductivity dependence of NiZn ferrite nanoparticles filled TPNR magnetic nanocomposite on temperature is given in Fig. 13b. TPNR was prepared from NR, LNR and high-density polyethylene (HDPE) in 20:10:70 ratio by melt blending [126]. Tremendous enhancement of thermal conductivity with increasing filler content up to 8 % of NiZn ferrite is observed, and thereafter, κ decreased slightly with filler content. The calculated enhancement in κ for 4, 8, and 12 wt% NiZn–TPNR nanocomposite samples was about 38, 108, and 81 %. The increase of thermal conductivity with filler rates could be attributed to the nature of NiZn ferrite nanoparticles, which are assumed to have a higher number of phonons' vibrational modes as well as higher mean free path due to its crystalline structure compared to the amorphous TPNR matrix. Also, the incorporation of NiZn ferrite nanoparticles creates a percolated network facilitating the phonon conduction inside the nanocomposite and therefore enhances κ [127]. Moreover, the T_g value of the TPNR matrix increases with filler content, suggesting an enhancement in the thermal stability of the TPNR matrix. Thus, 8 wt% filler content remarks the optimum filler concentration in the formation of the percolated network, above which κ value decreases. The decreasing trend in κ value for all nanocomposite samples with temperature is due to the increase of structural defects existing in the amorphous behavior of TPNR matrix, which damp the phonons' vibrational amplitudes.

2.2.6 Shape Memory Materials

Shape memory polymers (SMPs) are smart, responsive materials having numerous potential applications in various fields and in electronics [133]. Shape memory natural rubber (SMNR) has much importance among other SMPs in exhibiting exceptional properties such as strain storage of 1000 %, cold storage, cold programmability, and mechanical and thermal triggers tunable both during and after programming. They also possess energy storage capability and mechanical properties. SMNR decreases the melting temperature of crystals up to 40 K by the application of external mechanical stress. The mechanical stress parallel to stretching direction stabilizes the crystals indicated by linear increase in the trigger temperature, whereas the stress applied in the transverse direction weakens the shape-stabilizing crystals and resulted in decrease in the trigger temperature after programming (stress-induced melting). The reinforcing effect of parallel stress on the shape-stabilizing crystals is called stress-induced stabilization. A significant amount of crystals still exist in the stretched SMNR above the trigger temperature as evident from the XRD measurements. Above 100 °C temperature, this completely disappeared. Fukahori proposed the existence of partially continuous cross-linked phase and a continuous uncross-linked phase in vulcanized NR by which the two different melting regimes of the crystals were explained [134]. The strain-induced crystallization appears mainly in the cross-linked phase followed by thermal crystallization in the uncross-linked phase. The crystals formed in the cross-linked phase should be more sensitive to parallel stress than the crystals formed in the uncross-linked phase. This should then result in the two melting regimes.

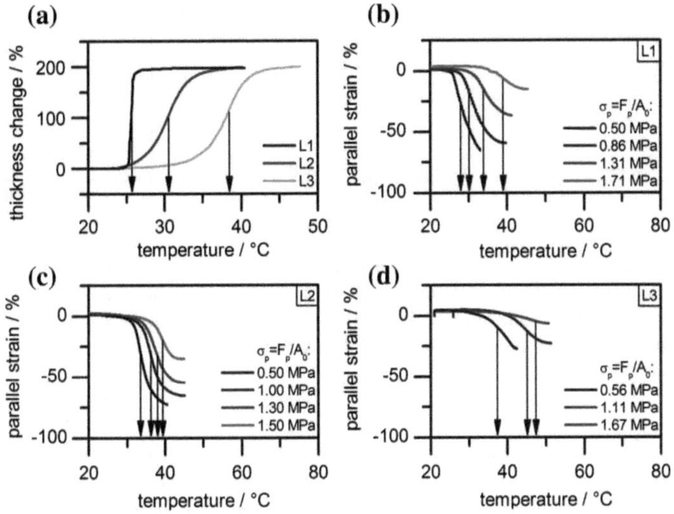

Fig. 14 **a** Determination of unstressed trigger temperatures of samples L 1–3 using TMA measurements. *Arrows* indicate different trigger temperatures (*T* trig, L1 258 °C; L2 305 °C; L3 384 °C, in all cases, the perpendicular strain increase was measured). Determination of trigger temperatures of **b** L1, **c** L2, and **d** L3 in dependence on different applied parallel stresses (in all cases, the parallel strain decrease was measured) [135]. Copyright 2013. Reprinted with permission from John Wiley & Sons

The shape memory property of two series of samples, one based on vulcanized NRL (degree of cross-linking: 0.2 %, abbreviated by L) and the other prepared by vulcanizing compacted rubber (degree of cross-linking: 0.1 %, abbreviated by NR), was analyzed at different stretching temperatures. Figure 14 shows the resulting trigger temperatures obtained from TMA measurements, which show the perpendicular strain increase during recovery for L 1–3. The average trigger temperatures (*T* trig) obtained for the samples L1, L2, and L3, respectively, were 25.8, 30.5, and 38.4 °C.

Figure 14b–d, respectively, illustrates the influence of constant stress applied parallel to prior programming direction on the samples L1, L2, and L3 [135]. All plots of parallel strain against temperature show a rising trigger temperature with increasing stress. This applied stress was significantly smaller than the stored stress, and as a result, the sample retracts to some extent upon increasing the temperature. This retraction continues until either the samples reach the limit of the apparatus or the samples rupture. The retraction process of L1 and L2 under different loads is triggered much more discrete than that of L3, which can be due to the stress-free trigger temperature ranges of the unstressed samples. The vulcanized samples NR 1–3 also follow similar results. The trigger temperature strongly depends on the stress of the surrounding amorphous phase on the strain-stabilizing crystals [136], and the parallel stress causes a stabilizing effect on the polymer crystals, resulting in an increase of the trigger temperature.

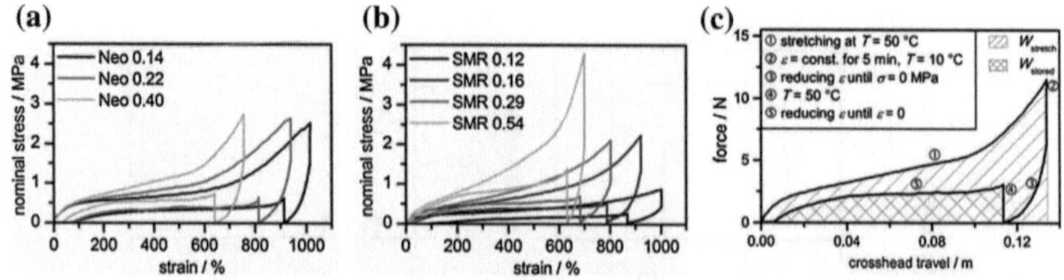

Fig. 15 Exemplary plots of nominal stress versus strain during a shape memory cycle of **a** latex-based and **b** SMR-based samples, plot of force versus crosshead travel of sample Neo 040 (dimensions: 178 mm × 382 mm × 11 mm) during a shape memory cycle [137]. Copyright 2013. Reprinted with permission from John Wiley & Sons

Figure 15a, b shows the variation of nominal stress with strain during the shape memory cycle of latex-based (Neo) and SMR-based samples [137]. The values of stored energy were obtained in a range from 1.34 to 4.88 J g^{-1} with efficiencies (ratio of stored to stretching energy) between 26.22 and 48.32 %. It is also found that the stored energies and efficiencies increase with increasing degree of cross-linking for both SMNR sources. The stored energy also depends on applied strain during programming.

The SMNR showed fixity ratios of up to 94 % and recovery ratios of up to 100 % combined with high stored strains in a range of 540–1180 % after 10 thermomechanical cycles. Energies of up to 4.88 Jg^{-1} could be stored with up to 53.30 % efficiency. This material also has excellent Young's modulus (two orders of magnitude upon triggering or programming). All these properties are observed in the case of latex-based samples as well (Fig. 15b); however, the masticated samples show superior shape memory and energy storage properties. The complete shape memory cycle of the latex-based sample Neo 0.40 is given in Fig. 15c for detailed information. The integral of the programming path 1 corresponds to the energy *W* stretch needed for stretching the rubber at 50 °C to a strain of 750 %. The elastically stored energy *W* during shape recovery at 50 °C is also equal to integral of the recovery path 5 [138].

The shape memory of NR in the presence of solvent vapors is illustrated in Fig. 16. For this, SMNR is kept in contact with toluene vapors and thus immediately triggered, resulting in complete shape recovery [139]. The programming steps showing the stretching (1) and fixation steps (2) are clear from the stress–strain plots of SMNR in Fig. 16a. Triggering the sample under unconstrained conditions resulted in strain answer (3), whereas triggering under constrained conditions resulted in stress answer (4). The solvent caused complete recovery, and so the qualitative and quantitative information on the trigger capability of the vapors is lost. The same effect was also observed in the case of other solvents such as chloroform, n-heptane, and tetrahydrofuran (THF). By mounting the programmed SMNR between a fixed clamp at the bottom and a spring scale, pulling force is determined as shown in Fig. 16b.

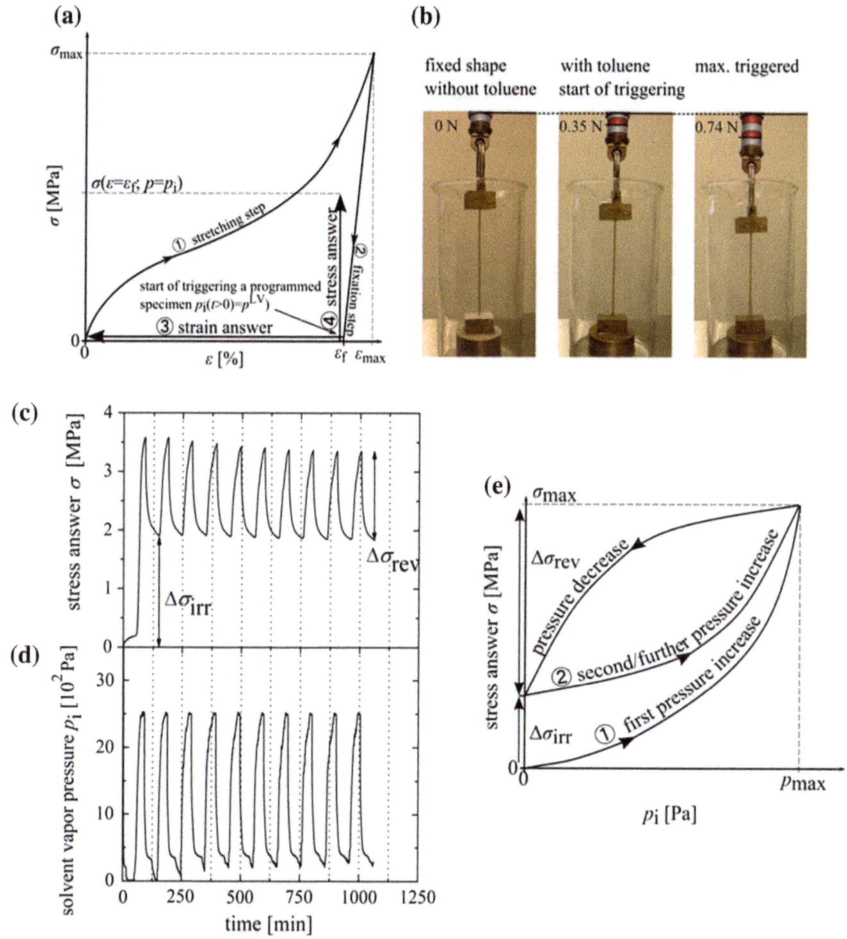

Fig. 16 **a** Stress–strain plots of the programming process of SMNR, **b** photographs illustrating the stress built up by exposure to toluene vapor temporal progress of **a** stress answer σ of constrained SMNR with **b** changing vapor pressure pi plotted, and **c** schematic illustration of the irreversible and reversible stress answers caused by the solvent vapor pressure during the first (Δσirr) and all following cycles (Δσrev) [139]. Copyright 2013. Reprinted with permission from American Chemical Society

In order to investigate the effect of solvent on triggering, the constrained SMNR sample was subjected to a force of about 1.7 N within 15 min, and thereafter, the unclamped sample was brought to atmospheric pressure following vapor evacuation. SMNR was not fully triggered as it did not recover its original shape, but stabilized 800 % strain. Cyclic experiment in which the stress is continuously measured by adding and removing toluene vapor pressure was further performed at constant time intervals to confirm partial triggering or triggering and reprogramming after solvent removal. As shown in Fig. 16c, the sample stress becomes

3.6 MPa upon increasing the toluene vapor pressure to 2500 Pa in the first cycle. When the vapor was evacuated below 300 Pa after 15 min, the sample stress decreased to 1.8 MPa. This two-way stress response is due to reprogramming of the SMNR sample. The results were reproducible. By comparing Fig. 16c and d, the direct dependence of stress answer on the vapor pressure increase is derived. The stress answer did not return to its initial value even after 24-h evacuation at which the stress was 1.7 MPa. Thus, the solvent-sensitive stress answer of SMNR on toluene vapor is composed of an irreversible stress answer $\Delta\sigma_{irr}$ that remains after the first cycle and a reversible stress answer $\Delta\sigma_{rev}$, as represented in Fig. 16e [140].

2.2.7 Actuators

Electroactive polymers (EAPs) convert electrical energy into mechanical energy and widely applied in actuators, sensors, motors, pumps, etc. [141]. Such dielectric and electrostrictive materials possess advantages of rapid response rates depending upon ionic diffusion, dipole moment, electrostatic forces, low operating voltage, high force generation capabilities, and high energy density per cycle, and thus, they are useful in fabricating catheters, automotive door locks, and artificial muscles for robotic and prosthetic devices and in bioengineering. Typically, the uses of actuators include adaptive optic applications to move an optical element, to connect thermal or electric pathways, and to provide agonist and antagonist controls in a linear motion, etc.

As NR composites are flexible and easily fabricable, they are also EAP. The application of an external electric stimulus to EAP induces dipole–dipole moments in molecules and thus induces electrostatic interactions between them. Higher force is required to deform such materials, which indicates that the mechanical properties of the material have been altered [141]. Niamlang et al. synthesized poly(p-phenylene vinylene) (PPV)/NR blend (10 %w/w PPV) in the presence of a cross-linker. The addition of a conducting polymer to the elastomer enhances the dipole moment of the conductive polymer and thus the interference of the dipole–dipole interactions between the cross-linking points. When external electric field strength is applied, the NR gel deflects toward anode, the functional group and strand are polarized, and ionic interaction between generated dipole moment and electrode is occurred. If the strength of the applied field is higher, stronger interaction occurs between dipole moment and electrode and causes higher degree of bending [141].

Blending NR with polythiophenes is also practiced by Puvanatvattana et al. by which they analyzed the electrorheological response of the samples to make it applicable in actuators. Electric field strength of 0–2 kV/mm was applied, and the corresponding storage and loss modulus values were noted. For PI of cross-linking ratios 2, 3, 5, and 7, the storage modulus sensitivity increased with applied field strength and attained maximum values of 10, 60, 25, and 30 %, respectively, at 2 kV/mm. For the PI/polythiophene blend, containing 5, 10, 20, and 30 vol.%, the moduli were higher than those of cross-linked PI. The storage modulus sensitivity of such blends also increased with field strength and reached maximum values of 50, 35, 110, and 45 %, respectively, at 2 kV/mim [142].

Orthorhombic and noncentrosymmetric perovskite lead zirconate ($PbZrO_3$) capable of spontaneous polarization was synthesized at a temperature below the Curie temperature and dispersed in NR. At a volume fraction of 0.007306, the electrical conductivity of the composite increased dramatically by two orders of magnitude at 500 kHz electrical frequency. The obtained composite sample showed variations in storage modulus with applied electrical field. When the field strength was 1 kV/mm, the temporal characteristic of the neat NR increased from 9.4 kPa to a steady-state value of 9.6 kPa, whereas for the composite, the increase was abrupt (from 9.4 to 12.6 kPa). This value decreased, but not went to the original value when the field was turned off. This behavior indicates that there are some irreversible interactions, possibly due to the dipole bonding between small lead zirconate particles at this volume concentration and some residual dipole moments remaining [143].

Based on the deflection responses of a standard Thai rubber (STR) composite of Al_2O_3 cross-linked with dicumyl peroxide, Tangboriboon et al. proposed a three-stage mechanism for the electromechanical cycle of the actuator. This is shown in Fig. 17a where both electrodes have the same concentration of dopant (Al^{3+}) when the cantilever is undistorted, and the electrochemical transfer of the Al^{3+} between electrodes causes bending either to the right or to the left. The STR/ alumina samples were cured using sulfur as well. They have checked the effect of cross-link agents between the sulfur/accelerator and the DCP/stabilizer on the deflection response of the composites under applied electrical field strength of 0–650 V/mm at room temperature. In the presence of applied field, the freely suspended end of the dielectric elastomer actuator bends toward one of the electrodes due to the interaction between the electrode and the nonsymmetrical dipole moments set up in the matrix. The maximum bending is observed for the DCP cured sample containing 60 phr alumina at 200 V/mm in which the dipole moment generated causes a nonsymmetrical negative charge distribution and as a result it

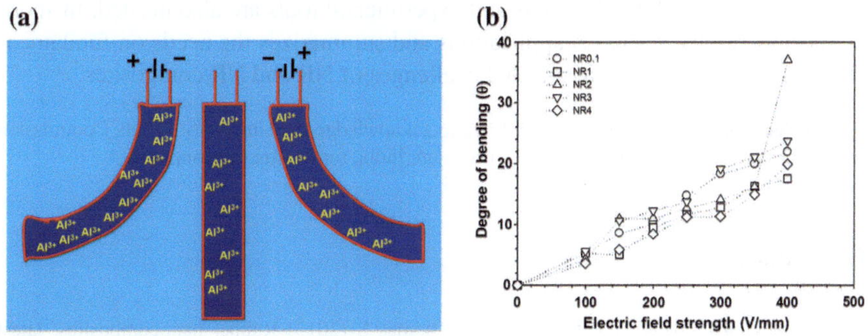

Fig. 17 **a** Three states during the electromechanical cycle of a dielectric elastomer actuator and **b** degree of bending of cross-linked natural rubber at various electric field strengths [145]. Copyright 2013. Reprinted with permission from Elsevier

bends toward anode. In addition, the alumina particles induce polarization within the composite and the dipole moments under applied electric field which results in high dielectric constant for the composites with maximum alumina content [144]. Also, the DCP curing induces better actuation effect when compared to sulfur curing [145] as per the results shown in Fig. 17b (all samples contain DCP at various percentages).

3 Conclusions

Nanotechnology offers technological advantages in various important areas, such as production, processing, storage, transportation, safety, and security. As presented in this review, the products based on NR derived from nanotechnology are applicable in electronics, aerospace science, valuable household items, value-added products, and industry. NR is an excellent polymer and obtained from nature, which has shown its own importance in the literature for long years. Nanofillers are capable of imparting superior properties such as gas permeability (impermeable tyres), electrical (conductive wires, sensors) and thermal conductivity, dielectric properties (capacitors), mechanical strength (cables, sheets, etc.), flame retardancy, shape memory, and microwave absorption properties to NR. A comprehensive as well as state-of-the-art review on NR and its composite applications has been made in this paper. Materials for piezoelectrics and low-friction lightweight devices with high durability can get at low cost. Major advantages of nanocomposites in comparison with their micro counterparts are addressed in most parts of this study. However, the fabrication of nanocomposites is often associated with filler–polymer compatibility, interfacial interaction, nanolevel dispersion, etc. Because of this, practical applications of NR composites are rather limited so far. Therefore, further developments to achieve nanofiller-reinforced composites with relatively high volume fractions and with maximum filler–polymer and filler–filler interactions are still required. For this, acquisition of fundamental knowledge and development of analytical models and experimental tools are also needed. In short, this review briefly discusses applications and summarizes the needs on fundamental research as well as the associated challenges of NR and NR composites.

Acknowledgement The authors would like to acknowledge the University Grants Commission and the Department of Atomic Energy Consortium, India, for the financial support.

References

1. Ponnamma D, Maria HJ, Chandra AK, Thomas S (2013) Rubber nanocomposites: latest trends and concepts. Adv Struct Mater 12:69–107
2. Harris J, Stevenson A (2011) On the role of nonlinearity in the dynamic behavior of rubber components. Rubber Chem Technol 59:740–764

3. Angellier H, Molina-Boisseau S, Dufresne A (2005) Mechanical properties of waxy maize starch nanocrystal reinforced natural rubber. Macromolecules 38:9161–9170
4. Sato S, Honda Y, Kuwahara M, Kishimoto H, Yagi N, Muraoka K, Watanabe T (2004) Microbial scission of sulfide linkages in vulcanized natural rubber by a white rot basidiomycete, ceriporiopsis s ubvermispora. Biomacromolecules 5:511–515
5. Sanguansap K, Suteewong T, Saendee P, Buranabunya U, Tangboriboonrat P (2005) Composite natural rubber based latex particles: a novel approach. Polymer 46:1373–1378
6. Ponnamma D, Sung SH, Hong JS, Ahn KH, Varughese KT, Thomas S (2014) Influence of non-covalent functionalization of carbon nanotubes on the rheological behavior of natural rubber latex nanocomposites. Eur Polymer J 53:147–159
7. Trabelsi S, Albouy PA, Rault J (2003) Effective local deformation in stretched filled rubber. Macromolecules 36:9093–9099
8. Rault J, Marchal J, Judeinstein P, Albouy PA (2006) Stress-induced crystallization and reinforcement in filled natural rubbers: 2H NMR study. Macromolecules 39:8356–8368
9. Poompradub S, Tosaka M, Kohjiya S, Ikeda Y, Toki S, Sics I, Hsiao BS (2005) Mechanism of strain-induced crystallization in filled and unfilled natural rubber vulcanizates. J Appl Phy 97:103529/1–103529/9
10. Ozbas B, Toki S, Hsiao BS, Chu B, Register RA, Aksay IA, Prud'homme RK, Adamson DH (2012) Strain-induced crystallization and mechanical properties of functionalized graphene sheet-filled natural rubber. J Polym Sci Part B Polym Phys 50:718–723
11. Brydson JA (1988) Rubbery materials and their compounds. Elsevier, Essex
12. Bode HB, Kerkhoff K, Jendrossek D (2001) Bacterial degradation of natural and synthetic rubber. Biomacromolecules 2:295–303
13. Schwerin M, Walsh D, Richardson D, Kisielewski R, Kotz R, Routson L, Lytle CD (2002) Biaxial flex-fatigue and viral penetration of natural rubber latex gloves before and after artificial aging. J Biomed Mater Res 63:739–745
14. Walsh DL, Schwerin MR, Kisielewski RW, Kotz RM, Chaput MP, Varney GW, To TM (2004) Abrasion resistance of medical glove materials. J Biomed Mater Res B 68:81–87
15. Kurian JK, Peethambaran NR, Mary KC, Kuriakose B (2000) Effect of vulcanization systems and antioxidants on discoloration and degradation of natural rubber latex thread under UV radiation. J Appl Polym Sci 78:304–310
16. Abad L, Relleve L, Aranilla C, Aliganga A, Diego CS, Rosa AD (2002) Natural antioxidants for radiation vulcanization of natural rubber latex. Polym Degrad Stab 76:275–279
17. Wu YP, Wang YQ, Zhang HF, Wang YZ, Yu DS, Zhang LQ, Yang J (2005) Rubber–pristine clay nanocomposites prepared by co-coagulating rubber latex and clay aqueous suspension. Compos Sci Technol 65(7):1195–1202
18. David JK, Hull TR (2012) A review of candidate fire retardants for polyisoprene. Polym Degrad Stab 97:201–213
19. Busfield JJC, Deeprasertkul C, Thomas AG (2000) The effect of liquids on the dynamic properties of carbon black filled natural rubber as a function of pre-strain. Polymer 41:9219–9225
20. Cai HH, Li SD, Rian TG, Wang HB, Wang JH (2003) Reinforcement of natural rubber latex film by ultrafine calcium carbonate. J Appl Polym Sci 87:982–985
21. Arroyo M, Lopez-Manchado MA, Herrero B (2003) Organo-montmorillonite as of carbon black in natural rubber compounds. Polymer 44:2447–2453
22. Jose L, Joseph R (1993) Study of the effect of polyethylene-glycol in field natural-rubber latex vulcanizates. Kaut Gummi Kunstst 46:220–222
23. Afiq MM, Azura AR (2013) Effect of sago starch loadings on soil decomposition of natural rubber latex. Int Biodeter Biodegrad 85:139–149
24. Kong LX, Peng Z, Li SD, Bartold PM (2000) Nanotechnology and its role in the management of periodontal diseases. Periodontol 40:184–196
25. Ranimol S, Thomas S (2010) Nanocomposites: state of the art, new challenges and opportunities In: Ranimol S, Thomas S (eds) Rubber nanocomposites: preparation, properties, and applications. Wiley, Singapore

26. Dufresne A (2010) Natural rubber green nanocomposites In: Ranimol S, Thomas S (eds) Rubber nanocomposites: preparation, properties, and applications. Wiley, Singapore
27. Ajayan PM, Schadler LS, Giannaris C, Rubio A (2000) Single-walled carbon nanotube–polymer composites: strength and weakness. Adv Mater 12:750–753
28. Thostenson ET (2001) Advances in the science and technology of carbon nanotubes and their composites: a review. Compos Sci Technol 61:1899–1912
29. Sadasivuni KK, Ponnamma D, Thomas S, Grohens Y (2014) Evolution from graphite to graphene elastomer composites. Prog Polym Sci 39:749–780
30. Novoselov KS (2004) Electric field effect in atomically thin carbon films. Science 306:666–669
31. Sadasivuni KK, Saiter A, Gautier N, Thomas S, Grohens Y (2013) Effect of molecular interactions on the performance of poly (isobutylene-co-isoprene)/graphene and clay nanocomposites. Colloid Polym Sci 291:1729–1740
32. Allen MJ, Tung VC, Kaner RB (2010) Honeycomb carbon: a review of graphene. Chem Rev 110:132–145
33. Zhu Y, Murali S, Cai W, Li X, Suk JW, Potts JR, Ruoff RS (2010) Graphene and graphene oxide: synthesis, properties, and applications. Adv Mater 22:3906–3924
34. Compton OC, Nguyen ST (2010) Graphene oxide, highly reduced graphene oxide, and graphene: versatile building blocks for carbon-based materials. Small 6:711–723
35. Mukhopadhyay P, Gupta RK (2011) Trends and frontiers in graphene-based polymer nanocomposites. Plast Eng 32:32–42
36. Prud'Homme RK, Ozbas B, Aksay I, Register R, Adamson D (2010) Functional graphene-rubber nanocomposites. US Patent No 7745528
37. Varghese S, Karger-Kocsis J (2004) Melt-compounded natural rubber nanocomposites with pristine and organophilic layered silicates of natural and synthetic origin. J Appl Polym Sci 91:813–819
38. Varghese S, Karger-Kocsis J (2003) Natural rubber-based nanocomposites by latex compounding with layered silicates. Polymer 44:4921–4927
39. Hambir S, Bulakh N, Jog JP (2002) Polypropylene/clay nanocomposites: effect of compatibilizer on the thermal, crystallization and dynamic mechanical behavior. Polym Eng Sci 42:1800–1807
40. Kodgire P, Kalgoannkar R, Hambir S, Bulukh N, Jog JP (2001) PP/clay nanocomposites: effect of clay treatment on morphology and dynamic mechanical properties. J Appl Polym Sci 81:1786–1792
41. Sadasivuni KK, Castro M, Saiter A, Delbreilh L, Feller JF, Thomas S, Grohens Y (2013) Development of poly(isobutylene-co-isoprene)/reduced graphene oxide nanocomposites for barrier, dielectric and sensing applications. Mater Lett 96:109–112
42. Ponnamma D, Sadasivuni KK, Grohens Y, Guo Q, Thomas S (2014) Carbon nanotubes based elastomer composites-an approach towards multifunctional materials. doi: 10.1039/C4TC01037J
43. Lin N, Yu J, Chang PR, Li J, Huang J (2011) Poly (butylene succinate)-based biocomposites filled with polysaccharide nanocrystals: structure and properties. Polym Compos 32:472–482
44. Carretero-Gonzalez J, Verdejo R, Toki S, Hsiao BS, Giannelis EP, López-Manchado MA (2008) Real time crystallization of organoclay nanoparticles filled natural rubber under stretching. Macromolecules 41:2295–2298
45. Carretero-González J, Retsos H, Verdejo R, Toki S, Hsiao BS, Giannelis EP, López-Manchado MA (2008) Effect of nanoclay on natural rubber microstructure. Macromolecules 41:6763–6772
46. Qu L, Huang G, Liu Z, Zhang P, Weng G, Nie Y (2009) Remarkable reinforcement of natural rubber by deformation-induced crystallization in the presence of organophilic montmorillonite. Acta Mater 57:5053–5060

47. Nie YJ, Huang GS, Qu LL, Wang XA, Weng GS, Wu JR (2011) New insights into thermodynamic description of strain-induced crystallization of peroxide cross-linked natural rubber filled with clay by tube model. Polymer 52:3234–3242
48. Nie YJ, Huang G, Qu L, Zhang P, Weng G, Wu JR (2011) Structural evolution during uniaxial deformation of natural rubber reinforced with nano-alumina. Adv Technol 22:2001–2008
49. Jiang HX, Ni QQ, Natsuki T (2010) Tensile properties and reinforcement mechanisms of natural rubber/vapor-grown carbon nanofiber composite. Polym Compos 31:1099–1104
50. Angles MN, Dufresne A (2000) Plasticized starch/tunicin whiskers nanocomposites 1 structural analysis. Macromolecules 33:8344–8353
51. Kim JT, Oh TS, Lee DH (2004) Curing and barrier properties of NBR/organo–clay nanocomposite. Polym Int 53:406–411
52. Putaux JL, Molina-Boisseau S, Momaur T, Dufresne A (2003) Platelet nanocrystals resulting from the disruption of waxy maize starch granules by acid hydrolysis. Biomacromolecules 4:1198–1202
53. Youssef H, Lucian AL, Orlando JR (2010) Cellulose nanocrystals: chemistry, self-assembly, and applications. Chem Rev 110:3479–3500
54. Abraham E, Deepa B, Pothan LA, Jacob M, Thomas S, Cvelbar U, Anandjiwala R (2011) Extraction of nanocellulose fibrils from lignocellulosic fibres: a novel approach. Carbohydr Polym 86:1468–1475
55. Schurz J (1999) Trends in polymer science—a bright future for cellulose. Prog Polym Sci 24:481–483
56. Teh PL, Ishak ZAM, Hashim AS, Karger-Kocsis J, Ishiaku US (2004) Effects of epoxidized natural rubber as a compatibilizer in melt compounded natural rubber–organoclay nanocomposites. Eur Polym J 40:2513–2521
57. Magaraphan R, Thaijaroen W, Lim-Ochakun R (2003) Structure and properties of natural rubber and mont morrilonite nanocomposites. Rubber Chem Technol 76:406–418
58. Nair KG, Dufresne A (2003) Crab shell chitin whisker reinforced natural rubber nanocomposites 1 processing and swelling behavior. Biomacromolecules 4:657–665
59. Nair KG, Dufresne A (2003) Crab shell chitin whisker reinforced natural rubber nanocomposites 2 mechanical behavior. Biomacromolecules 4:666–674
60. Nair KG, Dufresne A (2003) Crab shell chitin whiskers reinforced natural rubber nanocomposites 3 effect of chemical modification of chitin whiskers. Biomacromolecules 4:1835–1842
61. Peng Z, Kong LX, Li SD (2005) Non-isothermal crystallisation kinetics of self-assembled polyvinylalcohol/silica nano-composite. Polymer 46:1949–1955
62. Peng Z, Kong LX, Li SD (2005) Thermal properties and morphology of a poly (vinyl alcohol)/silica nanocomposite prepared with a self-assembled monolayer technique. J Appl Polym Sci 96:1436–1442
63. Peng Z, Kong LX, Li SD, Spiridonov P (2006) Poly (vinyl alcohol)/silica nanocomposites: morphology and thermal degradation kinetics. J Nanosci Nanotechnol 6:3934–3938
64. Li SD, Peng Z, Kong LX, Zhong JP (2006) Thermal degradation kinetics and morphology of natural rubber/silica nanocomposites. J Nanosci Nanotechnol 6:541–546
65. Saito R, Dresselhaus G, Dresselhaus MS (1998) Physical properties of carbon nanotubes. London Imperial College Press, London
66. Huczko A (2002) Synthesis of aligned carbon nanotubes. Appl Phys A Mater Sci Process 74:617–638
67. Berber S, Kwon YK, Tomanek D (2000) Unusually high thermal conductivity of carbon nanotubes. Phys Rev Lett 84:4613–4616
68. Ponnamma D, Sadasivuni KK, Strankowski M, Guo Q, Thomas S (2013) Synergistic effect of multi walled carbon nanotubes and reduced graphene oxides in natural rubber for sensing application. Soft Matter 9:10343
69. Hone J, Llaguno MC, Nemes NM, Johnson AT, Fischer JE, Walters DA, Casavant MJ, Schmidt J, Smalley RE (2000) Electrical and thermal transport properties of magnetically aligned single wall carbon nanotube films. Appl Phys Lett 77:666–668

70. Thomas PS, Abdullateef AA, Al-Harthi MA, Atieh MA, De SK, Rahaman M, Chaki TK, Khastgir D, Bandyopadhyay S (2012) Electrical properties of natural rubber nanocomposites: effect of 1-octadecanol functionalization of carbon nanotubes. J Mater Sci 47(7):3344–3349
71. Weng GS, Huang GS, Qu LL, Nie YJ, Wu JR (2010) Large-scale orientation in a vulcanized stretched natural rubber network: proved by in situ synchrotron X-ray diffraction characterization. J Phys Chem B 114:7179–7188
72. Tonpheng B, Andersson O (2008) Crosslinking, thermal properties and relaxation behaviour of polyisoprene under high-pressure. Eur Polym J 44:2865–2873
73. Liang J, Huang Y, Ma Y, Liu Z, Cai J, Zhang C, Gao H, Chen Y (2009) Electromagnetic interference shielding of graphene/epoxy composites. Carbon 47:922–925
74. De Rosa M, Mancinelli R, Sarasini F, Sarto MS, Tamburrano A (2009) Electromagnetic design and realization of innovative fiber-reinforced broad-band absorbing screens. IEEE Trans Electromag Compat 51:700–707
75. De Rosa M, Dinescu A, Sarasini F, Sarto MS, Tamburrano A (2010) Effect of short carbon fibers and MWCNTs on microwave absorbing properties of polyester composites containing nickel coated carbon fibers. Compos Sci Technol 70:102
76. De Bellis G, De Rosa IM, Dinescu A, Sarto MS, Tamburrano A (2010) Proceedings of the 2010 IEEE international symposium on electromagnetic compatibility, Fort Lauderdale 202
77. Angellier H, Molina-Boisseau S, Belgacem MN, Dufresne A (2005) Surface chemical modification of waxy maize starch nanocrystals. Langmuir 21:2425–2433
78. Trovatti E, Carvalho AJF, Ribeiro SJL, Gandini A (2013) Simple green approach to reinforce natural rubber with bacterial cellulose nanofibers. Biomacromolecules 14(8):2667–2674
79. Avérous L, Halley PJ (2009) Biocomposites based on plasticized starch. Biofuels Bioprod Bioref 3:329
80. Avérous L (2004) Biodegradable multiphase systems based on plasticized starch: a review. J Macromol Sci C Polym Rev C 44:231–274
81. Ray S, Bousmia M (2005) Biodegradable polymers and their layered silicate nanocomposites: in greening the 21st century materials world. Prog Mater Sci 50:962–1079
82. Roy N, Sengupta R, Bhowmick AK (2012) Modifications of carbon for polymer composites and nanocomposites. Prog Polym Sci 37(6):781–819
83. Dufresne A, Cavaille JY, Helbert W (1996) New nanocomposite materials: microcrystalline starch reinforced thermoplastic. Macromolecules 29:7624–7626
84. Baek JB, Lyons CB, Tan LS (2004) Grafting of vapor-grown carbon nanofibers via in-situ polycondensation of 3-phenoxybenzoic acid in poly (phosphoric acid). Macromolecules 37:8278–8285
85. Lu YL, Ye FY, Mao LX, Li Y, Zhang LQ (2011) Micro-structural evolution of rubber/clay nanocomposites with vulcanization process. Express Polym Lett 5:777–787
86. Coleman JN, Khan U, Blau WJ, Gunko YK (2006) Small but strong: a review of the mechanical properties of carbon nanotube–polymer composites. Carbon 44:1624–1652
87. Du JH, Bai J, Cheng HM (2007) The present status and key problems of carbon nanotube based polymer composites. eXPRESS Polym Lett 1:253–273
88. Jiang C, He H, Jiang H, Ma L, Jia D M (2013) Nano-lignin filled natural rubber composites: preparation and characterization. eXPRESS Polym Lett 7:480–493
89. Majdzadeh-Ardakani K, Ardakani Sh Sadeghi- (2010) Experimental investigation of mechanical properties of starch/natural rubber/clay nanocomposites. Digest J Nanomater Biostructures 5:307–316
90. Lopez-Manchado MA, Herrero B, Arroyo M (2003) Preparation and characterization of organoclay nanocomposites based on natural rubber. Polym Int 52:1070–1077
91. Andrews R, Jacques D, Minot M, Rantell T (2002) Fabrication of carbon multiwall nanotube/polymer composites by shear mixing. Macromol Mater Eng 287:395–403

92. Breuer O, Sundararaj U (2004) Big returns from small fibers: a review of polymer/carbon nanotube composites. Polym Compos 25:630–645
93. Liu Q, Zhang D, Fan T, Gu J, Miyamoto Y, Chen Z (2008) Amorphous carbon-matrix composites with interconnected carbon nano-ribbon networks for electromagnetic interference shielding. Carbon 46(3):461–465
94. Zhang CS, Ni QQ, Fu SY, Kurashiki K (2007) Electromagnetic interference shielding effect of nanocomposites with carbon nanotube and shape memory polymer. Compos Sci Technol 67:2973–2980
95. Yakuphanoglu F, Al-Ghamdi AA, El-Tantawy F (2014) Electromagnetic interference shielding properties of nanocomposites for commercial electronic devices. Microsyst Technol 1–9
96. Tanrattanakul V, Bunchuay A (2007) Microwave absorbing rubber composites containing carbon black and aluminum powder. J Appl Polym Sci 105:2036–2045
97. Kong I, Ahmada S, Abdullah MH, Hui D, Yusoff AN, Puryanti D (2010) Magnetic and microwave absorbing properties of magnetite-thermoplastic natural rubber nanocomposites. J Mag Mag Mater 322:3401–3409
98. Al-Hartomy OA, Al-Ghamdi A, Dishovsky N, Shtarkova R, Iliev V, Mutlay I, El-Tantawy F (2012) Dielectric and microwave properties of natural rubber based nanocomposites containing graphene. Mater Sci Appl 3:453
99. Arief PT, Kean CA, Jadranka T (2012) A novel polypyrrole and natural rubber based flexible large strain sensor. Sens Actuators B 20:166
100. He Q, Yuan T, Zhang X, Guo S, Liu J, Liu J, Liu X, Sun L, Wei S, Guo Z (2014) Heavy duty piezoresistivity induced strain sensing natural rubber/carbon black nanocomposites reinforced with different carbon nanofillers. Mater Res Express 1(3):035029
101. Tadakaluru S, Thongsuwan W, Singjai P (2014) Stretchable and flexible high-strain sensors made using carbon nanotubes and graphite films on natural rubber. Sensors 14:868–876
102. Knite M, Tupureina V, Fuith A, Zavickis J, Teteris V (2007) Polyisoprene—multi-wall carbon nanotube composites for sensing strain. Mater Sci Eng C 27:1125
103. Herculano RD, Brunello CA, Graeff CFO (2007) Optimization of a novel nitric oxide sensor using a latex rubber. J Appl Sci 7:3801
104. John H, Joseph R, Mathew KT (2007) Dielectric behavior of natural rubber composites in microwave fields. J Appl Polym Sci 103:2682–2686
105. Makled HM (2012) Dielectric properties of high coercivity barium ferrite–natural rubber composites. J Appl Polym Sci 126:969
106. Haseena AP, Unnikrishnan G, Kalaprasad G (2007) Dielectric properties of short sisal/coir hybrid fibre reinforced natural rubber composites. Compos Interf 14:763–786
107. Popielarz R, Chiang CK, Nozaki R, Obrzut J (2001) Dielectric properties of polymer/ferroelectric ceramic composites from 100 Hz to 10 GHz. Macromolecules 34:5910–5915
108. Marzinotto M, Santulli C, Mazzetti C (2007) Dielectric properties of oil palm-natural rubber biocomposites. IEEE Electr Insul Dielectr Phenom CEIDP 9777934:584–587
109. Jacob M, Varughese KT, Thomas S (2006) Dielectric characteristics of sisal—oil palm hybrid biofibre reinforced natural rubber biocomposites. J Mater Sci 41:5538–5547
110. Jacob M, Thomas S, Varughese KT (2004) Mechanical properties of sisal/oil palm hybrid fiber reinforced natural rubber composites. Compos Sci Tech 64:955–965
111. Tangboriboon N, Uttanawanit N, Longtong M, Wongpinthong P, Sirivat A, Kunanuruksapong R (2010) Electrical and electrorheological properties of alumina/natural rubber (STR XL). Compos Compos Mater 3:656–671
112. Elimat ZM, Zihlif AM, Ragosta G (2008) Study of ac electrical properties of aluminium–epoxy composites. J Phys D Appl Phys 41:165408
113. Musameh SM, Abdelazeez MK, Ahmad MS, Zihlif AM, Malinconico M, Martuscelli E, Ragosta G (1991) Some electrical properties of aluminum-epoxy composite. Mater Sci Eng B 10:29–33
114. Sadasivuni KK, Ponnamma D, Kumar B, Strankowski M, Cardinaels R, Moldenaers P, Thomas S, Grohens Y (2014) Dielectric properties of modified graphene oxide filled

polyurethane nanocomposites and its correlation with rheology. Compos Sci Technol 104:18–25
115. Dang ZM, Nan CW, Xie D, Zhang YH, Tjong SC (2004) Dielectric behavior and dependence of percolation threshold on the conductivity of fillers in polymer-semiconductor composites. Appl Phys Lett 85:97–99
116. Sinclair DC, Adams TB, Morrison FD, West AR (2002) CaCu3Ti4O12: one-step internal barrier layer capacitor. Appl Phys Lett 80:2153–2155
117. Jana PK, Sarkar S, Sakata H, Watanabe T, Chaudhuri BK (2008) Microstructure and dielectric properties of NaxTiyNi1 − x−yO (x = 0.05–0.30, y = 0.02). J Phys D Appl Phys 41:65403
118. Pohl HA (1978) Dielectrophoresis. Cambridge University Press, London
119. Jamal EMA, Joy PA, Kurian P, Anantharaman MR (2009) Synthesis of nickel–rubber nanocomposites and evaluation of their dielectric properties. Mater Sci Eng B 156:24–31
120. Psarras GC, Gatos KG, Karahaliou PK, Georga SN, Krontiras CA, Karger-Kocsis J (2007) Relaxation phenomena in rubber/layered silicate nanocomposites. eXPRESS Polym Lett 1:837–845
121. Kornev A, Bukanov A, Sheverdiaev O (2005) Technology of elastomeric materials, in Russian. Istek, Moscow
122. Banerjee P, Biswas S (2011) Dielectric properties of EVA rubber composites at microwave frequencies theory, instrumentation and measurements. J Micro Power Electromag Energ 45:24–29
123. Hernández M, Bernal MM, Verdejo R, Ezquerra TA, López-Manchado MA (2012) Overall performance of natural rubber/graphene nanocomposites. Compos Sci Technol 73:40–46
124. Yu J, Andersson O (2009) Thermal conductivity, heat capacity, and cross-linking of polyisoprene/single-wall carbon nanotube composites under high pressure. Macromolecules 42:9295–9301
125. Wei C, Srivastava D, Cho K (2002) Thermal expansion and diffusion coefficients of carbon nanotube-polymer composite. Nano Lett 2:647–650
126. Flaifel MH, Ahmad SH, Hassan A, Bahri S, Tarawneh MA, Yu L (2013) Thermal conductivity and dynamic mechanical analysis of NiZn ferrite nanoparticles filled thermoplastic natural rubber nanocomposites. Compos Part B 52:334–339
127. Gojny FH, Wichmann MHG, Fiedler B, Kinloch IA, Bauhofer W, Windle AH, Schulte K (2006) Evaluation and identification of electrical and thermal conduction mechanisms in carbon nanotube/epoxy composites. Polymer 47:2036–2045
128. Potts JR, Shankar O, Du L, Ruoff RS (2012) Processing–morphology–property relationships and composite theory analysis of reduced graphene oxide/natural rubber nanocomposites. Macromolecules 45:6045–6055
129. Zakaria MZ, Ahmad SH (2013) Investigation on thermal conductivity and mechanical properties of thermoplastic natural rubber filled with alumina and boron carbide nanocomposites. Energ Environ Eng J 2:11–14
130. Zhan Y, Wu J, Xia H, Yan N, Fei G, Yuan G (2011) Dispersion and exfoliation of graphene in rubber by an ultrasonically-assisted latex mixing and in situ reduction process. Macromol Mater Eng 296:590–602
131. Meera AP, Tlili R, Boudenne A, Ibos L, Poornima V, Thomas S, Candau Y (2012) Thermophysical and mechanical properties of TiO_2 and silica nanoparticle-filled natural rubber composites. J Elast Plast 44:1–14
132. Zhamu A, Bor JZ (2011) Pristine nano graphene-modified tyres. US Patent 2011/0046289A1
133. Ponnamma D, Sadasivuni KK, Strankowski M, Moldenaers P, Thomas S, Grohens Y (2013) Interrelated shape memory and Payne effect in polyurethane/graphene oxide nanocomposites. RSC Adv 3:16068
134. Fukahori Y (2010) Mechanism of the self-reinforcement of cross-linked NR generated through the strain-induced crystallization. Polymer 51:1621–1631

135. Heuwers B, Quitmann D, Hoeher R, Reinders FM, Tiemeyer S, Sternemann C, Tolan M, Katzenberg F, Tiller JC (2013) Stress-induced stabilization of crystals in shape memory natural rubber. Macromol Rapid Commun 34:180–184
136. Katzenberg F, Heuwers B, Tiller JC (2011) Superheated rubber for cold storage. Adv Mater 23:1909–1911
137. Heuwers B, Beckel A, Krieger A, Katzenberg F, Tiller JC (2013) Shape-memory natural rubber: an exceptional material for strain and energy storage. Macromol Chem Phys 214:912
138. Bruns N, Tiller JC (2006) Nanophasic amphiphilic conetworks with a fluorophilic phase. Macromolecules 39:4386
139. Quitmann D, Gushterov N, Sadowski G, Katzenberg F, Tiller JC (2013) Solvent-sensitive reversible stress-response of shape memory natural rubber. ACS Appl Mater Interf 5:3504
140. Jincheng W, Yan G (2011) Hyperbranched intumescent flame-retardant agent: application to natural rubber composites. J Appl Polym Sci 122:3474
141. Niamlang S, Thongchai S, Pawananant N, Sirivat A (2013) The electromechanical properties of crosslinked natural rubber. Energy Procedia 34:697–704
142. Puvanatvattana T, Chotpattananont D, Hiamtup P, Niamlang S, Sirivat A, Jamieson AM (2006) Electric field induced stress moduli in polythiophene/polyisoprene elastomer blends. React Funct Polym 66:1575–1588
143. Sirivat A, Petcharoen K, Pornchaisiriarun Y, Phansa-Ard C, Tangboriboon N (2014) Lead zirconate (PbZrO3) embedded in natural rubber as electroactive elastomer composites. J Innovative Opt Health Sci 7:1450016
144. Tangboriboon N, Datsanae S, Onthong A, Kunanuruksapong R, Sirivat A (2012) Electromechanical responses of dielectric elastomer composite actuators based on natural rubber and alumina. J Elastomers Plast 45:143–161
145. Niamlang S, Thongchai S, Pawananant N, Sirivat A (2013) The electromechanical properties of crosslinked natural rubber. Energy Procedia 34:697–704

Electronic Applications of Polymer Electrolytes of Epoxidized Natural Rubber and Its Composites

Fatin Harun and Chin Han Chan

Abstract The hope for composite polymer electrolytes (CPE) to fulfill future electronic applications has encouraged researchers to look up for wider range of polymer candidates, which are suitable for this purpose. Epoxidized natural rubber (ENR) is one of the candidates with polar epoxy group that helps in compatibilization with other polymers and provides coordination sites for fillers and organic or inorganic salt. Besides, low glass transition temperature, good dimensional stability and impact strength, and satisfactory stickiness finally provide sufficient contact between electrolytic layer and electrode for electronic devices. The desirable properties of ENR have earned researchers' attention to further explore its potential. Hence, this work will attempt to compile and review the ENR, ENR-based, and composite ENR-based electrolytes for electronic applications, especially in ionic conductivity.

Keywords Epoxidized natural rubber · Polymer electrolytes · Ionic conductivity · Glass transition temperature · Selective localization

1 Introduction to Epoxidized Natural Rubber

1.1 Structure and Composition

Pummere and Burkard [1] were the first to report the epoxidation reaction of natural rubber (NR). Epoxidized natural rubber (ENR) is a modified form of NR in which some of the double bonds in the *cis*-polyisoprene chain have been converted to oxirane groups. The epoxy groups are randomly distributed along the backbone

F. Harun · C.H. Chan (✉)
Faculty of Applied Sciences, Universiti Teknologi MARA (UiTM), 40450 Shah Alam, Selangor, Malaysia
e-mail: cchan@salam.uitm.edu.my

D. Ponnamma et al. (eds.), *Flexible and Stretchable Electronic Composites*, Springer Series on Polymer and Composite Materials,
DOI 10.1007/978-3-319-23663-6_2

of NR [2]. The chemistry of epoxidation of NR and the subsequent ring-opening reactions have been extensively studied by Swern [3]. Epoxidation is a stereo-specific reaction, and the rate of epoxidation is governed by the substituents on the double bond [3, 4]. Only in the mid-1980s, the pure samples of ENR were prepared and their potential applications were realized [5–7].

ENR is composed of 2-methylbut-2-ene and 2,3-epoxy-2-methylbutane units. Upon epoxidation, ENR can still undergo strain crystallization since oxygen is small enough to fit into the crystal lattice with only minor geometrical rearrangements. Therefore, it maintains the superior tensile properties of NR [8, 9]. Epoxidation is a highly exothermic reaction (60 kcal g^{-1} mol^{-1}), but under carefully controlled conditions of reaction temperatures and acid concentration below 20 w/w%, NR latex can be epoxidized to over 75 mol% without the formation of secondary ring-opened structures [7]. Nevertheless, most of the studies are limited to commercially available ENR, namely ENR-25 and ENR-50, where 25 and 50 denote 25 and 50 mol% of epoxide group. This type of chemical modification offers excellent scope of potential applications considering low cost of NR latex and simple nature of ENR production [10].

The properties of ENR change gradually with the epoxide level. Hysteresis, swelling resistance, and gas impermeability improve with increasing epoxide level, while resilience, on the other hand, decreases [5, 11]. The physical properties of ENR could be governed by increase in glass transition temperature (T_g), solubility, and ability to undergo strain crystallization [8].

The commercial production of ENR in the early time and after the year 1969 has several differences in their properties. Before 1969, the "old ENR" was known to have high Mooney viscosities of about 120, high gel contents (60–80 %) and acidic in nature (pH 6). This results in difficulty of mixing and filler dispersion in the rubber. Mixing can be optimized by premastication of the rubber in the presence of a base, but even so, filler could not be incorporated into the highly cross-linked macrogel phase causing heterogenous mixture [12].

In 1989, a different technological production of ENR known as Epoxyprene was launched by Kumpulan Guthrie Berhad, Malaysia. Epoxyprene has lower Mooney viscosities of about 80, more consistent composition, lower gel content (30–40 %), and alkaline in nature (pH 7–10), leading to easy processing of rubber. The advantages of the epoxyprene are that it can be compounded without the need of premastication, and homogenous carbon black filler distribution can be obtained [13, 14].

1.2 Synthesis Pathways

Epoxidation reaction can be carried out in solution or latex form (refer Table 1) but only latter is of commercial values. *In situ* epoxidation is preferred to minimize processing and costs [15].

Table 1 Epoxidation routes of NR

	Description
(1) Solution form	(a) Perbenzoic acid [16, 17]
	(b) Perphthalic acid [18]
	(c) Peracetic acid [19, 20]
	(d) Hydrogen peroxide/acetic mixtures catalyzed by *p*-toluene sulfonic acids [17, 19]
(2) Latex form	(a) Bromohydrin intermediates, hydrogen peroxide-catalyzed system [15] • Dibromide and bromohydrin formed during bromination, consumed bromine that was intended to react with rubber • Low efficiencies of epoxidation may be related to the difficulty in transferring water-soluble intermediate to the rubber particles
	(b) Peracetic acid [21–23] • Easy preparation using acetic anhydride and hydrogen peroxide • Low cost of reagents • Peracetic acid is stable and compatible with both aqueous and hydrocarbon phase of latex, uniform reaction • Quantitative conversion results in excellent efficiency of epoxidation • No evidence of ring-opened epoxide groups at moderately low modification level • Negligible side products when carried out under ambient temperature and in the absence of strong acid catalyst
	(c) Performic acid [15, 24, 25] • Prepared from hydrogen peroxide and formic acid • Epoxidation is accompanied by extensive amounts of secondary products • High acidity accelerates ring-opening reactions

Comparing the choices of epoxidation routes of NR, it can be concluded that epoxidation in latex using peracetic acid (refer Fig. 1) is one of the best methods of all, for laboratory scale as well as standard industrial production. To prepare latex epoxidation, it is necessary to first stabilize the latex by adding nonionic surfactant. The surface of the rubber particles surrounded by proteins is anionic charged, while peracetic acid to be added is cationic charged. Adding different charge surfactants to the NR latex would cause the rubber particles to destabilize. Therefore, nonionic surfactant must be used to stabilize NR latex against acid coagulation. After epoxidation completed, the ENR can be precipitated either by

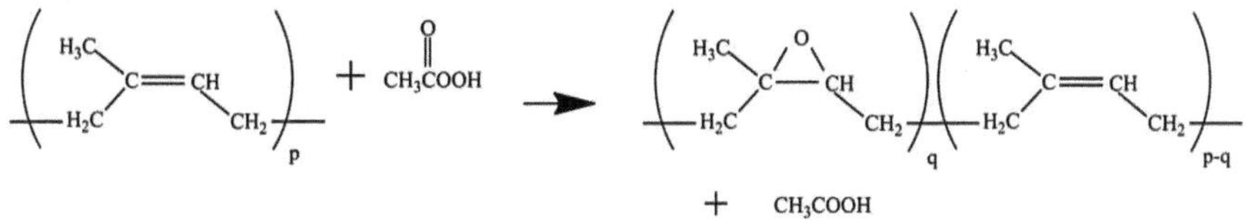

Fig. 1 Epoxidation of NR using peracetic acid

using alcohol or acetone, adding salt, or heating to temperature above the cloud point of the surfactant. Lastly, the ENR can be washed with base to remove all unreacted acid [26].

1.3 Characteristics

ENR consists of two parts namely, sol (soluble in organic solvent) and gel (insoluble in organic solvent). The sol was deduced to be intramolecular ether with linear structures, whereas gel is intermolecular ether with net-like structures [5, 27]. High cross-link structure of gel content was believed to be attributed by higher degree of epoxidation [5], minute structural changes (free radical) in ENR that may occur during milling [28], thermal effect during molding of rubber [29], interchain interactions by H-bonding through ring-opened products of ENR [29], and temperature [30]. Mild reaction conditions can avoid the ring-opening reactions. Nevertheless, the longer preparation time required is not favorable for scaled production of ENR.

1.3.1 ^{1}H-Nuclear Magnetic Resonance (^{1}H-NMR)

The determination of ENR structure using ^{1}H-NMR method (refer Table 2) was developed by Durbetaki and Miles [31]. For NR, the characteristic signals of

Table 2 Chemical shift assignment of ^{1}H-NMR for ENR

Chemical shift (ppm)	Assignment	Reference
1.7	Methyl CH_3–C=C	[32]
2.1	Methylene –CH_2–	[32]
5.1	Unsaturated methyne/olefinic hydrogen CH=C	[33]
1.3	Methyl of epoxy group CH_3–CO	[20]
2.7	Methyne of epoxy group	[20]
3.4	Diol	[2, 34]
3.9	Furan	[34]

Note ^{1}H-NMR spectrum for ENR-20

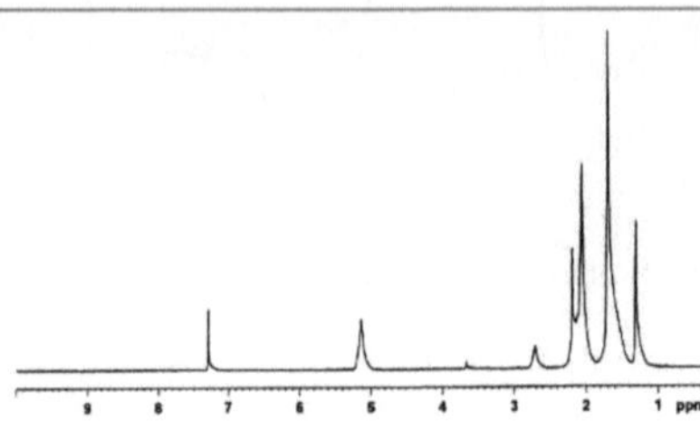

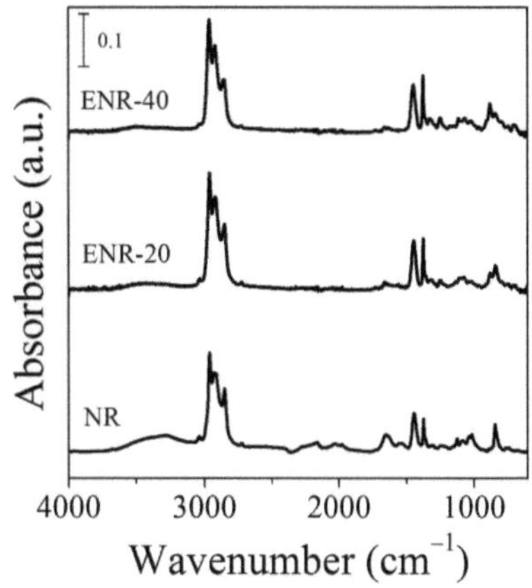

Fig. 2 FTIR spectra of NR and ENR

methyl, methylene, and unsaturated methyne protons of *cis*-1,4-isoprene units appeared at 1.7, 2.1 and 5.1 ppm, respectively. However, ENR spectrum shows two additional signals that appeared at 1.3 and 2.7 ppm which represent the methyl and methyne protons of the epoxy group, respectively.

1.3.2 Fourier Transform Infrared Spectroscopy (FTIR)

The FTIR spectra of NR and ENR are given in Fig. 2. The presence of C=C stretching (1654 cm^{-1}), CH olefin wagging (842 cm^{-1}), and CH_3 deformation (1376 cm^{-1}) confirm the identity of NR. On the other hand, ENR shows additional peak at 880 and 1250 cm^{-1} corresponding to oxirane ring and stretching vibration of C–O–C, respectively.

1.3.3 Thermal Properties

T_g of ENR increases as the amount of epoxide groups increases (refer Fig. 3). The presence of bulky epoxide group lowers the rotational freedom of the modified segment [15]. For every 1 mol% epoxidation, the T_g increases by approximately 1 °C [15, 35]. The presence of proteins in NR does not interfere with the formation of epoxide and segmental mobility of the chain [36].

The thermal stability of ENR in nitrogen atmosphere and in air exhibits different trend by thermal gravimetry analysis (TGA). In nitrogen atmosphere, ENR shows single-step decomposition. Presence of epoxy group was claimed to provide higher thermal stability in inert environment (refer Fig. 4) due to epoxide ring opening that occurs during the test which consequently slows down the rubber degradation [37]. On the other hand, presence of epoxy group destabilizes thermal

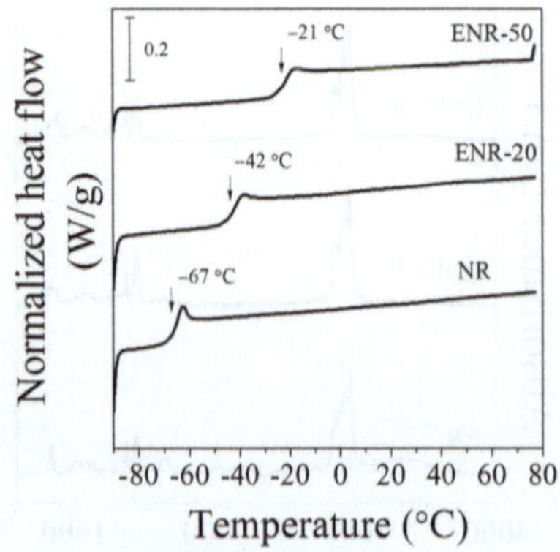

Fig. 3 DSC traces of reheating cycle for NR and ENRs

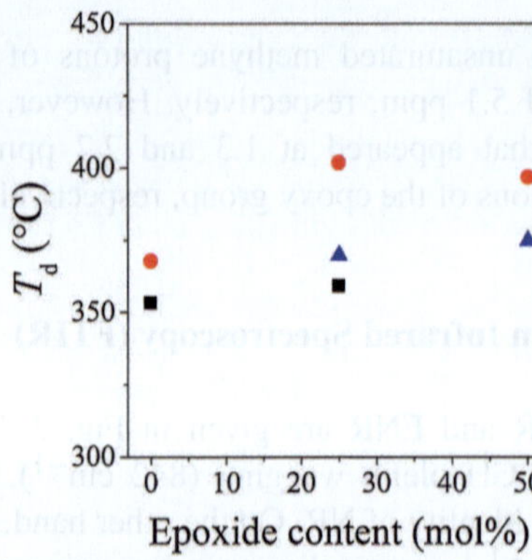

Fig. 4 Onset of thermal degradation temperature of ENR in nitrogen atmosphere, *square* [39], *circle* [40], and *triangle* [unpublished result]

stability of ENR in air. Thermal degradation of ENR in air shows two-step decomposition [27, 38]. The steps are in the range of 200–450 °C (approaches first-order reaction) and 450–550 °C, respectively.

1.3.4 X-ray Diffraction

NR can undergo strain-induced crystallization where the unit cell was reported to be orthorhombic (92°) or monoclinic (90°) [41, 42]. Meanwhile, ENR also displays the same unit cell where the volume increases with degree of epoxidation (refer Table 3). Beyond 50 mol%, the degree of crystallinity reduces drastically which may imply that the basic unit of NR can accommodate two epoxide groups [7, 8].

Table 3 Unit cell volume and degree of crystallinity estimated by X-ray diffractometer for ENR vulcanizates (2 phr sulfur) at 400 % strain. Copyright 1987 reprinted with permission from [13]

ENR	Unit cell volume (nm^3)	Degree of crystallinity
NR	0.955	11
ENR-25	0.985	11
ENR-50	0.999	10
ENR-75	1.036	4
ENR-90	1.036	2

Table 4 Mechanical properties of NR, ENR-25, and ENR-50

	NR	ENR-25	ENR-50
Modulus at 100 % elongation (MPa)	0.25 [39]	0.20 [39]	0.33 [44]
Elongation at break (%)		322 [14]	482 [14] 267 [44]
Green strength (MPa)[a]		0.34 [14]	0.49 [14]
Wallace plasticity	45 [14]	33 [14]	40 [14]
Plasticity retention index	>50 [14] 66 [45] 3 (after solvent extraction) [14]	45 [14]	45 [14] 22 [45]
Tensile strength (kPa)			548 [44]
Shore A hardness			15 [44]

[a]Green strength can be defined as resistance to deformation and fracture before vulcanization

1.3.5 Mechanical Properties

ENR becomes mechanically harder with increasing epoxide contents. This is related to the increased T_g of this modified NR [15]. Despite that, below 25 mol% modification, ENR is as elastic as NR [43]. Resilience of ENR decreases with increasing amount of epoxide, improving its wet grip performance. Some mechanical properties of ENR are summarized in Table 4.

1.3.6 Air Permeability

The air permeability of ENR is dependent on the level of epoxidation (refer Table 5). Epoxidation lowers the air permeability due to reduced segmental rotation rates and higher ENR density [11, 15].

Table 5 Air permeability of ENR. Copyright reprinted with permission from [46]

	Air permeability ($m^4 s^{-1} N^{-1} \times 10^{18}$)
NR	28.1
ENR-10	14.8
ENR-25	5.9
ENR-50	1.9

1.3.7 Aging Behavior

Wong and Ong [14] studied the aging behavior of NR, ENR-25, and ENR-50 at 50, 70, and 90 °C, respectively. As temperature increases, reduction in Wallace plasticity can be observed in ENR where the effect is more significant in higher epoxide rubber and at higher temperature (refer Fig. 5). Green strength, elongation, and modulus of stress–strain properties of ENR show no significant increase upon aging (refer Table 6). The amount of gel and epoxide content in ENR was found to remain consistent even when they were aged at 70 °C.

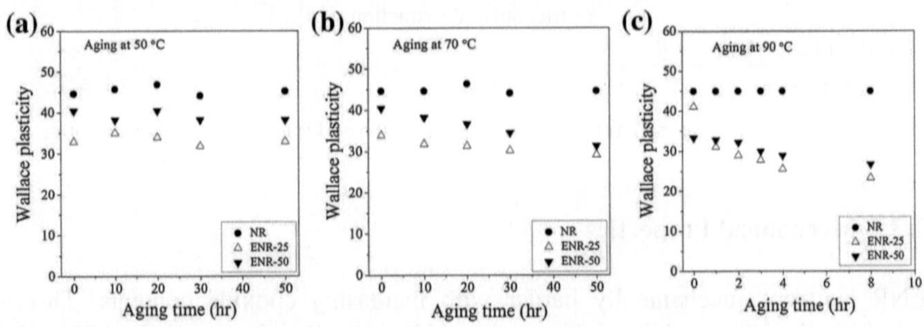

Fig. 5 Changes in Wallace plasticity at different temperatures, **a** 50 °C, **b** 70 °C, **c** 90°C. Copyright reprinted with permission from [14]

Table 6 Green strength and elongation at break of aging ENR. Copyright reprinted with permission from [14]

	ENR-25				ENR-50			
	Unaged	Aged at 50 °C			Unaged	Aged at 50 °C		
		4 h	16 h	32 h		4 h	16 h	32 h
Green strength (MPa)	0.34	0.21	0.26	0.24	0.49	0.42	0.40	0.37
Elongation at break (%)	322	327	307	310	482	513	493	447

1.4 Techniques to Determine Epoxide Content

There are many techniques to determine epoxide content of ENR. The most convenient and reliable method would be using ^{1}H-NMR and DSC. The advantages and disadvantages of each technique are summarized (refer Table 7).

Table 7 Various methods in determining epoxide content of ENR

<table>
<tr><th>Method</th><th>Techniques to determine epoxide content</th></tr>
<tr><td>1. ^{1}H-NMR [47]</td><td>Ratio of integrated areas of the olefinic and epoxy methane protons
$\text{Mol\% epoxide} = \left(\frac{A_{2.70}}{A_{5.14}+A_{2.70}}\right)$
In the range of 25–70 mol%, reproducibility was found to be ±1.5 %</td></tr>
<tr><td>2. ^{13}C-NMR [20]</td><td>$\text{Mol\% epoxide} = \frac{100(A_{64.5})}{A_{64.5}+A_{124.4,125.0,125.7}}$
In the range of 25–70 mol%, reproducibility was found to be ±0.8 %</td></tr>
<tr><td>3. Elemental analysis</td><td>%C, %H and %O
Large divergent of reproducibility: 12.9 %
Possibility of sample inhomogeneity which would be accentuated by small samples required for microanalysis</td></tr>
<tr><td>4. HBr titration [48]</td><td>Direct titration of oxirane ring with HBr is essentially quantitative and with few interferences for ENR at low modification level (<5 mol%)
Rapid technique being appropriate
<15 mol% where spectroscopic techniques are less accurate
Limited solubility at higher modification level as well as cyclization side reactions between adjacent oxirane rings</td></tr>
<tr><td>5. Differential scanning calorimetry (DSC) [49]</td><td>T_g of ENR increases linearly with epoxy content
$\text{Mol\% epoxide} = \frac{T_g+70}{0.92}$
Reproducibility: 0.25 %
Presence of significant side products derived from cyclization, hydrolysis, or esterification would nullify validity of calibration</td></tr>
<tr><td>6. Density [15]</td><td>T_g value is directly related to polymer density (density vs. T_g plot)
<table>
<tr><th>Epoxide content (%)</th><th>T_g (°C)</th><th>Density (g/cm^3)</th></tr>
<tr><td>0</td><td>−66.4</td><td>0.898</td></tr>
<tr><td>10</td><td>−60.3</td><td>0.906</td></tr>
<tr><td>20</td><td>−53.0</td><td>0.917</td></tr>
<tr><td>35</td><td>−42.2</td><td>0.937</td></tr>
<tr><td>50</td><td>−30.1</td><td>0.965</td></tr>
<tr><td>65</td><td>−16.6</td><td>0.986</td></tr>
</table>
Convenient low-cost measure of epoxy content
The progressive increase in T_g parallels with increase in polymer density due to reduction in volume of the chain segments</td></tr>
</table>

(continued)

Table 7 (continued)

Method	Techniques to determine epoxide content
7. FTIR [50, 51]	(a) Lambert–Beer law to calculate percentage of epoxy groups of ENR $C_e = \frac{100k_1A_2}{A_1+k_1A_2+k_2A_3}$ $C_d = \frac{C_2A_1}{k_1A_2}$ $C_o = 100 - C_d - C_d$ $A_1 = A_{835}$ $A_2 = A_{870}–0.14A_{835}$ $A_3 = A_{3460}–0.019A_{1375}$ where C_e, C_d, and C_o are the molar percentage of epoxy groups, carbon–carbon double bonds, and ring-opening products, respectively. A_{835}, A_{870}, A_{1375}, and A_{3460} are the absorbencies corresponding to 835, 870, 1375, and 3460 cm^{-1}. The values of k_1 and k_2, calculated from the C_e and C_d values determined from NMR method, are 0.77 and 0.34, respectively (b) Mol% epoxide = $\left(\frac{A_{874}}{A_{840}+A_{874}}\right)$

1.5 Applications of ENR

ENR has been utilized in various fields since the past such as automotives [52], substitute of acrylonitrile butadiene and butyl rubber products [50], coating [39, 53], clamping [54], wire and cabling [55], adhesions and sealants [56, 57], compatibilizer [58], and toughening agent in plastics [59, 60]. Recently, ENR is being featured as autonomous healing materials [61, 62] and for electronic applications [63–70] something uncommon and outside the traditional uses of rubber. In the next section, we will focus on the electronic applications of composite ENR-based polymer electrolytes, and in particular, we concentrate on impedance results.

2 Electronic Applications of ENR

2.1 Solid Polymer Electrolytes (SPEs)

SPE is made up of metal salt that dissolves in the polar matrix polymer to form a solid salt solution [71]. It has favorable mechanical strength, ease in fabrication of thin films into desirable size, and its ability to form effective electrode–electrolyte contacts [72]. Over many years, SPEs still remain the potential of interest for various applications in electrochemical devices, such as separators or electrolyte membranes in high energy density lithium-ion rechargeable batteries, fuel cell membrane, electrochromic displays, photoelectrochemical devices, smart windows, data storage, sensors, transistors, supercapacitors, and many more [71, 73, 74]. Therefore, it is desirable for SPEs to show acceptable conductivities

(~10^{-3} to 10^{-4} S cm^{-1}) at room temperature, dimensional stability, and elastomeric properties [73, 75].

The common polymer hosts used are thermoplastics such as poly(ethylene oxide) (PEO) [76–78], polyacrylonitrile (PAN) [79], poly(vinyl alcohol) (PVA) [80, 81], and poly(vinylidene fluoride) (PVDF) [82, 83]. Despite that, their conductivity is not high enough at room temperature. While ionic transport takes place mainly in amorphous phase, high degree of crystallinity in the semicrystalline polymer hosts (e.g., PEO) at room temperature is not favorable.

In view of that, ENR could be a potential candidate for SPE system. It has polar epoxy oxygen, low T_g, and amorphous nature that are suitable for ionic percolation at room temperature. In addition, ENR also offers flexibility and good contact between electrolytic layer and electrode for electronic devices [70].

2.1.1 ENR

Idris et al. [70] were the first to study ENR-based SPEs. They suggest that complexation between lithium cation (Li^+) and epoxy oxygen of ENR is similar to the mechanism of PEO and Li^+, the oldest and widely studied SPE system. The formation of transient cross-links of intra- and intermolecular coordination of oxygen atom in ENR with Li^+ restricts the segmental motion of ENR chains which in turn increase the T_g values. Above certain salt concentration, T_g value fluctuates when the level of transient cross-link is high enough to reduce Li^+ and polymer chain mobility, and distribution of salt becomes inhomogeneous. Under this condition, the salt–salt interactions are preferred than salt–polymer interactions [84].

As mentioned in previous section, properties of ENR changes gradually with epoxidation degree. Nevertheless, this effect was seldom considered and discussed. Yusoff et al. [85] have well demonstrated that ENR with more epoxy group will have stronger interaction with salt due to more coordination sites for Li^+ transport (see Fig. 6). The similar T_g trend is observed by Klinklai et al. [86] as

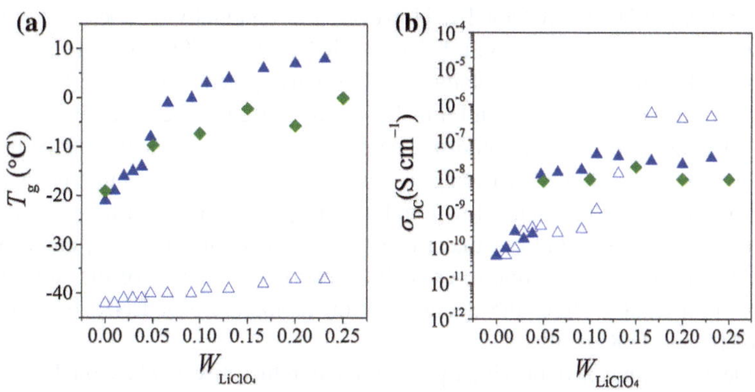

Fig. 6 **a** Glass transition temperature and **b** ionic conductivity of ENR/$LiClO_4$. *Open triangle* [85] indicates ENR-25; *filled triangle* [85] and *filled diamond* [67] indicate ENR-50

Table 8 Ionic mobility and diffusion coefficient of ENR/$LiClO_4$ ranging from $W_s = 0$–0.13 estimated from impedance spectrometer. Copyright reprinted with permission from [85]

	ENR-25	ENR-50
Ion mobility, $\alpha\mu$ ($cm^2\ V^{-1}\ s^{-1}$)	3.3×10^{-11}	4.4×10^{-10}
Diffusion coefficient, D ($cm^2\ s^{-1}$)	8.5×10^{-13}	1.1×10^{-11}

well. From FTIR study, a stronger ion–dipole interaction at C–O–C of ENR-50 with $LiClO_4$ was observed as compared to ENR-25. It is noted that Li^+ ions prefer to coordinate with epoxy oxygen than with the C=C bonds. ENR-50 has higher conductivity than ENR-25 when the weight percent of salt (W_s) < 0.15, which could be due to higher ion mobility, better salt molecular-chain segment correlation, and higher charge carrier diffusion rate (refer Table 8). When $W_s > 0.15$, the conductivity of ENR-50 is lower than ENR-25. This is because at high salt concentration, conductivity of ENR is dominantly governed by segmental motion rather than charge carrier density. The study suggests that polarization relaxations cause formation of stable percolation network, leading to significant increase in conductivity at $W_s = 0.05$–0.11 for ENR-50 and $W_s = 0.11$–0.17 for ENR-25. The results disagree with initial findings by Idris et al. [70] that report that ENR-25 has higher conductivity than ENR-50 at relatively high salt content.

The dielectric study done by Chan et al. [87] on ENR-25 doped with $LiClO_4$ indicates a non-Debye relaxation. The stable resistor networks only occur at $W_s \geq 0.2$. Molar conductivity indicates Kohlrausch-like variation over a certain range of concentration. The nonlinear deviation of molar conductivity at low salt concentration and unaffected polarizability of ENR-25 signifies the weak interaction between ENR-25 chains and salt molecules. For ENR-50 doped $LiClO_4$, Chan and Kammer [88] revealed that the values of diffusion coefficient and degree of dissociation (10^{-5}) remain constant over a wide range of $LiClO_4$ concentration ($W_s \leq 0.17$).

Tan et al. [67] studied the influence of various anions of lithium salts in ENR-50. The order of interaction strength between ENR and Li salt is $LiBF_4 < LiClO_4 < LiI < LiCF_3SO_3 < LiCOOCF_3$. However, the conductivity does not correlate well with the interaction strength: $LiBF_4 > LiCF_3SO_3$~$LiCOOCF_3 > LiI > LiClO_4$. In this work, lattice energy explains the conductivity behavior where lower lattice energy salt dissociates easily and thus has more charge carrier and higher conductivity. Despite that, $LiClO_4$ could not be explained using the same factor because $LiClO_4$ was said to have high ion pair formation which causes number of charge carrier to decrease and finally low conductivity performance. In terms of thermal stabilization, it was found out that $LiClO_4$ destabilizes while $LiBF_4$ enhances the thermal stability of ENR (due to cross-link reaction). The rest of the salts studied do not significantly affect the decomposition temperature of ENR, in agreement with FTIR and DSC results.

Table 9 summarized previously reported conductivity studies on ENR-based SPE. The values are too low for application. It is interesting to note that the strong interaction strength between salt and ENR does not necessarily ensure a

Table 9 Previously reported conductivity on various ENR-based SPE at room temperature

System	σ_{max} (S cm^{-1})	Reference
ENR-50/$LiN(SO_2CF_3)_2$	4.6×10^{-5}	[91]
ENR-50/$LiBF_4$	2.7×10^{-5}	[67]
ENR-25/$LiCF_3SO_3$	6.2×10^{-5}	[70]
ENR-50/$LiCF_3SO_3$	2.3×10^{-5}	[70]
	9.3×10^{-7}	[67]
ENR-50/$LiCOOCF_3$	2.0×10^{-6}	[67]
ENR-50/LiI	2.7×10^{-7}	[67]
ENR-25/$LiClO_4$	5.7×10^{-7}	[85]
ENR-50/$LiClO_4$	4.1×10^{-8}	[85]
	1.8×10^{-8}	[67]

high conductivity. There are many other factors to be considered such as ion salvation (lattice energy of salt), deformation of polymer structure due to inclusion of salt, and segmental movement of the polymer chains [89, 90]. Moreover, same ENR-based SPE system may not have similar ionic conductivity, for example, ENR-50/$LiCF_3SO_3$, as shown in Table 9. Those two separate studies on the same system show one order of conductivity difference. This could be due to difference in molecular weight of ENR, effect of non-rubber component (gel, protein), sample preparation, etc. Therefore, variation from ENR itself should be minimized to have fair comparison.

2.1.2 ENR Composites

Composite polymer electrolytes (CPE) introduced a filler which is dispersed to SPE as a third component. Inorganic fillers have been used since the 1980s where micrometer-sized ceramic particles were added to improve the mechanical properties. Later, nano-sized particles such as Al_2O_3, SiO_2, TiO_2, and ZrO_2 were used in attempt to enhance the conductivity as well. Although many studies were conducted about CPE, the conductivity behavior is still not well understood until now. Using effective medium theory (EMT) models, Wieczorek and Siekierski [92] explained that the dispersed filler particles are assumed to be surrounded by interface regions with an enhanced conductivity over the bulk electrolyte. However, this theory does not explain the reasons of conductivity differences in the interfacial regions. Several authors pointed out the relationship of Lewis acid–base interactions in describing the conductivity increase. Wieczorek et al. demonstrated the validity of the Lewis acid–base approach for various semicrystalline and amorphous systems [93, 94]. In 2001, Best et al. [95] described the conductivity behavior using a simple electrostatic model where he explained that the lithium cations experience the same potential at TiO_2 filler surfaces as at the polymer. This means that the lithium ions can move between these sites with a lower activation energy barrier. Later, they propose that the conductivity increase is due to the formation of percolating interface regions which allows the formation of conduction pathway interface near filler particles [96].

Tan and Bakar [97] studied thermal properties of ENR-50/$LiCF_3SO_3$ added with nano-sized Fe_3O_4. Trend of T_g follows as: ENR < ENR/Fe_3O_4 < ENR/$LiCF_3SO_3$/Fe_3O_4 < ENR/$LiCF_3SO_3$. Fe_3O_4 has negligible interaction with ENR in composite. However, in CPE, competing interactions occur between salt and filler or with ENR chains. In nitrogen atmosphere degradation, the trend of activation energy is: ENR/Fe_3O_4 < ENR/$LiCF_3SO_3$/Fe_3O_4 < ENR < ENR/$LiCF_3SO_3$. In CPE, salt suppresses catalytic effect of filler on the degradation of ENR. No conductivity study of the system was reported here.

2.1.3 ENR-Based Blends

Polymer blends are a mixture of two or more different polymers together. It can be classified into two main categories which are miscible and immiscible blends. Blending ENR with another polymer host, SPE may enhance the conductivity and mechanical properties.

An example of immiscible ENR-based SPE blends is PEO/ENR/$LiClO_4$. Chan and Kammer [98] found two T_g(s) with respect to PEO and ENR-50 over the entire composition range where both increase with addition of salt. Findings from T_g values show almost equal salt solubility in both polymers, but slower rates of crystallization and conductivity enhancement suggest that salt has higher solubility in PEO of the blend. The same PEO/ENR blends were studied by Noor et al. [99, 100], using $LiCF_3SO_3$ salt. The systems obey Arrhenius rule for temperature dependence impedance study. The salt-added PEO/ENR 70/30 blend has lower activation energy and higher conductivity pre-exponential factor as compared to PEO/$LiClO_4$. This indicates that the energy required to dissociate cations from anions and the mobility of charge carrier in the blend electrolyte is smaller. $LiCF_3SO_3$ decreases the thermal stability of PEO/ENR. Studies of Noor et al. [99, 100] and Chan et al. [87, 98] point toward preferential localization of lithium salt in PEO phase of the immiscible blends of PEO/ENR (when PEO is the major component) after T_g analysis using DSC, ion–dipole interaction using FTIR, etc. This leads to concentration of salt in PEO that becomes more concentrated in blends as compared to the neat PEO. Hence, the conductivity of salt-containing blends shall be higher than that of salt-containing PEO with addition of equivalent amount of salt.

It is important to stress that incorporation of ENR as minor component in other immiscible blends such as PVDF/ENR-50/$LiCF_3SO_3$ and PMMA/ENR-50/$LiCF_3SO_3$ also leads to very similar results where the polymer blend electrolytes exhibit higher conductivity as compared to PVDF/$LiCF_3SO_3$ and PMMA/$LiCF_3SO_3$, respectively. PVDF/ENR-50/$LiCF_3SO_3$ blend ratio of 60/40 at $W_s = 0.20$ shows highest conductivity (4.0×10^{-5} S cm^{-1}) for the system. The appreciable increase in T_g of PVDF demonstrates that lithium salt tends to interact more with PVDF than ENR [101]. For PMMA/ENR-50/$LiCF_3SO_3$

system, blend composition of 90/10 at $W_s = 0.60$ exhibits highest conductivity (5.1×10^{-5} S cm^{-1}). The ion–dipole interaction in FTIR characterization hints toward selective localization of lithium salt in PMMA [102].

PVC/ENR/$LiClO_4$ system also reveals higher conductivity than PVC-doped $LiClO_4$. Despite that, highest conductivity enhancement was obtained when ENR was employed as the major phase, which is different than the examples mentioned above. At constant salt content ($W_s = 0.40$), PVC/ENR 30/70 blend shows higher conductivity (6.8×10^{-8} S cm^{-1}) as compared to 70/30 (8.6×10^{-9} S cm^{-1}). Parameters other than impedance study were not discussed much in the literature; hence, localization of salt in these blends could not be concluded [103].

Besides achieving higher conductivity, polymer blends were also employed to provide mechanical strength. Mohammad et al. [104] studied ENR-50/polyethyl methacrylate (PEMA)/$NH_4CF_3SO_3$. The purpose of adding PEMA as a second component is to give mechanical strength where 60/40 blend ratio produces free standing film. Highest conductivity for this blend ratio (5.4×10^{-6} S cm^{-1}) is obtained when $W_s = 0.40$. This value is lower than PEMA/$NH_4CF_3SO_3$ (1.0×10^{-5} S cm^{-1}) [104, 105]. Only one T_g value (~5 °C) was observed in which the value is nearer to the T_g of ENR-50 (−21 °C) as compared to PEMA (65 °C). A more detailed study should be carried out to determine the miscibility of the blends.

Many assumptions and suggestions have been made in attempt to explain the conductivity enhancement in ENR-blend SPE. Most polymer blends are immiscible from the thermodynamic standpoint. The effect of phase separation of SPE binary blends due to preferential localization of salt in different phases is not so well explained except in Refs. [87, 98, 106]. Table 10 summarized the reported

Table 10 Previously reported conductivity on various ENR-blend SPE at room temperature

System	Composition	σ_{max} (S cm^{-1})	Reference
PEO/$LiClO_4$ PEO/ENR-50/$LiClO_4$ $T_{g,\,PEO} = -54$ °C $T_{g,\,ENR\text{-}50} = -19$ °C	$W_s = 0.15$ 60/40; $W_s = 0.11$	7.9×10^{-6} 8.0×10^{-5}	[107] [93]
PEO/$LiCF_3SO_3$ PEO/ENR-50/$LiCF_3SO_3$	$W_s = 0.15$ 70/30; $W_s = 0.20$	2.1×10^{-6} 1.4×10^{-4}	[99]
PVdF/$LiCF_3SO_3$ PVdF/ENR-50/$LiCF_3SO_3$ $T_{g,\,PVdF} = -40$ °C [108]	Not mentioned 60/40; $W_s = 0.20$	1.4×10^{-5} 2.7×10^{-5}	[101]
PMMA/$LiCF_3SO_3$ PMMA/ENR-50/$LiCF_3SO_3$ $T_{g,\,PMMA} = 120$ °C	$W_s = 0.40$ 90/10; $W_s = 0.60$	4.9×10^{-8} 5.1×10^{-5}	[102]
PEMA/$NH_4CF_3SO_3$ PEMA/ENR-50/$NH_4CF_3SO_3$ $T_{g,\,PEMA} = 65$ °C	$W_s = 0.35$ 40/60; $W_s = 0.40$	1.0×10^{-5} 5.4×10^{-6}	[105] [104]
PVC/$LiClO_4$ PVC/ENR/$LiClO_4$ $T_{g,\,PVC} = 82$ °C [109]	$W_s = 0.20$ 30/70; $W_s = 0.30$	8.8×10^{-10} 8.5×10^{-7}	[109] [103]

Table 11 Ionic conductivity of PEO/ENR-50/$LiCF_3SO_3$ in 70/30 blend at $W_s = 0.30$. Copyright reprinted permission from [110]

Weight percent of ZnO (%)	Ionic conductivity ($S\ cm^{-1}$)
0	1.39×10^{-4}
2	1.02×10^{-5}
4	1.25×10^{-5}
6	4.60×10^{-6}
8	2.54×10^{-6}

conductivity for ENR-blend SPE. Interestingly, when ENR is blended with another polymer host, sometimes ENR must be retained as minor phase (blends with PEO, PVdF, PMMA) and sometimes as major phase (blends with PVC) to achieve higher conductivity. In general, the use of ENR as dispersed phase to lead salt to become more concentrated in the major phase that governed the ionic conductivity in the immiscible blends allows lower percolation threshold that contributes to the enhancement of conductivity. Nevertheless, the preferential localization of salt is yet to be fully resolved. Detailed T_g study using Nernst distribution coefficient [87, 98, 106] could provide some insight on the distribution of salt among the blend constituents.

2.1.4 ENR-Blend Composites

Noor et al. [110] investigated the PEO/ENR-50/$LiCF_3SO_3$/nano-ZnO system. Surface morphology showed that ZnO was uniformly dispersed at low concentration, but the filler made the surface rough. Addition of ZnO causes reduction of ionic conductivity by one to two order of magnitude (refer Table 11). They suggest that filler causes blocking effect on ion transport, reduce effective interaction between polymer blends and salt, and unsuccessful in increasing number of ion dissociation. This explanation calls for future investigation.

2.2 Gel Polymer Electrolytes (GPEs)

GPE is a liquid electrolyte immobilized in polymer matrix. Normally, it has ionic conductivity about 10^{-3} S cm^{-1} at room temperature. However, GPE has its own drawback where it lacks mechanical stability to produce thin films. The liquid electrolyte can permeate into the polymer which ends up softening the polymer [111, 112].

2.2.1 ENR

Idris et al. [70] found that plasticized ENR-25/$LiCF_3SO_3$ electrolytes show higher conductivity than plasticized ENR-50/$LiCF_3SO_3$. Incorporation of plasticizer causes tremendous increase in conductivity and reduction in T_g. Mohamed et al.

Table 12 Ionic conductivity of ENR-based GPE at room temperature

System	Composition (w/w/w)	σ (S cm^{-1})	Reference
ENR-50/$LiCF_3SO_3$/EC	100/35/20	2.4×10^{-4}	[113]
ENR-50/$LiCF_3SO_3$/PC	100/35/10	4.9×10^{-4}	
ENR-25/$LiCF_3SO_3$/EC + PC	100/5/100	1.4×10^{-4}	[70]
ENR-25/$LiCF_3SO_3$/EC + PC	100/5/200	2.9×10^{-4}	
ENR-50/$LiCF_3SO_3$/EC + PC	100/5/100	4.0×10^{-5}	[70]
ENR-50/$LiCF_3SO_3$/EC + PC	100/5/200	1.3×10^{-4}	
ENR-50/$LiN(SO_2CF_3)_2$/EC + PC	100/5/50	5.8×10^{-4}	[91]
	100/5/100	8.4×10^{-4}	
	100/5/150	2.6×10^{-3}	

[113] studied ENR-50/$LiCF_3SO_3$ plasticized with EC and PC. The system obeys Vogel–Tamman–Fulcher rule, which describes conductivity based on the free-volume model. The presence of plasticizer causes the entropy configuration of ENR-50 to favor the provision of more free volume where ions can move easily through plasticizer-rich phase. The dielectric study shows that presence of salt and plasticizer at certain concentrations causes an increase in dielectric constant at lower frequencies that support that high conductivity is achieved for samples that have highest concentration of free ions. PC plasticized system has higher conductivity and better electrochemical performance than EC. Table 12 summarized reported studies for ENR-based GPE.

2.2.2 ENR Blend

Latif et al. [114] studied PMMA/ENR-50/$LiCF_3SO_3$/EC. The conductivity increases after addition of 20 % EC but decreases when plasticizer amount reached 40 % and above. Only morphology and impedance study were reported here. Therefore, detailed characterizations are required to compare the plasticized and unplasticized system.

Rahman et al. [115] studied PVC/liquid ENR-50 (LENR)/$LiClO_4$/PC. The conductivity increases with addition of plasticizer and reaches maximum when 70 % of PC was added. Not much information could be drawn out from the conductivity, morphology, and FTIR studies reported. The ionic conductivity of ENR blend-based GPE is summarized in Table 13.

Table 13 Ionic conductivity of ENR blend-based GPE at room temperature

System	Composition	σ (S cm^{-1})	Reference
PVC/LENR-50/$LiClO_4$	$W_s = 0.30$	2.3×10^{-8}	[116]
PVC/LENR-50/$LiClO_4$/EC	30/70/70;	2.1×10^{-7}	[115]
	$W_s = 0.30$		
PMMA/ENR-50/$LiCF_3SO_3$	90/10;	5.1×10^{-5}	[102]
PMMA/ENR-50/$LiCF_3SO_3$/EC	$W_s = 0.60$	4.1×10^{-7}	[114]
	Not mentioned		

3 Conclusion

The research for electronic applications in ENR-based systems is still scant. The improvement of ionic conductivity through utilization of ENR as a second component has become one of the promising ways to overcome the mechanical strength and low conductivity problem at room temperature. Until then, many challenges must be addressed to fully explore its potential. It has been noted that composition of ENR itself may introduce many variations of properties as compared to NR.

The effort to achieve acceptable conductivity while maintaining sufficient dimensional stability is not an easy task. The fundamental understanding of localization of salt in the different phase of immiscible binary polymer blend host and conductivity mechanism would provide a great advantage in controlling the properties. Phase separation of the multicomponent in the blends can be used as an effective tool to control the preferential localization of salt or filler in different phases, which maintain the percolation pathways. In general, the salt percolation threshold can be decreased by preferential localization of salt in the multiphase polymer materials.

With limiting studies in ENR-based CPEs, it could not be concluded at the moment on the criteria of the filler needed for the improvement of the conductivity properties of ENR or ENR blends. The problem that calls for attention is to choose inert or functionalized filler, where the functionalized filler may enhance the interaction of salt with polymer or cause repulsion of salt and polymer to be used as a polymer electrolyte. Interactions of filler with polymer, filler with salt, and salt with polymer; morphology; T_g(s) of polymer(s); location of salt; and filler in the multiphase system have led to huge challenge in generalization of selection criteria for filler.

References

1. Pummere R, Burkard PA (1922) Uber Kautschuk. Ber Dtsch Chem Ges 55:3458–3472
2. Davey JE, Loadman MJR (1984) A chemical demonstration of the randomness of epoxidation of natural rubber. Br Polym J 16:134–138
3. Swern D (1947) Electronic interpretation of the reactions of olefins with organic peracids. J Am Chem Soc 69:1692–1698
4. Rosowsky A (1964) Ethylene oxides. In: Weissberger A (ed) Chemistry of heterocyclic compounds. Wiley, Hoboken
5. Baker CSL, Gelling IR, Newell R (1985) Epoxidized natural rubber. Rubber Chem Technol 58:67–85
6. Gelling IR, Morrison NJ (1985) Sulphur vulcanization and oxidative ageing of epoxidized natural rubber. Rubber Chem Technol 58:243–257
7. Gelling IR (1988) Chemistry, structure and properties of epoxidised natural rubber. In: Proceedings international rubber technology conference, Penang, Malaysia, 17–19 October 1988, pp 415–427
8. Davies CKL, Wolfe SV, Gelling IR, Thomas AG (1983) Strain crystallization random copolymers produced by epoxidation of *cis*-1,4-polyisoprene. Polymer 24:107–113

9. Gelling IR, Porter M (1988). Natural rubber science and technology. In: Roberts AD (ed), Oxford University Press, Oxford, p 359
10. Cizravi JC, Naginlal C, Lim SL (1999) Moisture-sorption properties and studies on the dilute solution behavior of epoxidized rubber. J Appl Polym Sci 73:1633–1644
11. Roux C, Pautrat R, Cheritat R, Ledran F, Danjard JC (1967) Modification de Polymères Insaturés par Époxydation. J Polym Sci Part C Polym Symp 16:4687–4693
12. Abu A, Sidek D (1989) Easy processing epoxidised natural rubber. J Nat Rubber Res 4:119–132
13. Gelling IR (1987) Epoxidized natural rubber. Nat Rubber Technol 18:21–29
14. Wong ST, Ong LM (1992) Storage stability of epoxyprene. Kautsch Gummi Kunstst 45:284–286
15. Burfield DR, Lim KL, Law KS (1984) Epoxidation of natural rubber lattices. J Appl Polym Sci 29:1661–1673
16. Saffer A, Johnson BL (1948) Measurement of internal double bonds in polymers by perbenzoic acid addition. Rubber Chem Technol 21:821–829
17. Kolthoff IM, Lee TS, Mairs MA (1973) Use of perbenzoic acid in analysis of unsaturated compounds. J Appl Polym Sci Poly Chem Ed 2:206–220
18. Roux C, Pautrat R, Cheritat R, Pinazzi C (1964) Epoxidation of 1,4-*cis*-polyisoprenes. Comptes Rendus 258:5442–5445
19. Mairs JA, Todd J (1932) The oxidation of caoutchouc, Gutta-percha and Balata with hydrogen peroxide. J Chem Soc 29:386–399
20. Gemmer RV, Golub MA (1978) ^{13}C NMR spectroscopic study of epoxidized 1,4-polyisoprene and 1,4-polybutadiene. J Polym Sci Polym Chem Ed 16:2985–2990
21. Gelling IR, Smith JF (1979) Controlled viscosity by natural rubber modification. In: Proceedings of the international rubber conference, Venice, Italy, 3–6 October 1979, pp 140–149
22. Saito T, Klinklai W, Yamamoto Y, Kawahara S, Isono Y, Ohtake Y (2010) Quantitative analysis for reaction between epoxidized natural rubber and poly(L-lactide) through ^{1}H-NMR spectroscopy. J Appl Polym Sci 115:3598–3604
23. Abd El Rahman AMM, El-Shafie M, El Kholy SA (2012) Modification of local asphalt with epoxy resin to be used in pavement. Egypt J Petrol 21:139–147
24. Ng SC, Gan LH (1981) Reaction of natural rubber latex with performic acid. Eur Polym J 17:1073–1077
25. Xu K, He C, Wang Y, Luo Y, Liao S, Peng Z (2012) Preparation and characterization of epoxidized natural rubber. Adv Mater Res 396:478–481
26. Bradbury JH, Perera MCS (1988) Advances in the epoxidation of unsaturated polymers. Ind Eng Chem Res 27:2196–2203
27. Yu H, Zeng Z, Lu G, Wang Q (2008) Processing characteristics and thermal stabilities of gel and sol of epoxidized natural rubber. Eur Polym J 44:453–464
28. Gelling IR (1985) Modification of natural rubber latex with peracetic acid. Rubber Chem Technol 58:86–96
29. Kumar NR, Roy S, Gupta BR, Bhomic AK (1992) Structural changes in rubber during milling. J Appl Polym Sci 45:937–945
30. Ivan G, Giurginca M, Jipa S, Tavaru E, Setnescu T, Setnescu R (1993) Thermo-oxidative behavior of epoxidised natural rubber. J Nat Rubber Res 8:31–36
31. Durbateki AJ, Miles CM (1965) Near infrared and nuclear magnetic resonance spectrometry in analysis of butadiene polymers. Anal Chem 37:1231–1235
32. Hallensleben ML, Schmidt HR, Schuster RH (1995) Epoxidation of poly(*cis*-1,4-isoprene) microgels. Die AngewandteMakromolekulareChemie 227:87–99
33. Duch MW, Grant DM (1970) Carbon-13 chemical shift studies of the 1,4-polybutadienes and the 1,4-polyisoprenes. Macromolecules 3:165–174

34. Jayawardena S, Reyx D, Durand D, Pinazzi CP (1983) Synthesis of macromolecular antioxidants by reaction of aromatic amines with epoxidized polyisoprenes, 1. Model reactions of aniline with 3,4-epoxy-4-methylheptane and with 1,2-epoxy-3-ethyl-2-methylpentane. Die Makromolekulare Chemie Rapid Communications 4:449–453
35. Saendee P, Tangboriboonrat P (2006) Latex interpenetrating polymer networks of epoxidised natural rubber/poly(methyl methacrylate): an insight into the mechanism of epoxidation. Colloid Polym Sci 284:634–643
36. Eng AH, Tanaka Y, Gan SN (1997) Some properties of epoxidized deproteinized natural rubber. J Nat Rubber Res 12:82–89
37. Heping Y, Sidong L, Zheng P (1999) Preparation and study of epoxidized natural rubber. J Therm Anal Calorim 58:293–299
38. Roy S, Gupta BR, Chak TK (1993) Studies on the ageing behaviour of gum epoxidized natural rubber. Kautsch Gummi Kunstst 46:293–296
39. Jorge RM, Lopes L, Benzi MR, Ferreira MT, Gomes AS, Nunes RCR (2010) Thiol addition to epoxidized natural rubber: effect of the tensile and thermal properties. Int J Polym Mater 59:330–341
40. Tanrattanakul V, Wattanathai B, Tiangjunya A, Muhamud P (2003) In situ epoxidized natural rubber: improved oil resistance of natural rubber. J Appl Polym Sci 90:261–269
41. Morss HA (1938) An X-ray study of stretched rubber. J Am Chem Soc 60:237–241
42. Bunn CW (1942) Molecular structure and rubber-like elasticity I: the crystal structure of *β* Gutta-Percha, rubber and polychloroprene. Proc R Soc Lond Ser A 180:40
43. Bac NV, Huu CC (1996) Synthesis and applications of epoxidized natural rubber. J Macromol Sci Pure Appl Sci A33:1949–1955
44. Ratnam CT, Nasir M, Baharin A, Zaman K (2000) Electron beam irradiation of epoxidized natural rubber. Nucl Instrum Methods Phys Res B 171:455–464
45. Perera MCS (1990) Reaction of aromatic amines with epoxidized natural rubber latex. J Appl Polym Sci 39:749–758
46. Rahman HA (1984) Air permeability on various natural rubber grades. PED Internal Report No. 10. Rubber Research Institute of Malaysia, Kuala Lumpur
47. Hayashi O, Takahashi T, Kurihara H, Ueno H (1981) Monomer unit sequence distribution in partly-epoxidized *trans*-1,4-polyisoprene. Polym J 13:215–223
48. Burfield DR, Gan SN (1975) Non oxidative crosslinking reactions in natural rubber I: determination of crosslinking groups. J Polym Sci Polym Chem Ed 13:2725–2734
49. Davies CKL, Wolfe SV, Gelling IR, Thomas AG (1983) Strain crystallization random copolymers produced by epoxidation of *cis*-1,4-polyisoprene. Polymer 24:107–113
50. Zeng ZQ, Yu HP, Wang QF, Lu G (2008) Effects of coagulation processes on properties of epoxidized natural rubber. J Appl Polym Sci 109:1944–1949
51. Nakason C, Sainamsai W, Kaesaman A, Klinpituksa P, Songklanakarin J (2001) Preparation, thermal and flow properties of epoxidised natural rubber. J Sci Technol 23:415–424
52. Ravanbakhsh M, Khorasani SN, Khalili S (2015) Blending of NR/BR/ENR/EPDM-*g*-GMA by reactive processing for tire sidewall applications: effects of grafting and ENR on curing characteristics, mechanical properties, and dynamic ozone resistance. J Elastomers Plast. doi:10.1177/0095244315580453
53. Gelling IR, Porter M (1990) Chemical modifications of natural rubber. Oxford University Press, Oxford, p 359
54. Saito T, Klinklai W, Kawahara S (2007) Characterization of epoxidized natural rubber by 2D NMR spectroscopy. Polymer 48:750–757
55. Alwaan IM, Hassan A, Piah MAM (2015) Effect of zinc borate on mechanical and dielectric properties of metallocene linear low-density polyethylene/rubbers/magnesium oxide composite for wire and cable applications. Iran Polym J 24:279–288
56. Poh BT, Soo KW (2015) Effect of blend ratio and testing rate on the adhesion properties of pressure-sensitive adhesives prepared from epoxidized natural rubber 25/acrylonitrile–butadiene rubber blend. J Elastomers Plast. doi:10.1177/0095244314568471

57. Johnson T, Thomas S (2000) Effect of epoxidation on the transport behaviour and mechanical properties of natural rubber. Polymer 41:7511–7522
58. Sam ST, Hani N, Ismail H, Noriman N, Ragunathan S (2014) Investigation of epoxidized natural rubber (ENR 50) as a compatibilizer on cogon grass filled low density polyethylene/soya spent flour. Mater Sci Forum 803:310–316
59. Nakason C, Wannavilai V, Kaesaman A (2006) Effect of vulcanization system on properties of thermoplastic vulcanizates based on epoxidized natural rubber/polypropylene blends. Polym Test 25:34–41
60. Margaritis AG, Kalfoglou NK (1987) Miscibility of chlorinated polymers with epoxidized poly(hydrocarbons) 1: epoxidized natural rubber/poly(vinyl chloride) blends. Polymer 28:497–502
61. Rahman MA, Penco M, Peroni I, Ramorino G, Janszen G, Landro LD (2012) Autonomous healing materials based on epoxidized natural rubber and ethylene methacrylic acid ionomers. Smart Mater Struct 21:035014
62. Lin T, Ma S, Lu Y, Guo B (2014) New design of shape memory polymers based on natural rubber crosslinked via Oxa-Michael reaction. ACS Appl Mater Interfaces 6:5695–5703
63. Mahmood WAK, Azarian MH (2015) Thermal, surface, nanomechanical and electrical properties of epoxidized natural rubber (ENR-50)/polyaniline composite films. Curr Appl Phys 15:599–607
64. Abdullah Z, Ibrahim KMYK (2014) Electrical tracking performance of thermoplastic elastomer nanocomposites material under high voltage application. Int J Sci Eng Res 5:708–711
65. Rahman MYA, Ahmad A, Lee TK, Farina Y, Dahlan HM (2012) $LiClO_4$ salt concentration effect on the properties of PVC-modified low molecular weight LENR50-based solid polymer electrolyte. J Appl Polym Sci 124:2227–2233
66. Noor SAM, Ahmad A, Talib IA, Rahman MYA (2011) Effect of ZnO nanoparticles filler concentration on the properties of PEO-ENR50-$LiCF_3SO_3$ solid polymeric electrolyte. Ionics 17:451–456
67. Tan WL, Bakar MA, Bakar NHHA (2013) Effect of anion of lithium salt on the property of lithium salt-epoxidized natural rubber polymer electrolytes. Ionics 19:601–613
68. Mohammad SF, Zainal N, Ibrahim S, Mohamed NS (2013) Conductivity enhancement of (epoxidized natural rubber 50)/poly(ethyl methacrylate)–ionic liquid-ammonium triflate. Int J Electrochem Sci 8:6145–6153
69. Zainal N, Idris R, Sabirin MN (2011) Characterization of (ENR-50)-ionic liquid based electrolyte system. Adv Mater Res 287:424–427
70. Idris R, Glasse MD, Latham RJ, Linford RG, Schlindwein WS (2001) Polymer electrolytes based on modified natural rubber for use in rechargeable lithium batteries. J Power Sour 94:206–211
71. Armand MB (1987) Current state of PEO-based electrolyte. In: MacCallum JR, Vincent CA (eds) Polymer electrolyte reviews, vol 1. Elsevier, London, pp 1–21
72. Rajendran S, Kannan R, Mahendran O (2001) Ionic conductivity studies in poly(methyl methacrylate)-polyethylene oxide hybrid polymer electrolytes with lithium salts. J Power Sour 96:406–410
73. Murata K, Izuchi S, Yoshihisa Y (2000) An overview of the research and development of solid polymer electrolyte batteries. Electrochim Acta 45:1501–1508
74. Polu AR, Kumar R, Rhee HW (2015) Magnesium ion conducting solid polymer blend electrolyte based on biodegradable polymers and application in solid-state batteries. Ionics 21:125–132
75. Armand MB (1986) Polymer electrolytes. Annu Rev Mater Sci 16:245–261
76. Wright PV (1975) Electrical conductivity in ionic complexes of poly(ethylene oxide). Br Polym J 7:319–327
77. Chan CH, Sim LH, Kammer HW, Tan W, Abdul Nasir NH (2011) Ionic transport and glass transition temperature of polyether-salt complexes: dependence on molecular mass of polymer. Mater Res Innov 15:14–17

78. Klongkan S, Pumchusak J (2015) Effects of nano alumina and plasticizers on morphology, ionic conductivity, thermal and mechanical properties of PEO-$LiCF_3SO_3$ solid polymer electrolyte. Electrochim Acta 161:171–176
79. Bushkova OV, Animitsa IE, Lirova BI, Zhukovsky VM (1997) Lithium conducting solid polymer electrolytes based on polyacrylonitrile copolymers: ion solvation and transport properties. Ionics 3:396–404
80. Every HA, Zhou F, Forsyth M, MacFarlane DR (1998) Lithium ion mobility in poly(vinyl alcohol) based polymer electrolytes as determined by ^{7}Li NMR spectroscopy. Electrochim Acta 43:1465–1469
81. Yang JM, Wang SA (2015) Preparation of graphene-based poly(vinyl alcohol)/chitosan nanocomposites membrane for alkaline solid electrolytes membrane. J Membr Sci 477:49–57
82. Ohno H, Matsuda H, Mizoguchi K, Tsuchida E (1982) Demonstration of solid-state cell based on poly(vinylidene fluoride) system containing lithium perchlorate. Polym Bull 7:271–275
83. Ataollahi N, Ahmad A, Lee TK, Abdullah AR, Rahman MYA (2014) Preparation and characterization of PVDF-MG49-$NH_4CF_3SO_3$ based solid polymer electrolyte. E-Polym 14:115–120
84. Le Nest JF, Gandini A, Cheradama H (1988) Crosslinked polyethers as media for ionic conduction. Br Polym J 20:253–268
85. Yusoff SNHM, Sim LH, Chan CH, Hashifudin A, Kammer HW (2013) Solid solution of polymer electrolytes based on modified natural rubber. Polym Res J 7:159–169
86. Klinklai W, Kawahara S, Marwanta E, Mizumo T, Isono Y, Ohno H (2006) Ionic conductivity of highly deproteinized natural rubber having various amount of epoxy group mixed with lithium salt. Solid State Ionics 177:3251–3257
87. Chan CH, Kammer HW, Sim LH, Yusoff SNHM, Hashifudin A, Tan W (2014) Conductivity and dielectric relaxation of Li salt in poly(ethylene oxide) and epoxidized natural rubber polymer electrolytes. Ionics 20:189–199
88. Chan CH, Kammer HW (2015) Polymer electrolytes-relaxation and transport properties. Ionics 21:927–934
89. Gray FM (1991) Solid polymer electrolyte—fundamentals and technological applications. VCH, Weinheim 245
90. Gray FM (1997) Polymer electrolytes. The Royal Society of Chemistry, UK
91. Razali I, Nor WAHWS (2007) Characterization of plasticized and non-plasticised epoxidised natural rubber based polymer electrolyte systems. Solid State Sci Technol 15:147–155
92. Wieczorek W, Siekierski M (1994) A description of the temperature dependence of the conductivity for composite polymeric electrolytes by effective medium theory. J Appl Phys 76:2220–2226
93. Wieczorek W, Zalewska A, Raducha D, Florjanczyk Z, Stevens JR (1998) Composite polyether electrolytes with Lewis acid type additives. J Phys Chem B 102:352–360
94. Wieczorek W, Florjanczyk Z, Stevens JR (1995) Composite polyether based solid electrolytes. Electrochim Acta 40:2251–2258
95. Best AS, Adebahr J, Jacobsson P, MacFarlane DR, Forsyth M (2001) Microscopic interactions in nanocomposite electrolytes. Macromolecules 34:4549–4555
96. Adebahr J, Best AS, Byrne N, Jacobsson P, MacFarlane DR, Forsyth M (2003) Ion transport in polymer electrolytes containing nanoparticulate TiO_2: the influence of polymer morphology. Phys Chem Chem Phys 5:720–725
97. Tan WL, Bakar MA (2014) The effects of magnetite particles and lithium triflate on the thermal behavior and degradation of epoxidized natural rubber (ENR-50). Am-Eurasian J Sustain Agric 8:111–122
98. Chan CH, Kammer HW (2008) Properties of solid solutions of poly(ethylene oxide)/epoxidized natural rubber blends and $LiClO_4$. J Appl Polym Sci 110:424–432

99. Noor SAM, Ahmad A, Rahman MYA, Abu Talib I (2010) Solid polymeric electrolyte of poly(ethylene oxide)-50 % epoxidized natural rubber-lithium triflate (PEO-ENR50-$LiCF_3SO_3$). Nat Sci 2:190–196
100. Noor SAM, Ahmad A, Talib IA, Rahman MYA (2010) Effects of ENR on morphology, chemical interaction and conductivity of PEO-$LiCF_3SO_3$ solid polymer electrolyte. Solid State Sci Technol Lett 18:115–125
101. Aziz M, Chee LC (2005) Preparation and characterization of PVDF/ENR50 polymer blend electrolyte. Solid State Sci Technol 13:126–133
102. Latif F, Aziz M, Katun N, Ali AMM, Yahya MZ (2006) The role and impact of rubber in poly(methyl methacrylate)/lithium triflate electrolyte. J Power Sour 159:1401–1404
103. Ahmad A, Rahman MYA, Ali MLM, Hashim H, Kalam FA (2007) Solid polymeric electrolyte of PVC-ENR-$LiClO_4$. Ionics 13:67–70
104. Mohammad SF, Idris R, Mohamed NS (2010) Conductivity studies of ENR based porton conducting polymer electrolytes. Adv Mater Res 129:561–565
105. Anuar NK, Zainal N, Mohamed NS, Subban RHY (2012) Studies of poly(ethyl methacrylate) complexed with ammonium trifluoromethanesulfonate. Adv Mater Res 501:19–23
106. Chan CH, Kammer HW, Sim LH, Harun MK (2011) Blends of epoxidized natural rubber and thermoplastics. In: Rubber: types, properties and uses. Nova Science Publishers, New York, pp 305–336
107. Sim LH, Gan SN, Chan CH, Kammer HW, Yahya R (2009) Compatibility and conductivity of $LiClO_4$ free and doped polyacrylate-poly(ethylene oxide) blends. Mater Res Innov 13:278–281
108. Mohajir BEE, Heymans N (2001) Changes in structural and mechanical behavior of PVDF with processing and thermomechanical treatments 1: Change in structure. Polymer 42:5661–5667
109. Rahman MYA, Ahmad A, Wahab SA (2009) Electrical properties of a solid polymeric electrolyte of PVC-ZnO-LiClO4. Ionic 15:221–225
110. Noor SAM, Ahmad A, Talib IA, Rahman MYA (2011) Effect of ZnO particles filler concentration on the properties of PEO-ENR50-$LiCF_3SO_3$ solid polymeric electrolyte. Ionics 17:451–456
111. Feuillade G, Perch P (1975) Ion-conductive macromolecular gels and membranes for solid lithium cells. J Appl Electrochem 5:63–69
112. Abraham KM, Alamgir M (1990) Li^+-Conductive solution polymer electrolytes with liquid-like conductivity. J Electrochem Soc 137:1657–1659
113. Mohamed SN, Johari NA, Ali AMM, Harun MK, Yahya MZA (2008) Electrochemical studied on epoxidised natural rubber-based gel polymer electrolytes for lithium-air cells. J Power Sour 183:351–354
114. Latif F, Aziz M, Ali AMM, Yahya MZA (2009) The coagulation impact of 50 % epoxidised natural rubber chain in ethylene carbonate-plasticized solid electrolytes. Macromol Symp 227:62–68
115. Rahman MYA, Ahmad A, Lee TK, Farina Y, Dahlan HM (2011) Effect of ethylene carbonate (EC) plasticizer on poly(vinyl chloride)-liquid 50 % epoxidised natural rubber (LENR50) based polymer electrolyte. Mater Sci Appl 2:818–826
116. Rahman MYA, Ahmad A, Lee TK, Farina Y, Dahlan HM (2012) $LiClO_4$ salt concentration effect on the properties of PVC-modified low molecular weight LENR50-based solid polymer electrolyte. J Appl Polym Sci 124:2227–2233

Electronic Applications of Ethylene Vinyl Acetate and Its Composites

Soumen Giri and Chaoying Wan

Abstract Ethyl vinyl acetate (EVA) copolymers exhibit diverse properties ranging from semicrystalline polymer to rubber-like elastomer, which highly depends on the VA percentage and molecular weight. It can be used as membrane materials for electronic applications due to its polarity, flexibility, good processability, and low cost. Insulating EVA can be used for cable sheath, automotive sound damping, or sound barrier sheets. Electrical-conductive EVA with the addition of conductive fillers is suitable for electromagnetic interference shielding applications. Thermal-conductive EVA is a popular material for encapsulation of photovoltaic modules. This chapter covers the recent development of EVA materials in supercapacitors (SCs), actuators, electromagnetic shield interference, sensors, and photovoltaic modules. The flame-retardant properties of EVA composites are discussed. The challenges and perspectives of EVA and EVA-based composites in future electronics are discussed.

Keywords Ethyl vinyl acetate (EVA) copolymers · Electrical conductivity · Supercapacitors · Composites

1 Introduction of EVA

The ongoing research interest to create new properties from the combination of various polymers has stimulated various copolymers to be produced. Ethyl vinyl acetate (EVA), a class of random copolymers of ethylene and vinyl acetate (VA),

S. Giri
Materials Science Center, Indian Institute of Technology Kharagpur, Kharagpur 721302, West Bengal, India

C. Wan (✉)
International Institute for Nanocomposites Manufacturing, WMG, University of Warwick, Coventry CV4 7AL, UK
e-mail: chaoying.wan@warwick.ac.uk

D. Ponnamma et al. (eds.), *Flexible and Stretchable Electronic Composites*, Springer Series on Polymer and Composite Materials,
DOI 10.1007/978-3-319-23663-6_3

Table 1 Examples for commercially available EVA copolymers

Trade name	Producer
Apifive®	API SpA
Appeel®, Brynel®, Elvax®	DuPont Packaging & Industrial Polymers
Cosmothene®	TPC, The Polyolefin Company (Singapore)
ELEVATE®	Westlake Chemical
Escorene™Ultra	Exxon Mobil Chemical
Evatane®	Arkema
Evathene	USI Corporation
Greenflex®	Polimeri Europa
Hanwha®	Hanwha Chemical
Honam®	Honam Petrochemical Corporation, South Korea
J-REXEVA	Japan Polyolefins Co., Ltd.
Levapren®	Lanxess Deutschland GmbH
MDI EVA	Modern Dispersions, Inc.
Nipoflex®	Tosoh Corporation
Novatec®	Japan Polychem Corporation

exhibits intermediate properties between polyethylene (PE) and poly(vinyl acetate) (PVAc) that are dependent on the proportion of both monomers [1–4]. Their molecular structures can be readily tailored by varying the VA concentration, macromolecular weight and crystallinity, so that producing excellent properties such as impact strength, puncture resistance, crack resistance, adhesion, and optical transparency. EVA has found broad industry applications in adhesive, sealants, packaging, tubing, footwear, wire and cable insulation, photovoltaic encapsulates, and sheets. Some commercial grades of EVA copolymers are listed in Table 1.

1.1 Synthesis and Structure of EVA Copolymers

Ethylene copolymers including EVA were firstly synthesized by ICI (Great Britain) in the 1930s, in the same laboratories where PE was discovered [1]. The chemical structure of EVA is given in Fig. 1. Low levels of VA comonomer were initially used for modifying PE. Since the 1950s to present, a broad range of EVA copolymers has been developed [3–6].

The copolymerization process of ethylene with VA was unknown until the mechanism was first described by M.W. Perrin and patented in US Patent 2,200,429 [2]. Till today, different types of catalysts, reaction medium, and synthesis methods have been proposed. The hydrolysis or partial hydrolysis of the EVA copolymers was patented in US Patents 2,386,347 and 2,399,653. The VA incorporation into the polymer backbone maintains from 0 up to 90 mol%, which has considerable influence on both thermal and physical properties of EVA. The

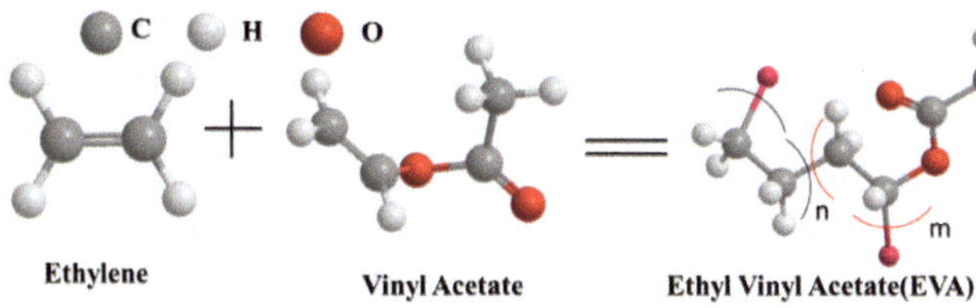

Fig. 1 Polymerization reaction of EVA

copolymerizable monomers may be used during the EVA synthesis are α-olefins which include propylene, isobutylene, α-octene, and α-dodecene. Unsaturated acids are acrylic acid, methacrylic acid, crotonic acid, maleic acid, and itaconic acid; nitriles and amides can also be alternative to acrylonitrile, methacrylonitrile, acrylamide, and methacrylamide, respectively. Alkyl vinyl ethers, vinyl ketone, N-vinylpyrrolidone, vinyl chloride, and vinylidene chloride may also be the good candidates as monomer [6]. These additional monomers can be used to improve the overall properties of EVA copolymers. For example, acrylic acid, methacrylic acid, crotonic acid, maleic acid, and itaconic acid act as adhesion prompters; *N*-methylol acrylamide as cross linker; and acrylamide, versatic acid, and vinyl esters as semicrystalline polymers.

There are different types of polymerization applicable for synthesis of EVA, namely solution polymerization, suspension polymerization, emulsion polymerization, and bulk polymerization. Among these processes, solution polymerization is more preferred. This consists of either continuous process or batch process and generally uses methanol as the solvent [6]. There are various types of azonitrile-based initiators, but more preferable initiators are 2,2′-azobisisobutyronitrile, 2,2′-azobis-(2,4-dimethylvaleronitrile), 2,2′-azobis-(4-methyl-2,4-dimethylvaleronitrile), 2,2′-azobis-(4-methoxy-2,4-dimethylvaleronitrile), and 2,2′-azobis-(2-cyclopropyl propionitrile). Organic peroxide-based initiators, namely isobutyryl peroxide, cumyl peroxyneodecanoate, diisopropyl peroxycarbonate, di-*n*-propyl peroxydicarbonate, *t*-butyl peroxyneodecanoate, lauroyl peroxide, benzoyl peroxide, and *t*-butyl hydroperoxide, can be used as initiators for the solution polymerization process [6].

1.2 Relationship of Structure and Properties

The structure and properties of EVA copolymers are mainly determined by the VA concentration (VA%), molecular weight, melt flow index (MFI), and crystallinity.

The VA% in the copolymer chains affects its degree of crystallinity (X_c) as shown in Table 2 [7]. The presence of VA unit hinders the polymer chain from forming crystals, and the VA-containing chain segments are mainly reside in the amorphous domain of the polymer. With the increase of VA%, the polymer

Table 2 Physical properties of LDPE and PVA homopolymers and the EVA copolymers [7]

Sample	PVAc	EVA-40	EVA-33	EVA-25	EVA-18	EVA-9	LDPE
VA(%) nominal	100	40	33	25	18	9	0
VA(%) experimental	97	40	33	28	20	10	0
T_g (K)	303	256	–	257	259	261	–
T_m (K)	–	–	337	343	357	377	383
M_n 10^4	2.3	2.4	2.2	1.6	2.1	3.7	2.1
M_n/M_w	2.9	2.8	2.9	2.5	4.5	6.0	16.2
X_{DSC} (%)	–	–	–	5	17	25	40

transforms from LDPE which has around 40 % crystallinity, to completely amorphous EVA elastomer above 40 % of VA. The glass transition temperature (T_g), melting point (T_m), and freezing point [8] of copolymers are decreased with the VA% increasing. LDPE containing 0 % VA has a melting point of about 110 °C and a freezing temperature of 89 °C. As the VA content increases to 40 %, the melting point and freezing temperature are decreased to 47 and 27 °C [8], respectively. The EVA copolymer with VA > 40 % has an even lower freezing point. The range of X_c and VA% greatly affects the properties such as stiffness, clarity, solubility, polarity, and permeability. EVA copolymers (15–30 wt% VA) are comparable with plasticized PVC or PE wax. EVA (30–40 wt% VA) is soft, elastic, and highly fillable and is desirable for coatings and adhesives. EVA (40–50 wt% of VA) is rubber-like polymer and can be cross-linked as cable insulation by either peroxide or radiation. Copolymers with 70–95 wt% of VA are used for manufacturing of emulsion paints, adhesives, and coatings.

The mechanical flexibility and transparency of EVA are enhanced with VA% increasing. EVA copolymers with up to 10 wt% VA are more transparent, flexible, and tougher than LDPE. EVA rubber (EVM, VA > 40 wt%) has been used to modify polyamide [9, 10]. Wu et al. [9] prepared EVM-modified ternary polyamide copolymer (tPA) thermoplastic elastomers via a dynamic vulcanization process in the presence of dicumyl peroxide (DCP). A phase transition of EVM/tPA blend was observed at weight ratio of 60/40. The addition of ethylene-acrylic acid copolymer or maleic anhydride-grafted EVM (EVM-g-MAH) as a compatibilizer led to finer EVM phase dimension in tPA matrix, and improvement in both tensile strength and elongation at break of the blends. A high-performance EVM/tPA (70/30) thermoplastic elastomer with Shore A hardness of 75, tensile strength of 24 MPa, elongation at break of 361 %, and set at break of 20 % was obtained at 5 wt% of EVM-g-MAH and 3.5 phr DCP. Furthermore, reactive compounding of PA6/EVM blends was studied and proved the ester-amide exchange reaction between PA6 and EVM during the melt-blending process [10]. In the presence of dibutyltin oxide (DBTO) as a catalyst, PA6-grafted EVM copolymer (EVM-g-PA6) was quantified as 6.8 wt% for PA6/EVM/DBTO (60/40/1) blend at 230 °C for 60 min. The reaction rate constant was dependent on the catalyst concentration, PA6/EVM ratio, and shearing condition.

MFI is measured as the grams of polymer flowing in ten minutes through a capillary of a specific diameter and length by a pressure applied at 190 °C (or 230 °C). It is an indirect measure of the molecular weight and viscosity of polymers. Low MFI value reflects high molecular weight and viscosity of polymers. Higher concentration of VA results broader molecular weight distribution (MWD) and affects melt flow characteristics. The long chains in broad MWD polymers tend to be entangled more than shorter length chains which attribute to the difference in melt flow and elastic response [1]. Therefore, broad MWD EVA copolymers exhibit relatively high viscosity at very low shear stress.

EVA copolymers reacted with most of the organic solvents and oxidizing agents of high concentration such as >30 % nitric and >50 % chromic. The chemical properties depend on the presence of VA%, i.e., this effect increases with increasing content of VA in the EVA copolymer. Aqueous salt solutions do not have any effect on the EVA copolymers [1]. EVA is resistant to dilute mineral acids, alkaline substances, alcohols, fats, oils, and detergents but not to concentrated mineral acids, ketones, and aromatic or chlorinated hydrocarbons. The increase of VA% increases the polarity of the copolymer and thus provides a number of interesting and important properties such as adhesion, printability, dielectric loss factor, and compatibility with polar polymers.

The presence of polar VA group in EVA restricts its application as high-dielectric material. This means EVA cannot be used in telephone or television cable industry where high frequencies are encountered. On the contrary, it ensures the use in radio frequency sealing applications where EVA films containing more than 15 wt% VA are required. The polar nature of EVA helps to incorporate high level of carbon black and flame-retardant additives. In addition, EVA can be used in low- to medium-voltage low-frequency applications considering the dielectric strength and volume resistivity of EVA [11, 12].

2 Electronic Application of EVA Copolymers

Conductive EVA composites can be made by the incorporation of electrical conducting fillers through solution-mixing or melt-compounding process. The electrical resistivity of EVA can be significantly reduced when the fillers formed a conductive network in the EVA matrix, so-called electrical-conductive percolation threshold. The conductivity of the composites (σ) near the conductor–insulator transition is described using the laws of power as follows: $\sigma = (\varphi - \varphi_c)^t$, where φ is the volume fraction of conducting filler, φ_c is the percolation threshold, and t is the critical exponent beyond the percolation threshold. Conductive carbon fillers, such as carbon black (CB), carbon nanotube (CNT), carbon fiber (CF), graphene, and metal particles, are often used for this purpose. Due to the good flexibility, resistance to environmental stress cracking, and processibility [13–15], conductive EVA materials have played an important role in applications in flexible electrodes, electromagnetic radiation shielding, strain sensors, photovoltaic modules, and cable sheathing.

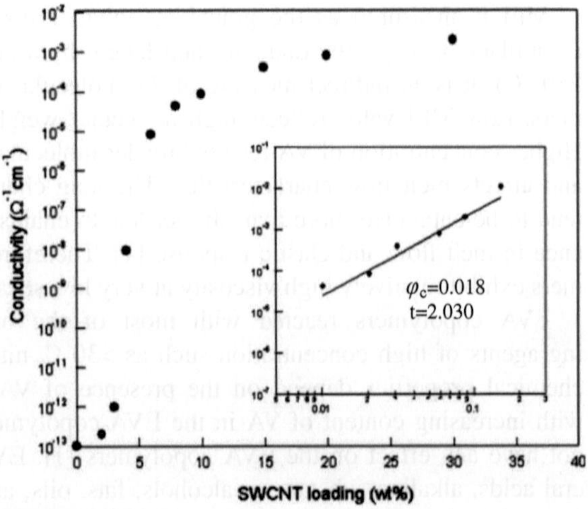

Fig. 2 Electrical conductivity of EVA filled with SWCNT, *inset* shows a log–log plot of the conductivity as a function of $\varphi - \varphi_c$ [18]. Copyright 2008, Reprinted with permission from Springer

For different carbon fillers-filled EVA (28 wt% VA) composites, the percolation thresholds of electrical conductivity were reported to be $\varphi_{CB} = 0.14$ vol.%, $\varphi_{CF} = 0.07$ vol.%, and $\varphi_{MWCNT} = 0.03$ vol.% [16]. The electrical conductivities at the percolation threshold for EVA/CB, EVA/CF, and EVA/CNT composites were 2.5×10^{-6}, 5.0×10^{-5}, and 2.5×10^{-4} S/cm, respectively [16]. For expanded graphite (EP)-reinforced EVA (19 wt% VA), the φ_c was 8.2 vol.% and the maximum electrical conductivity was 0.002 S/cm [17]. For single-walled carbon nanotube (SWCNT)-filled EVA (28 wt% VA), the electrical conductivity was reached 10^{-4} S/cm at $\varphi_c = 3.5$ wt% [18], as shown in Fig. 2. Higher conductivity of 12.1 S/cm was achieved in EVA/CNT composite paper (containing 40 wt% CNT) which was prepared via a dipping coating method [19]. Extruded polyaniline/EVA blends showed high conductivity of 1–2 S/cm with 30 wt% of polyaniline and gallic acid compatibilizers [20].

Electrical conductivity of polymer composites is mainly dependent on the filler intrinsic conductivity, dimension, aspect ratio, distribution, and concentration. Environment temperature also affects conductivity [21]. The dependence of electrical conductivity of temperature, so-called positive-temperature-coefficient (PTC) effect, generally happens at above the T_m of polymers. The melt and thermal expansion of polymer break the conductive network in the polymer matrix, thus leading to a large resistivity jump of polymer composite. In addition, above T_m, a negative-temperature-coefficient (NTC) phenomenon appears because of the rearrangement of the fillers [22]. The NTC effect refers to the decrease of the resistivity with increasing temperature. To increase the interface interactions, CF was chemically oxidized before adding to EVA. In comparison with those untreated CF-filled EVA, the modified CF/EVA showed higher electrical reproducibility after thermal cycles and independence of the conductivity on time. As shown in Fig. 3, the large drop of the resistivity with the increase of time for the untreated CF-filled EVA composites indicates that there is weak interaction

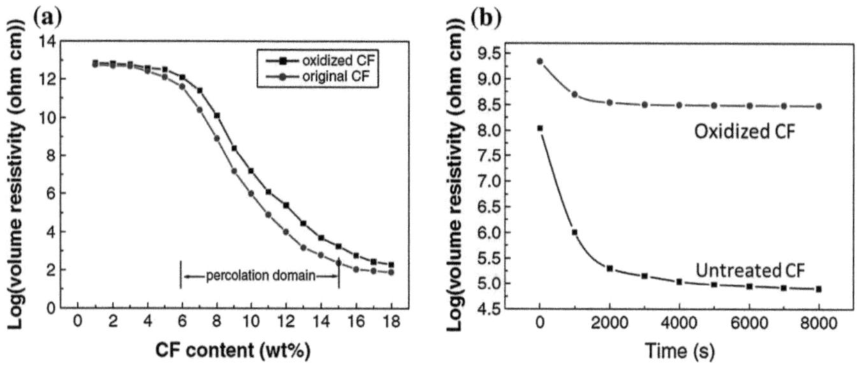

Fig. 3 **a** The log resistivity as a function of CF content at room temperature. **b** Log resistivity as a function of time for EVA/CF (14 wt%) at 100 °C [23]. Copyright 2004. Reprinted with permission from John Wiley & Sons

between EVA chains and CF surface, so that the EVA chain segments have more mobility and lead to more CF rearrangement and agglomeration. For oxidized CF-filled EVA composite, a slight decrease of resistivity with time indicates stronger interfacial interaction between EVA and CF, which prevents the mobility of EVA chains and CF fillers [23]. Therefore, the physical/chemical interaction at the matrix/filler interface helps the stabilization of the conductive network.

2.1 Supercapacitors

Flexible supercapacitors (SCs) are a new class of energy storage devices due to its lightweight and high flexibility, high specific/volumetric energy, and power densities [24–26]. Carbon-based bulky papers show good electrochemical performance and have been directly used as electrodes in flexible SCs [27]. However, the bulky papers become brittle under harsh conditions and are only limited to small scale applications. Recently, several other types of flexible and conductive materials have been developed as electrode substrates [28, 29]. For example, fluorine-doped SnO_2 (FTO) glass, flexible polyethylene terephthalate (PET)/indium tin oxide substrates, carbon cloth, conductive polymers, and CNT- or Au-coated commercial papers. With the deposition of electroactive nanomaterials on the substrates, the performance of charge collection and transport is highly improved due to the improved integrity and flexibility of the electrodes [26]. EVA-based conductive composites represent significant advantages over the above flexible substrates, due to its good flexibility, mechanical properties, chemical resistance, good processibility, and low cost [19, 29].

In the case of EVA/CB/CNT composites ($E_{49.5}B_{49.5}C_1$, where 49.5, 49.5, and 1 are the mass ratios of EVA, CB, and CNT, respectively) prepared by solution-mixing process [30], the addition of CNT to EVA/CB reduced the elongation at

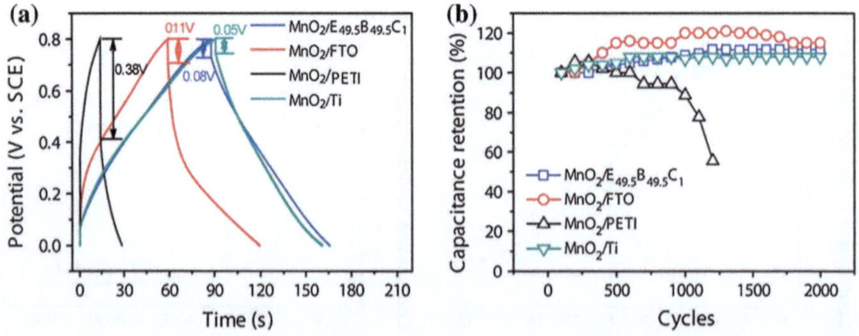

Fig. 4 **a** Galvanostatic charge/discharge curves at a current density of 0.67 mA/cm^2. **b** Cycle stability examined by repeated galvanostatic charge/discharge tests at a current density of 1.33 mA/cm^2 with the potential window ranging from 0 to 0.8 V [30]. Copyright 2014. Reprinted with permission from Royal Society of Chemistry

break but increased the Young's modulus and tensile strength. The elongation at break remained at 22.0 ± 3 % indicating excellent flexibility. The electrical conductivity of $E_{49.5}B_{49.5}C_1$ was 0.17 S/cm. The deposition of 2 wt% of MnO_2 onto the $E_{49.5}B_{49.5}C_1$ substrate provided a high specific capacitance of 214.6 F/g and good cycle performance (>2000), high specific power density (10,950 Wh/kg), and high energy density (30.4 Wh/kg) (Fig. 4). The capacitive behavior as tested at 50 mV/s was not affected much under different bending angles. Additionally, this device was also tested under stress and destress conditions, and there was no capacitance loss after 1500–2000 stress cycles [30].

EVA/CNT papers were prepared by dipping coating of lens papers into EVA/CNT/xylene solutions, and the mechanical properties and electrical conductivity can be controlled by adjusting the ratio of EVA/CNT/xylene and the number of impregnation (Fig. 5) [19]. The obtained conductive papers show higher

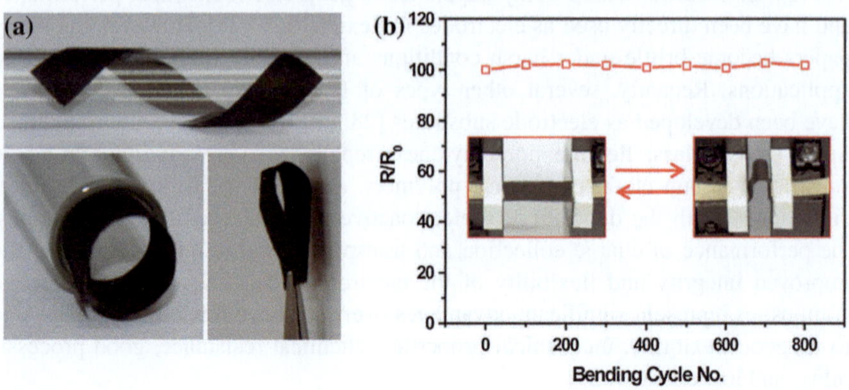

Fig. 5 **a** Flexible 45 s-MnO_2/EVA/40 wt% CNT paper. **b** Resistance variation of the as-prepared electrode as a function of bending–releasing cycles [19]. Copyright 2014. Reprinted with permission from Elsevier Ltd.

performance than the EVA/CB/CNT composites [30] and graphene film [31, 32]. For the EVA/CNT paper containing 40 wt% of CNT, the Young's modulus, tensile strength, and elongation at break of the paper substrate were achieved at 372.9, 22.3 MPa and 55.3 %, respectively. The further treatment by deposition of MnO_2 onto the EVA/CNT substrate for 120 s provided the as-prepared electrodes with the highest areal capacitance of 0.083 F/cm^2 (0.36 mg/cm^2) at scan rate of 10 mV/s, which is higher than the MnO_2/graphene composites (0.081 F/cm^2, 0.42 mg/cm^2, 10 mV/s) [33] and MnO_2/CNT/textiles (0.076 F/cm^2, 0.40 mg/cm^2, 10 mV/s) [34]. The performance of the 120 s-MnO_2/EVA/CNT electrode is accelerated by the good electrical conductivity of EVA/CNT substrate (12.1 S/cm) and high accessible area by the coarse EVA/CNT surface. The amorphous MnO_2 layer directly grown on the substrate ensures the highly redox active sites which facilitates fast and reversible redox reaction. If two and three 45 s-MnO_2/EVA/40 wt% CNT-SC devices connected in series, it enables to lighten up a red light-emitting diode for about 5 min after charging at 1 mA cm^{-2} with an enhanced voltage window of 1.6 and 2.4 V.

Furthermore, 3D co-continuous conductive composites formed of PANI-impregnated EVA/CNT (40 wt%) paper were tested for SCs [35]. The flexible composites of 0.3 mm in thickness achieved a high areal-specific capacity of 575 mF/cm^2, a specific capacitance of 1105 F/g at a scan rate of 5 m/Vs, and capacity retention of 100 % after 45,000 cycles. The capacitance was increased with the EVA/CNT substrate thickness increasing. The PANI-integrated EVA/CNT with continuous network structure shows substantially higher capacity than other reported PANI-based electrodes, such as PANI/CNT/PDMS with a capacitance retention of 95.6 % after 500 cycles [36], and graphene–PANI kept a capacitance retention of 82 % after 1000 cycles [37]. This is because the integration of PANI into the continuous EVA/CNT structures facilitates the interfacial charge transfer as well as prevents the PANI from structure pulverization.

The melting temperature of EVA can be varied from 60 to 100 °C, depending on the VA%. EVA has been used as a thermoresponsive material to coat on separators offering thermal shutdown protection functions for lithium ion batteries [38]. About 5.2-μm-thick EVA microsphere layer was coated onto PP/PE separator membrane to make the separator substrate impermeable. At 90 °C, the coating layer melted then formed a barrier on the separator surface (Fig. 6), thus cutting off the ion conduction between the electrodes and shutting down the battery reactions. The thin EVA barrier only produces a marginally negative effect on the ionic transport and has no adverse effect on the normal cycling performance of the cells.

The tailored polarity, mechanical flexibility, and good processibility of EVA open great opportunities for flexible electronics applications. As shown in Table 3, EVA-based CNT or PANI composites show comparable electrochemical performance as compared to cotton cloth, PDMS and CNT or graphene bulky papers, but with higher flexibility and lower cost.

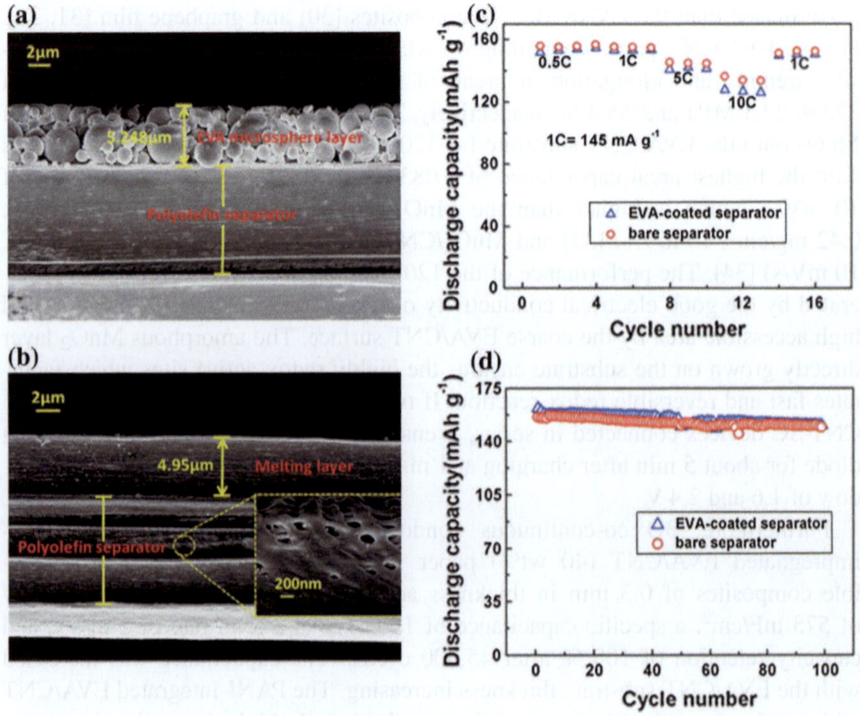

Fig. 6 SEM images of the EVA microspheres-coated separator before (**a**) and (**b**) after heated at 90 °C. **c** Rate capability and (**b**) cycling performance of Li/$LiCoO_2$ coin cells using both conventional bare separator and EVA-coated separator at ambient temperatures [38]. Copyright 2014. Reprinted with permission from Royal Society of Chemistry

2.2 Actuators

Actuator materials have the properties to change their dimension upon the applied external stimulus, such as light, heat, and electric voltage. It is being used in various applications, such as robotics, piezo-drives, shape-memory alloys, switches, drug delivery systems, and optical displays.

One of the great challenges in fabrication of actuators is the filler dispersion and alignment in the polymer matrix. The lack of understanding of the mechanism hinders the development of actuator materials. EVA is attractive in developing reversible shape-memory polymer (SMP) due to its low cost. Large mechanical response was observed upon the addition of MWCNT fillers into EVA. This composite showed enhanced light absorption and thermal capacity, yielding strains on the order of 1–15 %, with sharp onset responses and relaxation times on the order of tens of seconds. The elastic EVA/MWCNT composites can contract or expand upon stimulation depending on prestrains. The extent of prestrain is responsible for the bimodal behavior of the composites (contraction or expansion) although the mechanism is still not clear. The scenario was explained as light being rapidly

Table 3 Comparison of electrochemical performance of flexible supercapacitors

Flexible SCs	Energy density (Wh/kg)	Power density (Wh/kg)	Capacitance	Retention	Reference
45 s-MnO_2/EVA/40 wt% CNT	9.4	6780	0.126 F/cm^2 at 0.5 mA/cm^2	97 % after 5000 cycles	[19]
MnO_2-EVA/CB/CNT	30.4	10,950	214.6 F/g at 10 mV/s	100 % after 2000 cycles	[30]
EVA/CNT/PANI composites	–	–	1106 F/g at 5 mV/s	100 % after 45,000 cycles	[35]
Graphene/cotton cloth	12.3	1.5 Kw/kg	326.8 F/g in 6 M KOH solution	93.8 % after 1500 cycles	[62]
PET/ITO/CNT-PANI	–	–	233 F/g	100 % after 1000 cycles	[63]
PANI/CNT/PDMS	0.15 mWh/cm^3 at 2 mA/cm^2	–	159 F/g at 5 mV/s	94 % after 2000 cycles	[36]
SWCNT–MnO_2	73.6	14.6	529.9 F/g	99.9 % after 2000 cycles	[64]
CNT/MnO_2	–	–	167.5 F/g at 77 mA/g	96 % after 1000 cycles	[65]

absorbed by MWCNTs, producing a thermal perturbation to the surrounding polymeric chains, consequently affecting the prestrained distribution throughout the matrix. Polymeric chains are then able to relax and contract, resulting in mechanical output. This mechanism is confirmed by in situ near-edge X-ray absorption fine-structure (NEXAFS) spectroscopy [12].

The photo-actuation is also observed in EVA with dispersed and aligned MWCNTs or SWCNTs composites. The using of a specially synthesized compatibilizer, cholesteryl 1-pyrenecarboxylate (PyChol), promotes the dispersion of MWCNTs in EVA (Fig. 7) [39]. It is found that the uniaxial orientation of MWCNTs is an essential precondition to ensure the photo-actuating behavior of MWCNTs in polymeric matrices [40].

Figure 8 depicts the carbon K-edge NEXAFS spectra at various temperatures. The inverse variations in intensity with temperature between the emissions at 284.8 and 288.5 eV confirm the 1s → π*C=C and 1s → π*C=O transitions. This temperature-variable NEXAFS spectra confirms the synchronous interplay between π*C=C and π*C=O resonances (albeit with different intensities) suggesting conformational change of polymer chains and MWCNTs as a result of latching through the CH-π interactions. It is also reported that intensities in prestrained composites are linearly correlated by a factor of 10 across the investigated temperature range. These results predict the strong conformational coupling between C=C of CNTs and C=O of EVA, possibly involving CNT torsional effects. The mechanism behind the photo-actuation process is further explained by the following schematic diagram (Fig. 9) [41].

Another study of the photo-actuation behavior of the EVA/PyChol/CNTs has been reported by the using of light-emitting diode (LED). The samples are tested

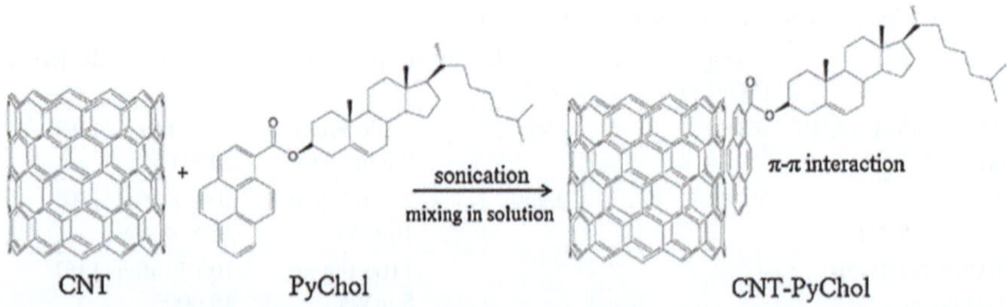

Fig. 7 Schematic depiction of non-covalent modification of MWCNT using PyChol [39]. Copyright 2013. Reprinted with permission from Elsevier Ltd.

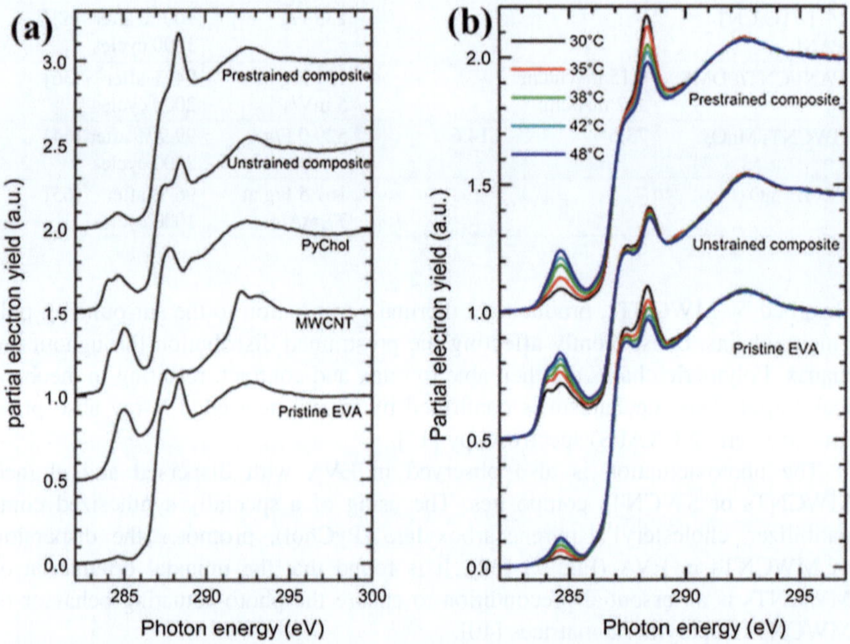

Fig. 8 **a** Carbon K-edge NEXAFS spectra of pristine EVA, unstrained (PyChol, EVA/0.7 wt% MWCNT), and prestrained EVA/0.7 wt% MWCNT composites (*bottom* to *top*) were studied at magic angle and room temperature. **b** Carbon K-edge NEXAFS spectra of pristine EVA, unstrained, and prestrained composites were obtained at various temperatures [12]. Copyright 2014. Reprinted with permission from American Chemical Society

under a dynamic mechanical analyzer equipped with a red LED at a wavelength of 627 nm and an accurate dynamometer using two types of light sources: a white LED at various diode power settings and a red laser diode with $\lambda = 670$ nm [39].

The composites are used to develop Braille element by specially designed molds [40, 42]. It is observed that the height of the Braille element increases temporarily after illumination and the Braille element returns to its original shape with height after switching of the light. It is mainly due to its photo-actuation properties

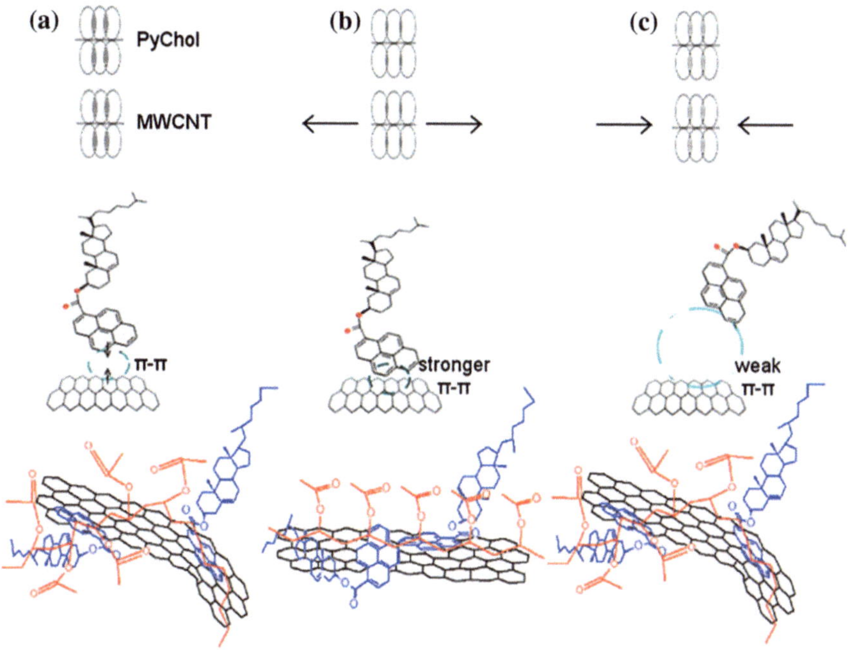

Fig. 9 Schematic diagrams of interactions among the EVA/PyChol/MWCNT composites. For the simplicity, only vinyl acetate is considered instead of EVA (*red*). **a** PyChol molecules (*blue*) attached to MWCNTs (*black*) through π–π interactions, and EVA adhesion occurred through CH–π interactions leads to the formation of isotropic C=O groups. **b** Prestrained MWCNTs (aligned along plan) and C=O groups from EVA exist in isotropic configuration. **c** By the aging of the composites, π–π interactions between PyChol and MWCNTs are vanished which results as PyChol detachment from the surface of MWCNTs and terminating communication between EVA and MWCNT. Above figures depict the changes of π–π interactions through three stages [41]. Copyright 2014. Reprinted with permission from AIP Publishing LLC.

of the composites. The above phenomenon indicates the process is completely reversible. This photo-actuation property is characterized by the atomic force microscopy and nanoindentor [40]. From Fig. 10, it is observed that the height changes take place of a Braille element with time upon illumination and after switching off the red/blue LED. The effect of blue/red LED on the composites at the current 200 mA and 300 mA is listed in Table 4. The blue LED light is more effective for the photo-actuation as compared with red LED light.

2.3 EMI Shielding

Electromagnetic interference (EMI) of electronic devices, which is conducted and/or radiated electromagnetic signals, may interact to other electronic devices causing the degradation of device performances. Electrically conductive polymer

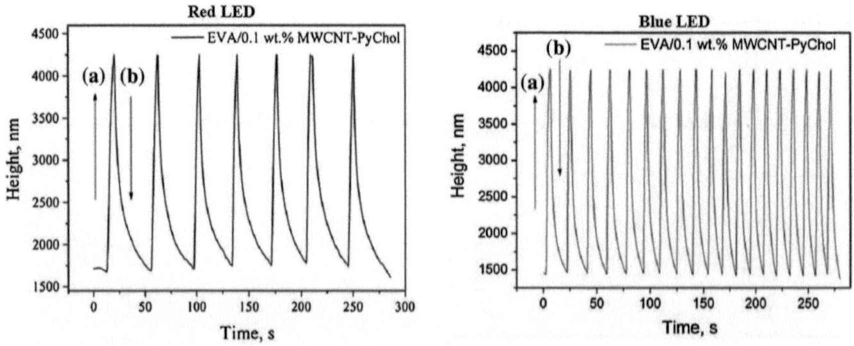

Fig. 10 An AFM recording of the height changes over time for a Braille element based on an EVA/0.1 wt% MWCNT-PyChol composite **a** upon illumination (*red/blue* LED, at an applied current of 300 mA) and **b** after switching off the *red/blue* LED [40]. Copyright 2014. Reprinted with permission from Springer

Table 4 The height changes with power of the red and blue diodes are measured at different applied current (i.e., 200 and 300 mA) [40]

Applied current (mA)	Power of red LED (mW)	Height changes (μm)	Power of blue LED (mW)	Height changes (μm)
200	4.4	6.4	5.8	9.8
300	6.6	9.7	9.6	15.2

composites have shown great potential in replacing metal materials in EMI shielding applications due to its lightweight and low cost.

Electrical-conductive EVA composites have shown EMI shielding effectiveness (SE); for example, an optimum SE was achieved at 45–55 dB for EVM filled with 18 phr MWCNT and 75 phr polypyrrole: PPy(Cl-) loadings. The formation of conductive network by MWCNT and the enhancement in interface and conductivity for those PPy(Cl-)-coated MWCNT could effectively enhance the EMI shielding of composites [43].

By construction of sandwich-structured composites with EVA/graphene composite film and paraffin layers (Fig. 11), the EMI shielding performance was reached up to 27 dB [44].

EVA filled with conductive PANI showed effective EMI shielding over the microwave range of 200–2000 MHz and X-band frequency range of 8–12 GHz. Here, the PANI was doped with DBSA and blended with the EVA. The electrical resistivity of EVA/PANI-DBSA blends was decreased with the PANI-DBSA content increasing, and reached from 10^{13} to 10^{3} Ω cm when the PANI-DBSA increased to 33.3 wt% [45].

If P_{inc} denotes the measured incident power density at a measuring point before the shield put in the place, P_{trans} denotes the measured transmitted power density at the same measuring place after the shield is kept in place. Hence, the

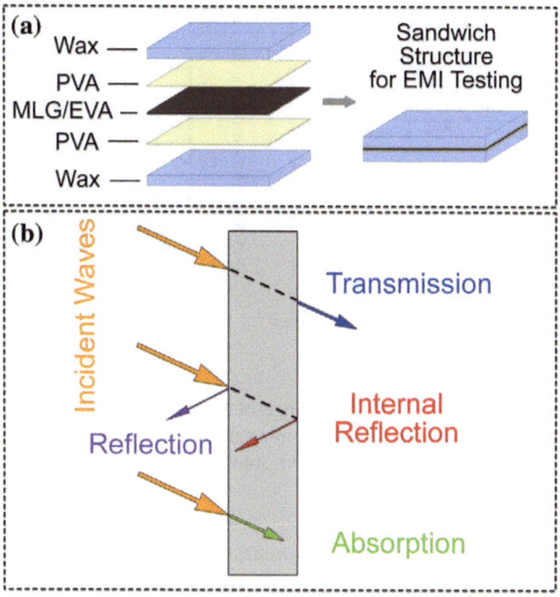

Fig. 11 **a** Sandwich structures coated with G-E films. **b** Mechanisms for EMI shielding [44]. Copyright 2014. Reprinted with permission from Elsevier Ltd.

effectiveness measurement of EMI shielding is basically an insertion loss which is expressed in decibels (dB) unit.

$$SE = -\log\left(P_{inc}/P_{trans}\right) \tag{1}$$

The SE of EVA composites containing different concentrations of PANI-DBSA in the X-band frequency range is shown in Fig. 12. It is found that the SE depends on PANI-DBSA concentration in the composites and SE increases gradually with the increasing concentration of PANI-DBSA. It is well established that the SE of a conductive composite is related to its conductivity. PANI-DBSA acts as conductive filler in blends with EVA, contributing to formation of conductive paths.

Another study of EVA/SWCNT composites demonstrates the enhancement of conductivity as well as EMI shielding of the composite with the increase content of SWCNT [18]. Here, the percolation threshold was obtained at 3.5 wt% of SWCNT. The investigation showed that the EMI shielding of the composites with 15 wt% SWCNT is ~22–23 dB in 8–12 GHZ frequency range. As the conductivity of the EVA/SWCNT composites increases with the increase content of SWCNT, 4 wt% of SWCNT present in the composite takes higher lifting of the conductivity in the order of 10^{-5} Ω cm^{-1} (Fig. 13). This is due to the formation of interconnected network of SWCNTs. On the other hand, the low level of SWCNTs (<2 wt%) embedded in the polymer matrix resulted decrease in conductivity due to the isolation of SWCNT from each other by the insulated polymer matrix, but increase in the content of SWCNTs at certain levels such as 3 and 4 wt%; the

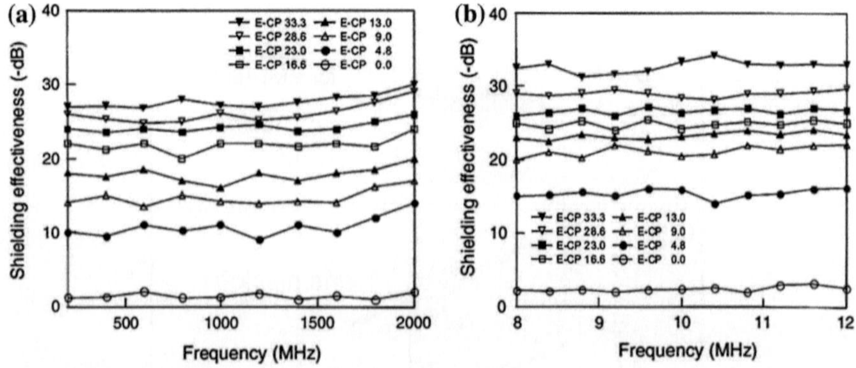

Fig. 12 Shielding effectiveness with the change of frequency in the **a** 100–2000 MHz region and **b** X-band region of EVA composites containing different concentration of PANI-DBSA [45]. Copyright 2004. Reprinted with permission from John Wiley & Sons

steeply rise of conductivity clearly demonstrates the conductor–insulator transition as the SWCNTs form a conductive network throughout the matrix. This uniformly distributed conductive network interacts with the interrupting electromagnetic waves and contributes to the EMI SE. The percentage of SWCNT in the matrix also reflects the SE properties of the materials. Figure 13 indicates the gradual rise of the SE curve with the increase addition of the SWCNT.

EMI SE increases with increasing electrical conductivity because of the reflection effect. When multiple reflections are ignored, the EMI SE can be calculated by the following equation:

$$SE = 50 + 10\log\left(\sigma/f\right) + 1.7t\sigma f^{1/2} \tag{2}$$

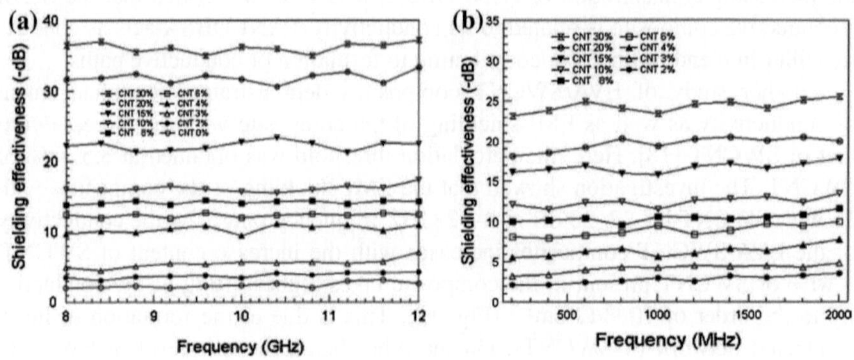

Fig. 13 EMI SE with the change of frequency measured in **a** 8.0–12.0 GHz and **b** 200–2,000 MHz range for SWCNT/EVA composites containing various SWCNT loadings [18]. Copyright 2008. Reprinted with permission from Springer

where SE is in dB, σ is the conductivity ($\Omega^{-1}cm^{-1}$) at room temperature, t is the thickness of the sample (cm), and f is the measurement frequency (MHz), respectively. This empirical relationship does not consider the effect of multiple reflections. The composite with higher percentage of SWCNTs showed lower return loss value (RL) against frequency that means the composite with higher SE has a lower RL. The variation of RL with frequency is nonlinear because of the non-uniform distribution of the SWCNT into the polymer matrix that results the formation of voids with different sizes in the conducting domain. These voids affect RL because of their effect on the external reflection. Equation (2) indicates that the SE depends not only on the conductivity, but also on the frequency and thickness of the sample. The composites exhibit EMI SE of ~11 dB at 4 wt% SWCNT, while ~37 dB for 30 wt% SWCNT.

EVA-based microwave-absorbing materials are generally made by blending with conductive polymers such as PPy, PANI, and polyalkylthiophene, in order to generate dielectric and magnetic losses. In comparison with the conductive fillers, conductive polymers offer the benefits of weight reduction, especially in the applications requiring a good compromise between absorption bandwidth and surface mass [46]. Dielectric constant (permittivity) and loss of EVA/PANI composites were found to increase with the PANI content increasing. The dielectric constant and loss were almost frequency independent for neat EVA matrix and were found to decrease for the filled composite systems with the increase in frequency. A new model was proposed to predict the permittivity of highly dielectric composite systems and shown in Eq. (3) [47].

$$\log \varepsilon_c = \log \varepsilon_f + \frac{\log \varepsilon_m - \log \varepsilon_f}{e^{\frac{v_f}{M*\log F}}} \tag{3}$$

where F is the measurement of frequency; M is the morphological factor/fitting parameter in the range of 0–1; ε_c, ε_m, and ε_f are the dielectric constant of polymer composite, polymer matrix, and fillers, respectively; and v_f is the volume fraction of fillers.

2.4 Sensors

Polymeric materials have attracted great attention in the chemical and biomedical fields for the application as sensors. The material properties of these sensors depend on their electrical properties which are controlled upon interaction with a variety of chemicals. They are used to detect solvent leakage or detection of organic/inorganic gases or liquids such as hydrocarbons-chlorinated solvents, gasoline, ammonia, and nitrogen dioxide. EVA copolymers could contribute for sensors when blending with conducting fillers. In the EVA/high impact polystyrene (HIPS)/CB system, CB particles are preferentially distributed within the polar EVA phase and along the interface at the higher CB concentrations. The electrical conductivity of the blend was

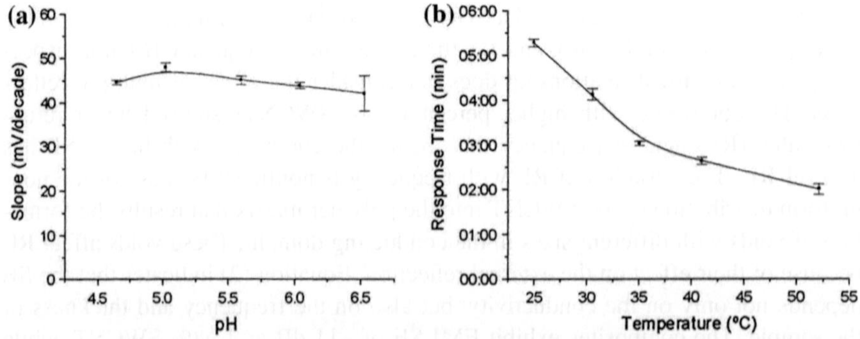

Fig. 14 a The response of biosensor is measured with change of pH, temperature: (25.0 ± 0.1) °C. Ionic strength: 0.1 mol^{-1} and **b** temperature effect on the response time of the biosensor for ascorbate, pH: 5.0. Ionic strength: 0.1 mol^{-1} [49]. Copyright 1999. Reprinted with permission from Elsevier Ltd.

increased due to the attraction of CB to EVA [48]. The interface properties of HIPS/EVA play a crucial role in the sensing process, and a double-percolation structure is preferable for sensor materials. It is necessary to measure the relationship of resistivity–shear rate for sensor analysis since the sensitivity depends on the filament's production rate. In addition, the DC electrical resistivity at different shearing levels showed the sensitivity to organic liquid solvents.

The relative resistivity of filaments HIPS/EVA/CB (55/45/5) was measured at different exposure time to the liquid and vapor solvents benzene and n-heptane [48]. It is found that the filaments are sensitive to the liquid phase of benzene and heptane but not to vapor phase. The bicontinuous interpenetrating structure of the composites is likely to powerfully sense the liquids which have great affinity to each component.

EVA is also used for potentiometric biosensor for *L*-ascorbic acid based on ascorbate oxidase of natural source. The enzyme is collected from the epicarp of *Cucumissativus* L. fruit and immobilized in a graphite/epoxy electrode by occlusion in EVA 40 wt%. Previously, it was tried to immobilize the ascorbate oxidase extract by covalent attachment with glutaraldehyde on the graphite/epoxy electrode, but the attempt was not so successful as it fails to show good selectivity. The problem is overcome by the use of polymeric membrane made of EVA copolymer. The observed sub-Nerntian behavior between 8.0×10^{-6} and 4.5×10^{-4} mol/l for ascorbate ion in 0.1 mol/l KH_2PO_4 buffer, pH 5.0, and at (25.0 ± 0.1) °C of this potentiometric biosensor exhibited with a detection limit of 4.2×10^{-6} mol/l. The electrode worked well in the $4 < pH < 6$ and temperature of above 60 °C (Fig. 14) [49].

2.5 Photovoltaic Modules

Photovoltaics (PVs) represent a clean and renewable alternative energy source in reducing greenhouse emission. To effectively utilize the solar energy, hybrid

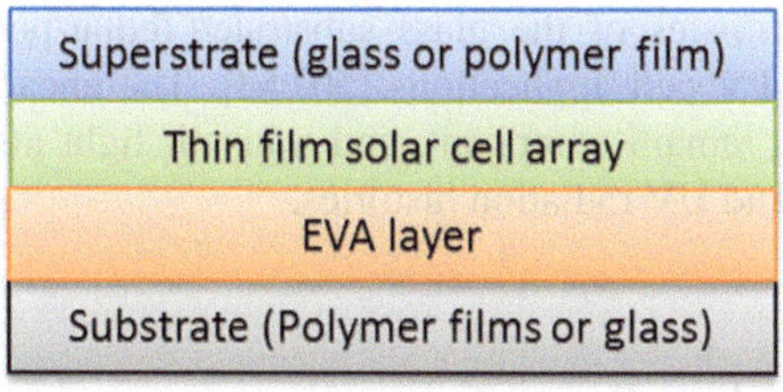

Fig. 15 A typical PV module encapsulation configuration

photovoltaic/thermal solar system and jet impingement cooling device have been developed in order to disseminate the heat quickly and reduce the thermal degradation of solar cells. PV modules are generally encapsulated into a "sandwich" structure as shown in Fig. 15, for example. This is to provide multiple purposes, including physically hold components in place and provide electrical insulation, reduce moisture ingress, optically couple superstate materials (e.g., glass) to PV cells, protect components from mechanical stress by mechanically decoupling components via strain relief, and protect materials from corrosion.

Encapsulation materials include EVA, polyvinyl butyral (PVB), polydimethylsiloxane (PDMS), ionomer, thermoplastic polyurethane (TPU), and polyolefin. EVA is now becoming the dominant PV encapsulate materials replacing PDMS due to its low cost and easy to form films via melt-extrusion. Typically, EVA containing 27–33 wt% VA is used to balance its characteristics, including low crystallinity/high light transmittance, low melting temperature (45–65 °C), and easy melting for processing. Incorporating EVA into PV backsheets can improve dielectric properties and adhesion to the encapsulates, thus reducing the Tedlar film (a PET-PVF composite film) usage in backsheets, and increase the panel power through light reflection accordingly. EVA film can be thermally cross-linked with peroxide (such as tertbutyl-2-ethylhexyl-peroxycarbonate) at elevated temperatures during lamination. A trialkoxysilane is used to promote adhesion between EVA and inorganic surfaces [50]. The effects of UV radiation are mitigated by the inclusion of a UV absorber such as a benzotriazole or benzophenone. In addition, a hindered amine light stabilizer and possibly a phenolic phosphonite may be added as antioxidants. For example, the EVA layer in encapsulated Si solar cells is about 0.5 mm in thickness with thermal conductivity (K) of 0.23 W/mK, Tedlar film is about 0.1 mm in thickness with $K = 0.36$ W/mK, and Si layer is 0.25–0.4 mm in thickness with $K = 148$ W/mK [51]. The EVA layer requires high heat-dissipating ability and electrical resistivity. The addition of 60 vol.% of SiC, ZnO, or boron nitride could improve the K of EVA to 2.85, 2.26, and 2.08 W/mK, respectively [51, 52]. The improved heat-dissipating ability and PV efficiency of EVA expanded its applications in encapsulating rear films for solar cells.

The adhesion of EVA laminates in PV backsheets determines the performance of the PV cells. The synergistic increase in the interfacial hydrophobicity, siloxane density, and cross-linking density is key to the improvement in the EVA adhesion strength for the laminates of the glass substrate/(Tedlar/polyester/EVA tri-laminate) backsheet in PV cell applications [50, 53]. The encapsulates must adhere well to all surfaces, remain compliant, and transmit light after exposure to temperature, humidity, and UV radiation histories.

2.6 Flame-Retardant Properties

Low smoke and halogen-free flame-retardant properties are crucial in development of wire and cable sheathing materials in order to achieve better safety for electrical equipment and devices. High loadings (>60 wt%) of metal hydrate fillers, such as magnesium hydroxide (MH) and aluminiumtrihydrate (ATH), are required for improving the flame retardancy of polymers. The effects are evaluated by cone calorimeter test (CCT), thermogravimetric analysis (TGA), limiting oxygen index (LOI), and UL-94 test. Halogen-free flame-retardant EVA composites have been studied extensively. The dilution of fuel gases by water released by ATH at above 200 °C and the associated endothermic effect are the most important mechanisms that improve the flame retardancy of EVA [54, 55]. However, the high loadings of fillers cause a reduction of flexibility and processability. Recently, the use of nanofillers in combination with the hydrated minerals has been developed in order to leverage the usage of hydrated minerals, mechanical properties, and flame retardancy of the composites [54–60].

In EVA/EG/MH composites, the EG with bigger particle size and higher expansion ratio could increase the thermal stability, LOI value, and improve the UL-94 flammability properties of EVA [54]. The heat release rate, effective heat of combustion, and mass loss rate of the composites decrease remarkably with increasing the particle size and expansion ratio of EG. Similar effects were found in EVA/silica/MH composites, and the addition of silica not only reduced the heat release and mass loss rates, but also depressed the smoke release during the combustion of the composites. This synergistic flame-retardant mechanism of fumed silica in the EVA/MH materials is mainly due to the physical process of fumed silica acting as enhanced char/silica layers in the condensed phase which prevents the heat and mass transfer in the fire.

In EVA/organo-montmorillonite (OMMT)/ATH composites, EVA containing 3 wt% OMMT and 47 wt% ATH revealed the best elongation and fire resistance (LOI = 28) [57]. The deacetylation of the EVA is accelerated in the presence of clay, revealing the catalytic effect of β-elimination of the sites containing strongly acid group. It reduces the rate of thermal degradation and delays weight loss due to the labyrinth effect of the silicate layers in the polymer matrix [58]. Layered double hydroxide (LDH) particles have been used to increase the flame retardancy of EVA, such as MgAl–CO_3, ZnMgAl–CO_3, MgAlFe–CO_3, MgAl-borate, ZnAl-borate, MgAl–PO_4, MgAl-oleate, ZnAl-oleate, MgAl-N-(2-(5,5-dimethyl-1,3,2-dioxaphosphinyl-2-ylamino)-hexyl)acetamide-2-propyl acid, and MgAl-2-aminotoluene-5-sulfonic acid [59]. It is observed that the EVA filled with 60 wt% of MgAl–PO_4 provides lower peak of heat release rate (pHRR) reduction properties as compared to that with MgAl–CO_3, showing better flame-retardant properties. MgAl-borate and ZnAl-borate LDHs modified EVA composites exhibited significant pHRR reductions. The pHRR reduction increased with an increase in the LDH loadings. Organic-modified MgAl LDHs, those containing PAHPA, oleate, or ASA, exhibit high flame-retardant properties, where the PHRR reduction was 43 % (5 wt% loading), 36 % (10 wt% loading), and 39 % (3 wt% loading), respectively. So it is clear that EVA copolymer containing ZnAl, MgAl, and ZnMgAl LDHs are all promising flame-retardant materials [59].

The thermal degradation and flammability of EVA composites filled with ultrafine kaolinite or ATH were evaluated with up to 60 wt% filler loadings [60]. TGA and cone calorimeter analyses showed a higher decomposition temperature range and an improved FR performance for EVA/kaolinite composites in comparison with EVA/ATH composites. For a loading of 35 wt% kaolinite, the pHRR of EVA/kaolinite was reduced by 55 % compared to EVA/ATH. Moreover, a significant intumescent behavior was found in EVA/kaolinite composites during cone calorimeter tests.

3 The Challenges and Perspectives of EVA and EVA-Based Composites

The increasing demand for portable and flexible electronics including photovoltaic cells, SCs, actuators, and wearable devices has initiated extensive efforts to explore flexible, lightweight, and environmentally friendly energy storage devices. EVA copolymers have great potential in the applications of flexible electronics due to its versatile properties that can be adjusted by varying the VA concentration, molecular weight, and crystallinity, thus offering a wide range of materials with properties from crystalline polymers to elastomers and from adhesives to rubbers.

There still remain some challenges to use the EVA in practice. In case of PV devices, EVA copolymers are typically used as encapsulated materials to provide mechanical support, optical coupling, electrical isolation, and protection against environmental exposure. When EVA comes under exposure to atmospheric water and/or ultraviolet radiation, it decomposes with excretion of acetic acid resulting of the lower pH and enhances the surface corrosion rates of the inserted devices. Although the decomposition rate of acetic acid is slow that may not lead to rapid deterioration of modules by any catalyze reaction. Another limitation of EVA is its low T_g of about −15 °C, because in some region, the temperature can be reached below the T_g and it can cause the more vulnerable damages of the material when some mechanical load is applied. As compared to LDPE, the use of EVA reduces chemical resistance, barrier properties, and creep resistance of the composites [61]. However, the polarity and good processibility of EVA make it ready to accommodate various conductive nanofillers, so that expand its applications in electromagnetic interference shielding, thermal-conductive encapsulation, and substrates for SCs.

References

1. Henderson AM (1993) Ethylene-vinyl acetate (EVA) copolymers: ageneral review. IEEE Electr Insul Mag 9:30–38
2. Perrin MW, Fawcett EW, Williams EG (1940) Interpolymerization of ethylene. USP 2200429
3. Roedel MJ, Del W (1955) Ethylene/vinyl acetate polymerization process. USP 2703794
4. Reyes JD (2014) Innovative uses of ethylene vinyl acetate polymers for advancing health-care. In: SPE polyolefins conference, USA
5. Integrated Design Engineering Systems (DES), The Plastics Web®, IDES Inc (2006) 209 Grand Avenue Laramie, WY 82070 USA. http://www.ides.com/prospector/
6. Fink JK (2010) Ethylene vinyl acetate copolymers. In: Handbook of engineering and specialty thermoplastics: polyolens and styrenics. Wiley, Hoboken, pp 187–209
7. Yamaki SB, Prado EA, Atvars TDZ (2002) Phase transitions and relaxation processes in ethylene-vinyl acetate copolymers probed by fluorescence spectroscopy. Euro Polym J 38: 1811–1826
8. http://www.dupont.com//content/dam/dupont/products-and-services/packaging-materials-and-solutions/packaging-materials-and-solutions-anding/documents/elvax_thermal_properties.pdf. Accessed March 2015
9. Wu W, Wan C, Zhang Y (2013) Morphology and mechanical properties of ethylene-vinyl acetate rubber/polyamide thermoplastic elastomers. J Appl Polym Sci 130:338–344
10. Wu W, Wan C, Zhang H, Zhang Y (2014) In situ ester-amide exchange reaction between polyamide6 and ethylene-vinyl acetate rubber during melt blending. J Appl Polym Sci. doi:10.1002/APP.40272
11. Amrollahi H, Borhani M (2015) Effect of EVA content upon the dielectric properties in LDPE-EVA Films. Int J Eng Res 4:69–72
12. Winter AD, Larios E, Alamgir FM et al (2014) Thermo-active behavior of ethylene-vinyl acetate multiwall carbon nanotube composites examined by in situ near-edge X-ray absorption fine-structure spectroscopy. J Phys Chem C 118:3733–3741
13. Zhang Z, Zhai T, Lu X et al (2013) Conductive membranes of EVA filled with carbon black and carbon nanotubes for flexible energy-storage devices. J Mater Chem A1:505–509
14. Stark W, Jaunich M (2011) Investigation of ethylene/vinyl acetate copolymer (EVA) by thermal analysis DSC and DMA. Polym Test 30:236–242
15. Kuila T, Khanra P, Mishra AK et al (2012) Functionalized-graphene/ethylene vinyl acetate co-polymer composites for improved mechanical and thermal properties. Polym Test 31:282–289
16. Sohi NJS, Bhadra S, Khastgir D (2011) The effect of different carbon fillers on the electrical conductivity of ethylene vinyl acetate copolymer-based composites and the applicability of different conductivity models. Carbon 49:1349–1361
17. Yousefzade O, Hemmati F, Garmabi H et al (2014) Thermal behavior and electrical conductivity of ethylene vinyl acetate copolymer/expanded graphite nanocomposites: effects of nanofiller size and loading. J vinyl Addi Tech. doi:10.1002/vnl.21428
18. Chandra Das N, Maiti S (2008) Electromagnetic interference shielding of carbon nanotube/ethylene vinyl acetate composites. J Mater Sci 43:1920–1925
19. Zhang Z, Wang W, Li C et al (2014) Highly conductive ethylene-vinyl acetate copolymer/carbon nanotube paper for lightweight and flexible supercapacitors. J Power Sour 248:1248–1255

20. Dopico-Garcia MS, Lasagabaster-Latorre AAA, Garcia X et al (2014) Extruded polyaniline/EVA blends: enhancing electrical conductivity using gallate compatibilizers. Synth Met 189:193–202
21. Nakis M, Ram A, Flashner F (1978) Electrical properties of carbon black filled polyethylene. Polym Eng Sci 18:649–653
22. Narkis M, Vaxman A (1984) Resistivity behavior of filled electrically conductive crosslinked polyethylene. J Appl Polym Sci 29:1639–1652
23. Di W, Zhang G, Zhao Z et al (2004) Ethylene-(vinyl acetate) copolymer/carbon fiber conductive composite: effect of polymer-filler interaction on its electrical properties. Polym Int 53:449–454
24. Yuan L, Xiao X, Ding T et al (2012) Paper-based supercapacitors for self-powered nanosystems. Angew Chem 124:5018–5022
25. Yuan C, Yang L, Hou L et al (2012) Growth of ultrathin mesoporous Co_3O_4 nanosheet arrays on Ni foam for high-performance electrochemical capacitors. Energy Environ Sci 5:7883–7887
26. Wang X, Lu X, Liu B et al (2014) Flexible energy-storage devices: design consideration and recent progress. Adv Mater 26:4763–4782
27. Yang X, Zhu J, Qiu L et al (2011) Bioinspired effective prevention of restacking in multilayered graphene films: towards the next generation of high-performance supercapacitors. Adv Mater 23:2833–2838
28. Lu X, Wang G, Zhai T et al (2012) Stabilized tin nanowire arrays for high-performance and flexible supercapacitors. Nano Lett 12:5376–5381
29. Lu X, Zhai T, Zhang X et al (2012) WO3–x@Au@MnO2 core-shell nanowires on carbon fabric for high-performance flexible supercapacitors. Adv Mater 24:938–944
30. Zhang Z, Zhai T, Lu X et al (2013) Conductive membranes of EVA filled with carbon black and carbon nanotubes for flexible energy-storage devices. J Mater Chem A 1:505–509
31. Zhang LL, Zhao X, Stoller MD et al (2012) Highly conductive and porous activated reduced graphene oxide films for high-power supercapacitors. Nano Lett 12:1806–1812
32. Wu Q, Xu Y, Yao Z et al (2010) Supercapacitors based on flexible graphene/polyaniline nanofiber composite films. ACS Nano 4:1963–1970
33. Cheng Y, Lu S, Zhang H et al (2012) Synergistic effects from graphene and carbon nanotubes enable flexible and robust electrodes for high-performance supercapacitors. Nano Lett 12:4206–4211
34. Choi BG, Hong J, Kang HW et al (2011) Facilitated ion transport in all-solid-state flexible supercapacitors. ACS Nano 5:7205–7213
35. Zhang Z, Hang Y, Yang K et al (2015) Three-dimensional carbon nanotube/ethylvinylacetate/polyaniline as a high performance electrode for supercapacitors. J Mater Chem A 3:1884–1889
36. Yu MH, Zhang YF, Zeng YX et al (2014) Water surface assisted synthesis of large-scale carbon nanotube film for high-performance and stretchable supercapacitors. Adv Mater 26:4724–4729
37. Cong HP, Ren XC, Wang P et al (2013) Flexible graphene–polyaniline composite paper for high-performance supercapacitor. Energy Environ Sci 6:1185–1191
38. Ji W, Jiang B, Ai F et al (2015) Temperature-responsive microspheres-coated separator for thermal shutdown protection of lithium ion batteries. RSC Adv 5:172–176
39. Czaniková K, Torras N, Esteve J et al (2013) Nanocomposite photoactuators based on an ethylene vinyl acetate copolymer filled with carbon nanotube. Sens Actuat B 186:701–710
40. Czaniková K, Omastová M, Krupa I et al (2014) Chemistry: the key to our sustainable future, Chapter 1. In: Elastomeric actuators based on ethylene-vinyl acetate and carbon nanotubes. Springer, Heidelberg, pp 1–14
41. Winter AD, Jaye C, Fischer D et al (2014) Prestrain relaxation in non-covalently modified ethylene-vinyl acetate |PyChol| multiwall carbon nanotube nanocomposites. APL Materials 2:066105

42. Czaniková K, Krupa I, Račko D et al (2015) In situ electron microscopy of Braille microsystems: photo-actuation of ethylene vinyl acetate/carbon nanotube composites. Mater Res Express 2(2):02561. doi:10.1088/2053-1591/2/2/025601
43. Huang C, Wu J, Tsao K et al (2011) The manufacture and investigation of multi-walled carbon anotube/polypyrrole/EVA nanopolymeric composites for electromagnetic interference shielding. Thin Solid Films 519:4765–4773
44. Song W, Cao M, Lu M et al (2014) Flexible graphene/polymer composite films in sandwich structures for effective electromagnetic interference shielding. Carbon 66:67–76
45. Das NC, Yamazaki S, Hikosaka M (2005) Electrical conductivity and electromagnetic interference shielding effectiveness of polyaniline–ethylene vinyl acetate composites. Polym Int 54:256–259
46. Olmedo L, Hourquebie P, Jousse F (1993) Microwave absorbing materials based on conducting polymers. Adv Mater 5:373–377
47. Rahaman M, Chaki TK, Khastgir D (2012) Consideration of interface polarization in the modeling of dielectric property for ethylene vinyl acetate (EVA)/polyaniline conductive composites prepared through in-situ polymerization of aniline in EVA matrix. Euro Poly J 48:1241–1248
48. Narkis M, Srivastava S, Tchoudakov R et al (2000) Sensors for liquids based on conductive immiscible polymer blends. Syn Met 113:29–34
49. Fernandes JCB, Kubota LT, Neto GDO (1999) Potentiometric biosensor for l-ascorbic acid based on ascorbate oxidase of natural source immobilized on ethylene–vinylacetate membrane. Analytica Chimica Acta 385:3–12
50. Pern FJ (2005) Enhanced adhesion of EVA laminates to primed glass substrates subjected to damp heat exposure. In: Conference record of the thirty-first IEEE photovoltaic specialists conference, 3–7 Jan, pp 495–498
51. Kempe M (2011) Overview of scientific issues involved in selection of polymers for PV applications. In: Presented at the 37th IEEE photovoltaic specialists conference (PVSC 37), Seattle, Washington, 19–24 June
52. Pern J (2008) Module encapsulation materials, processing and testing. APP international PV reliability workshop, Shanghai, China
53. Lee B, Liu J, Sun B, et al (2008) Thermally conductive and electrically insulating EVA composite encapsulants for solar photovoltaic (PV) cell. Express Polym Lett 2:357–363
54. Li Z, Qu B (2003) Flammability characterization and synergistic effects of expandable graphite with magnesium hydroxide in halogen-free flame-retardant EVA blends. Polym Degrad Stab 81:401–408
55. Fu M, Qu B (2004) Synergistic flame retardant mechanism of fumed silica in ethylene-vinyl acetate/magnesium hydroxide blends. Polym Degrad Stab 85:633–639
56. Beyer G (2002) Carbon nanotubes as flame retardants for polymers. Fire Mater 26:291–293
57. Chang M, Hwang S, Liu S (2014) Flame retardancy and thermal stability of ethylene-vinyl acetate copolymer nanocomposites with alumina trihydrate and montmorillonite. J Indus Eng Chem 25:1596–1601
58. Hull TR, Price D, Liu Y, et al (2003) An investigation into the decomposition and burning behavior of Ethylene-vinyl acetate copolymer nanocomposite materials. Polym Degrad Stab 82:365–371
59. Gao Y, Wu J, Wang Q, et al (2014) Flame retardant polymer/layered double hydroxide nanocomposites. J Mater Chem A 2:10996–11016
60. Batistella M, Otazaghine B, Sonnier R, et al (2014) Fire retardancy of ethylene vinyl acetate/ultrafine kaolinite composites. Polym Degrad Stab 100:54–62
61. Kempe MD, Jorgensen GJ, Terwilliger KM, et al (2006) Ethylene-vinyl acetate potential problems for photovoltaic packaging. Conference paper, NREL/CP-520-39915
62. Liu W, Yan X, Lang J et al (2012) Flexible and conductive nanocomposite electrode based on graphene sheets and cotton cloth for supercapacitor. J Mater Chem 22:17245–17253
63. Lin H, Li L, Ren J, Cai Z et al (2013) Conducting polymer composite film incorporated with aligned carbon nanotubes for transparent, flexible and efficient supercapacitor. Sci Rep 3. doi:10.1038/srep01353

64. Shen J, Liu A, Tu Y (2012) Asymmetric deposition of manganese oxide in single walled carbon nanotube films as electrodes for flexible high frequency response electrochemical capacitors. Electrochim Acta 78:122–132
65. Chou SL, Wang JZ, Chew SY et al (2008) Electrodeposition of MnO_2 nanowires on carbon nanotube paper as free-standing, flexible electrode for supercapacitors. Electrochem Commun 10:1724–1727

Electronic Applications of Polyurethane and Its Composites

Seema Ansari and M.N. Muralidharan

Abstract One of the current and future challenges in electronics is to develop suitable multifunctional materials which can address simultaneously several parameters such as flexibility, lightweight, conductivity, environmental impact, and production cost. Polymer electronics forms a new and high-potential technological field, which may pave the way to many novel applications and products. This chapter focuses on the electronic applications of polyurethane (PU) and its composites filled with different conductive fillers including carbon black (CB), graphite, metal particles (MP), carbon nanotube (CNT), and graphene. This chapter also covers PU-conducting polymer blends for various electronic applications. PU composites and blends have wide applications in electronics and optoelectronics. The applications of these include sensors, actuators, EMI shielding, electrolytes for supercapacitors and batteries, electrostatic dissipation, and shape memory applications. The potential global market for printed electronics is huge. PU and its composites-based adhesives have found applications in flexible and printed electronics.

Keywords Polyurethane · Composites · Electronics · Blends · Graphene

1 Introduction

In the modern electronics era, multifunctional devices are gaining more and more momentum. Multifunctionality is impossible within the confines of rigid, planar substrates of the conventional electronic manufacturing. Plentiful attentions have been made to find an alternate way to make electronic components on any desired

---

S. Ansari (✉) · M.N. Muralidharan
Center for Materials for Electronics Technology, Athani P.O, Thrissur, Kerala 680 581, India
e-mail: seema@cmet.gov.in; seemacmet@gmail.com

D. Ponnamma et al. (eds.), *Flexible and Stretchable Electronic Composites*, Springer Series on Polymer and Composite Materials,
DOI 10.1007/978-3-319-23663-6_4

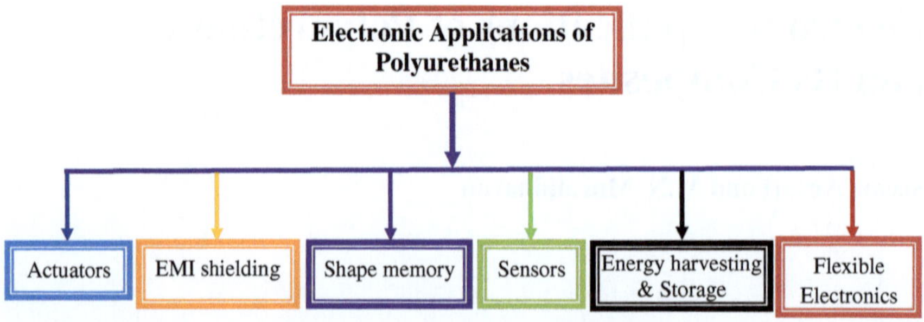

Fig. 1 Applications of polyurethane and its composites in electronics

substrates, especially those of flexible ones. The integration of electronic functionality into everyday objects can improve their benefit to the user and also give rise to new applications such as flexible displays, [1] flexible and conformal antennas, [2, 3] thin-film transistors, [4, 5] sensors, [6] actuators, electronic solar cell arrays, [7, 8] flexible energy storage devices, [9, 10] electronic clothes [11] and so forth.

Polyurethanes (PU) are widely used high-performance materials with unique properties including excellent elongation and high-impact strength, good elasticity, excellent low temperature resistance, and biocompatibility. [12–14]. They have applications ranging from rigid foam as insulation in walls, roofs, and appliances to flexible foam in upholstered furniture, as sealants, adhesives and coatings, as medical devices, and in foot wear [15]. PU are thermoplastic or thermoset polymers formed from polyisocyanates with at least one other species containing active hydrogen such as polyols, polyamines, hydroxyl-terminated polyesters, polyethers, or polycarbonates.

In this chapter, we will be dealing with PU and its composites filled with different conductive fillers including carbon black (CB), graphite, metal particles (MP), carbon nanotube (CNT), and graphene. We will be also covering PU-conducting polymer blends for various electronic applications. PU composites and blends have wide applications in electronics and optoelectronics. The applications of these include sensors, actuators, EMI shielding, electrolytes for supercapacitors, electrostatic dissipation, and shape memory applications. Printed electronics are gaining momentum in electronics industry. PU and its composites-based adhesives have found applications in flexible and printed electronics. The applications of the PU and its composites in electronics are summarized in Fig. 1.

2 Shape Memory Applications

Shape memory polymers (SMPs) are stimuli-responsive smart materials with the ability to undergo a large recoverable deformation upon the application of an external stimulus [16]. Widespread research is going on in SMPs with various

$$\text{—R'—NHCOO—R''—OOCHN—R'—NHCOO————R—O—[—CO—(CH_2)_5—O]_n—H}$$

(Hard segment, Diisocyanate + extender) **(Soft segment, Polyol)**

Fig. 2 Polyurethane with microphase separation structure [28]. Copyright 2011. Reprinted with permission from Elsevier Ltd.

external stimulus such as heat, electricity, light, magnetism, moisture, and even a change in pH value. International research interest in the shape memory polymers have been rapidly growing since its discovery in 1980 [16–23]. SMPs must consist of two segments, of which one should be highly elastic and another should be able to reduce its stiffness upon a particular stimulus which can either be a molecular switch or a stimulus-sensitive domain. When exposed to a suitable stimulus, the switching/transition is triggered and strain energy stored in the temporary shape is released, which consequently results in the shape recovery. SMPs and their composites can recover their original shapes after large deformation when subjected to any suitable external stimulus.

SMPs, SMP composites, or SMP structures have applications including space deployable structures (e.g., hinges, trusses, mirrors, and reflectors), morphing skins used for folding or variable camber wings in aircraft, biomedical, and bio-inspirational instruments (e.g., a polymer vascular stent with shape memory as a drug delivery system, smart surgical sutures, and laser-activated SMP microactuator to remove a clot in a blood vessel), SMP textiles, automobile actuators, and self-healing systems.

The SMPs generally consist of netpoints or "hard segments" and switches or "soft segments." The hard segment determines the permanent shape while the soft segment allows the material to deform and take on a fixed temporary shape. To trigger the shape memory effect, different external stimuli such as heat, light, magnetism, or moisture can be used.

In PU, netpoints can be formed either by physical cross-links or by chemical cross-links through covalent bonds [24]. Chemically cross-linked SMPs (thermosets) exhibit less creep when compared to physically cross-linked SMPs (thermoplastics); and hence, the irreversible deformation during shape recovery is less. Thermoplastic PU (TPU) SMPs have many advantages over other available thermoplastic SMPs, including higher shape recoverability (maximum recoverable strain > 400 %), a wider range of shape recovery temperature (from −30 to 70 °C), better biocompatibility, and better processing ability [21, 25, 26]. In TPU SMPs, the hard segments are formed either from a long-chain macrodiol with a higher thermal transition temperature, or from diisocyanates and chain extenders. The soft segment is formed from the polyols [27]. A schematic representation of PU structure is given in Fig. 2.

2.1 Thermoresponsive Shape Memory Effect

Among all types of SMPs, the thermoresponsive systems are the most common systems. Typical thermomechanical cycle of a thermoresponsive SMP consists of three steps. In the first step, the SMP is fabricated into an original required shape. Then in the second step, the deformed shape is created by a thermomechanical procedure. In this step, the SMP is heated above the thermal transition temperature (T_{trans}). The thermal transition temperature can be either a glass transition temperature, T_{g}, or a melting temperature, T_{m}. Then, the temporary predeformed shape is created by applying an external force followed by cooling well below T_{trans} and removal of the constraint. In the last step, when ever required, the predeformed SMP is heated above the T_{trans} to recover the original shape [28]. The typical shape memory effect during thermomechanical cycle can be illustrated as in Fig. 3.

PU SMPs can be deformed into a temporary shape and can be brought back to its original shape by applying a suitable stimulus. A thermomechanical procedure, which is often called programming, is used to create the temporary shape and to trigger the recovery effect. Programming can be done in a stress- or strain-controlled mode. The thermomechanical conditions of the programming affect the shape memory properties. Santiago et al. studied the effect of stress-holding time on the shape memory properties of a PU sample under successive thermomechanical cycles. They conducted the experiments with two different programming temperatures near the glass transition temperature. The shape recovery ratio was found to decrease considerably with cycling even with shorter holding time. The effect of stress-holding time on the shape recovery properties was found to be dependent on the temperature at which it has been applied. However, neither the stress-holding time nor the successive cycles were found to affect the shape fixity ratio and switching temperature, but a higher programming temperature resulted in higher shape fixity ratio and switching temperature. The study also demonstrated that shape recovery was drastically affected when the samples were subjected to stress-holding times at high temperature. The degradation of the shape recovery was found to be more prominent when the programming temperature was less than the glass transition temperature. The studies suggest that, for potential applications where an SMP needs to work under a constant stress and over successive cycles,

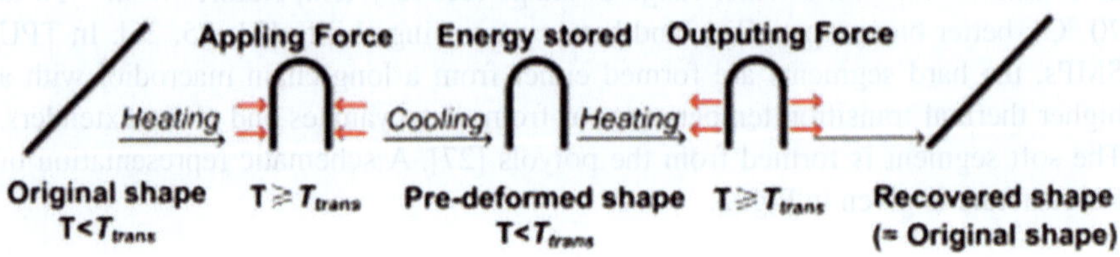

Fig. 3 Shape memory effect during a typical thermomechanical cycle [28]. Copyright 2011. Reprinted with permission from Elsevier Ltd.

it is better to use the programming temperature above the glass transition temperature and reduce the stress-holding time by quenching the temperature as fast as possible [29].

In the recent years, the "cold hibernated elastic memory" (CHEM) technology based on SMPs has received tremendous attention from the scientific community. The material used in the CHEM technology is either SMPs in foam structure or sandwich structures with SMP foam cores and polymeric composite skins [30]. The basic material used was TPU foams with a wide range of T_g. For the simple demonstration of the technology in Earth's atmosphere, a PU foam of $T_g = 63$ °C was used, whereas for Mars application, a PU with $T_g = 4$ °C was used. The studies demonstrated that PU foams with uniformly distributed micron-sized cells are suitable for ultra-lightweight porous membrane or thin-film space applications. The influence of long-term storage in cold hibernation on the recovery strain–stress of PU SMP foam was studied and found to exhibit excellent stability [31, 32].

Normally, shape memory effect works in one direction. Studies were also successful in the development of PU shape memory systems which can work in two directions [33]. For this operating method, two PUs with two different glass transition temperatures are used (T_{g1} and T_{g2} where $T_{g1} < T_{g2}$). The two kinds of SMPs have reverse memory shapes with respect to each other. On heating, when the temperature reaches T_{g1}, the memory shape related to T_{g1} appears. On continued heating, when the SMPs reach T_{g2}, the first memory shape is canceled by the memory shape related to T_{g2} and it moves in opposite direction. The main advantage of this method is that it requires only a simple heating and cooling cycle to operate a two-directional shape recovery. This method was also demonstrated by fabricating two prototypes of cantilever and diaphragm design using two kinds of PUs with glass transition temperatures 35 and 55 °C.

Completely shape recoverable PU fibers can be prepared by adjusting the hard–soft segment content [34]. The mechanical properties of fibers prepared by wet spinning of prepolymerized PU were comparable to other commercial fibers such as Lycra. From the shape memory characterization, it was also found that the thermal setting temperature has a huge influence on the mechanical properties and shape memory property. A higher thermal setting temperature can lead to lower tensile modulus and tenacity. An optimal thermal setting treatment can achieve a complete heat responsive recovery of the shape memory PU fiber. Similarly, polycaprolactone (PCL)-based PU fiber prepared by wet spinning process exhibited a shape recovery ratio of more than 95 % and shape fixity ratio of more than 80 % [35].

In addition to pure PU and its blends, particle-reinforced or fiber-reinforced composites of PU can be used as SMPs. Compared to pure PU, PU composites are multifunctional materials which possess superior properties including high electrical conductivity, magnetic-responsive performance, and high stiffness.

Conducting particles such as CB, carbon nanoparticles, CNTs, graphene, and even metal powders can be used to improve the shape memory performance of the PU. Similar to the particle-reinforced PU, continuous fiber-reinforced SMPs also offer good improvement in properties such as strength, stiffness, and resistance against relaxation and creep. Carbon fibers and nanopapers made of carbon fibers are extensively used to improve the shape memory performance. Other than carbon fibers, glass fibers, and Kevlar fibers can also be used for improving the performance of the SMPs.

CB/PU composites attracted the interest of many researchers due to the economical advantage of the filler. The studies of Yang on the thermomechanical properties of carbon powder-loaded PU demonstrated that effective increase in the recovery stress can be achieved by the incorporation of carbon powders [36].

The PU/CB composites prepared by solution–precipitation and melt compression molding were found to exhibit enhanced shape memory effect [37]. Since the CB fillers were aggregating, the percolation threshold was achieved at a CB concentration of 20 wt%. The CB fillers were found to decrease the crystallinity of PCL soft segments of the PU. The fillers did not affect the stable physical cross-link structure which is necessary to store the elastic energy. Even though the response temperature of the shape memory effect was not improved very much, the strain fixation was improved.

Due to the unique electrical, thermal, and mechanical properties of graphene, enormous interest was raised in utilizing graphene/PU composites for shape memory applications. The introduction of very small amount (0.1 wt%) of high-quality graphene in PU enabled to achieve a shape recovery of 1.8 MPa cm^{-3} [38]. Geometrically large but highly crystallized few layer graphene was utilized for this purpose. It was also observed that for an effective stress transfer from the graphene to PU, the oxygen-containing functional groups at the edge plane of graphene played an important role.

The shape memory properties of functionalized graphene/PU nanocomposites were studied by Han and Chun [39]. Graphene nanosheets (GNs) functionalized with phenethyl alcohol groups were covalently incorporated into PU matrix. The shape memory tests were performed by thermomechanical cycles. Shape fixity of 98 % and shape recovery ratio of 94 % after 4 cycles were obtained for these composites at an extremely low loading of 0.5 %. The hysteresis loss was as low as 2 %. The cross-linked structures formed by the covalent bonding between functionalized GNs and the NCO terminated PU resulted in extremely homogeneous dispersion which in turn helped to improve the shape memory properties.

Chemically modified graphene oxide using 3-aminopropyltriethoxysilane incorporated into silanized PU through sol–gel reactions exhibited novel shape memory performance [40]. When modified graphene oxide composites of PU of two different molecular weights formed into a cohesive layer by interpenetrating polymer network, two undisturbed glass transitions were exhibited by which an intermediate plateau region was formed, and hence, triple-shape memory performance was demonstrated.

The shape memory properties of PU were also improved by the incorporation of nanoclay particles. Nanoclay-tethered PU nanocomposites prepared by the bulk polymerization of PCL diol, methylene diisocyanate, and butane diol with reactive nanoclay were reported to have improved the shape recovery to a great extend [41]. Even though the well-exfoliated clay particles in the polymer decreased the crystallinity of the soft segment phase, the remaining crystallinity was enough for shape recovery performance. An increase of 20 % in magnitude of shape recovery stress was achieved with a loading of 1 wt% nanoclay.

ZnO nanorods were used as fillers to improve the shape memory properties of TPU [42]. Alkyl thiol-modified ZnO nanorods were dispersed in a low-hard-segment-density PU. The effect of nanoparticle aspect ratio, morphology, and interfacial interactions responsible for trapping the elastic strain for shape memory applications were also studied. Even though the enhancement in the modulus was modest and shape memory performance was not improved very much under low loading, it was considerably improved on higher filler loading. The improvement in shape recovery performance on ZnO nanorod addition was not as great as for carbon nanofibers, which is explained due to the difference in their impact on strain-induced crystallization of the PU matrix.

When PUs were incorporated with glass fibers, it was found that the creep properties of the shape memory PU such as rupture strain and rupture time was considerably improved due to the reinforcing effect of the glass fibers [43]. It was also observed that there is an optimum fiber weight fraction between 10 and 20 wt% to have an extremely low residual strain during cyclic loading. However, if the glass fiber weight fraction is increased beyond a limit, it can affect the shape memory properties [44].

2.2 *Electrically Triggered Shape Memory Effect*

A lot of research has been performed to the development of electrically triggered PU SMPs. Electrical stimulus-based PU SMPs can work either due to electrostrictive effect or due to resistive heating. A number of PU-based material systems have emerged as electrically triggered SMPs.

Electroactive polymer blend nanocomposites based on TPU elastomer, polylactide (PLA), and surface-modified CNTs were reported by Raja et al. [45]. Surface modification of CNTs was done by ozonolysis. Loading of the surface-modified CNT in the PU/PLA polymer blends resulted in the significant improvement of the mechanical, thermal, and electrical properties of the polymer blend. The shape recoverability of the PU/PLA/modified CNT composites were found to be excellent at lower applied DC voltages, when compared to pure or pristine CNT-loaded system. The response time was also significantly improved to 10 s with modified CNT composite than with pristine CNT composite which took 40 s for the shape recovery.

In a similar study, MWCNTs of 10–20 nm diameter and 20 mm length were incorporated into PU matrix [46]. They have also optimized the conditions for surface modification of the MWCNTs in mixed solvents of nitric acid and sulfuric acid which further improved the mechanical and electrical properties of the composites. An energy conversion efficiency of 10.4 % was achieved for the electro-active shape recovery of these composites.

The electro-active properties of PU/CB composites were studied by incorporating Ni powder chains into the composite [47]. In this approach, an additional small amount of randomly distributed Ni microparticles (0.5 vol.%) in the PU/CB composite significantly reduced the electrical resistivity. With the alignment of these Ni into chains by applying a magnetic field during preparation, the electrical conductivity was significantly enhanced. The electroactive properties were also improved.

The role of graphene as a reinforcing filler for memorizing the original shape and as a conductive filler by actuating the shape recovery by resistive heating was studied by hydroxyl-functionalized graphene in PU [48]. The shape memory performance studies revealed a maximum shape recovery of 98.6 % and a shape fixity of 99.1 %. The resistive heating and hence the shape recovery was more effective with higher voltage.

The shape memory properties of thermoresponsive PU filled with micron-sized Ni powders were investigated [49]. Under a weak static magnetic field (0.03 T), the filler particles formed conductive Ni chains, thereby improving the electrical conductivity of the composite in the chain direction. This makes it more suitable for Joule heat-induced shape recovery. It was observed that single chains start to form at 1 % volume fraction of Ni. As the Ni content increases, it shows a bundling nature. The Ni chains were found to reinforce the composites significantly but their influence on the glass transition temperature is about the same as that of the randomly distributed Ni powders. After five stretching–shape recovery cycles, the Ni chains still existed, which indicates the possibility of using chained SMPs for cyclic actuation.

Carbon nanofibers and oxidized carbon nanofibers were used to improve the shape memory performance of PU composites on resistive heating [50]. The soft segment crystallinity was found to be reduced in the presence of carbon nanofibers and oxidized carbon nanofibers. Out of the two types of nanofibers, the reduction in crystallinity of soft segment was low for the oxidized carbon nanofiber as the filler. When compared to CB filler, the reduction in crystallinity was low for fiber fillers, and hence, they did not show any positive temperature coefficient (PTC) effect. The shape memory properties of PU were improved with the incorporation of carbon nanofibers.

In a conductive composite prepared by the incorporation of silver nanowires (AgNWs) into the surface layer of shape memory polyurethane (SMPU), it was demonstrated that it is possible to recover not only the shape but also the conductivity of the composite under a thermal stimulation [51]. The developed AgNW/SMPU composites are stretchable and electrically conductive, and moreover, they exhibited reversible strain-dependent conductivity and can find applications in smart polymer materials for flexible electronics.

PU composites, incorporated with multilayered carbon nanopaper developed by the self-alignment of carbon nanofibers, exhibited improved shape recovery behavior [52]. The PU matrix was engineered to have a T_g of 50 °C to which the carbon nanofiber papers were incorporated. This significantly improved the T_g as well as electrical properties and thereby improving the Joule heating triggered shape recovery. The electrically actuated shape recovery behavior of the composites was demonstrated at a voltage of 30 V.

2.3 Light-Responsive Shape Memory Effect

Light-induced SMPs are of great importance from a practical perspective since it can be remotely controlled using light radiation. Light-triggered SMPs can be produced by incorporating reversible photoreactive molecular switches which react to a particular wavelength region of light [53–56]. References [16, 53] studied the shape memory performance of light-induced SMP. The polymer is initially stretched into strained and coiled polymer segments. Then, the polymer segments are irradiated with ultraviolet light with a wavelength >260 nm and by cross-linking to create new covalent netpoints so that the strain is fixed to get the temporary shape. When required, the sample in the temporary shape is irradiated with UV light of $\lambda < 260$ nm, through cleavage of the cross-links the permanent shape can be recovered.

The synergistic effect of the CNTs and reduced graphene oxide (RGO), on the shape memory performance of hyperbranched PU composites in response to a photothermal stimulation, was studied using a near-infrared (IR) laser as the stimulus [57]. Composites with a 7/3 of CNT/RGO ratio and a 1 wt% of nanocarbon content exhibited the best shape recovery performance under laser illumination. The studies demonstrated that for an IR laser-triggered shape memory PU composite, the use of hybrid-type fillers enhanced the effective conversion from light and thermal energy to mechanical work. The better performance is explained due to an enhanced three-dimensional interconnection between CNTs and graphene in the composites.

Due to the unique heating features and remote controllability of IR-induced shape memory effect, they have the potential to find many smart application including those in biomedical field where electrical stimuli are least preferred. In the studies on the capability of absorbing IR light in CB-filled SMPs, it was found that the composite covers a wide range of absorption spectrum and exhibits an increased absorptivity than pure SMP [58]. Due to the higher absorption of IR light by SMP/CB composites, they are able to produce remarkable heating effect when irradiated with light having wavenumbers between 500 and 3000 cm^{-1}. As a result, the shape recovery ability and shape recovery speed for the SMP/CB composite are remarkably enhanced than that of pure SMP.

2.4 Magnetically Triggered Shape Memory Effect

SMPs can be triggered indirectly either by magnetically induced heating or by radiant thermal energy of IR light. In magnetically induced SMPs, the polymer is filled with magnetic particles and the inductive Joule heating is achieved through an alternating magnetic field [59, 60].

Magnetically induced shape recovery of SMP composites can be achieved by the incorporation of magnetic nanoparticles such as Fe_2O_3 and Fe_3O_4 [59, 60].

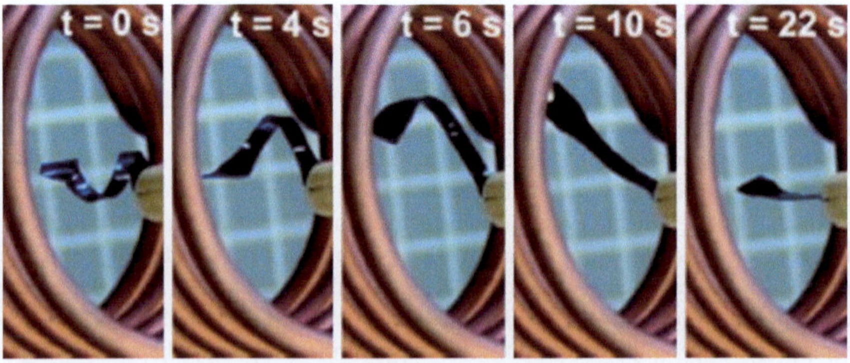

Fig. 4 Shape memory effect of PU composite in a magnetic field. The transition from a temporary shape of a corkscrew-like spiral to permanent shape of a plane strip took 22 s [60]. Copyright 2006. Reproduced with permission from National Academy of Sciences, U.S.A.

An alternating magnetic field is used to produce inductive Joule heating. In magnetically induced SMPs, the shape memory effect can be controlled remotely and selective location heating can be achieved. These types of SMPs can find potential application in biomedical field like in tumor therapy (magnetic fluid hyperthermia) and as contactless microactuators for clot removal in blood vessels.

Magnetically induced shape memory performance of SMP/Fe_2O_3 nanocomposites was studied by Mohr et al. [60]. A polyether urethane and a biodegradable copolymer of poly(*p*-dioxanone) and poly(ε-caprolactone) were used as the matrix and iron (III) oxide core in silica matrix was used as the filler. The shape recovery of PU composites was achieved by inductive heating in an alternating magnetic field ($f = 258$ kHz; $H = 30$ k Am^{-1}). It was observed that the sample geometry and the nanoparticle content have great influence on the maximum temperatures achieved by inductive heating in a specific magnetic field. The magnetically induced shape memory effect of the composites is shown in Fig. 4.

2.5 *Moisture-/Solvent-Driven Shape Memory Effect*

The shape memory effect can also be triggered by lowering T_{trans} with the uptake of water. When immersed in water, solvent molecules diffuse into the polymer sample and act as plasticizers, leading to a reduction in the transition temperature and resulting in shape recovery. In the PU polymers and their composites, T_g is lowered from the transition temperature to ambient or below. The lowering of T_g is directly dependent on the moisture uptake, or it indirectly depends on the immersion time. When the water uptake is adjusted between 0 and 4.5 wt%, a decrease in the T_g between 0 and 35 K can be achieved [61, 62].

Investigations were also carried out on polyetherurethane–polysilesquisiloxane block copolymers for water-actuated shape memory effect [63, 64]. A low molecular weight poly(ethylene glycol) (PEG) was used as the polyether segment. When immersed in water, the PEG segment is dissolved, which results in the disappearance of T_m, thereby recovering the permanent shape.

Similar to water-driven shape memory effect, other solutions can also trigger the shape memory effect. It can be mainly in two ways: through chemical interaction such as hydrogen bonding or interaction resulting in volume change or through swelling. The response of PU SMP to water or to solution is due to the plasticizing effect of the solution molecule on the polymeric materials. This increases the flexibility of the macromolecular chains and thus reduces the transition temperature of the materials until shape recovery occurs.

2.6 Multistimuli-Responsive Shape Memory Effect

PU-based SMPs which can be driven by more than one type of stimulus have more versatility for practical device applications. Multistimuli-responsive shape memory materials which respond to both water and electrical signals were demonstrated by incorporating MWCNTs into shape memory PU by a transfer method [65]. The hydrophilicity of the composites was enhanced by the surface modification of CNT. Faster and more complete shape recovery was observed for higher CNT content under both the water and the electrical stimulations.

Crystalline shape memory PU/graphene composites with various modifications to improve compatibility were also studied for shape memory applications [66]. Due to better compatibility of methanol-modified graphene and grafted PU, the shape recovery was considerably improved, but the resistive heating efficiency was decreased since the grafted PU hindered the intimate contact between graphene. The shape recovery by IR radiation was better for methanol-modified graphene composite than for pristine graphene composite since the IR absorption bands of graphene were strengthened with methanol modification.

Electroactive composites using multiwalled carbon nanotubes (MWCNTs)-filled hyperbranched polyurethane (HPU) were explored for shape memory applications [67]. The branched structure of HPU was found to be very effective for the homogeneous dispersion of MWCNTs without any surfactant or surface modification. The shape recovery studies on these composites under thermal as well as electrical triggers proved that these composites are excellent shape memory materials. They exhibited more than 98 % shape recovery and a rapid recovery time of 9 s in both thermal triggering and electrical actuating shape memory behavior. The work also suggested that the use of hyperbranched polymer-assisted MWCNT dispersion is a convenient and economical way to significantly improve the actuation as well as the mechanical properties of polymer composites.

Multiresponsive PU SMPs, due to their excellent biocompatibility, can be used as different clinical devices which are in contact or implanted in the human body

[68, 69]. Thermo-/moisture-responsive PU SMPs also extend possibilities to fabricate micro-/nanodevices for surgery/operation at the cellular level. Many polymer micromachines, which are similar or even smaller in size to that of a cell, have already been developed and demonstrated for various applications. [70, 71] These can be operated from outside the cell by triggering with a laser beam. In the case of a thermo-/moisture-responsive PU SMP, a piece of original curved SMP is straightened and then inserted into a living cell. When it absorbs moisture inside the cell, the SMP recovers its original shape. Since the recovery strain of SMPs is near to hundred percent, it makes it possible to make machines in the size similar or smaller than cells and insert them into the cell which can be driven for useful applications.

3 Sensors

As the physical and chemical characteristics of the polymers can be tailored over a wide range, the application of polymers in sophisticated electronic measuring devices such as sensors is increasing day by day. During last few decades, polymers have gained tremendous recognition in the field of artificial sensors by mimicking natural sense organs. By replacing classical sensor materials such as semiconductors, semiconducting metal oxides, solid electrolytes, ionic membranes, polymers and its composites have taken the lead role in gas sensors, pH sensors, ion-selective sensors, humidity sensors, biosensor devices, etc. These sensors have better selectivity and rapid measurements. Recently, PU and its composites are being used in the sensor devices either as active components (participate in sensing mechanisms) or as passive components (immobilize the component responsible for sensing the analyte).

3.1 Gas Sensors

Conductive polymer composites have received significant attention as gas sensors for various toxic gases or volatile organic compounds (VOCs). These composites have found promising applications in industries because of their advantages such as good processability, chemical stability, cost-effectiveness, design capability, and easy regulation of electrical conductivity within a wide range compared to intrinsic conductive polymer materials. The principle of these sensor devices relies on changes in conductivity of the used composite when exposed to gases. The conductivity of the composites may decrease or increase depending on the nature of the conductive filler used and the nature of the toxic vapor. Due to percolation mechanism, there is a drastic variation in electrical resistances in response to vapor exposure. Upon exposure to an organic vapor, the composites will swell to some degrees depending on the polymer–vapor interactions. It is believed that this swelling results in redistribution of the conductive fillers in the matrix and hence a change in the resistance of composites, which provides measurable signals characterizing the surrounding vapor-phase analytes [72, 73]. This transition is a reversible process and when the composites are transferred to air, it reaches its original structure.

Conductive composites with PU matrix are found to be potential candidates for wide-spectrum gas sensing element [74, 75]. PU-conductive composites are able to respond to many solvents including nonpolar and polar ones due to the coexisting nonpolar and polar segments on PU chains.

Waterborne polyurethane (WPU)-/CB-conductive composites having low percolation threshold (0.7–0.95 wt%) were synthesized by Chen et al. [76, 77]. These composites are quite sensitive to organic solvent vapors regardless of their polarities and showed drastic changes in their conductivity. With increase in CB content, negative and positive vapor coefficient phenomena of the composites were observed especially in the polar solvents. The vapor response habit of the composites in polar solvent atmosphere changes from negative vapor coefficient (NVC) to positive vapor coefficient (PVC) with a rise in CB content. However, in the case of weakly polar or nonpolar solvent vapors, the composites showed no responses because the filler loading is lower than the percolation threshold. But when the CB loading exceeds the percolation threshold, these composites showed remarkable responsivity. The composites behaved differently in different type of vapors due to the different interactions among the matrix polymer, the fillers, and the solvents.

In the case of nonpolar solvent, the solvent molecules prefer to immigrate into the microphase of the WPU soft segments, leading to volumetric expansion of the composites. The CB particles are also tend to be dispersed inside the microphase of the WPU soft segments, even a small amount of the solvent vapor is sufficient to directly disconnect the conducting paths formed by the contacted CB. In the case of a polar solvent, the solvent molecules prefer to diffuse into the microphase of the WPU hard segments. The localized swelling of the hard segments has to release first their restriction on the soft segments and then indirectly damages the CB conducting networks that are selectively located in the microphase of the WPU soft segments. Therefore, the higher equilibrium polar solvent absorbance on the composites is not bound to result in higher maximum electrical responsivity, as compared to the response to nonpolar vapor.

In order to reduce CB loading and to broaden the applicability as gas sensors, PU/CB-based composites were prepared by latex blending [75]. By this technique, conductive composites having low electrical threshold and improved processability were made. The composite with 3.5 wt% of CB exhibits significant increase of its electrical resistance in the environment of organic solvent vapors, regardless of the solvent's polarities and the maximum variation in resistance decreases linearly with decreasing vapor concentration. The slopes of the linear dependences change with the solvent species and hence allow quantifying vapor concentration and species. These composites show high reproducibility by withstanding over 5000 times of repeated sensing tests. The response habit of the CB/WPU composites as a function of temperature and vapor pressure were also studied [77]. Under saturated vapor test conditions, elevated temperature increases the rate of response but results in diverse dependences of the maximum magnitude of response on temperature due to the simultaneous variation in the vapor pressure. Under a constant vapor pressure, however, the composite's responsivity is found to be proportional to the inverse temperature, revealing the nature of thermally activated desorption. Heat-induced variation in the resistance of composites or the filler particles is ruled out and probable possibility is the maximum responsivity of the composites decreases with decreasing vapor pressure at a given temperature. Cross-linking treatment of the matrix polymer was found to be an effective way to weaken the unwanted NVC effect and to improve reproducibility of CB/WPU composites [78]. The cross-linking agent was added to the composite latexes, forming intramolecular cross-linked networks among the matrix polymer of the composites. The method greatly increased the filler/matrix interfacial interaction and reduced the mobility of CB particles. Due to this, during vapor absorption in the composites, the reconstruction of conduction paths through reaggregation of the disconnected filler particulates became difficult. Hence, the unwanted NVC effect was significantly weakened and the gas sensitivity and the performance reproducibility were enhanced.

3.2 Biosensors

The use of PU and their composites as biosensors has remarkable contribution in the sensors field. A lot of research has been performed to the fabrication of PU-based biosensors including glucose sensors, protein sensors, and urea sensors with considerable success.

Shin et al. [79] reported potentiometric biosensors using asymmetric-type PU membrane system which consists of a plasticized PU-based ion-selective membrane with a coating of thin hydrophilic polyurethane (HPU) film directly mixed with an enzyme at the top. Since the HPU:enzyme layer does not significantly influence the intrinsic potentiometric response of the basic ion-selective PU membrane and the addition of these two layers is very easy, this system is well suited for solid-state-type potentiometric biosensors. For instance, they have used glucose oxidase (GOD) as one of the enzyme for glucose sensing. The GOD oxidizes the glucose to gluconic acid, and the resulting pH change is detected. It was also demonstrated that with a careful selection of the organic solvent for formulating the HPU:enzyme layer, the proposed approach may prove valuable in the design and fabrication of other types of sensor devices including amperometric biosensors. They have also demonstrated the sensing of urea and adenosine with these systems by varying the enzymes. A schematic drawing of different types of HPU:enzyme-based biosensors is given in Fig. 5.

An amperometric glucose biosensor based on an acrylated-PU photocurable polymer membrane was reported [80]. The enzyme was immobilized on to the membrane by entrapment. An epoxy–graphite-conductive composite was used

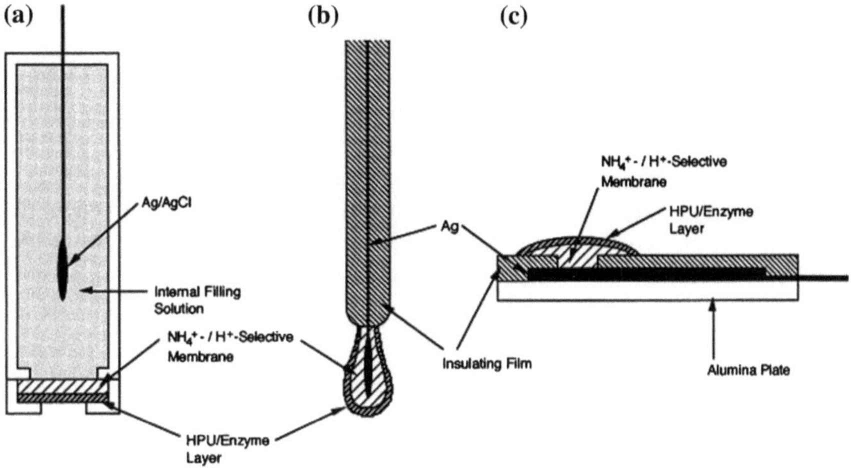

Fig. 5 Schematic drawing of different types of HPU-based biosensors [79]. Copyright 1998. Reprinted with permission from Elsevier Ltd.

as the amperometric transducer which showed a good adhesion to the membrane surface. The use of UV-curable polymer resulted in a robust membrane with a long-term stability of around 3 months. The process can also be atomized since the membranes can be deposited by photolithography. The effect of introduction of graphite powder into the polymer matrix was also studied and an increase of response of about three times was reported. A flexible glucose biosensor for long-term subcutaneous implantation was developed using epoxy-enhanced PU membrane [81]. These porous membranes consisted 30–40 % epoxy resin adhesive and 50–70 % PU and exhibited excellent durability. It was reported that the epoxy enhanced the structural strength and adhesive nature of the sensor element. The in vivo performance of the sensors proved that the sensors are highly durable and can work at least 56 days when implanted subcutaneously in rat.

A high-performance glucose sensor for the continuous monitoring of glucose levels was demonstrated by Han et al. [82] The sensor design consisted of a non-conducting poly(*p*-aminophenol) (pAP) film on a plasma-deposited platinum electrode with a HPU membrane mixed with polyvinyl alcohol/vinyl butyral copolymer (PVAB) as an outer membrane. The non-conducting layer of pAP was used to immobilize the enzyme, glucose oxidase. The *p*-aminophenol monomer with an optimum concentration of enzyme (10 mg/mL) was electropolymerized to form the inner enzyme layer which displayed better sensor performance when compared to the layers formed by conventional cross-linking method. An outer membrane with an optimized ratio of HPU to PVAB (3:2) was having high oxygen permeability and good ability to properly control glucose diffusion. The optimized glucose sensor with good sensitivity and fast response (<5 s) was capable of continuous monitoring for 20 days in a buffer solution, while in human serum, the lifetime was only about 3 days.

PU-grafted calcium alginate hydrogel microspheres were reported by Li et al. [83] for the selective imprinting of bovine serum albumin (BSA). The grafting of hydrogel with PU side chain was beneficial for the specific recognition of protein. It was also found that the PU-modified imprinted hydrogel had higher affinity to template molecules. The effect of grafting ratio on specific rebinding was also studied. The formation of BSA-specific binding sites and molecular selectivity in the single and binary competitive protein was also verified. The samples were found capable of recognizing and separating target proteins and hence can be used for chemical sensing and bio-separation applications.

3.3 Strain Sensors

PU is also widely used as strain sensors. A high-deformation strain-sensing element using multiwalled CNTs/TPU composite was reported [84]. The composite was prepared by enmeshing a nonwoven PU filter membrane with CNTs and melding them together. The studies revealed that these composites can be elongated up to 400 % with an increase in resistance of about 270 times. In terms of gauge factor, it increases linearly with strain with values changing from 4 at start of deformation to 69 at 403 % strain. This proves that the material is well suited for a high-sensitive strain gauge. They also suggest potential applications as a strain sensor in a human knee flexion and its cyclic movement. Strain sensors were also developed by coating TPU/CNT-conductive polymer composite on to a commercially available spandex multifilament yarn [85]. Very good strain sensitivity was achieved with a CNT concentrations as low as 0.015 wt%. Even though the CNT loading has an effect on the resistance, it was observed that more significance should be given to homogeneous, continuous coating. Under cyclic loading tests, partial recovery of resistance was observed in the first loading cycle, while good reversibility was observed on subsequent cycles, proving these materials have good potential as sensors for smart textiles.

Flexible strain sensor arrays on PU substrates prepared by patterning different conductive solutions through a tilted drop process were reported by Chang et al. [86]. Conductive polymer solutions were prepared by incorporating various fillers such as carbon nanofibers and MWCNT into PU solution. Gauge factors from 0.34 to 7.98 were obtained for 3–7 % strain. The sensor arrays were also able to withstand bending tests at 150° more than 50 times without any deformation. The best sensitivity with a gauge factor of 7.98 was obtained when a conductive polymer solution with 36.6 % carbon nanofibers, 10.9 % sodium dodecyl sulfate, and 52.5 % PU was used. The method proposed for the fabrication of sensors is important from an industrial perspective since it can be scaled up to mass production.

3.4 Ion Sensors

Ion-selective field-effect transistors are one of the important applications of PU membranes in the field of sensors. Bratov et al. [87] developed a Ca^{2+}-ion-sensitive ISFETs using a photocured PU-based polymer membrane with alkylphosphate exchanger (DOPPCa) and three neutral carrier ionophores (ETH 1001, ETH 129, and ETH 5234). They used dioctylsebacate as plasticizer. They studied the Ca^{2+} sensitivity of the sensor both in pure $CaCl_2$ solutions and in background solutions containing Na^+, K^+, and Mg^{2+} ions. The effect of the ionophore on the selectivity of Ca^{2+} was studied. ISFETs with membranes containing ETH 129 exhibited high selectivity and long lifetime in constant contact with a solution and high precision of ionized calcium determination. The estimation of Ca^{2+} ions in whole milk was also reported.

4 Polyurethane-Based Actuators

Actuators are materials or devices which can undergo shape or mechanical change in response to an appropriate external stimulus [88]. Polymer actuators have attracted the attention of researchers in the last few decades because of their low cost, ease of processing, large displacements obtained, and their resemblance to the properties of human muscles. PU and its composites in particular have sealed their place as the most promising polymer actuators in many applications including artificial muscles and robotics. Depending on the stimulus used for actuation, they can work as electrical, optical, thermal, chemical, and pneumatic actuators.

4.1 Electrical Actuators

Electromechanical properties of PU elastomers have created enormous interest in researchers due to its remarkable characteristics such as large electric field-induced strain, high specific energy, and fast speed of response [89]. Electromechanical actuation in PU can be mainly of two mechanisms: Maxwell's effect and the actual electrostrictive effect. The pure electrostriction is the direct coupling between an electric polarization and mechanical strain response, while the Maxwell effect is mainly due to an attractive force between the opposite charges on the electrodes [89]. However, investigations have already revealed that the major contribution (up to 85 %) to the electromechanical activity is from electrostrictive effect due to a phase separation in the PU. The minor factor (less than 15 %) was attributed to the Maxwell stress, due to the electrostatic attraction under applied electric field [90].

Virgin PU has been widely studied as an electroactive polymer in the last few decades. Studies on the effect of chemical structure of PU on its actuation behavior revealed that polyester-based PU is a better electromechanical actuator than polyether-based PU [91, 92]. The electromechanical actuation of PU can be controlled by controlling the chemical and polymeric nature of the PU as well as the film thickness [93]. A monolayer bending actuator, which can act as a bimorph, was demonstrated by Watanabe et al. using PU film [94]. A deformation curvature (1/R) of 36 m^{-1} under an electric field of 10 mV/m within 3 s was achieved. In a study regarding electromechanical response of fiber actuators prepared using PU tubes, it was found that the blocking force obtained for PU tubes was significantly higher when compared to silicone tubes [95]. The effect of applied prestrain on the electromechanical performance was also studied. For uniaxially prestrained samples, the axial actuation strain decreased with increasing prestrain, whereas the radial actuation strain exhibited an increasing trend.

Jung et al. [96] studied the performance of a diaphragm-type polymer actuator fabricated using segmented PU (SPU). They have studied the relationship between the elastic modulus and maximum deflection as a key property of the Maxwell stress effect and the relationship between the dielectric constant and maximum deflection as a key property of the electrostriction effect. The SPU-based diaphragm-type actuator was used to formulate theoretical calculations which were compared with experimental values.

Even though virgin PU is comparatively better electromechanical actuator material than many other polymers, efforts to reduce the voltage required to drive the actuators have lead to the introduction of PU composite actuators. Basically, two types of fillers are used for low-voltage-driven PU composite actuators: ferroelectric fillers and conductive fillers. When ferroelectric fillers such as PZT particles are used, the relative permittivity ε_r of the polymer matrix is small enough to be ignored compared to the filler material. In this case, the strain obtained is directly related to the permittivity of the filler material. The required strain can be tuned by the volume of the filler in the composite. Unlike in ferroelectric fillers where each particle is supposed to become a dipole when an electric field is given, in the case of conductive filler the electrons stay on the surface of each conductive particle during electric field induction. Consequently, a smaller electric field is sufficient to drive such conductive composite actuators [97].

The most widely explored conductive fillers in PU composite actuators are carbonaceous materials such as carbon particles, CNTs, fullerene derivatives, and graphene. Carbon nanopowder/PU (C/PU) composites were studied as actuator materials. It was observed that when compared to some classical actuator material, the C/PU composites required very low electric field to get maximum strain, but the generative force was comparatively low [98].

MEMS actuators were developed using PU/CB nanocomposites through a typical microsystem fabrication process [99]. The actuator performance at the fundamental resonance frequency was studied using a sinusoidal signal (electrostriction mode). The performance in the pseudo-piezoelectric mode was also studied by applying an additional DC bias in order to induce permanent polarization and create a prestress in the material. The performance of the actuators in the pseudo-piezoelectric mode was larger compared to the electrostrictive mode due to the constant direction of electric field in the pseudo-piezoelectric mode. The actuation of the nanocomposite membranes was found to be more efficient than that of pure PU. However, for the development of high-frequency PU-based microactuators, the time constant needs to be improved.

A diaphragm-type actuator was demonstrated using an electrostrictive CB/PU composite membrane [100]. A theoretical model for the diaphragm actuator was developed and found that experimental results fitted well with the developed model. When the deflection at the center of a 50-μm composite film with 1.25 wt% CB concentration is compared with pure PU membrane of same thickness under a 9 V/μm electric field at 700 Hz, a fourfold increase was achieved which suggests significant enhancement in using such devices in MEMS systems. Moreover, the performance of the diaphragm operating in a pseudo-piezoelectric mode at the resonant frequency (700 Hz) demonstrates the possibility of high-frequency actuators. The diaphragm deflections were found to be enhanced by loading the polymer with CB particles. A CB loading of 1.25 % was reported as the optimum.

Under electrical stimulus, CNT/PU composites were found to exhibit shape memory behavior which can be controlled by the concentration of CNT in the composite. At lower concentrations of CNT, the composites showed a delayed shape recovery, whereas the shape recovery was accelerated with increasing CNT content [101]. Electrothermal bimetallic actuators with MWCNT/WPU and MWCNT/silicone rubber composites were demonstrated [102]. Very large bending displacements of 28 mm or a curvature of 0.29 cm^{-1} under a very low driven voltage of 7 V was achieved. The actuator exhibited considerable controllability, large mechanical output, and long lifetime. They were also able to work under a reduced working power of 25 MW/mm^3 in air atmosphere. In addition to the contribution due to the mismatch in the coefficients of thermal expansion, the unique negative temperature coefficient effect of nanotube composites also contributed to the actuation mechanism. Furthermore, the bimetallic actuators based on various polymer matrices are supposed to provide revelations for electrothermal bimorph actuators. The bimetallic actuator using two different polymer matrices open up opportunities for fabricating a class of actuators with extensive prospects in applications such as biomedical devices, miniature or bio-mimetic robotics, artificial muscles, switches, and microsensors.

In order to improve the performance of CNT/PU composite actuators, several modifications to CNT have been attempted by many researchers. The electromechanical actuation behavior of PU nanocomposites with metal nanoparticles (Ag and Cu)-decorated MWCNTs was studied by Raja et al. [103]. The introduction of the metal nanoparticles-decorated CNTs in PU resulted in significant improvement of mechanical, thermal, and electrical properties. PU/MWCNTs nanocomposites exhibited remarkable strain recovery at lower applied DC voltages. PU/metal-decorated CNTs showed high strain recovery ability after several cycles compared to PU/pristine CNT composites.

With the discovery of graphene in 2004, the studies on graphene-based PU composites for actuator applications attracted the attention of many researchers. However, the dispersibility of graphene was a major problem. Many attempts were made to functionalize graphene to make homogeneously dispersed polymer composites. PU composites were prepared using poly(methylmethacrylate)-functionalized graphene [104]. Graphene was functionalized using poly(methylmethacrylate) via atom transfer radical polymerization and was used as a conducting filler in PU. The measurement of electromechanical properties proved that the electric field-induced strain behavior of poly(methylmethacrylate)-functionalized graphene/PU (MGPU) composites were significantly improved when compared to pure PU films. For instance, 1.50 wt% MGPU film exhibited an electric field-induced strain of 32.8 % which was nearly two times than that of pure PU (17.6 %).

Ionic liquids can also be incorporated into the polymer matrix in order to improve the actuation performance of PU. Transparent ionic liquid/polyurethane (IL/PU) gels were synthesized by the addition reaction of polyol and diisocyanate in the presence of 1-ethyl-3-methylimidazolium bis(trifluoromethylsulfonyl)imide [105]. The IL/PU gels were sandwiched between two conductive polymer films made of poly(3,4-ethylenedioxythiophene) doped with poly(4-styrenesulfonate) (PEDOT:PSS) as flexible electrodes. Under electric field, the IL/PU/PEDOT:PSS composites exhibited fast and intensive bending toward anode. A bending displacement of 3.8 mm, corresponding to a strain of 0.32 %, was attained for 40 wt% of ionic liquid at 2 V. Moreover, the IL/PU/PEDOT:PSS composite actuator worked at frequencies higher than 10 Hz.

Similar to the conductive PU composites, conductive PU blends were also investigated by some researchers for electromechanical actuator application. The actuation behavior of PU/polyaniline blends were studied by Jaaoh et al. [106]. They observed that the solvent used for casting the blend films has an effect on the electrostriction behavior of the blends. In NMP (1-methyl-2-pyrrolidone) solvent, polyaniline component effectively enhanced the electrostriction of PU/polyaniline blends, whereas opposite effect was observed when DMF (N,N-dimethyl formamide) was used as the solvent. They attributed this to the formation of hard interfacial domains between the dispersed polyaniline and the PU matrix which contributes to the enhanced electrostriction which in turn improves the actuator performance.

4.2 Thermoresponsive and pH-Sensitive Actuators

Thermally responsive PUs are basically used for shape memory application which is dealt as a separate section in this chapter. In shape memory application, the heating is done through different methods including resistive heating or magnetically induced heating. In certain PU actuators, direct heating is used to trigger the actuation mechanism. The bending actuation characteristics of a PU/TiNi wire composite belt by direct heating and cooling revealed improved actuation properties. The rate of force recovery was about 93–94 % of maximum force [107]. Two-way actuation of a poly(esterurethane) was achieved between +60 and −20 °C. The driving force for the two-way actuation was the soft segment crystallization and soft segment melting. Multiple-cycling tests revealed that poly(esterurethane) is suitable for thermoresponsive actuators which can work under constant stress conditions [108].

PU-based pH-sensitive actuators for drug delivery have attained considerable attention in the biomedical field. A pH-sensitive actuator for drug delivery was developed by introducing pyridine rings to PU backbone. The mechanism of pH-stimulated actuation is due to the formation of a hydrogen bond between nitrogen atom on the pyridine ring and H–N of urethane in neutral or alkaline environments. The breakage of hydrogen bonds in acidic medium due to the protonation of the pyridine ring reverses the shape change. The actuation behavior is well suited for drug delivery because it is independent of temperature and only depends on the pH [109]. In a similar work, PU nanoparticles which can change shape in response to pH and temperature were used for the controlled delivery of drugs. These PUs exhibited faster release at lower pH value, and the drug release was increased with temperature [110].

4.3 Optical Actuators

Optically triggered actuators have attracted considerable interest in the recent years due to their unique features such as wireless actuation and remote controllability. Optical actuators are extremely useful in biomedical field where stimulus other than electricity is preferred. PU nanocomposites which can be actuated by IR light have enormous potential in this field. The incorporation of conductive fillers such as CNTs or graphene can combine the IR absorption characteristics of the filler and the unique mechanical properties of PU. The fillers act as an "energy transfer unit" which absorbs the light energy and transfers to the PU which acts as the "molecular switch unit" where the mechanical deformation takes place.

Due to the excellent IR absorption characteristics of the CNTs, many studies have been performed to the optical actuation of CNT-filled PU. Improved optical actuation was demonstrated in remotely actuated PU/CNT composite prepared using commercial TPU with low glass transition temperature [111]. PU composite with 1 wt% of CNT which was prestrained to 300 % contracted on IR irradiation and exerted almost 19 J to lift a weight of 60 g to 3.3 cm with a force of ~588 N, whereas neat PU, which is transparent to IR light, showed no actuation on IR irradiation. On a comparison of the CNT-filled PU with CB-filled PU, it was observed that addition of four times more CB than CNT is required to impart comparable properties. The dispersed CNT in PU can alter the strain dependence of polymer crystallite formation and increase the strain set and stored energy density. These composites could demonstrate remotely controlled actuators which have great potential for many practical applications.

In the past decade, graphene and its derivatives have emerged as a competitor for CNTs in polymer nanocomposite actuators. Graphene with its unique electrical, thermal, mechanical, and optical properties has proved its efficiency to be used as an energy transfer unit in light-triggered polymer actuators. Repeatable IR-triggered actuation performance was demonstrated using a TPU/sulfonated graphene nanocomposite [112]. With 1 wt% filler content, when exposed to IR light, the composite actuators could lift a weight of 21.6 g to a height of 3.1 cm with 0.21 N force. An estimated energy density of more than 0.33 J/g was achieved. The mechanical properties of the composite were also improved significantly. The IR-triggered actuation of isocyanate-functionalized graphene and reduced graphene oxide as fillers in PU matrix were also studied. However, the actuation performance of sulfonated graphene composite was much better than the other two fillers due to the better dispersibility and better IR absorption of sulfonated graphene.

The photomechanical actuation properties of thermally reduced graphene oxide (TRGO)/TPU composites were studied under IR light illumination [113]. The composites exhibited fast photomechanical response. The photomechanical strain was found to increase with increasing prestrain. An optimum amount of 2 wt% of TRGO was found to give maximum photomechanical properties of TPU, but further increase in TRGO concentration resulted in reduction in properties. With the increase in TRGO content, the response time was also improved. A photomechanical strain of 50.2 % with a stress of 1680 kPa was obtained for 2 wt% TRGO/TPU composites at 220 % prestrain. It was demonstrated that these composites are capable of finding potential application as light-triggered actuators in many fields including robotics.

The studies on the effect of a hybrid filler-containing sulfonated reduced graphene oxide (SRGO) and sulfonated CNT on the optically triggered actuation of PU revealed improved actuation behavior [114]. Interconnected SRGO/sulfonated CNT hybrid fillers dispersed in PU matrix exhibited excellent IR absorption characteristics. Even a low loading of 1 % of fillers improved the crystallization of soft segments for obtaining larger strain. The SRGO/sulfonated CNT ratio was found to have an effect on the output force, energy density, and recovery time. PU composites with SRGO: sulfonated CNT weight ratio of 3:1 exhibited maximum IR-triggered actuation. It could lift a weight of 107.6 g up to 4.7 cm in 18 s. A high energy density of 0.63 J/g and a shape recovery force of 1.2 MPa were obtained. This improvement was mainly due to the increase in thermal conductivity of the composite. In order to get high mechanical output, a trade-off between stiffness of the PU composite and the heat transfer should be achieved through proper control of the SRGO and sulfonated CNT content.

4.4 Pneumatic Actuators

In some specific applications such as nursing care devices or artificial muscles for aged humans which need lightweight and flexible actuators, pneumatic actuators are well preferred over electrically driven actuators. PUs definitely have carved their own space in these applications as well. PU-based flexible pneumatic actuators which could be operated even if deformed by any external force were reported for use in nursing care systems. In a pneumatic actuator, two types of pneumatic cylinders were used, one with a ball-type gasket and another with a Y-ring-type gasket to seal the cylinder head [115]. The flexible cylinder was made of two different PUs, a hard PU and a soft PU. The relation between input pressure and generated force was studied with the cylinder curved at different angles. A linear relationship was obtained even if the cylinder tube is bent. When compared to other pneumatic actuators, a larger generative force to weight ratio was obtained. For instance, output force of 20 N for a less than 0.1 kg of actuator was obtained with a stroke of 1 m.

A similar type of pneumatic artificial muscle actuator was developed by using a coating of a shape memory PU [116]. When compared to conventional pneumatic actuators, these types of shape memory PU-based actuators exhibited several advantages. At normal temperatures, these can be functioned as conventional pneumatic actuators. But we can change the initial position of the actuator by simple heating by utilizing the shape memory effect of PU coating. By changing the area of heating, we can also change the direction of bending.

A bionic hand which works on pneumatic actuators which uses PU of different stiffness was demonstrated to have very good ability for grasping various objects [117]. The finger sacks and palm sheets were made up of relatively stiffer PU, whereas the whole-hand structure was covered with soft PU. The adaptive grasping ability of this bionic hand using pneumatic actuators was tested for various objects using a controlled air input. The stability of manipulation for grasping different objects was improved by the use of multilayer structure of PU material with different stiffness for different parts.

Refreshable Braille displays are used to convert electronic text signals to Braille dots for helping visually challenged people to read. Due to the rapid response, large force generation, and low cost, pneumatic actuators are promising candidates for Braille cell actuation. A refreshable Braille cell using PU-based pneumatic microbubble actuators was demonstrated to have satisfactory performance for static and vibrational tactile display applications [118]. The PU-based pneumatic microbubble arrays were actuated through solenoid valves as per the received signals. The displacement of PU-based Braille dot was about 0.56 mm at 0.2 Hz operating frequency and generated a force of 66 mN at 100 kPa applied pressure. A lightweight compact device was possible through this technology.

PU composite-based morphing wings for aircraft application which is operated by a PU-based pneumatic actuator were studied by Peel et al. [119]. The upper and lower skins of the wing were prepared by PU composites. For upper skin of the wing, carbon fiber laminate impregnated with a semirigid PU elastomer was used, whereas for lower skin of the wing, carbon fiber laminate with less stiff PU was employed. The pneumatic actuator required for the wings were also fabricated using carbon tow, carbon fiber, and carbon cloth composites with PU elastomer. The wings operated with the pneumatic actuators exhibited excellent angles of deflection. The PU composite-based braided actuators generated higher force than normal filament wound actuators and possess potential for practical applications.

5 EMI Shielding

Electromagnetic interference (EMI) shielding using radio-frequency and microwave frequency absorbing materials are in high demand in commercial, military, and scientific electronic devices and communication instruments. In conventional EMI systems, magnetic particles or MP are used as absorbing materials; however, high specific gravity and difficult formulation have limited their practical applications. There is a need for lightweight, flexible and structurally sound EMI shielding material to protect the workspace and environment from radiation coming from computers and telecommunication equipments as well as for protection for sensitive circuits.

Recently electrically conducting PU composites have found application in EMI shielding because of their lightweight, resistance to corrosion, flexibility, and processing advantages [120–123]. The EMI shielding efficiency (SE) of a material is the ratio of the incident field strength (V/m) to that fraction, which is transmitted, expressed in a decibel scale as 20 log 10 (E_i/E_t) (dB) [15]. E_i and E_t are the intensity of electric field of incident and transmitted electromagnetic waves through a shielding material, respectively. Total SE is the sum of three parts: reflection, absorption, and rereflection [15, 121]. SE of a composite material depends on many factors, including the filler's intrinsic conductivity, dielectric constant, and aspect ratio.

5.1 Carbon Nanotube/Polyurethane Composites

Liu et al. [14] reported well-dispersed single-walled carbon nanotube/PU (PU/SWCNT) composite EMI shielding materials prepared by a simple physical blending method. An EMI SE up to 17 dB at the band range of 8.2–12.4 GHz (X-band) was obtained for PU/SWCNT composites with 20 wt% SWCNT loading. The composites showed a percolation threshold as low as 0.2 wt%. They also reported that at low CNT-loading EMI mechanism is dominantly reflection, while with increased SWCNT loading and frequency, absorption was the predominant mechanism. Using EMI shielding theory, this trend is explained with the intrinsic properties of the components. At high SWCNT loadings, the intrinsic properties favored the absorption ability rather than the reflecting one. Processable composites of single-walled CNTs with soluble cross-linked polyurethane (SCPU) show the best microwave absorption of 22 dB at 8.8 GHz [122]. A solution or hot pressing technique processable SCPU was mixed with various loadings of SWCNT (0–25 wt%) to prepare composites and the microwave absorption characteristics were studied in the range of 2–18 GHz. The absorbing peak position moves to lower frequencies with increasing SWCNT.

Partially conductive polymer composites were made with poly(urea urethane) (PUU) with a poly(dimethylsiloxane) soft segment and different types of conductive fillers—CNT, silver-coated CNT (CNT–Ag), and nickel-coated CNT

(CNT–Ni) [120]. CNT concentration was varied from 3, 4, and 5 parts per hundred parts of resin (phr), in the PUU, and these composites possessed SE of 8.5, 28.4, and 26.0 dB, respectively, at electromagnetic wave frequencies of 12.3, 16.2, and 15.9 GHz, respectively. SE of the composites that contained CNT–Ni and CNT–Ag increased with the filler loading. At the same modified-CNT loading, the CNT–Ni-filled composites had a higher SE than those filled with CNT–Ag. Multiwalled CNT (MWCNT)/PUU nanocomposites have been studied for EMI shielding, surface resistance, microwave absorption, and mechanical properties [123]. The microwave absorption and tensile strength of MWNT/PUU were also improved by the uniform dispersion of MWCNT in PUU matrix. MWCNT end-capped PUU system with 5 phr of functionalized MWCNT showed the maximum microwave absorption of 19.41 dB at about 16 GHz. The EMI SE of pristine PUU, silane end-capped PUU, phenyltrimethoxysilane end-capped PUU, and MWNT end-capped PUU at 500 MHz were −1.25, −1.28, −0.97, and −1.03 dB, respectively. The pristine PUU or silane end-capped PUU exhibits higher EMI SE than that of P-silane-capped PUU or MWCNT end-capped PUU because of the lower surface resistance.

Long-length MWCNTs-reinforced PU composites were prepared by solvent casting followed by compression molding technique [121]. EMI SE of these composites was studied in the frequency range of 8.2–12.4 GHz (X-band) and EMI SE of the composites is achieved up to −41.6 dB at 10 wt% loading of MWCNT. The major cause for improved SE has been attributed to the significant improvement in the electrical conductivity of the composites by 13 orders of magnitude, i.e., from 10^{-14} for pure PU to 7.9 S cm^{-1} for MWCNT-PU composites.

5.2 Graphene/Other Fillers—Polyurethane Composites

Flexible and lightweight GN/waterborne polyurethane (WPU) composites which exhibited high electrical conductivity and electromagnetic shielding performance were reported [124]. Covalently modified GN with aminoethyl methacrylate (AEMA-GNs) exhibited high compatibility with a WPU matrix which caused the AEMA-GNs to homogeneously disperse in the WPU matrix. 5 vol.% AEMA-GN-loaded WPU composites exhibited electrical conductivity of 43.64 S/m and EMI shielding effectiveness of 38 dB over the frequency of 8.2–12.4 GHz.

Ferrite–PU composites prepared with different ferrite ratios of 50, 60, 70, and 80 % in PU matrix have been investigated in X-band frequency range [125]. The M-type hexaferrite composition $BaCo^{+2}_{0.9}Fe^{+2}_{0.05}Si^{+4}_{0.95}Fe^{+3}_{10.1}O_{19}$ was prepared by solid-state reaction technique. The composite with 80 % ferrite content has shown a minimum reflection loss of −24.5 dB (>99 % power absorption) at 12 GHz with the −20 dB bandwidth over the extended frequency range of 11–13 GHz for an absorber thickness of 1.6 mm. These composites can fruitfully be utilized for suppression of EMI and reduction of radar signatures. Conducting carbon in PU matrix for EMI shielding was reported by Abbas et al. [126].

5.3 Polyurethane Blends

Polyaniline–polyurethane (PANI–PU) blends were also used for EMI shielding applications [127, 128]. The microwave absorption, microwave reflection, and EMI shielding properties of PANI–PU blends at S-band and X-band frequencies were reported. It was found that the conducting blends showed a very low percolation threshold and their electronic properties can be tuned easily with the mass fraction of polyaniline in the blends or through chemical process. EMI SE from 30 to 90 dB at 8.2–18 GHz band was found for mono- and three-layer samples with a very lightweight and good mechanical properties.

6 Polymer Electrolytes

Polymeric solid electrolytes have received considerable attention for application in lithium-ion batteries (LIB), fuels cells, super capacitors, electrochemical sensors, etc. [129–132]. They are having many advantages over conventional liquid electrolytes and solid inorganic electrolytes in terms of their processability, higher energy density, mechanical strength, no-leakage, flexibility for cell design, non-combustible reaction products at the electrode surface, etc. [133, 134]. Various polymers including poly(vinylidenefluoride) (PVDF), polyacrylonitrile (PAN), polyvinyl chloride (PVC), poly(methyl methacrylate) (PMMA) and PU, have been used as the host polymers for preparation of polymer electrolytes [134–136]. Polymeric solid electrolytes consist of either a polymer having electrolytic properties by itself, or composites of an inorganic salt incorporated into a polymer [137–139].

For the current battery applications, polymeric solid electrolytes are required to have high ionic conductivity (>10^{-5} S/cm at ambient temperature), good thermal dimensional stability, good electrochemical stability, easy processability, etc. [140]. The mobility of the ions in polymer electrolyte is dependent on the flexibility of the polymer segments and the dimensional stability of the polymeric material depends on the rigidity of polymer segments. Considering this, a polymer composed of hard segments and soft segments is good for the matrix for the polymeric solid electrolytes and in recent years PU-based polymer electrolytes have attracted great attention [141–144]. In the polyether-based PU, the hydrogen bonding between the carbonyl group and N–H group in the hard segments of TPU leads to the formation of physically cross-linked hard domains and responsible for the dimensional stability [145, 146]. The soft segments will act as solvents to solvate the cations favoring the transportation of the ions and contribute to the conductivity of the polymer electrolyte. The advantage of TPU is that its property can span a wide range by tailoring the structure and the ratio of soft segment and hard segment [147].

6.1 Polyurethane-Based Electrolytes

TPU with a mixture of soft segments [poly(ethylene glycol) (PEG) and poly(tetramethylene glycol) (PTMG)] was prepared as an ion-conducting polymer electrolyte by Chen et al. [142]. The changes in the morphology and ion conductivity of the phase-segregated TPU-based solid and gelatinous polymer electrolytes as a function of the lithium perchlorate concentration were reported. The conductivity changes revealed the combined influence of PTMG and PEG units in TPU. The room-temperature conductivity of approximately 1×10^{-4} S/cm was found for TPU containing 50 wt% liquid electrolytes. The swelling characteristics in a liquid electrolyte and the dimensional stability were evaluated for these TPUs. TPU system containing both PEG and PTMG as soft segments showed good dimensional stability and ionic conductivity and was found to be suitable for electrolyte applications.

Polymer electrolytes were prepared by the addition of mixture of ionic liquid N-ethyl (methylether)-N-methylpyrrolidinium trifluoromethanesulfonimmide ($PYRA_{12O1}TFSI$), lithium bis(trifluoromethanesulfoneimide) salt, and propylene carbonate into TPU matrix [148]. These electrolytes have good self-standing characteristics, and also a sufficient level of thermal stability and a fairly good electrochemical window. The ionic conductivity increases with increasing amount of mixture, and the character of temperature dependence of conductivity indicates decoupling of ion transport from polymer matrix. The maximum ionic conductivity measured at room temperature was 10^{-4} S cm^{-1}.

A series of comb-like cationic PUs were synthesized by quaternizing different bromoalkanes (C_2H_5Br, $C_8H_{17}Br$, and $C_{14}H_{29}Br$) with PU [149]. Solid polymer electrolytes were prepared by doping TPU with $LiClO_4$. Alkyl quaternary ammonium salts in the polymer backbone are the inherent plasticizers, which make the electrolytes exhibit liquid-like behavior. The plasticizing effects of PU electrolytes containing higher alkyl chain lengths are more effective than that of PU electrolytes containing lower alkyl chain length. Maximum ionic conductivity of 1.1×10^{-4} S cm^{-1} at room temperature was obtained for PU-C8 electrolytes containing 50 wt% $LiClO_4$.

Gel PU electrolyte was prepared by the UV curing of a liquid electrolyte containing an aliphatic urethane acrylate [150]. By controlling the concentration of TPU, the ionic conductivities of the gel electrolytes were varied from 9.8×10^{-3} to 4.7×10^{-3} S/cm at room temperature. The quasi-solid-state DSSC fabricated with the gel electrolyte given an overall energy conversion efficiency of 6.1 % under irradiation of 100 MW/cm^2 at room temperature. The fabrication process of DSSC and DSSC cell fabricated is given in Figs. 6 and 7. From Fig. 6, it can be observed that PU electrolyte can be integrated into DSSC through simple and cost-effective processes such as printing and UV curing. This makes it more commercially viable. Figure 7 shows the photographs of a DSSC fabricated using PU electrolyte. Four different TPUs were made from poly(ethylene glycol) (PEG), 4,4-methylenebis(phenyl isocyanate) (MDI), and 1,4-butanediol (1,4-BDO) with

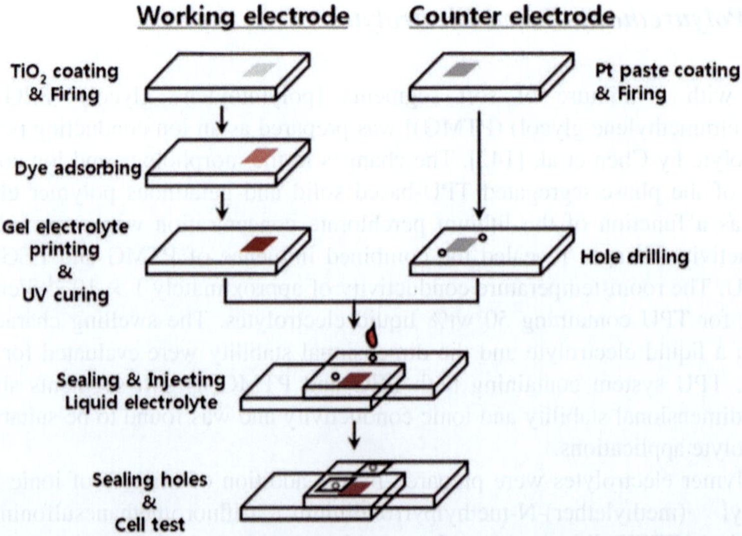

Fig. 6 Fabrication process of a DSSC with UV-curable electrolyte [150]. Copyright 2010 reprinted with permission from Elsevier Ltd.

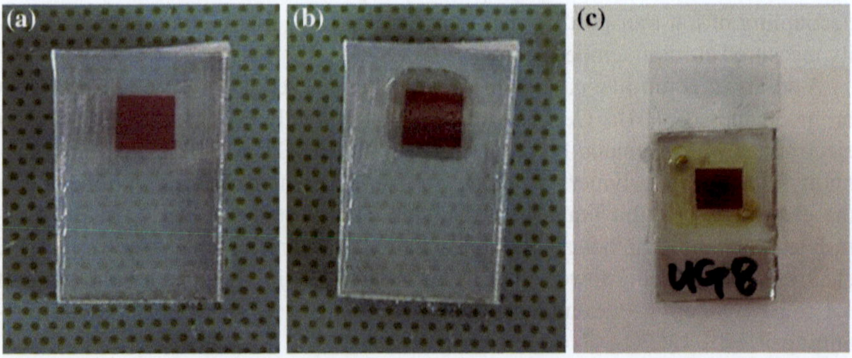

Fig. 7 Photographs of **a** TiO_2 electrode, **b** TiO_2 electrode coated with gel electrolyte, and **c** a DSSC assembled with UG8 [150]. Copyright 2010. Reprinted with permission from Elsevier Ltd.

different 1,4-BDO/PEG ratios [151]. The effect of polymer structure on the conductivity of the polymer electrolytes was investigated. The conductivity depends on the soft segment concentration and on the degree of phase separation exhibited by these materials, and ionic conductivity as high as 3×10^{-4} S/cm at room temperature was obtained.

To study the effects of the anion structure on the properties of PU electrolytes, TPU electrolytes with different lithium salts such as LiCl, $LiClO_4$, $LiN(SO_2CF_3)_2$ (LiTFSI) were made [152]. The effect of the anion structure on monomer (PTMG) prior to polymerization and on the properties of TPU electrolytes on post-polymerization was studied. It was found that the anion structure of lithium salt has a significant influence on the ionic conductivity, thermal stability, and tensile property of TPU electrolytes. The TPU electrolytes with LiTFSI have a high ionic conductivity up to 10^{-5} S/cm at 300 K. The ionic conductivity of the electrolytes with lithium salts is in the order as follows: LiCl < $LiClO_4$ < LiTFSI. This is due to the fact that the lithium salts with larger anions were easily dissociated in TPU and had stronger interaction with TPU, which provided more charge carriers and gave higher ionic conductivity.

A bio-based PU film was synthesized from palm kernel oil-based monoester-OH, and the solid electrolyte was prepared by doping PU with lithium iodide salts [153, 154]. Photoelectrochemical cell was developed from this bio-based PU solid electrolyte. From the X-ray diffraction studies of PU electrolytes with 25 wt% LiI, it was found that LiI reduced the semicrystalline characteristics of plasticized PU. The ionic conductivity was increased with the addition of 25 wt% LiI and a highest conductivity of 7.6×10^{-4} S cm^{-1} was obtained. A dye-sensitized solar cell of FTO/TiO_2-dye/PU-LiI-I_2/Pt was prepared. Under a light intensity of 100 MW cm^{-2}, it indicated a photovoltaic effect with the short-circuit current density of 0.06 mA cm^{-2} and open-circuit voltage of 0.14 V, respectively.

6.2 PU/PVDF Blends-Based Electrolytes

Hybrid microporous gel polymer electrolytes (HMGPEs) based on the polyvinylidene fluoride (PVDF)/PU lithium salt (PLS) were fabricated by thermal phase separation technique [155]. PLS was synthesized via condensation copolymerization of polyethylene glycol 800 and diphenylmethane-diisocyanate and then neutralized with LiOH. It was reported that the incorporation of PLS in PVDF matrix can enhance ionic conductivities and electrochemical stabilities of the HMGPEs. The enhancement in the ionic hopping conductivity is due to the formation of dendrite of PLS onto the walls of PVDF micropores that constructs the ionic pathway through the polymer host's sites. The enhanced electrochemical stability of HMGPEs is attributed to the exceptional affinity of PLS to liquid electrolyte, and the electrolyte solution tightly trapped into the pores of hybrid microporous membrane. LIB were made using HMGPEs and the analysis shows that HPGPEs with weight ratio 80:20 of PVDF and PLS delivers the highest charge–discharge capacity (about 175 mA h g^{-1}), and the cell based on HMGPEs with weight ratio 90:10 of PVDF and PLS shows the best stability in cycle performance. TPU/PVDF (1:1 wt/wt)-based microporous GPEs prepared by electrospinning technique have also been reported [156]. This polymer electrolyte had a high ionic conductivity about 3.2×10^{-3} S cm^{-1} at room temperature and exhibited a high electrochemical stability up to 5.0 V versus Li^+/Li, good mechanical strength and stability. The Li/GPE/$LiFePO_4$ cell delivered a high discharge capacity at 0.1 °C rate at 25 °C (167.8 mA h g^{-1}) and very stable cycle performance.

Electrospun GPE consisting of TPU and poly(vinylidene fluoride-co-hexafluoropropylene) (PVDF-HFP) was also investigated by researchers [157]. The GPE shows a maximum ionic conductivity of 4.1×10^{-3} S cm^{-1} with electrochemical stability up to 5.5 V versus Li^+/Li at room temperature. It shows a first charge–discharge capacity of 168.8 mA h g^{-1} when the GPE is evaluated in a Li/PE/$LiFePO_4$ cell under 0.1 °C rate at the first cycle. There is microscale attenuation (1 %) in the 30 cycles of charge and discharge tests. The TPU/PVDF-HFP membrane has a high tensile strength (8.4 ± 0.3 MPa) and elongation at break (118.7 ± 0.2 %).

6.3 Polyurethane Composites-Based Electrolytes

PU composites are also used as electrolytes. The effect of silica on the electrochemical characteristics of TPU-based electrolyte was studied [158]. MCM-41 mesoporous silica was added in proportions ranging from 5 to 20 wt% with respect to TPU. The MCM-41 addition to the system was found to improve the electrochemical stability of the membranes and reduce the gel electrolyte/metallic Li interfacial resistance.

GPE films based on the TPU/PVDF with and without in situ ceramic fillers (SiO_2 and TiO_2) are prepared by electrospinning [159]. The electrospun TPU–PVDF blending membrane with 3 % in situ TiO_2 shows a highest ionic conductivity of 4.8×10^{-3} S cm^{-1} with electrochemical stability up to 5.4 V versus Li^+/Li at room temperature and has a high tensile strength (8.7 ± 0.3 MPa) and % elongation at break (110.3 ± 0.2).

Electrochemical behavior of PU ether electrolytes/CB composites and application to double-layer capacitor was reported by Furtado et al. [160]. The performance of the capacitors was characterized by a variety of electrochemical techniques, including impedance spectroscopy, cyclic voltammetry, galvanostatic charge–discharge methods, and long-term cycling tests. The electrochemical double-layer capacitors prepared using composite electrodes containing 20 wt% of CB showed a specific capacitance of 0.7 F cm^{-3} under 2 V providing an energy density of 0.4 W h L^{-1}.

7 Flexible and Printed Electronics

In modern era, there have been intensive studies into flexible electronic devices involving low processing temperature, low material cost, high fabrication speed, and amenability to mass manufacturing (e.g., by reel-to-reel printing) [161–165]. Being the basic components, printed electrical conductors and interconnect components are one of the major area where tremendous research is going on in the flexible electronics so as to cater to the next generation of consumable flexible electronic devices [166–169]. Breakthroughs in these areas will not only enable the development of a broad range of devices and applications that are not possible today but also benefit the development of many current consumer electronics including flexible displays [164, 170, 171], flexible and conformal antenna arrays and thin-film transistors [1, 4, 172], membrane keyboards and electronic solar cell arrays [7, 8], radio-frequency identification (RFID) tags [173], flexible batteries [9, 10], electronic circuits fabricated in clothing and biomedical devices [11, 168, 174, 175].

In all the lightweight conductive materials, flexible and stretchable conductive polymers and their composites have gained extensive interest of researchers due to their hi-tech applications in the fields of flexible electronic displays, smart textiles, medical electronics, sensors, and actuators. These composites/nanocomposites have the advantages of being able to be stretched, folded, and deformed into the required shape. In comparison with the normal rigid conductors, stretchable conductive polymers are a new concept and have the potential applications in the arbitrary curved surfaces and movable parts of large-area electronics.

Flexible conducting composites are prepared by loading conductive fillers into polymers [176–178]. Generally, the conductive fillers have good electronic performance and stability, but poor mechanical robustness, while polymers have good mechanical properties, but poor electronic properties. Therefore, the conductive polymer composites have excellent mechanical robustness and electronic performance and it is the key to realize stretchable electronics.

Araki et al. [179] reported extremely stretchable electrical conductors compounded with microsized silver flakes and PU printed on a variety of stretchable or foldable substrates. The high electrical conductors were fabricated by filling half of the volume of a PU elastomer with microsized silver flakes under a low temperature of 70 °C. These PU-based conductors could be folded without peeling off the polymeric or cellulosic substrates because the matrix substances of PU-enhanced robust adhesion of the conductors with the substrates. Furthermore, PU-based conductors printed on the PU substrates maintained their low resistivity up to a strain of 600 % and they demonstrated the illumination of a LED above a strain of 400 % because of the high affinity of PU not only with the silver flakes but also with their substrates. The stretchable and foldable wiring systems enable continuous roll-to-roll processing under low temperature, resulting in relatively simple and inexpensive processing that is suitable for forthcoming future technologies such as flexible displays, solar cells, electronic paper, panel sensors, and actuators.

Elastic conductive nanocomposite was developed with MWCNTs and PU. In order to form a well-developed conducting network of MWCNT in the PU matrix, an ionic liquid was used as dispersant and uniaxial tension was given for the alignment of MWCNT [180]. The obtained conductive nanocomposite has the initial conductivity greater than 5.3 S cm^{-1}, and the stretchability of more than 100 %. The glass transition temperature (T_g) and melting temperature (T_m) of nanocomposites increased when compared with the pure PU. The stretchable conductive PU/MWCNT nanocomposite can be uniaxially stretched up to 100 % without significant variation in electrical conductivity (Fig. 8). Further, these composites

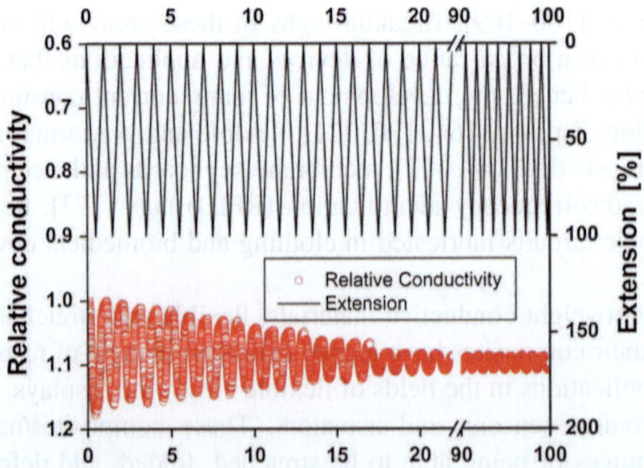

Fig. 8 The changes of relative conductivity of the nanocomposite during cyclically elongating by 100 % and relaxing to their unstretched state for 100 times [180]. Copyright 2011. Reproduced with permission from the Royal Society of Chemistry

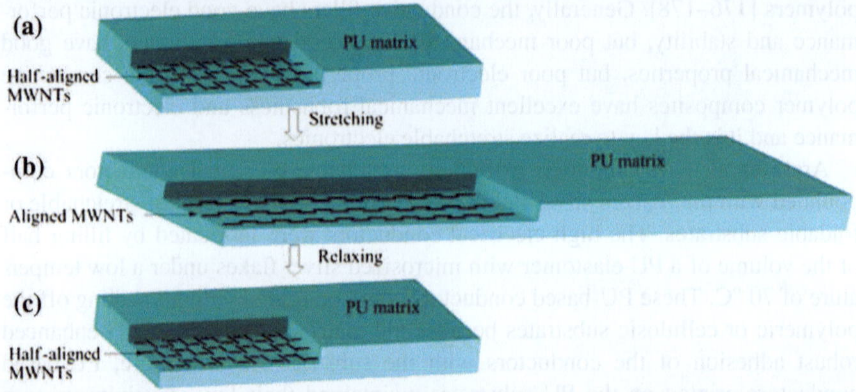

Fig. 9 Schematic representation of the variation of the MWCNTs' conductive network within the PU matrix under the condition of stretching/relaxing. **a** Image of MWCNTs/PU film with half-aligned MWCNTs in the PU matrix. **b** The MWNTs rearrangement in composite under tensile strain. **c** The conductive network of MWNTs recovers to its original state caused by the elastic recovery of PU [180]. Copyright 2011. Reproduced with permission from the Royal Society of Chemistry

remained functionally uniform even up to 100 times cyclically straining and relaxing, indicating excellent durability. The measurement of temperature-dependent conductivity indicates that a 3D hopping mechanism dominates the conductivity of MWCNTs. It is reported that the rearrangement of MWCNTs in the PU matrix is contributing to the resistance variations. Schematic representation of the variation of the MWCNTs' conductive network within the PU matrix under the condition of stretching/relaxing is given in Fig. 9. Under the condition of tensile strains, MWCNTs can slide against each other along the tensile direction, and the half-aligned MWCNTs will be rearranged, which leads to a high degree of alignment without damage of the connection among MWCNTs and the conductive network. These highly stretchable conductive polymer nanocomposites can be applied in high-performance flexible and stretchable electronic circuits and in the flexible mechanical actuation system, such as in flexible electrical artificial skin and in integrated actuators. Further, these stretchable composites enable the construction of electronic integrated circuits to spread over any surface, including arbitrary curved surfaces and movable parts.

Flexible electronics requires stringent properties for interconnect materials [164, 181, 182]. Interconnects for flexible electronics are required to maintain excellent mechanical robustness and electrical interconnectivity during mechanical deformation in addition to the conventional functions of providing ground, sufficient power, and signal transmission. There are considerably fewer reports on adhesive materials to support the flip-chip interconnects or surface-mount technology (SMT) for flexible electronics. Conventional silicone-based electrically conductive adhesives (ECAs) usually have limited conductivity (resistivity $> 2.0 \times 10^{-4}\ \Omega$ cm), while epoxy-based ECAs suffer from the rigidity [183]. ECAs with high electrical conductivity, good mechanical compliance, low process temperature, and strong adhesion are needed to enable SMT and flip-chip interconnection on flexible and low-cost substrates in flexible electronics. PU-based flexible ECA (PU-ECA) that can meet all of the above requirements was developed by Li et al. [184]. They demonstrated that by selecting the appropriate resin, the PU-ECA exhibits excellent electrical conductivity (resistivity $\approx 1.0 \times 10^{-5}\ \Omega$ cm) and the resistivity does not change when bending at a 2.64 % flexural strain, rolling at a radius of 8 mm or pressing under 250 kPa. They also report that the adhesion strength is much better than that of a commercial ECA (Abletherm 3188) and pure PU resin. The PU-ECA have many other advantages, including a low curing temperature (150 °C), which enables printing and curing on low-cost flexible substrates; simple and cost-effective processing, by eliminating the use of any expensive silver nanoparticles [185, 186] or additives [171, 187] to achieve high electrical conductivity; and environmental friendliness. The PU-ECA enables various interconnection techniques in flexible and printed electronics: it can serve as a die-attach material for flip-chip, as vertical interconnect access (VIA)—filling and polymer bump materials for 3D integration, and as a conductive paste for wearable radio-frequency devices.

Silver nanowire-embedded PU optical adhesive coating which can apply to rigid glass substrate or flexible plastic polyethylene terephthalate (PET) and elastomeric polydimethlysiloxane (PDMS) having transparency and conductivity similar to indium tin oxide (ITO) was reported [188]. These flexible and transparent coatings remain conducting even at high bending strains. Compared to ITO, they are more durable and resistant to scratches. These PU-based transparent conducting electrodes coated on PDMS can be used at a tensile strain as high as 76 %. They also exhibit high reliability withstanding 250 bending cycles of 15 % strain without much increase in its electrical resistance. In flexible light-emitting electrochemical cells, these coatings are considered as excellent candidates for transparent conducting electrodes, due to their smooth surfaces (root-mean-squared surface roughness < 10 nm). The flexible light-emitting devices developed using these coatings can emit light even at a bending radius as low as 1.5 mm. Studies on reliability also proved that, after 20 bending cycles of 25 % tensile strain, they can perform as good as unstrained devices.

Isotropically conductive adhesives (ICAs), which can be printed through many conventional printing methods and can be cured at 150 °C within a few minutes, have been reported by researchers [189–191]. Recently, highly conductive ICAs with biocompatible and environmentally benign resin materials have attracted more attention [185, 192–195].

Water-based ICAs (WBICAs) composed of silver microflakes, and water-based aliphatic PU resin dispersant was reported by Yang et al. [2]. Through combining surface iodination and in situ reduction treatment (by adding a small dose of sodium borontetrahydride), the electrical conductivity of the WBICAs is dramatically improved (resistivity: 8×10^{-5} Ω cm with 80 wt% of silver). The reliability studies (stable for at least 1440 h during 85 °C/85 % RH aging) showed that they meet the essential requirements for microelectronic applications. It was reported that when compared to the other traditional dispersants for ICAs, such as epoxy, polyester, and polyacrylate, water-based PU as the resin has a few advantages such as follows:

1. Environmental friendliness. Water being the solvent, the manipulation does not involve toxic volatile materials.
2. Variety of PU materials are possible, and hence, mechanical strength of PU can be tuned by selecting from a large pool of starting materials, even from the extractions of plants [196, 197].
3. High reliability for general electronic packaging applications due to strong urethane bond.
4. Economic. The curing of the ICAs can even be carried out at room temperature.
5. Compatible to many conventional printing techniques such as screen printing, gravure printing, and roll-to-roll printing

The development of PU-based flexible electronic materials has revolutionized the field of flexible and stretchable electronics. Prototype applications of PU-based flexible substrates were demonstrated in light-emitting diode (LED) arrays and radio-frequency identification (RFID) antennas, which showed satisfactory performances.

PU-based flexible electronics have already been demonstrated for practical applications by many researchers. Figure 10 shows a PU-based ECA substrate to which LED lights are attached. It can be noted that the brightness of the LED does not diminish even when rolled. Similarly, a flexible antenna made using PU-ECA is shown in Fig. 11. These flexible antennae can be bent or twisted without damaging the antenna pattern.

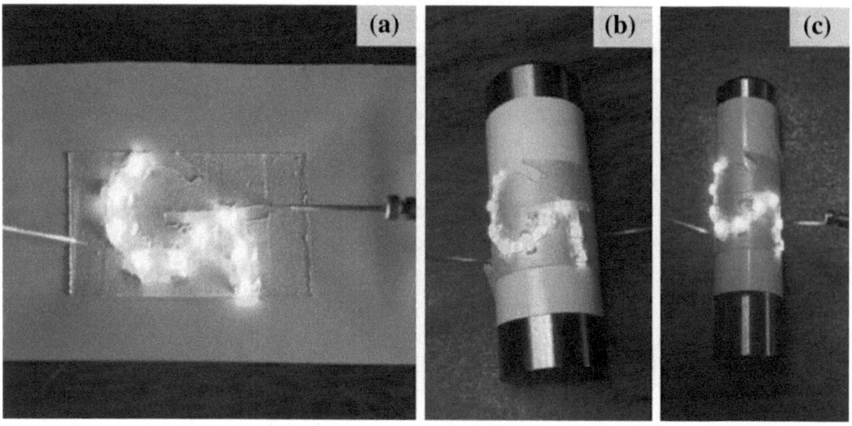

Fig. 10 **a** LED chips are attached to photopaper substrate by PU-ECA. The brightness of the LED chips does not change when rolling at radius of 18.5 mm (**b**) and 15 mm (**c**) [184]. Copyright 2013. Reprinted with permission from Wiley-VCH Verlag GmbH & Co.

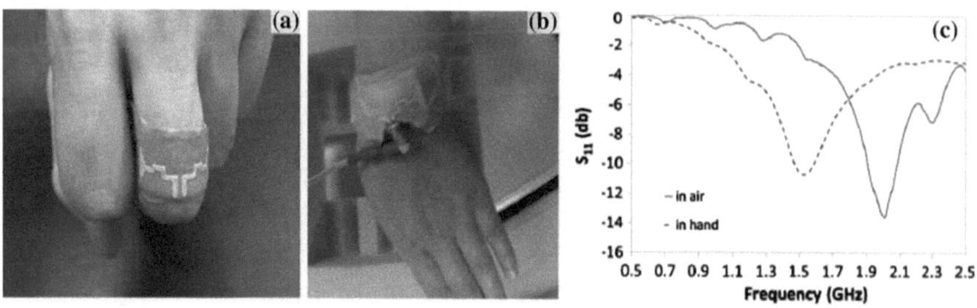

Fig. 11 **a** Flexible antenna made of PU-ECA can be printed on bandage and wrapped around a finger. **b** Testing an antenna on a wrist with a vector network analyzer (VNA). **c** The S_{11} parameter of the antenna in the frequency range of 0.5–2.5 GHz [184]. Copyright 2013. Reprinted with permission from Wiley-VCH Verlag GmbH & Co.

8 Energy Harvesting

With the advances in the area of wireless sensor networks and devices, the need for self-powered devices also grew in demand. In this context, Energy harvesting, a process of converting energy from ambient environment to usable electrical energy, has become a hot topic of research. In energy harvesting applications, in order to convert mechanical energy to electrical energy, electroactive materials including piezoelectrics and electrostrictive materials are used extensively. PU composites have also emerged as a potential candidate for energy harvesting applications.

PU composites with low concentrations of copper powder were used as an electrostrictive materials for mechanical energy harvesting by Putson et al. [198]. Both nanosized copper particles and micron-sized copper particles were studied as filler material. The measurement of the harvested current and harvested power also demonstrated an enhancement of the energy harvesting capability when nanofillers were employed. The harvested current and output power were found to increase with filler content. However, the increase in figure of merit for the harvested current or power was found to be more pronounced for nanosized filler. The energy harvesting performance of the composites was much better than pure PU. For instance, 3 % nanosized copper particle-filled PU composite exhibited almost tenfold gain in harvested output power than neat PU, whereas same amount of micron-sized filler resulted in threefold gain in harvested output power than pure PU.

A lot of electrostrictive polymers have been investigated for energy harvesting in the recent past. However, a static field is necessary for energy harvesting. In order to overcome this problem, electrostrictive polymers are hybridized with electrets. Belhora et al. [199] developed an analytical model based on electrostatic and electrostrictive equations, and it was verified with an experimental study using PU and P(VDF-TrFE-CFE) films, with and without carbon nanopowder as filler material. The energy harvesting performance was studied by employing PU, P(VDF-TrFE-CFE) and composites filled with carbon nanopowder. The hybridization of electrostrictive polymers and electrets were found to exhibit excellent energy harvesting performance. A layered structure of polymer and electrets was used. A thin layer of low-permittivity material such as CYTOP was used as the electret which was charged by a negative corona discharge. The electret was used to create an electric field in order to polarize the electrostrictive polymer. Figure 12 depicts the experimental setup used for the energy harvesting measurements. The electrostricive polymer and the electret were glued using conducting adhesive and mounted on a test rig with one part fixed and other part movable in one direction. The film was driven with a given strain profile and strained along one direction. The generated force was measured using a force sensor. The composites loaded with carbon nanopowder showed enhanced current harvesting and energy conversion. It was found that a combination of electrostrictive polymers and electrets seem to be more advantageous for practical applications which require low or variable frequencies, low cost, and large areas. Electrostrictive polymers and electrets were also found to have great potential as candidates for the development of μ-generators.

In a similar work by Eddiai et al. [200] cellular polypropylene electrets were used along with PU composites. After high-voltage corona poling of cellular polypropylene electrets on an electrostrictive PU/CB composite, the efficiency of electromechanical conversion for energy harvesting application was studied. In the energy harvesting experiments, it was possible to obtain harvested power up to 13.93 nW using a low electric field of 0.4 V/μm and a transverse strain of 3 % at a mechanical frequency of 15 Hz. The efficiency was 78.14 % at low frequency which is very significant compared to other structures. Thus, it was demonstrated

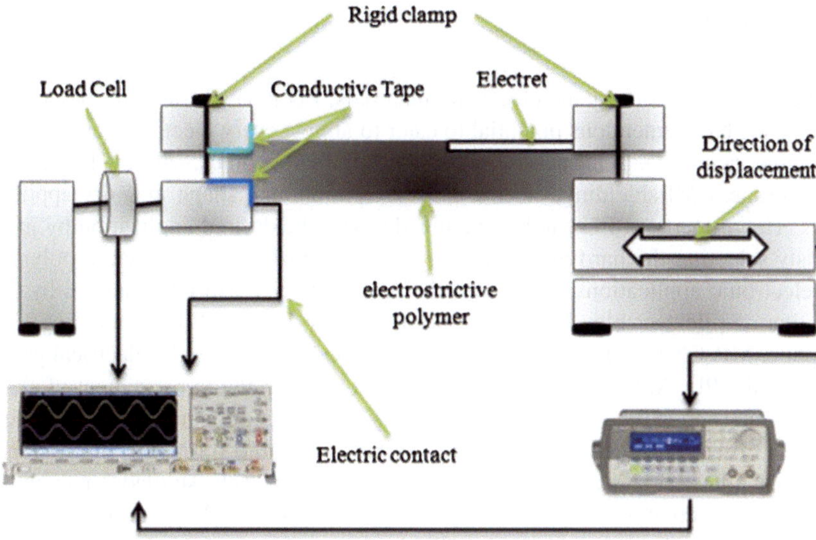

Fig. 12 Illustration of experimental setup for energy harvesting measurements [199]. Copyright 2013. Reprinted with permission from Elsevier Ltd.

that the use of polypropylene electrets with electrostrictive polymers was one of the best ways to decrease the power of polarization, thereby improving the efficiency of the electromechanical conversion in energy harvesting. The studies proved that electrostrictive polymers with cellular polypropylene electrets are good candidates which can compete with other piezoelectric materials such as PZT in energy harvesting applications.

Fiorido et al. [201] studied the energy harvesting and actuation behavior of PU/iron carbide nanocomposites fabricated by electrospinning. These materials were found to exhibit both electrostrictive and magnetoelectric (ME) properties. For a diaphragm-type actuator made up of PU/2.5 wt% nanoiron carbide composites, an almost sevenfold gain for the deflection strain under 17 V/μm was obtained with a significant increase in the dielectric permittivity. However, higher filler loading reduced the actuation performance due to aggregation of iron carbide, but the ME properties still showed an increase even above a filler content of 2.5 wt%. When subjected to a magnetic field, these composites generated a current at least 100 times higher than polymer-based systems studied for their ME behavior. As required for energy harvesting applications, the magnitude of ME current is independent of the applied DC bias magnetic field and is a linear function of the AC alternative field magnitude or applied frequency.

9 Conclusions

In the modern era of organic electronic materials, PU along with their composites and blends has tremendous potential to cater to almost every electronic application including actuators, sensors, dielectrics, flexible electronics, energy harvesting and storage, EMI shielding, electrostatic dissipation, and shape memory applications. The property of PU can be fine-tuned to suit the desired application by judiciously selecting the hard and soft segments. This makes PU a versatile polymer for electronic applications. Moreover, the properties can also be enhanced by the addition of fillers or blending with other polymers. Various fillers including CB, graphite, MP, CNT, and graphene have been studied to improve the electrical properties of the PU. Although much work has been done on the improvement of electrical properties of PU, considerable work is still needed to improve percolation threshold of the electrical conductivity further. From the foregoing discussions, it is clear that PU and its composites/blends are extensively studied for applications such as actuators, shape memory effects, and sensors. However, more works are being done all over the world in PU-based systems in other areas including flexible electronics and energy. In future, energy is going to be the major concern for mankind. Even though some effort has already been put in by researches in the energy harvesting and storage applications, there is ample scope for PU-based research for energy harvesting and storage. PU-based electrolytes for supercapacitors and batteries can emerge as a prominent area of research in the near future. PU-based flexible electronic systems can bring revolutionary changes in the electronics industry.

References

1. Park S-I, Xiong Y, Kim R-H, Elvikis P, Meitl M, Kim D-H, Wu J, Yoon J, Yu C-J, Liu Z, Huang Y, Hwang K-C, Ferreira P, Li X, Choquette K, Rogers JA (2009) Printed assemblies of inorganic light-emitting diodes for deformable and semitransparent displays. Science 325:977–981
2. Yang C, Lin W, Li Z, Zhang R, Wen H, Gao B, Chen G, Gao P, Yuen MMF, Wong CP (2011) Water-based isotropically conductive adhesives: towards green and low-cost flexible electronics. Adv Funct Mater 21:4582–4588
3. Russo A, Ahn BY, Adams JJ, Duoss EB, Bernhard JT, Lewis JA (2011) Pen-on-paper flexible electronics. Adv Mater 23:3426–3430
4. Sekitani T, Yokota T, Zschieschang U, Klauk H, Bauer S, Takeuchi K, Takamiya M, Sakurai T, Someya T (2009) Organic nonvolatile memory transistors for flexible sensor arrays. Science 326:1516–1519
5. Han L, Song K, Mandlik P, Wagner S (2010) Ultraflexible amorphous silicon transistors made with a resilient insulator. Appl Phys Lett 96:042111
6. Takei K, Takahashi T, Ho JC, Ko H, Gillies AG, Leu PW, Fearing RS, Javey A (2010) Nanowire active-matrix circuitry for low-voltage macroscale artificial skin. Nat Mater 9:821–826
7. Fan Z, Razavi H, Do J-W, Moriwaki A, Ergen O, Chueh Y-L, Leu PW, Ho JC, Takahashi T, Reichertz LA, Neale S, Yu K, Wu M, Ager JW, Javey A (2009) Three-dimensional nanopillar-array photovoltaics on low-cost and flexible substrates. Nat Mater 8:648–653

8. Yoon J, Baca AJ, Park S-I, Elvikis P, Geddes JB, Li L, Kim RH, Xiao J, Wang S, Kim T-H, Motala MJ, Ahn BY, Duoss EB, Lewis JA, Nuzzo RG, Ferreira PM, Huang Y, Rockett A, Rogers JA (2008) Ultrathin silicon solar microcells for semitransparent, mechanically flexible and microconcentrator module designs. Nat Mater 7:907–915
9. Hu LB, Wu H, La Mantia F, Yang YA, Cui Y (2010) Thin, flexible secondary li-ion paper batteries. ACS Nano 4:5843–5848
10. Hu L, Choi JW, Yang Y, Jeong S, La Mantia F, Cui L-F, Cui Y (2009) Highly conductive paper for energy-storage devices. Proc Natl Acad Sci 106:21490–21494
11. Yamada T, Hayamizu Y, YamamotoY Yomogida Y, Izadi-Najafabadi A, Futaba DN, Hata K (2011) A stretchable carbon nanotube strain sensor for human-motion detection. Nat Nanotech 6:296–301
12. Li YJ, Hanada T, Nakaya T (2006) Surface modification of segmented polyurethanes by grafting methacrylates and phosphatidylcholine polar head groups to improve hemocompatibility. Chem Mater 1:763–770
13. Shirasaka H, Inoue S, Asai K, Okamoto H (2000) Polyurethane urea elastomer having monodisperse poly(oxytetramethylene) as a soft segment with a uniform hard segment. Macromolecules 33:2776–2778
14. Liu Z, Bai G, Huang Y, Ma Y, Du F, Li F, Guo T, Chen Y (2007) Reflection and absorption contributions to the electromagnetic interference shielding of single-walled carbon nanotube/polyurethane composites. Carbon 45:821–827
15. Gurunathan T, Rao CRK, Narayan R, Raju KVSN (2013) Polyurethane conductive blends and composites: synthesis, and applications perspective. J Mater Sci 48:67–80
16. Nelson A (2008) Stimuli-responsive polymers: engineering interactions. Nat Mater 7:523–525
17. Behl M, Lendlein A (2007) Shape-memory polymers. Mater Today 10:20–28
18. Lendlein A, Kelch S (2002) Shape-memory polymers. Angew Chem Int 41:2034–2057
19. Liu YJ, Lv HB, Lan X, Leng JS, Du SY (2009) Review of electro-activate shape-memory polymer composite. Compos Sci Technol 69:2064–2068
20. Gunes IS, Jana SC (2008) Shape memory polymers and their nanocomposites: a review of science and technology of new multifunctional materials. J Nanosci Nanotechnol 8:1616–1637
21. Ratna D, Karger-Kocsis J (2008) Recent advances in shape memory polymers and composites: a review. J Mater Sci 43:254–269
22. Liu C, Qin H, Mather PT (92007) Review of progress in shape-memory polymers. J Mater Chem 17:1543–1558
23. Beloshenko VA, Varyukhin VN, Voznyak YV (2005) Shape memory effect in polymers. Russ Chem Rev 74:265–284
24. Hu JL (2007) Shape memory polymers and textiles. In: Characterization of shape memory properties in polymers. CRC Press, Boca Raton, pp 197–225
25. Yakacki CM, Willis S, Luders C, Gall K (2008) Deformation limits in shape-memory polymers. Adv Eng Mater 10:112–119
26. Kim BK, Lee SY, Lee JS, Baek SH, Choi YJ, Lee JO, Xu M (1998) Polyurethane ionomers having shape memory effects. Polymer 39:2803–2808
27. Leng JS, Du SY (2010) Shape memory polymer and multifunctional composite. In: Jiang HY, Schmidt A (eds) The structural variety of shape memory polymers. CRC Press, Boca Raton, pp 21–64
28. Leng J, Lan X, Liu Y, Du S (2011) Shape-memory polymers and their composites: stimulus methods and applications. Prog Mater Sci 56:1077–1135
29. Santiago D, Ferrando F, Flor SD (2014) Influence of holding time on shape recovery in a polyurethane shape-memory polymer. J Mater Eng Perform 23:2567–2573
30. Sokolowski WM, Tan SC (2007) Advanced self-deployable structures for space applications. J Spacecraft Rockets 44:750–754
31. Huang WM, Lee CW, Teo HP (2006) Thermomechanical behavior of a polyurethane shape memory polymer foam. J Intell Mater Syst Struct 17:753–760

32. Tey SJ, Huang WM, Sokolowski WM (2001) Influence of long-term storage in cold hibernation on strain recovery and recovery stress of polyurethane shape memory polymer foam. Smart Mater Struct 10:321–325
33. Imai S, Sakurai K (2013) An actuator of two-way behavior by using two kinds of shape memory polymers with different Tgs. Precis Eng 37:572–579
34. Zhu Y, Hu J, Yeung LY, Liu Y, Ji F, Yeung K (2006) Development of shape memory polyurethane fiber with complete shape recoverability. Smart Mater Struct 15:1385
35. Meng Q, Hu J, Zhu Y, Lu J, Liu Y (2007) Polycaprolactone-based shape memory segmented polyurethane fiber. J Appl Polym Sci 106:2515–2523
36. Yang B (2007) Influence of moisture in polyurethane shape memory polymers and their electrical conductive composites. PhD Thesis. School of Mechanical and Aerospace Engineering, Nanyang Technological University, Singapore
37. Li F, Qi L, Yang J, Xu M, Luo X, Ma D (2000) Polyurethane/conducting carbon black composites: structure, electric conductivity, strain recovery behavior, and their relationships. J Appl Polym Sci 75:68–73
38. Jung YC, Kim JH, Hayashi T, Kim YA, Endo M, Terrones M, Dresselhaus MS (2012) Fabrication of transparent, tough, and conductive shape-memory polyurethane films by incorporating a small amount of high-quality graphene. Macromol Rapid Commun 33:628–634
39. Han S, Chun BC (2014) Preparation of polyurethane nanocomposites via covalent incorporation of functionalized graphene and its shape memory effect. Compos Part A 58:65–72
40. Kim YJ, Park HC, Kim BK (2015) Triple shape-memory effect by silanized polyurethane/silane-functionalized graphene oxide nanocomposites bilayer. High Perform Polym. doi:10.1177/0954008314565398
41. Cao F, Jana SC (2007) Nanoclay-tethered shape memory polyurethane nano composites. Polymer 48:3790–3800
42. Koerner H, Kelley J, George J, Drummy L, Mirau P, Bell NS, Hsu JWP, Vaia RA (2009) ZnO nanorod—thermoplastic polyurethane nanocomposites: morphology and shape memory performance. Macromolecules 42:8933–8942
43. Ohki T, Ni QQ, Iwamoto M (2004) Creep and cyclic mechanical properties of composites based on shape memory polymer. Sci Eng Compos Mater 11:137–147
44. Ohki T, Ni QQ, Ohsako N, Iwamoto M (2004) Mechanical and shape memory behavior of composites with shape memory polymer. Compos Part A Appl Sci Manuf 35:1065–1073
45. Raja M, Ryu SH, Shanmugharaj AM (2013) Thermal, mechanical and electroactive shape memory properties of polyurethane (PU)/poly (lactic acid) (PLA)/CNT nanocomposites. Eur Polym J 49:3492–3500
46. Cho JW, Kim JW, Jung YC, Goo NS (2005) Electroactive shape-memory polyurethane composites incorporating carbon nanotubes. Macromol Rapid Commun 26:412–416
47. Leng JS, Huang WM, Lan X, Liu YJ, Du SY (2008) Significantly reducing electrical resistivity by forming conductive Ni chains in a polyurethane shape-memory polymer/carbon-black composite. Appl Phys Lett 92:204101
48. Choi JT, Dao TD, Oh KM, Lee H, Jeong HM, Kim BK (2012) Shape memory polyurethane nanocomposites with functionalized graphene. Smart Mater Struct 21:075017
49. Leng JS, Lan X, Liu YJ, Du SY, Huang WM, Liu N, Phee SJ, Yuan Q (2008) Electrical conductivity of shape memory polymer embedded with micro Ni chains. Appl Phys Lett 92:014104
50. Gunes IS, Jimenez GA, Jana SC (2009) Carbonaceous fillers for shape memory actuation of polyurethane composites by resistive heating. Carbon 47:981–997
51. Luo H, Li Z, Yi G, Wang Y, Zu X, Wang H, Huang H, Liang Z (2015) Temperature sensing of conductive shape memory polymer composites. Mater Lett 140:71–74
52. Lu H, Liang F, Yao Y, Gou J, Hui D (2014) Self-assembled multi-layered carbon nanofiber nanopaper for significantly improving electrical actuation of shape memory polymer nanocomposite. Compos Part B 59:191–195
53. Lendlein A, Jiang HY, Junger O, Langer R (2005) Light-induced shape-memory polymers. Nature 434:879–882

54. Baer GM, Small W, Wilson TS, Benett WJ, Matthews DL, Hartman J, Maitland DJ (2007) Fabrication and in vitro deployment of a laser-activated shape memory polymer vascular stent. Biomed Eng Online 6:43
55. Monkman GJ (2000) Advances in shape memory polymer actuation. Mechatronics 10:489–498
56. Yu YL, Ikeda T (2005) Photodeformable polymers: a new kind of promising smart material for micro- and nano-applications. Macromol Chem Phys 206:1705–1708
57. Yi DH, Yoo HJ, Mahapatra SS, Kim YA, Cho JW (2014) The synergistic effect of the combined thin multi-walled carbon nanotubes and reduced graphene oxides on photothermally actuated shape memory polyurethane composites. J Colloid Interface Sci 432:128–134
58. Leng JS, Wu XL, Liu YJ (2009) Infrared light-active shape memory polymer filled with nanocarbon particles. J Appl Polym Sci 114:2455–2460
59. Schmidt AM (2006) Electromagnetic activation of shape memory polymer networks containing magnetic nanoparticles. Macromol Rapid Commun 27:1168–1172
60. Mohr R, Kratz K, Weigel T, Lucka-Gabor M, Moneke M, Lendlein A (2006) Initiation of shape-memory effect by inductive heating of magnetic nanoparticles in thermoplastic polymers. PNAS 103:3540–3545
61. Yang B, Huang WM, Li C, Li L, Chor JH (2005) Qualitative separation of the effects of carbon nano-powder and moisture on the glass transition temperature of polyurethane shape memory polymer. Scripta Mater 53:105–107
62. Jung YC, So HH, Cho JW (2006) Water-responsive shape memory polyurethane block copolymer modified with polyhedral oligomeric silsesquioxane. J Macromol Sci Part B Phys 45:453–461
63. Yang B, Huang WM, Li C, Li L (2006) Effects of moisture on the thermomechanical properties of a polyurethane shape memory polymer. Polymer 47:1348–1356
64. Huang WM, Yang B, An L, Li C, Chan YS (2005) Water-driven programmable polyurethane shape memory polymer: demonstration and mechanism. Appl Phys Lett 86:114105
65. Luo H, Li Z, Yi G, Zu X, Wang H, Huang H, Wang Y, Liang Z, Zhang S (2014) Multi-stimuli responsive carbon nanotube–shape memory polymeric composites. Mater Lett 137(2014):385–388
66. Park JH, Dao TD, Lee H, Jeong HM, Kim BK (2014) Properties of graphene/shape memory thermoplastic polyurethane composites actuating by various methods. Materials 7:1520–1538
67. Mahapatra SS, Yadav SK, Yoo HJ, Ramasamy MS, Cho JW (2014) Tailored and strong electro-responsive shape memory actuation incarbon nanotube-reinforced hyperbranched polyurethane composites. Sensor Actuat B-Chem 193:384–390
68. Feninat FE, Laroche G, Fiset M, Mantovani D (2002) Shape memory materials for biomedical applications. Adv Eng Mater 4:91–104
69. Sokolowski W, Metcalfe A, Hayashi S, Yahia L, Raymond J (2007) Medical applications of shape memory polymers. Biomed Mater 2:S23–S27
70. Ikuta MS, Hayato K (2003) Submicron manipulation tools driven by light in a liquid. Appl Phys Lett 82:133–135
71. Ikuta MS, Hayato K (2003) Force-controllable, optically driven micromachines fabricated by single-step two-photon microstereolithography. J Microelectromech Syst 12:533–539
72. Lundberg B, Sundqvist B (1986) Resistivity of a composite conducting polymer as a function of temperature, pressure, and environment: applications as a pressure and gas concentration transducer. J Appl Phys 60:1074–1079
73. Marquez A, Uribe J, Cruz R (1997) Conductivity variation induced by solvent swelling of an elastomer-carbon black–graphite composite. J Appl Polym Sci 66:2221–2232
74. Hu JW, Chen SG, Zhang MQ, Li MW, Rong MZ (2004) Low carbon black filled polyurethane composite as candidate for wide spectrum gas sensing element. Mater Lett 58:3606–3609
75. Chen SG, Hu JW, Zhang MQ, Li MW, Rong MZ (2004) Gas sensitivity of carbon black/waterborne polyurethane composites. Carbon 42:645–651

76. ChenS G, Hu XL, Hu J, Zhang MQ, Rong MZ, Zheng Q (2006) Relationships between organic vapor adsorption behaviors and gas sensitivity of carbon black filled waterborne polyurethane composites. Sens Actuators B 119:110–117
77. Chen SG, Hu JW, Zhang MQ, Rong MZ (2005) Effects of temperature and vapor pressure on the gas sensing behavior of carbon black filled polyurethane composites. Sens Actuators B 105:187–193
78. Chen SG, Hu JW, Zhang MQ, Rong MZ, Zheng Q (2006) Improvement of gas sensing performance of carbon black/waterborne polyurethane composites: effect of crosslinking treatment. Sens Actuators B 113:361–369
79. Shin JH, Yoon SY, Yoon IJ, Choi SH, Lee SD, Nam H, Cha GS (1998) Potentiometric biosensors using immobilized enzyme layers mixed with hydrophilic polyurethane. Sens Actuators B 50:19–26
80. Puig-Lleixa C, JimeÂneza C, Bartroli J (2001) Acrylated polyurethane-photopolymeric membrane for amperometric glucose biosensor construction. Sens Actuators B 72:56–62
81. Yu B, Long N, Moussy Y, Moussy F (2006) A long-term flexible minimally-invasive implantable glucose biosensor based on an epoxy-enhanced polyurethane membrane. Biosens Bioelectron 21:2275–2282
82. Han JH, Taylor JD, Kim DS, Kim YS, Kim YT, Cha GS, Nam H (2007) Glucose biosensor with a hydrophilic polyurethane (HPU) blended with polyvinyl alcohol/vinyl butyral copolymer (PVAB) outer membrane. Sens Actuators B 123:384–390
83. Li L, Ying X, Liu J, Li X, Zhang W (2015) Molecularly imprinted polyurethane grafted calcium alginate hydrogel with specific recognition for proteins. Mater Lett 143:248–251
84. Slobodian P, Riha P, Saha P (2012) A highly-deformable composite composed of an entangled network of electrically-conductive carbon-nanotubes embedded in elastic polyurethane. Carbon 50:3446–3453
85. Zhang R, Deng H, Valenca R, Jin J, Fub Q, Bilotti E, Peijs T (2012) Carbon nanotube polymer coatings for textile yarns with good strain sensing capability. Sens Actuators A 179:83–91
86. Chang F-Y, Wang R-H, Yang H, Lin Y-H, Chen T-M, Huang S-J (2010) Flexible strain sensors fabricated with carbon nano-tube and carbon nano-fiber composite thin films. Thin Solid Films 518:7343–7347
87. Bratov A, Abramova N, Dom´ınguez C C, Baldi A A (2000) Ion-selective field effect transistor (ISFET)-based calcium ion sensor with photocured polyurethane membrane suitable for ionised calcium determination in milk. Anal Chim Acta 408:57–64
88. Ahir SV, Squires AM, Tajbakhsh AR, Terentjev EM (2006) Infrared actuation in aligned polymer-nanotube composites. Phys Rev B 73:085420
89. Diaconu I, Dorohoi D, Topoliceanu F (2006) Electrostriction of a polyurethane elastomer-based polyester. IEEE Sens J 6:876–880
90. Guillot FM, Balizer E (2003) Electrostrictive effect in polyurethanes. J Appl Polym Sci 89:399–404
91. Ueda T, Kasazaki T, Kunitake N, Hirai T, Kyokane J, Yoshino K (1997) Polyurethane elastomer actuator. Synth Met 85:1415–1416
92. Petcharoen K, Sirivat A (2013) Electrostrictive properties of thermoplastic polyurethane elastomer: effects of urethane type and soft-hard segment composition. Curr Appl Phys 13:1119–1127
93. Diaconu I, Dorohoi D, Ciobanu C (2008) Electromechanical response of polyurethane films with different thickness. Rom J Phys 53:91–97
94. Watanabe M, Hirai T, Ueda T, Suzuki M, Amaike Y (1999) Polyurethane actuators using bending piezoelectricity and bending electrostriction. J Intell Mat Syst Struct 10:100–104
95. Arora S, Ghosha T, Muth J (2007) Dielectric elastomer based prototype fiber actuators. Sens Actuator A-Phys 136:321–328

96. Jung Y, Park H, Jo N, Jeong H (2007) Fabrication and performance evaluation of diaphragm-type polymer actuators using segmented polyurethane according to chemical-hard-segment content. Sens Actuator A-Phys 136:367–373
97. Yuse K, Guyomar D, Kanda M, Seveyrat L, Guiffard B (2011) Development of large-strain and low-powered electro-active polymers (EAPs) using conductive fillers. Sens Actuator A-Phys 165:147–154
98. Petit L, Guiffard B, Seveyrat L, Guyomar D (2008) Actuating abilities of electroactive carbon nanopowder/polyurethane composite films. Sens Actuator A-Phys 148:105–110
99. Roussel M, Malhaire C, Deman AL, Chateaux JF, Petit L, Seveyrat L, Galineau J, Guiffard B, Seguineau C, Desmarres JM, Martegoutte J (2014) Electromechanical study of polyurethane films with carbon black nanoparticles for MEMS actuators. J Micromech Microeng 24:055011
100. Galineaua J, Guiffard B, Seveyrata L, Lallart M, Guyomar D (2013) Study and modeling of an electrostrictive polyurethane diaphragm loaded with conductive carbon black. Sens Actuator A-Phys 189:117–124
101. Chen J, Zhang Z, Huang W, Li J, Yang J, Wang Y, Zhou Z, Zhang J (2015) Carbon nanotube network structure induced strain sensitivity and shape memory behavior changes of thermoplastic polyurethane. Mater Des 69:105–113
102. Zeng Z, Jin H, Zhang L, Zhang H, Chen Z, Gao F, Zhang Z (2015) Low-voltage and high-performance electrothermal actuator based on multi-walled carbon nanotube/polymer composites. Carbon 84:327–334
103. Raja M, Shanmugharaj AM, Ryu SH, Subha J (2011) Influence of metal nanoparticle decorated CNTs on polyurethane based electro active shape memory nanocomposite actuators. Mater Chem Phys 129:925–931
104. Chen T, Qiu J, Zhu K, He X, Kang X, Dong E (2014) Poly(methylmethacrylate)-functionalized graphene/polyurethane dielectric elastomer composites with superior electric field induced strain. Mater Lett 128:19–22
105. Okuzaki H, Takagi S, Hishiki F, Tanigawa R (2014) Ionic liquid/polyurethane/PEDOT:PSS composites for electro-active polymer actuators. Sens Actuator Chem 194:59–63
106. Jaaoh D, Putson C, Muensit N (2015) Deformation on segment-structure of electrostrictive polyurethane/polyaniline blends. Polymer 61:123–130
107. Tobushi H, Hayashi S, Hoshio K, Makino Y, Miwa N (2006) Bending actuation characteristics of shape memory composite with SMA and SMP. J Intell Mater Syst Struct 17:1075–1081
108. Bothe M, Pretsch T (2012) Two-way shape changes of a shape-memory poly (ester urethane). Macromol Chem Phys 213:2378–2385
109. Chen H, Li Y, Liu Y, Gong T, Wang L, Zhou S (2014) Highly pH-sensitive polyurethane exhibiting shape memory and drug release. Polym Chem 5:5168–5174
110. Wang A, Gao H, Sun Y, Yang Y, Wu G, Wang Y, Fan Y, Ma J (2013) Temperature- and pH-responsive nanoparticles of biocompatible polyurethanes for doxorubicin delivery. Int J Pharm 441:30–39
111. Koerner H, Price G, Pearce NA, Alexander M, Vaia RA (2004) Remotely actuated polymer nanocomposites—stress-recovery of carbon-nanotube-filled thermoplastic elastomers. Nat Mater 3:115–120
112. Liang J, Xu Y, Huang Y, Zhang L, Wang Y, Ma Y, Li F, Guo T, Chen Y (2009) Infrared-triggered actuators from graphene-based nanocomposites. J Phys Chem C 113:9921–9927
113. Muralidharan MN, Ansari Seema (2013) Thermally reduced graphene oxide/thermoplastic polyurethane nanocomposites as photomechanical actuators. Adv Mat Lett 4:927–932
114. Feng Y, Qin M, Guo H, Yoshino K, Feng W (2013) Infrared-actuated recovery of polyurethane filled by reduced graphene oxide/carbon nanotube hybrids with high energy density. ACS Appl Mater Interfaces 5:10882–10888
115. Akagi T, Dohta S, Tani T (2001) Development of a flexible pneumatic actuator with a flexible tube. In: INTERMAC2001 joint technical conference, Tokyo, Japan OSP-3 (10933)

116. Takashima K, Noritsugu T, Rossiter R, Guo S, Mukai T (2012) Curved type pneumatic artificial rubber muscle using shape-memory polymer. J Robot Mechatron 24:472–479
117. Takamuku S, Fukuda A, Hosoda K (2008) Repetitive grasping with anthropomorphic skin-covered hand enables robust haptic recognition. IEEE/RSJ Int Conf Intell Robots Syst Nice. doi:10.1109/IROS.2008.4651175
118. Wu X, Kim SH, Zhu H, Ji CH, Allen MG (2012) A refreshable braille cell based on pneumatic microbubble actuators. J Microelectromech S 21:908–916
119. Peel LD, Mejia J, Narvaez B, Thompson K, Lingala M (2009) Development of a simple morphing wing using elastomeric composites as skins and actuators. J Mech Des 131:091003
120. Ma CCM, Huang YL, Kuan HC, Chiu YS (2005) Preparation and electromagnetic interference shielding characteristics of novel carbon-nanotube/siloxane/poly-(urea urethane) nanocomposites. J Polym Sci Part B Polym Phys 43:345–358
121. Gupta TK, Singh BP, Teotia S, Katyal V, Dhakate SR, Mathur RB (2013) Designing of multiwalled carbon nanotubes reinforced polyurethane composites as electromagnetic interference shielding materials. J Polym Res 20:169
122. Liu Z, Bai G, Huang Y, Li F, Ma Y, Guo T, He X, Lin X, Gao H, Chen Y (2007) Microwave absorption of single-walled carbon nanotubes/soluble cross-linked polyurethane composites. J Phys Chem C 111:13696–13700
123. Wu HL, Ma CCM, Yang YT, Kuan HC, Yang CC, Chiang CL (2006) Morphology, electrical resistance, electromagnetic interference shielding and mechanical properties of functionalized MWNT and poly(urea urethane) nanocomposites. J Polym Sci Part B Polym Phys 44:1096–1105
124. Hsiao S-T, Ma C-CM, Tien H-W, Liao W-H, Wang Y-S, Li S-M, Yang C-Y, Lin S-C, Yang R-B (2015) Effect of covalent modification of graphene nanosheets on the electrical property and electromagnetic interference shielding performance of a water-borne polyurethane composite. ACS Appl Mater Interfaces 7:2817–2826
125. Abbas SM, Dixit AK, Chatterjee R, Goel TC (2007) Complex permittivity, complex permeability and microwave absorption properties of ferrite–polymer composites. J Magn Magn Mater 309:20–24
126. Abbas SM, Chandra M, Verma A, Chatterjee R, Goel TC (2006) Complex permittivity and microwave absorption properties of a composite dielectric absorber. Compos Part A 37:2148
127. Lakshmi K, John H, Mathew KT, Joseph R, George KE (2009) Microwave absorption, reflection and EMI shielding of PU–PANI composite. Acta Mater 57:371–375
128. Wojkiewicz JL, Fauveaux S, Redon N, Miane JL (2004) High electromagnetic shielding effectiveness of polyaniline-polyurethane composites in the microwave band. Int J Appl Electromagn Mech 19:203–206
129. Scrosati B (1993) Applications of electroactive polymers. Chapman and Hall, London
130. Xi J, Qiu X, Tang X, Zhu W, Chen L (2006) PVDF–PEO blends based microporous polymer electrolyte: effect of PEO on pore configurations and ionic conductivity. J Power Sour 157:501–506
131. Ohno H, Matsuda H, Mizoguchi K, Tsuchida E (1982) Demonstration of solid-state cell based on poly(vinylidene fluoride) system containing lithium perchlorate. Polym Bull 7:271–275
132. Hirata M, Inoue S, Yosomiya R (1985) Humidity dependence of the electric characteristics of quaternized poly(vinylpyridine)-perchlorate complexes. Angew Makromol Chem 131:125–134
133. Stephan AM, Nahm KS (2006) Review on composite polymer electrolytes for lithium batteries. Polymer 47:5952–5964
134. Stephan AM (2006) Review on gel polymer electrolytes for lithium batteries. Eur Polym J 2006:21–42

135. Santhosh P, Vasudevan T, Gopalan A, Lee KP (2006) Preparation and properties of new cross-linked polyurethane acrylate electrolytes for lithium batteries. J Power Sour 160:609–620
136. Santhosh P, Gopalan A, Vasudevan T, Lee KP (2006) Preparation and characterization of conducting poly(diphenylamine) entrapped polyurethane network electrolyte. J Appl Polym Sci 101:611–617
137. Wright PV (1975) Electrical conductivity in ionic complexes of poly(ethylene oxide). Br Polym J 7:319–327
138. Wintersgill MC, Fortanella JJ, Greenbaum SG, Asamic KJ (1985) D.s.c., electrical conductivity, and n.m.r. studies of salt precipitation effects in PPO complexes. Br Polym J 20:195–198
139. Kobayashi N, Uchida M, Tsuchida E (1985) Poly[lithium methacrylate-co-oligo(oxyethylene)methacrylate] as a solid electrolyte with high ionic conductivity. Solid State Ionics 17:307–311
140. Gray FM (1997) Polymer electrolytes. The Royal Society of Chemistry, UK
141. Yoshimoto N, Nomura H, Shirai T, Ishikawa M, Morita M (2004) Ionic conductance of gel electrolyte using a polyurethane matrix for rechargeable lithium batteries. Electrochim Acta 50:275–279
142. Chen W, Chen H, Wen T, Digar M, Gopalan AJ (2004) Morphology and ionic conductivity of thermoplastic polyurethane electrolytes. Appl Polym Sci 91:1154–1167
143. Santhosh P, Gopalan A, Vasudevan T, Lee KP (2006) Evaluation of a cross-linked polyurethane acrylate as polymer electrolyte for lithium batteries. Mater Res Bull 41:1023–1037
144. Li Y, Wu F, Chen R (2009) Synthesis and characterization of mixing soft-segmented waterborne polyurethane polymer electrolyte with room temperature ionic liquid. Chin Chem Lett 20:519–522
145. Parnell S (2002) Doctoral dissertation, University of Akron, Akron
146. Jeung S (2005) Doctoral dissertation, University of Akron, Akron
147. Hepburn C (1982) Polyurethane elastomers. Applied Science Publishers, NY
148. Lavall RL, Ferrari S, Tomasi C, Marzantowicz M, Quartarone E, Magistris A, Mustarelli P, Lazzaroni S, Fagnoni M (2010) Novel polymer electrolytes based on thermoplastic polyurethane and ionic liquid/lithium bis(trifluoromethanesulfonyl) imide/propylene carbonate salt system. J Power Sour 195:5761–5767
149. Liu L, Wu X, Li T (2014) Novel polymer electrolytes based on cationic polyurethane with different alkyl chain length. J Power Sour 249:397–404
150. Lee H, Han C, Sung Y, Sekhon SS, Kim K (2011) Gel electrolyte based on UV-cured polyurethane for dye-sensitized solar cells. Curr Appl Phys 11:S158–S162
151. Wen T, Du Y, Digar M (2002) Compositional effect on the morphology and ionic conductivity of thermoplastic polyurethane based electrolytes. Eur Polym J 38:1039–1048
152. Wang S, Jeung S, Min K (2010) The effects of anion structure of lithium salts on the properties of in-situ polymerized thermoplastic polyurethane electrolytes. Polymer 51:2864–2871
153. Suait MS, Ahmad A, Badri KH, Mohamed NS, Rahman MYA, Azanza Ricardo CL, Scardi P (2010) The potential of polyurethane bio-based solid polymer electrolyte for photoelectrochemical cell application. Int J Hydrogen Energy 39:3005–3017
154. Daud FN, Ahmad A, Badri KH (2014) An investigation on the properties of palm-based polyurethane solid polymer electrolyte. Int J Polym Sci 326716
155. Xing Y, Wu Y, Wang H, Yang G, Li W, Xu L, Jianga X (2014) Preparation of hybrid polymer based on polyurethane lithium salt and polyvinylidene fluoride as electrolyte for lithium-ion batteries. Electrochim Acta 136:513–520
156. Wu N, Cao Q, Wang X, Chen Q (2011) Study of a novel porous gel polymer electrolyte based on TPU/PVDF by electrospinning technique. Solid State Ionics 203:42–46

157. Zhou L, Cao Q, Jing B, Wang X, Tang X, Wu N (2014) Study of a novel porous gel polymer electrolyte based on thermoplastic polyurethane/ poly(vinylidene fluoride-co-hexafluoropropylene) by electrospinning technique. J Power Sour 263:118–124
158. Lavall RL, Ferrari S, Tomasi C, Marzantowicz M, Quartarone E, Fagnoni M, Mustarelli P, Saladino ML (2012) MCM-41 silica effect on gel polymer electrolytes based on thermoplastic polyurethane. Electrochim Acta 60:359–365
159. Wu N, Cao Q, Wang X, Li S, Li X, Deng H (2011) In situ ceramic fillers of electrospun thermoplastic polyurethane /poly(vinylidene fluoride) based gel polymer electrolytes for Li-ion batteries. J Power Sour 196:9751–9756
160. Furtado CA, de Souza PP, Goulart Silva G, Matencio T, Pernaut JM (2001) Electrochemical behavior of polyurethane ether electrolytes: carbon black composites and application to double layer capacitor. Electrochim Acta 46:1629–1634
161. Yan H, Chen Z, Zheng Y, Newman C, Quinn JR, Dotz F, Kastler M, Facchetti A (2009) A high-mobility electron-transporting polymer for printed transistors. Nature 457:679–686
162. Sekitani T, Zschieschang U, Klauk H, Someya T (2010) Flexible organic transistors and circuits with extreme bending stability. Nat Mater 9:1015–1022
163. Zschieschang U, Ante F, Yamamoto T, Takimiya K, Kuwabara H, Ikeda M, Sekitani T, Someya T, Kern K, Klauk H (2010) flexible low-voltage organic transistors and circuits based on a high-mobility organic semiconductor with good air stability. Adv Mater 22:982–985
164. Gates BD (2009) Flexible electronics. Science 323:1566–1567
165. Service RF (2004) Lighting the way to speedier chips. Science 304:674–675
166. Yang C, Gu H, Lin W, Yuen MMF, Wong CP, Xiong M, Gao B (2011) Silver nanowires: from scalable synthesis to recyclable foldable electronics. Adv Mater 23:3052–3056
167. Ahn BY, Duoss EB, Motala MJ, Guo XY, Park SI, Xiong YJ, Yoon J, Nuzzo RG, Rogers JA, Lewis JA (2009) Omnidirectional printing of flexible, stretchable, and spanning silver microelectrodes. Science 323:1590–1593
168. Forrest SR (2004) The path to ubiquitous and low-cost organic electronic appliances on plastic. Nature 428:911–918
169. Crone B, Dodabalapur A, Lin YY, Filas RW, Bao Z, LaDuca A, Sarpeshkar R, Katz HE, Li W (2000) Large-scale complementary integrated circuits based on organic transistors. Nature 403:521–523
170. De S, Higgins TM, Lyons PE, Doherty EM, Nirmalraj PN, Blau WJ, Boland JJ, Coleman JN (2009) Silver nanowire networks as flexible, transparent, conducting films: extremely high DC to optical conductivity ratios. ACS Nano 3:767–1774
171. Chun KY, Oh Y, Rho J, Ahn JH, Kim YJ, Choi HR, Baik S (2010) Highly conductive, printable and stretchable composite films of carbon nanotubes and silver. Nat Nanotechnol 5:853–857
172. Fan ZY, Ho JC, Takahashi T, Yerushalmi R, Takei K, Ford AC, Chueh YL, Javey A (2009) Toward the development of printable nanowire electronics and sensors. Adv Mater 21:3730–3743
173. Nikitin PV, Lam S, Rao KVS (2005) Low cost silver ink RFID tag antennas. In: Antennas and propagation society international symposium, vol 2B, pp 353–356. doi:10.1109/APS.2005.1552015
174. Cohen-Karni T, Timko BP, Weiss LE, Lieber CM (2009) Flexible electrical recording from cells using nanowire transistor arrays. Proc Natl Acad Sci USA 106:7309. doi:10.1073/pnas.0902752106
175. Siegel AC, Phillips ST, Dickey MD, Lu NS, Suo ZG, Whitesides GM (2010) Flexible printed circuit boards on paper substrates. Adv Funct Mater 20:28–35
176. Khang DY, Jiang HQ, Huang Y, Rogers JA (2006) A stretchable form of single-crystal silicon for high-performance electronics on rubber substrates. Science 311:208–212

177. Kim DH, Ahn JH, Choi WM, Kim HS, Kim TH, Song JZ, Huang YGY, Liu ZJ, Lu C, Rogers JA (2008) Stretchable and foldable silicon integrated circuits. Science 320:507–511
178. Sekitani T, Takamiya M, Noguchi Y, Nakano S, Kato Y, Sakurai T, Someya T (2007) A large-area wireless power-transmission sheet using printed organic transistors and plastic MEMS switches. Natl Mater 6:413–417
179. Araki T, Nogi M, Suganuma K, Kogure M, Kirihara O (2011) Printable and stretchable conductive wirings comprising silver flakes and elastomers. IEEE Electron Device Lett 32:1424–1426
180. Shang S, Zeng W, Tao X (2011) High stretchable MWNTs/polyurethane conductive nanocomposites. J Mater Chem 21:7274–7280
181. LeMieux MC, Bao ZN (2008) Flexible electronics: stretching our imagination. Nat Nanotechnol 3:585–586
182. Rogers JA, Someya T, Huang Y (2010) Materials and mechanics for stretchable electronics. Science 327:1603–1607
183. Cong H, Pan T (2008) Photopatternable conductive PDMS materials for microfabrication. Adv Funct Mater 18:1912–1921
184. Li Z, Zhang R, Moon K-S, Liu Y, Hansen K, Le T, Wong CP (2013) Highly conductive, flexible, polyurethane-based adhesives for flexible and printed electronics. Adv Funct Mater 23:1459–1465
185. Zhang R, Lin W, Moon K-S, Wong CP (2010) Fast preparation of printable highly conductive polymer nanocomposites by thermal decomposition of silver carboxylate and sintering of silver nanoparticles. ACS Appl Mater Interfaces 2:2637–2645
186. Jiang H, Moon K-S, Wong CP (2006) Surface functionalized silver nanoparticles for ultrahigh conductive polymer composites. Chem Mater 18:2969–2973
187. Ma R, Kwon S, Zheng Q, Kwon HY, Kim JI, Choi HR, Baik S (2012) Carbon-nanotube/silver networks in nitrile butadiene rubber for highly conductive flexible adhesives. Adv Mater 24:3344–3349
188. Miller MS, O'Kane JC, Niec A, Carmichael RS, Carmichael TB (2013) Silver nanowire/optical adhesive coatings as transparent electrodes for flexible electronics. ACS Appl Mater Interfaces 5:10165–10172
189. Liu J (1999) Conductive adhesives for electronics packaging. Electrochemical Publications, Port Erin
190. Li Y, Moon KS, Wong CP (2005) Electronics without lead. Science 308:1419–1420
191. Li Y, Wong CP (2006) Recent advances of conductive adhesives as a lead-free alternative in electronic packaging: materials, processing, reliability and applications. Mater Sci Eng R 51:1–35
192. Li Y, Wong CP (2006) High performance anisotropic conductive adhesives for lead free interconnects soldering. Surf Mount Technol 18:33–39
193. Zhang RW, Moon KS, Lin W, Wong CP (2010) Preparation of highly conductive polymer nanocomposites by low temperature sintering of silver nanoparticles. J Mater Chem 20:2018–2023
194. Yang C, Yuen MMF, Xu B (2008) Using novel materials to enhance the efficiency of conductive polymer. In: The 58th IEEE electronic components and technology conference, vol 5. pp 213–218. doi:10.1109/ECTC.2008.4549972
195. Li Z, Zhang R, Liu Y, Le T, Wong CP (2012) Highly conductive, flexible, bio-compatible poly-urethane based isotropic conductive adhesives for flexible electronics. In: IEEE 62nd electronic components and technology conference (ECTC), pp 406–411. doi:10.1109/ECTC.2012.6248862
196. Petrovic ZS (2008) Polyurethanes from vegetable oils. Polym Rev 48:109–155
197. Guner FS, Yagci Y, Erciyes AT (2006) Polymers from triglyceride oil. Prog Polym Sci 31:633–670

198. Putson C, Lebrun L, Guyomar D, Muensit N, Cottinet P-J, Seveyrat L, Guiffard B (2011) Effects of copper filler sizes on the dielectric properties and the energy harvesting capability of nonpercolated polyurethane composites. J Appl Phys 109:024104
199. Belhora F, Cottinet P-J, Hajjaji A, Guyomar D, Mazroui M, Lebrun L, Boughaleb Y (2013) Mechano-electrical conversion for harvesting energy with hybridization of electrostrictive polymers and electrets. Sens Actuators A 201:58–65
200. Eddiai A, Meddad M, Guyomar D, Hajjaji A, Boughaleb Y, Yuse K, Touhtouh S, Sahraoui B (2012) Enhancement of electrostrictive polymer efficiency for energy harvesting with cellular polypropylene electrets. Synth Met 162:1948–1953
201. Fiorido T, Galineau J, Salles V, Seveyrat L, Belhora F, Cottinet P-F, Hu L, Liu Y, Guiffard B, Moortele AB-VD, Epicier T, Guyomar D, Brioude A (2014) Bifunctional organic/inorganic nanocomposites for energy harvesting, actuation and magnetic sensing applications. Sens Actuators A 211:105–114

Electronic Applications of Polyamide Elastomers and Its Composites

Paulina Latko and Anna Boczkowska

Abstract The aim of this chapter is to promote the usage of polyamide (PA) thermoplastic elastomers and its composites in the electronic sector. General knowledge about this group of materials and their applications will be presented in Sect. 1. The structure and synthesis methods will be briefly described in Sect. 2 as there are few books focusing on these issues. Properties of PA thermoplastic elastomers—especially from the perspective of electronic applications—are discussed in Sect. 3. The small market for PA elastomers including manufacturers, commercial products, and the properties of some of the grades of PA elastomers is described in Sect. 4. Finally, the application of PA elastomers in general sectors and in the electronic industry is included in Sect. 5, which is divided into 5 subsections. Such topic as cable insulating, antistatic blends, conductive composites and nanocomposites, sensors, and textile composites will be described as thermoplastic PA elastomers which are commonly used there.

Keywords Polyamide elastomer · Flexibility · Insulating · Conductive properties

Abbreviations

TPE	Thermoplastic elastomer
TPE-A, COPA, TPA	Thermoplastic polyamide elastomer
SBC	Styrenic block copolymer
SBS	Poly(styrene–butadiene–styrene)
SEBS	Styrene–ethylene/butylene–styrene

P. Latko · A. Boczkowska (✉)
Faculty of Materials Science and Engineering, Warsaw University of Technology,
Woloska 141, 02-507 Warsaw, Poland
e-mail: anna.boczkowska@inmat.pw.edu.pl

P. Latko
e-mail: paulina.latko@inmat.pw.edu.pl

D. Ponnamma et al. (eds.), *Flexible and Stretchable Electronic Composites*,
Springer Series on Polymer and Composite Materials,
DOI 10.1007/978-3-319-23663-6_5

TPO	Thermoplastic polyolefine
TPU	Thermoplastic polyurethane
COPE	Copolyether ester
PEBA	Polyether block amide
EMI	Electromagnetic interference
PEE	Poly(ether–ester)
CNT	Carbon nanotubes
MWCNT	Multi-walled carbon nanotubes
SWCNT	Single-walled carbon nanotubes

1 Introduction

In early 1972, the Paris-based company Atochem S.A (now known as ARKEMA) started to develop a new type of non-plasticized flexible polymer in order to bridge a gap in the engineering of thermoplastic resins. They found that mixing polyamide (PA) with polyether results in a material with good flexibility. Two years later, the first experiments with TPE-A synthesis started, but commercial production did not start due to the lack of a good catalyst. The pilot plant was successfully developed until 1980, when the first grade of polyether block amides (PEBA) with the trade name Pebax® was formally made available on the market. Subsequently, the development of various two-phase TPE-A manufacturing processes rapidly grew. As a result of the presence of rigid PA and soft polyether segments, TPE-A offers a high level of flexibility over a wide range of temperatures [1, 2]. Compared to other types of TPE, the TPE-A market is small and new. According to a 2003 report [3], the price of TPE-A is higher than other types of TPE, and this is still the case in 2015 [1]. Because the performance-to-price ratio is the deciding factor when choosing a material for a specific application, TPE-A is usually not chosen.

The high price of polymeric materials can be reduced with polymer-based composites. Reinforcing polymers with filler results in a material with better properties than pure polymer. The type, amount, and dispersion of fillers all affect the final properties of the composite. If a nanosize filler is used, the resulting composite is called a nanocomposite. However, filler decreases the flexibility of elastomers, which should be taken into account when designing composites based on TPE. This is especially true of TPE-A and is therefore a very important factor which affects further applications because TPE-A has the highest flexibility in all of TPEs across a broad range of temperatures.

Two trends have been identified on the basis of our own review of various industrial sectors and scientific literature. From a scientific perspective, the synthesis of new grades of TPE-A is deeply investigated as well as production of membranes for gas separation. From an industrial perspective, TPE-A is used for medical devices (e.g., catheters), sports goods (e.g., ski boots), and textiles. In the electronic industry, TPE-A is mainly used for covering cables and wires with insulation jacketing. The review also found that some grades of antistatic PEBA available on the market are mixed with various thermoplastic polymers, resulting in polymeric blends for electronic equipment and packaging [4].

There are many papers concerning the mixing of PEBA elastomer with PVC or PPS to produce films for gas separation [5, 6]. The main area in which TPE-A and its composites are used is filtration, due to its high permeability and polar gas selectivity and its suitability for fabricating thin membranes [7, 8]. Some membranes are produced from PEBA doped with a small amount of carbon black designed for separation of volatile organic compounds in water. In order to improve the occasionally insufficient permeability or gases separation, the addition of organically modified clay into TPE-E has been also studied [9]. Reinforcement of polymer with clay results in better thermal and mechanical properties, scratch resistance, and even flame retardancy [10, 11]. As TPE-A can be made into a thin film, it could be an ideal candidate for touch switch membranes with conductive particles [12].

Polymeric composites are found in 9 % of electronic applications of usage when the rest includes neat thermoplastic and thermosets. A few areas of the electronic industry can be distinguished as the most popular in terms of TPE: (i) wire and cable insulation and jacketing; (ii) connectors; (iii) integrated circuits, transistors, and diodes; and (iv) molded electrical devices. TPE-A is mainly used as an insulating material.

2 Structure and Synthesis of TPE-A

PA thermoplastic elastomers are a group of multi-block copolymers consisting of hard and soft segments connected by characteristic amide linkages (Fig. 1). The hard segments are always PAs, whereas the soft segments can be either polyesters or polyether-esters. The most important examples of TPE-A are polyesteramide, polyetheresteramide, polycarbonate esteramide, and polyether block amide [13], the latter of which is the most important and is produced on the largest scale.

Thermoplastic PA elastomers are an example of polymer synthesis by polycondensation reaction between functional groups in the initial components [14], although it is also possible to synthesize TPE-A via anionic ring-opening polymerization of lactams. In terms of the mechanism, the reaction seems easy, but carrying out the process is much more difficult. Generally, the choice of technique is dependent on the starting components as well as the desired properties of the final

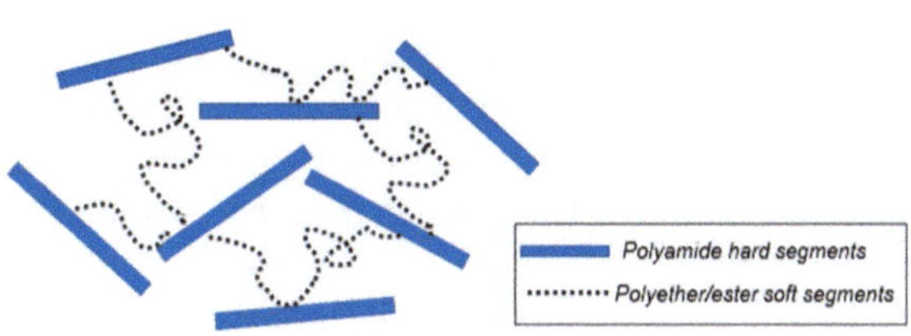

Fig. 1 Schematic representation of thermoplastic elastomers structure

material. From a technical point of view, the reaction can be carried out in a bulk, in a solution, or as an interfacial process. Moreover, a catalyst, high or low temperature, as well as one-step or two-step methods can be utilized [4, 14, 15]. All of these factors mean there are multiple ways to produce TPE-A; however, these will not be described in detail here. In this work, general information and ideas about the technical aspects of TPE-A fabrication will be given, as well as modern manufacturing processes of commercial TPE-A.

Bulk polymerization is a method carried out in the absence of any solvent or other medium which makes the process much easier. The main problem with TPE-A is the melting point of PAs (even 290 °C), and at such a high temperature, the polyether starts to degrade. In order to decrease the processing temperature, it is worth using PA salt, prepolymer, or an appropriate catalyst. According to the method described in [15], polyisobutylene (PIB)–PA multi-block copolymers were obtained by three-step process. Firstly, a prepolymer with a soft block of PIB was synthesized as a difunctional primary amine-terminated PIB. At the same time, the hard segmented prepolymer of PA was received as difunctional carboxylic acid-terminated PA. Afterward, they were mixed in bulk under nitrogen at 220 °C resulting in copolymers with different ratios between segments.

Polymerization in solution is utilized very often when high-viscosity monomers are used. Polar solutions such as N, N-dimethylacetamide, m-cresol, and tetramethylene sulfone are often utilized. As described in [16], a method of synthesizing thermoplastic elastomer PA with naphthalenedicarboxamide hard segments was achieved at a very low temperature (0–20 °C) by applying diacid dichlorides and N, N-dimethylacetamide as a solvent. The obtained polymer has a rather low molecular weight due to possible side reactions involving the solvent. The reaction time between components can be as much as 16 h, but this time can be reduced by adding a catalyst. The most often used catalyst is $Ti(OBu)_4$, which belongs to the group of transition metal catalysts. Al, Zr, Ge, Hf, Sn, and Sb can also be used instead of Ti. The last two are mostly used in a mixture [4]. It was shown in [17] that polymerization conducted in the presence of catalyst $Zr(OBu)_4$ leads to the formation of polymers with much higher molecular weights.

One way of carrying out the polymerization process for incompatible monomers is *interfacial polymerization*. In the case of TPE-A, a reaction between PA6.6 and PEO was conducted in chloroform at 60–90 °C. The main drawback of this process is the amount of solvent required [18].

Polycondensation method between dicarboxy PA and dihydroxy polyether is mainly used in the industry nowadays. By applying chain coupling, the reaction occurs at lower temperatures and in a shorter time. Moreover, there is lack of by-products and adjustment of both segment ratios is easy because the reaction between components occurs by polyaddition. Furthermore, the lack of toxic solvents and the simpler equipment and monomer preparation makes this process harmful [19]. This method is used by ARKEMA (France) to produce the diverse group of PEBA under the trade name Pebax®. This two-step process is shown schematically in Fig. 2. Firstly, lactam/amino and acid/diacid + diamine react with chain-terminating diacid, resulting in a PA with two carboxylic groups.

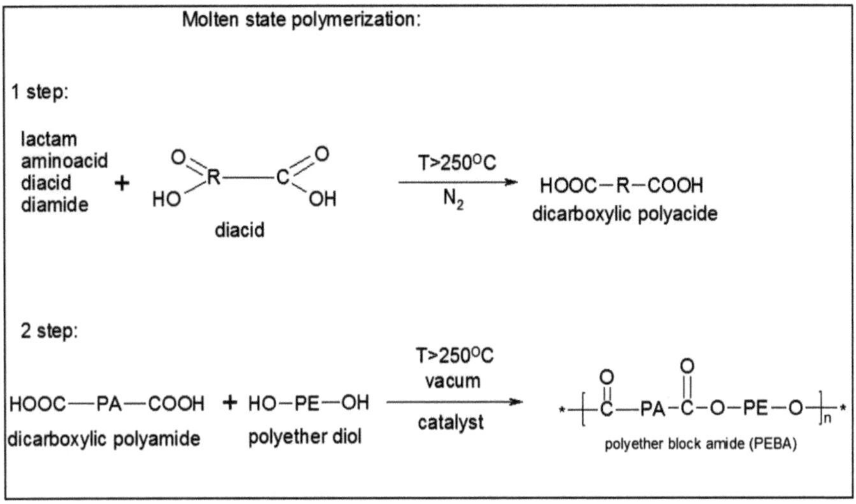

Fig. 2 Two-step process of PEBA synthesis

Then, according to simple esterification under vacuum and in the presence of a catalyst between HOOC-PA-COOH and HO-PE-OH, a TPE-A with an ester bond is created [4].

Currently, PA12 is mostly used as hard segments in Pebax®. It is obtained by polymerization of lactam-12, which in turn is obtained from cyclododecane in petroleum. There is a growing trend in bioplastics to synthesize a new class of TPE-A with the trade name Pebax®Rnew. This class of materials based on PA11 as a hard segment is synthesized indirectly from castor or seeds. According to the type of the functional end groups of the utilized starting components, the linkage between hard and soft segments can be an amide, an ester, a urethane, or a urea [4] (Fig. 3).

As mentioned in the introduction, the hard segments of TPE-A consist of one of the PAs, mainly aliphatic and alicyclic. They could be PA6, PA6.6, PA12, PA6.10, PA6.12, PA10.10, PA11.6, or PA4.6. However, the most important are PA12, PA11, and PA6. It is also possible to use aromatic, semiaromatic, or semicycling PAs. Therefore, PAs are characterized by a wide and high melting range which means that in PA elastomers, the hard segments have a melting point (T_m) in the range 120–275 °C. Above this melting point, the hard phase of elastomer becomes soft and fluid, which makes the polymer less stiff [2]. PA segments have two types of linkages which affect the properties of the material. The first is a polar amide group in the PAs which allows the creation of intra- and intermolecular hydrogen bonds which increase the strength of the elastomer. The second is an ester group which by easy hydrolytic degradation of ester bond leads to the reduction of crystallinity. As a soft phase of TPE-A, a poly(tetramethylene oxide) (PTMO/PTMG/PTHF), poly(ethylene glycol) (PEG/PEO), poly(propylene glycol) (PPG/PPO), or bisphenol A ethoxylate (BEO) are often selected [4, 17].

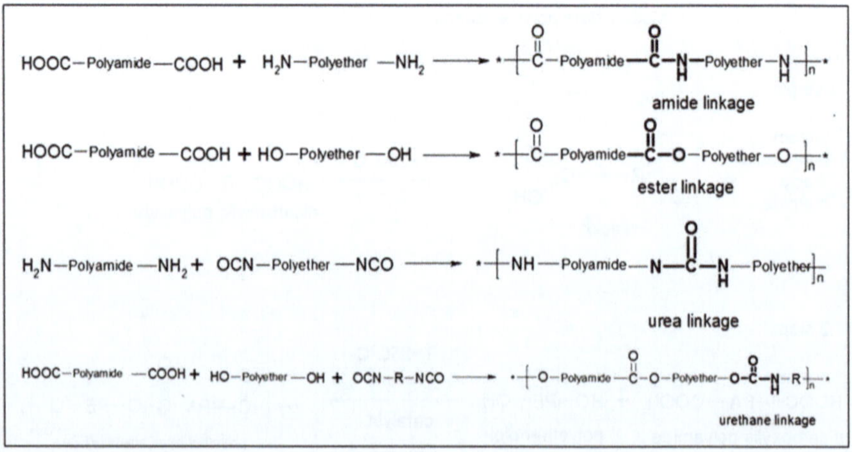

Fig. 3 Possible linkages between segments occur in polyamide elastomers

In contrast to hard segments, the compounds used as soft phase are characterized by low T_g in the range −60 to −40 °C. Below this temperature, the soft phase becomes hard, which results in stiffness higher than rubber [2].

On the one hand, the presence of two thermodynamically immiscible segments results in micro-phase-separated morphology of TPE-A, which is highly dependent on the PA/polyether ratio (PA/PE). Intuitively, the higher content of PA segment (PA/PE > 1) causes the higher Young's modulus (150–700 MPa) and the lower elasticity (app. 20 %). Contrary, when PA/PE < 1, the stiffness of the material decreases, but the elastic properties increase (100–300 %) [4]. On the other hand, these segments are not separated at the macroscopic level because of the physical cross-link. Below the melting point of PA segment, the soft phase is chemically bonded to the amide groups in the hard segments. The created hydrogen bonds result in a physical cross-linked network responsible for the elastomeric properties of the material. In turn, above T_m, the material becomes a common thermoplastic polymer because the previously created network is disrupted (Fig. 4) [10, 20].

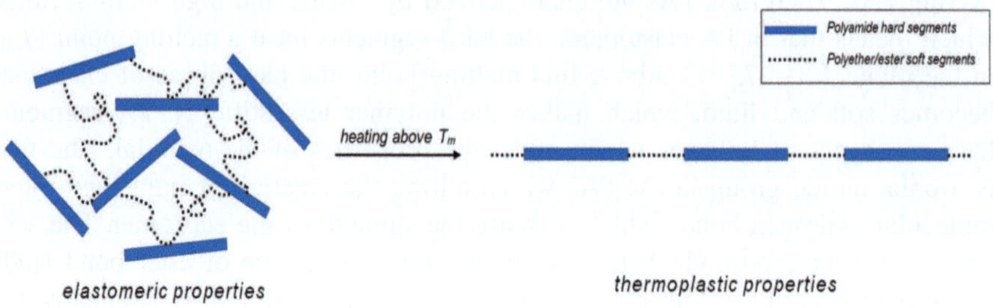

Fig. 4 Changing in TPE-A structure during heating

3 Properties

The chemical composition, molecular weight distribution, and proportion of hard and soft segments are the factors which affect the final properties of TPE-A. The main role of hard segments is to control the degree of crystallinity, melting point, and mechanical strength. In turn, soft segments are responsible for low temperature flexibility, chemical resistance, and thermal and hydrolytic stability [13]. The most characteristic properties of TPE-A are shown in Fig. 5.

TPE-A combines low density (lightweight) with outstanding properties which are especially desired in electronic devices. As previously mentioned, TPE-A belongs to a segmented group of polymers containing two types of domains which are well separated from each other. Due to this special morphology, TPE-A is characterized by versatile physical properties. It means that TPE-A can be used in extrusion, injection, melt spinning, etc., and also can be decorated or recycled [4].

The values of the characteristic temperatures of both segments determine the so-called service temperature range. For TPE-A, this range is in the range −65 to 275 °C as a result of the melting point of PA and glass transition of component used as soft segment [2]. Compared to other types of TPE, the similar service temperature range has COPE. Additionally, TPE-A maintains a high level of flexibility within a wide service temperature range, especially at very low temperatures, and is resistant to kinking when the right precursor for hard segments is chosen. In order to obtain a polymer with non-kinking behavior, the PA should have a high melting temperature, a high amount of crystallinity phase, and good mechanical properties [16]. The wide flexibility range even at low temperatures and non-kinking features are characteristics of TPE-A and do not occur for other TPEs.

Abrasion resistance is especially important in the cable industry. PA elastomers are resistant to scratch, which is dependent on the segment ratio. For instance, abrasion resistance decreases from 107 to 34 mm^3 when the percentage of PA12 segments falls from 80 to 23 %, respectively [4].

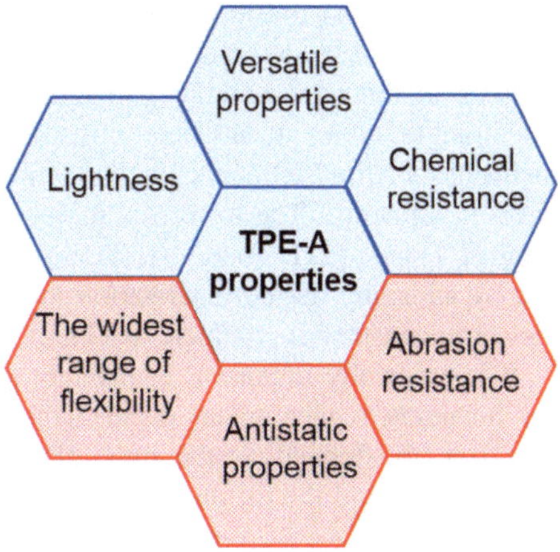

Fig. 5 The outstanding properties of TPE-A

Dependent on the type of PA and polyester used, the final polymer can be hydrophobic or hydrophilic [4]. Nevertheless, TPE-A has excellent resistance to many chemicals when compared to other thermoplastic elastomers. This is highly dependent on the type and amount of PA segments used because they are more sensitive to chemicals than polyether blocks. The polymer can swell in grades in which they are softer than hard segments. In other cases, the polar character of soft segments and the crystalline nature of the hard segments lead to poor affinity to hydrocarbon oils and solvents [2].

In electronic applications, the electrical properties of material play the main role. Similarly to most polymers, TPE-A is an insulator (dielectrics), which means that it does not have free charges which can move when current is applied. Apart from its insulating ability, its resistance to arcing, vibrations, and mechanical shocks which can both damage the structure of material is equally important. Furthermore, external factors such as temperature, moisture, radiation, and contaminates have a strong effect on its performance as an insulator. In order to determine TPE-A's suitability as an insulator, factors such as dielectric constant, dielectric strength, surface and volume resistivity and dissipation factor should be taken into account [21].

Dielectric constant (ε) is defined as the ratio between the capacitance of the capacitor with a dielectric layer between the plates (C) to the capacitance of the capacitor in a vacuum (C_0):

$$\varepsilon = \frac{C}{C_0} \tag{1}$$

By using dielectric spectroscopy, the dielectric permittivity of a material is expressed by the complex equation: $\varepsilon^* = \varepsilon' - i\varepsilon''(i = \sqrt{-1})$. In this case, the real part ε' refers to the dielectric constant, while the imaginary part ε'' corresponds to the dielectric loss. From a physical point of view, dielectric constant is understood as an electrical energy which can be stored in a material's structure leading to the insulation of each element from each other. The higher the dielectric constant, the higher the amount of energy that can be stored under applied voltage [4, 21]. According to Fig. 6, TPE-A has the highest TPU dielectric constant of all TPEs. It should be noted that these values are average. There are some grades of PEBA available on the market which differ significantly from each other in dielectric strength values. It has been observed that the molecular weight of hard segments has a strong influence on dielectric constant of TPE-A, while soft segments have no influence. The lower the molecular weight, the lower the dielectric constant of TPE-A [22]. Therefore, TPE-A can be obtained with the desired properties by adjusting the polymerization methods between monomers. Generally, the dielectric constant of commercial TPE-A is 4.00–9.5 [24, 27].

It is especially important in insulating applications to know when the electrical breakdown phenomena occur. This is determined by the value of dielectric strength, defined as the ratio of the maximum voltage in which material breaks down and the distance between the electrodes. Because of the still higher

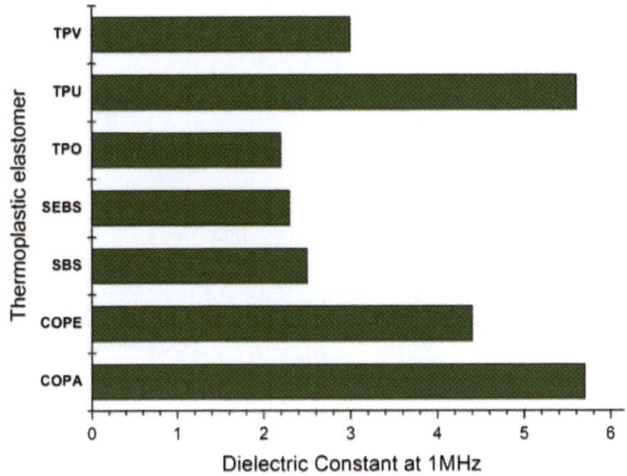

Fig. 6 Average values of dielectric constant for TPE

transmission voltages applying into the electronic devices, the dielectric strength should be high with simultaneous working reliability of the material. The average value of the dielectric strength of commercial TPE-A is 36.1 kV/mm [21, 23].

The level of volume resistivity of materials determines their application in the electronic industry. This is schematically represented in Fig. 7. Insulator materials have volume resistivity higher than 10^{14} Ω cm. TPE-A has volume and surface resistivity of 3.58×10^{15} Ω cm and 2.24×10^{15} Ω, respectively; therefore, it is used mainly as insulating layers in cable and wires.

ARKEMA offers two types of PEBA with antistatic properties which can reduce or eliminate buildup of static electricity or prevent dust attraction. The trade names are Pebax®MV 1074 SA and Pebax®MH 1657, with a surface resistivity of 3×10^9 and 1×10^9 Ω m, respectively. By adding TPE-A to a thermoplastic polymer, the charge decays much slower, which protects materials from the environment. Antistatic is an extremely important issue for electronic components; therefore, it will be described in-depth in Sect. 5.2.

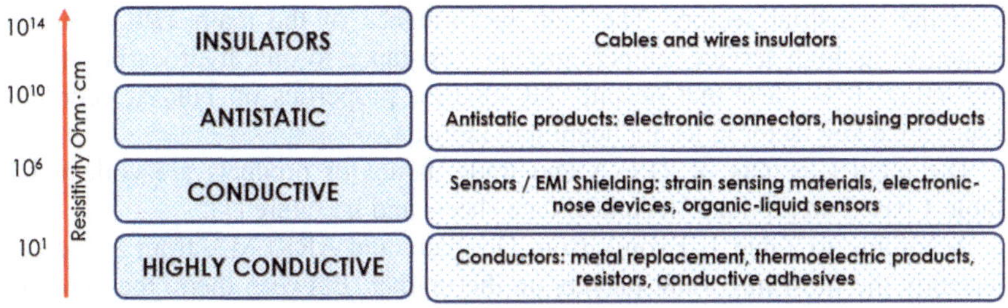

Fig. 7 Dependence between resistivity of materials and application sectors

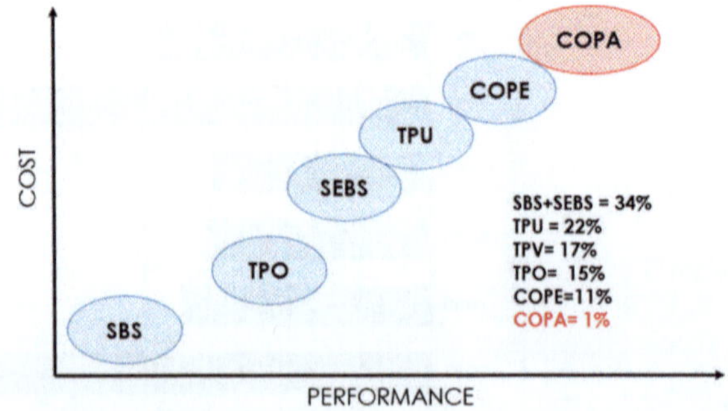

Fig. 8 Relation between cost and performance of commonly used types of TPE with global production level

Table 1 Commercial TPE-A

Supplier	Country	Trade name	Hard segments	Soft segments
ARKEMA	France	Pebax® Pebax®Rnew	PA12/PA6 PA11	PTMO/PPO PTMO
Evonik	Germany	Vestamid®E	PA12 PA612	PTMO
EMS-Grivory	Switzerland	Griflex Griltech®ELY	PA12 PA12	Polyether PTMO
Ube Industries	UK	Ubesta XPA	PA12	Uniquely designed polyether
Brüggemann Chemical	Germany	Nyrim®	PA6	–

4 TPE-A Market

According to the latest recent report, the global TPE market accounted for 5.6 % of total plastic production in 2013. It is predicted that TPE production will increase by about 0.7 % between 2014 and 2018. The main areas of application are the automotive industry (42 %), followed by the medical, sport, and electronic industries. Among the whole TPE group, the production of TPE-A is lowest, as shown in Fig. 8, which is due to the high price of TPE-A and the low demand from the industrial sectors. The high performance of the main TPE groups is related to its high service temperature and resistance to oils, hydrocarbon fluids, and aging. It is clear that despite the superior properties of TPE-A, economic issues are a key factor in its application in industry.

The main suppliers of TPE-A and trade names of products are collected in Table 1 based on products' technical data sheets and literature [1].

According to Table 1, Evonik from Germany and ARKEMA from France are currently the most important players on the market. Between them, they produce various grades of TPE-A with the trade names Vestamid and Pebax, respectively. Evonik offers various grades of Vestamid®E, a thermoplastic elastomer with PA12

Table 2 Applications of Vestamid® grades produced by Evonik

Vestamid®	Typical applications
E40-S3	Noiseless gears, seals, functional elements of sports shoes, processing aids in the extrusion of TPU, films
E47-S1 E47-S3	Sports shoe soles, packaging films, non-skid surfaces, sports glasses, protective goggles
E55-S1 E55-S3	Alpine ski boots, sports shoe soles, pneumatic lines, rolls, films
E58-S4	Specialty grade for transparent sports shoe soles
E62-S1	Alpine ski boots, noiseless gears, conveyor belts
E62-S3	Alpine ski boots, noiseless gears, conveyor belts, fiber optic cable
EX9200	Decorative and protective films

and polyether segments. These are listed in Table 2 with grades appropriate for the electronic industry highlighted.

It should be noted that the two-digit number after the letter E determines the Shore hardness in scale D. Additionally, when S1 is in the name, this denotes thermal and S2 and S3 thermal and UV light stabilization. With the increase of polyether content, such properties as flexibility and higher impact strength become more apparent [4]. According to Table 2, the main field of PEBA application is sport, alpine footwear, and mechanical parts of machines. However, the E62-S3 grade is designed for fiber optic cables and will be presented later in this chapter.

As a leader in PEBA production, ARKEMA offers five groups of PEBA with the trade names Pebax®, Pebax®Clear, Hydrophilic Pebax®, Pebax®MED, and Pebax®Rnew. According to Table 3, the first group is for general-purpose use. ARKEMA is also the first producer of PEBA from renewable resources based on castor oil. As has been already mentioned, one type of Pebax® offers permanent antistatic properties. There are two grades of Hydrophilic Pebax® specified as Pebax®MV 1074 SA and Pebax®MH 1657 with surface resistivity of 3×10^9 and 1×10^9 Ω/sq, respectively. This TPE-A is used as an antistatic agent and blended

Table 3 Applications of Pebax® grades produced by ARKEMA

Grade	Main properties	Typical applications
Pebax®	High performances independent of temperature	Power transmission belts and silent gears, transportation, sport, and leisure
Pebax®Clear	High transparency High UV resistance	Racket bumpers, sport, and ski shoes
Hydrophilic Pebax®	Permanent antistatic properties, high moisture and vapor transmission rate	Permanent antistatic additive, breathable film in construction wrapping, medical, sport clothing, food packaging
Pebax®MED	Biocompatibility sterilization feasibility	Angioplasty and urology catheters, connectors for catheters, medical gown
Pebax®Rnew	Bio-based origin and high performance	Air freshener, conveyor belts, fibers, power transmission belts, racket bumpers, sport shoes

with many thermoplastic polymers to produce new types of composites for electronic applications.

Similarly to Evonik, EMS-Grivory manufactures a bio-based Griflex PEBA with 10–100 % renewable content. These nylon 12-based materials, based on castor and canola oil, have properties similar to standard Grilflex and are intended for sporting goods.

The newest group of TPE-A is a material with the trade name Nyrim® based on PA6 produced by reactive injection molding. The grade E50-R2 is designed for permanently antistatic articles and has a surface resistivity between 10^6 and 10^8 Ω. Moreover, the high-impact toughness and light structures are appropriate for application Nyrim® in the electrical industry [4].

5 Applications

The main subject of this work is the applications of TPE-A and TPE-A composites in electronics. Therefore, the next part of the chapter will focus on areas where TPE-A is already used or has the potential for use.

5.1 Cable Insulators

Wire, cable, and jackets are the main applications of pure TPE-A and filler-reinforced TPE-A. This is due to both its outstanding insulation properties and its ability to respond differently to various voltages [21]. Electronic cables are used for a wide range of purposes, from common household applications to more sophisticated technical situations. The most typical types of cables distinguished by their applications are shown in Table 4.

Many types of electronic cable are available, and new products are constantly introduced to the market. The structure of a simple coaxial electronic cable includes conducting wire pairs running side by side, usually made of metal strands assembled together into various configurations by twisting or bonding. Each conductive wire has to be insulated and sheathed. In order to keep these single wires separated electrically and physically, the insulation materials should have excellent dielectric and transmission properties. The outer part of a cable jacket is made up of insulating materials. Insulation of cables is necessary to prevent electrical leakage and contact with other conductors as well as to protect cables from various environmental factors. Cable jackets are made of special polymer which fulfills some specific requirements, primarily the safe and efficient functioning of the cable.

Nowadays, polymeric materials are a popular choice for insulation jackets. There are specialty polymers whose excellent features fulfill the long list of requirements in the electronic sector. On the basis of market information from

Table 4 Examples of electronic cable applications

Cables type	Main requirements	Examples of application
Household	Low tension	Wiring of domestic equipments: ovens, washing machines, etc.
Lighting	High temperature resistance	Lamps and lighting equipments
Safety	Insulation of the electrical conductor	Emergency lamps, signal lamps for safety exits, audio and video signaling devices mainly in public area
Ignition and battery	High tear resistance High electrical breakdown resistance	Burners, boilers, test devices, sparking plugs, neons
Control and instrumentation	High notch or abrasion resistance High chemical resistance High operating temperatures	To connect electrical instrument circuits, alarm equipment for switchgear, in vehicle monitoring equipment for detection of traffic movements, inductive loop detectors
Railway	High tear resistance High mechanical strength High flame resistance UV stability	Trains, subways and the associated tunnel systems
Ship, marine, offshore	High dielectric strength resistance to water treeing and thermal stability	Signal transmission (TV, Internet) navigation system

various cable manufacturers, the main polymers used for jacket insulation are as follows: polyvinyl chloride (PVC), polyethylene (PE), polypropylene (PP), polyurethane (PU), PA/Nylon (PA), polytetrafluoroethylene (PTFE), fluoropolymers, silicone, fluorosilicone, and TPE. TPE-A is also used as a jacketing material due to its good insulating and high resistance properties. TPE-A can compete with other jacketing materials; however, it is often not used due to economical factors.

The list of an electronic cable jacket is long and is dependent on four groups of factors. The first group is the most reliable and is only partially connected to the other three groups. This is the *electrical performance* of the cable, which is strongly associated with the properties of chosen material, primarily its capacity to store (capacitance) and lose electrical energy (attenuation). Secondly, the velocity of signal transmission through the length of the cable, whose integrity can potentially be affected by electromagnetic interference or cross talk from undesired coupling of signals. All of these features are directly included and expressed by a dielectric constant. If the dielectric constant value is known, the appropriate material can quickly be chosen for a given application. A low dielectric constant equates to low capacitance and attenuation and high impedance, meaning that the material has good insulation properties. Figure 9 presents a comparison between dielectric constant values of different polymers used for cable insulation.

The last two features which affect the electrical performance of insulating material are dielectric strength, which determines the ability of a material to prevent voltage breakdown and working voltage, which is the maximum voltage that can be applied to a cable.

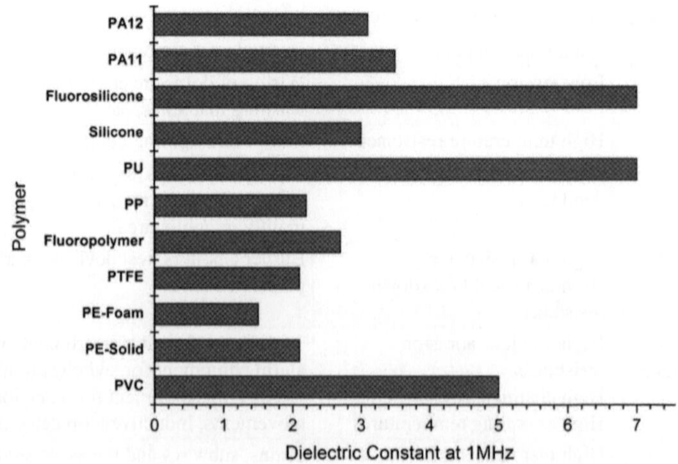

Fig. 9 Dielectric constant values for polymers used as cable's jacket

Apart from the electrical performance of jacket materials described above, there are three groups of so-called physical features that also affect cable performance. They destroyed action causes physical degradation of cable material and also affects the electrical properties of the material. The first group is *mechanical performance* related to all possible movement of cables during installation and further operations. Rolling or twisting produces kinetic energy which can damage the cable. Therefore, the material chosen for a cable's jacket should possess high elongation and mechanical strength. These properties are strongly associated with the material's flexibility, meaning how easy the material can be bent in each direction without breaking or destroying its structure. Also, cable movement, especially in repeated cycles, can cause jacket abrasion as a result of cable contact with sharp surfaces.

As mentioned previously, the cable's jacket is the first contact with the surrounding environment. Therefore, the protection of the inner cable construction is dependent on the behavior of the jacket and in sometimes on quickly changing working conditions. This is the second group of physical factors affecting cable functioning—*the environmental stress.* The most variable factor is obviously temperature, which can range from very low to very high values. Due to the strong dependence between polymer properties and working temperatures, this factor is especially problematic and sometimes not easy to monitor. At very low temperatures, the polymer is more brittle, while at high temperatures, the macromolecule chains of the polymer become softer than they should be. Hence, it is necessary that cable jackets operate in variable conditions while maintaining their flexibility and not degrading. Similarly, excessive pressure facilitates the penetration of liquids or gases into the jacket, which also accelerates the degradation process. Nowadays, cable environments are more demanding. Numerous liquids such as chemicals, cleaning fluids, fuel, lubricants, and water can easily come into contact with cable jackets; therefore, resistance to chemicals is a key property of materials chosen for jackets. It should be noted that radiation and contaminants such as mud and metal particles in the environment are also detrimental to the life span of a cable.

The last group of factors affecting cable jackets is *specific stress*, which is dependent on the type of environment in which the cables will be used. All the possible application types are listed in Table 4. For instance, cables intended for public areas such as transportation should work at low voltage because of safety issues. In turn, the key requirements for cables for marine applications are oil resistance at high temperature as well as the ability to maintain this property between −40 and 90 °C [24].

Beyond these four groups of factors affecting cable jackets, flammability is also an extremely important factor. All of the polymeric materials with the exception of chlorinated and fluorinated polymers combust. This may be caused by local overheating or electrical faults and can be spread rapidly from the affected area. Therefore, flame retardants are added to polymers to inhibit ignition and impede fire.

Mainly due to its excellent insulating properties, TPE-A is popular in the cable industry. Compared to other polymeric materials, TPE-A's dielectric constant is as high as fluorosilicones, PU, or PVC (Fig. 9). TPE-A is a better material for cable jacketing not only because of its good dielectric properties, but also primarily due to its flexibility, which is superior to that of the aforementioned polymers. Furthermore, these properties are maintained across a broad range of operating temperatures from −50 °C to almost 300 °C. Finally, its resistance to abrasion, kinking, and environmental factors increases the potential of TPE-A in the cable industry.

There are a few grades of ARKEMA Pebax® appropriate for cable jacketing, for example, Pebax® 7233, 7033, 6333, 5533, 4033, 3533. The electrical properties of these grades are listed in Table 5.

According to Table 5, surface resistivity and volume resistivity are in a range suitable for insulating materials. Besides, the bulk conductive properties expressed by dielectric strength are high enough to withstand an electrical field applied to the device. In terms of the mechanical properties, ARKEMA offers PEBA with

Table 5 Electrical properties of some of the PEBA grades

Grade	Electrical resistivity (Ω cm)	Surface resistance (Ω)	Dielectric strength (kV/mm)	Dissipation factor frequency 1e + 6 Hz
Pebax® 4033 SN 01	2.56×10^{12}	2.59×10^{13}	39.0	0.00047
Pebax® 5533 SN 01	3.20×10^{13}	1.64×10^{13}	38.5	0.00048
Pebax® 6333 SN 01	9.33×10^{13}	2.67×10^{13}	42.5	0.00039
Pebax® 7233 SN 01	7.74×10^{13}	1.84×10^{14}	44.5	0.00074

various hardnesses from 25 Shore D up to 72 Shore D, while remaining mechanically strong but flexible.

Evonik offers Vestamid®E for fiber optic jacketing. It is characterized by very high flexibility over a wide range on temperatures and therefore is commonly used for non-kinking cables [21]. The ability of PEBA to twist easily without mechanical breakage is ideal for medical vascular catheters, and its resistivity to abrasion makes it suitable for the electronic industry. Furthermore, Vestamid®E 62-S3 is resistant to heat and UV light so is recommended for fiber optic cables. Simply put, a fiber optic cable (Fig. 10) consists of a bundle of fibers inside a loose tube. As described previously, harsh external factors can reduce the efficiency or destroy a cable. Therefore, the fibers are protected by tube which can be made of PEBA like Vestamid®E 62-S3. This material is easily processed by extrusion processes with high stability and shows good resistance to hydrolysis. Evonik also offers one grade of PA12 TPE-A under the trade name Vestamid® *Terra* DS which is obtained from renewable sources and has good stress, cracking, and chemical resistance. Vestamid® *Terra* DS is intended for copper core insulation, cables, and plastic optical fiber sheathing as well as cable conduits. Protection against damage from rodents and termites is achieved by using Vestamid®L2170, especially in such countries such as Brazil or India, where this problem is particularly common. In turn, for fire protection applications, some grades of Vestamid® are offered as halogen- and phosphorus-free flame retardant layers of cables and wires. There are, for instance, Vestamid® LX9057, LX9104, L1604 or X7166.

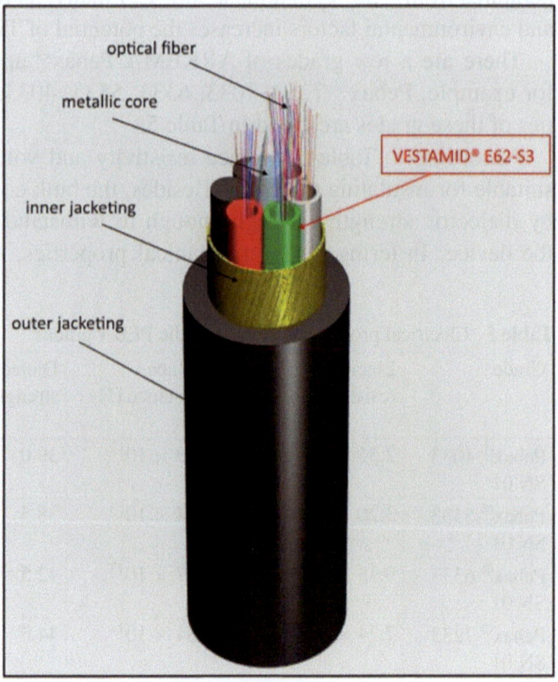

Fig. 10 Fiber optic cable with PEBA insulation layer

TPE-A elastomers possess all of the properties necessary to be used as insulation cables in the automotive industry. However, their commercial usage is still low mainly due to the lower price of other materials such as PVC, PA, or TPU and TPO in the TPE group. Similarly to those materials, TPE-A could be used for data transmission cables in a wide variety of telematic systems in vehicles. TPE-A-based optic fibers fulfill all of the necessary requirements for automotive hi-tech communication systems. Evonik offers the special dual-layer Vestamid® with high mechanical and fire stability designed to protect cables from environmental breakage especially in the automotive industry.

The addition of nanoclay is known to improve the abrasion resistance and hardness of TPE-A [25]. Moreover, such nanocomposites display better flame retardancy and therefore could be used in the cable industry as an innovative material which does not contain halogens [26]. Also, due to the good processability of TPE-A, melt mixing of organoclay with pellets of PEBA is common and produces satisfactory results such as higher thermal stability and solvent resistance enhanced by the clay nanoparticles [27]. Similarly, as has been briefly described, the extrusion of PEBA nanocomposites doped with functionalized alumina nanoparticles acts as a precursor for membrane fabrication [28]. Such organic–inorganic composites show higher thermo-oxidative stability with a degradation temperature between 300 and 400 °C. These composites based on PEBA with improved thermal behavior are promising for the cable industry, but there are relatively few new papers covering this topic. These were mentioned above, but the main point is that these works are the enhancement of gas permeability. The addition of clay particles to TPE-A leads to higher stiffness and improved scratch resistance, flame resistance, and glass transition, meaning they could be applied in the electronic industry. To the best of our knowledge, because TPE-A is a new group of materials, there are not many papers which address the aforementioned topic.

5.2 Antistatic Blends and Composites

Materials used in electronic devices should usually be electrically conductive. Therefore, the accumulation of electric charges on the surface of a device is a real problem which makes it necessary to reduce the volume or surface resistivity by accelerating the dissipation of charges. Very fast and slow charge decay can be prevented by keeping the resistivity at an antistatic performance level ($>10^{10}$ Ω m, Fig. 7). When the dissipation of charges is uncontrolled, this primarily causes dust contamination or short circuits which makes fire or even explosion highly probable. As has been already mentioned, TPE-A is an insulator. Therefore, doping it with conductive particles can lead to resistivity reduction resulting in electrically conductive material [21, 29]. However, the amount of conductive fillers required is sometimes as much as 30 %, which makes the processing of a composite into a final product difficult. Moreover, the tendency of conductive fillers to cause agglomerate formation results in a non-homogenous material structure. Sometimes, surface modification of particles is performed, but usually, this requires sophisticated methods and equipment. Besides, the relatively high price and occasional migration of fillers in a polymer matrix can hamper the movement of charges [10]. Hence, in order to gain antistatic polymer behavior, some grades of TPE-A are perfect antistatic agent candidates. This in mainly because of the semicrystalline poly(ethylene oxide) (PEO) segments in the chemical composition of TPE-A. Contrary to amorphous phase that causes conductive properties of material in the presence of PEO, the antistatic behavior is pronounced [4]. The dissipation of charges in those composites occurs due to the movement of polyether phase and charge transfer through PA chains. ARKEMA offers a special type of Pebax® with PEG groups for antistatic properties with the trade names MV1074 and MH1657. Merely 5–20 % of the antistatic agent is needed to decrease the surface resistivity by a few orders of magnitude even in dry conditions. Antistatic PEBA is suitable and widely mixed with various thermoplastic matrixes as PE, PP, ABS, PMMA, PVC, PC, PBT, PS, PA, or POM to create antistatic blends. It should be noted that changes in surface resistivity are strictly dependent on the ratio between thermoplastic polymer and the added PEBA [30]. Compared to conductive fillers, PEBA can maintain the good appearance and properties of the product. It can be easily fabricated by melt blending or compounding with the presence of a compatibilizer. These antistatic polymer blends are used, for example, in photocopier or printer components, computer, and mobile housing, as well as in technical packaging. Moreover, they are a promising type of packaging for the electronic industry as they protect sensitive parts (e.g., smart cards) from mechanical damage [4].

An interesting system was presented by Wang et al. [31] in which PEBA was modified by solid polymer electrolyte (SPE) containing one salt from $LiPF_6$, $LiAsF_6$, $LiClO_4$, or $LiBF_4$. This antistatic system was mixed with HIPS, and the surface resistivity was significantly decreased to $10^{8-9}\ \Omega\ cm^{-2}$.

As was highlighted in the introduction, one of the most investigated PEBA composites/blends is gas separation or filtration membranes. One example is the preparation of a highly selective membrane from PVC/Pebax®1657 for CO_2 capturing from a polar mixture [6]. This is similar to membranes fabricated by the casting of activated carbon/ Pebax-2533 on a poly(vinylidene fluoride) (PVDF) which has higher permittivity to CO_2 in the presence of carbon [32]. Additionally, TPE-A is added to a thermoplastic polymer to reduce the viscosity and improve the processing by increasing the flexibility [33]. Moreover, a blend consisting of 95 % PEBA and 5 % PEE, and not more than 3 % of triglycidyl isocyanurate has better mechanical performance [34].

5.3 Conductive Composites and Nanocomposites

Currently, electrically conductive composites based on polymers are of great interest of many researches. Plenty of different polymeric composites with conductive fillers have already been fabricated. Obviously, the practical usage of these materials is strictly dependent on the level of electrical conductivity achieved. As was shown in Fig. 7, the electrical resistivity determines the area of application [35]. For instance, the material with high resistance presented in Sect. 5.1 can be used as cable insulation, while decreasing its resistance a little makes the material suitable as an antistatic barrier.

Decreasing the resistivity of polymers is possible by incorporating conductive additives within the polymer matrix. There could be some metal particles (copper, silver, nickel) which do not only increase electrical conductivity significantly, but also increase the weight of the composite. Therefore, more attention is paid to composites doped with carbonaceous fillers such as carbon black, graphite, carbon fibers, or more recently carbon nanotubes and graphene [36]. These letters are increasingly studied nowadays because of their outstanding properties and ability to increase the electrical conductivity of nanocomposites under low loads. In the dependence on type of use of fillers, percentage and quality of dispersion, and distribution of fillers, the obtained electrical conductivity could be lower or higher. According to the obtained value, composites can be used just in antistatic applications or as EMI shielding structures when highly conductive [37, 38]. Despite the fact that many polymer composites and nanocomposites with conductive particles have already been synthesized and defined, those based on TPE-A are still in early stages of development. This is confirmed by a few scientific papers which describe conductive composites or nanocomposites produced from TPE-A in comparison with those produced from other TPE.

One example is synthesis of conductive membranes with 0–5wt% of MWCNT for mechanical property improvement [39]. Indeed, the mechanical performance was increased, but the rate of gas permeation was also faster in the presence of MWCNT. It was observed that the velocity of butanol removal is faster when CNT is added, compared to pure PEBA [40]. Similarly, membranes are fabricated with 5wt% of SWCNT and 5wt% of MWCNT and the selectivity of CO_2 in the mixture of CO_2/H_2 or CO_2/N_2 increased by 28 % (for SWCNT) and 13 % (for MWCNT) [41]. The same improvement in CO_2 selectivity was achieved for membranes prepared from PEBA/PEG /MWCNT [42]. Unfortunately, none of these authors focused on the electrical properties of the membrane PEBA + MWCNT, but they are probably conductive and could be used as a material for electronics.

The conductivity of nanocomposites prepared from PEBA and MWCNT was investigated by other authors [43]. In this case, Pebax®53 (ARKEMA, France) was mixed with 0.25–5 wt% of MWCNT produced by chemical vapor deposition. Nanocomposites were prepared by an extrusion process at 200 °C with a mixing time of 7 min. Then, electrical conductivity was measured using a 4-point probe to find the percolation threshold. It was observed that at 2wt% of MWCNT, a resistance decrease occurs and the measured electrical value was 2.12×10^{-6} s/cm. After further loading of MWCNT, the electrical conductivity increased to 10^{-4} s/cm for 5 wt% of MWCNT. The obtained values are promising, and such composites could be used as a touch switches.

The influence of the type and shape of the graphene plates on the electrical properties of porous PEBA films was recently investigated [44]. Authors evaluated the electrical resistance changes under applied pressure when the graphene content was 15 wt%. The electrical resistance was between 8.1×10^5 and 3.4×10^5 Ω after fabrication. In turn, when pressure was applied, there was no change in resistivity as the material is not sensitive. One nanocomposite showed a linear response to pressure which makes it promising as a piezoresistive material for electronics. It was also observed that—dependent on the type of graphene—the intrinsic structure of composites can be regular with a large range of interconnecting pores. Such materials can potentially be used in the electronic industry as insulators, chemical gas sensors, or smart fabrics [45]. Conductive nanocomposites based on TPE-A would be a good material for actuators as they belong to the group of block copolymers, which can convert external chemical or electrical energy into mechanical energy [46]. This is especially true because TPE-A has a high dielectric constant and flexible performance within wide range of temperatures. Generally, TPE is known to be used in actuators because it differs from conventional homopolymer dielectric elastomers (acrylics, silicones, rubber) in nanostructure morphology, shape memory property, and electric actuation mechanism. TPE-As have two glass transition temperatures and good low-temperature flexibility [47]. Among the entire group of nanocomposite, materials based on TPE are characterized by the highest actuation strain under lowest voltage. Therefore, they are the most promising when low voltage is required, for instance, in spacecraft components [48]. One type of actuators is dielectric elastomer actuator consisting of two electrodes with an isolating layer between them. When the voltage is applied, its thickness shrinks and returns to normal when voltage is off. Such actuators are used, for instance, for artificial muscles and mainly consist of acrylic polymers and carbon grease [49]. However, in some working conditions, especially temperatures, they are insufficient. Therefore, TPE-A doped with carbon nanotubes could be a good candidate to replace the compliant electrodes in dielectric actuators. The aforementioned MWCNT/graphene-doped PEBA could be used as compliant electrodes if the electrical conductivity is high enough. In turn, the insulating layer could be made from neat PEBA. However, these types of materials are rather new and no works about TPE-A composites for actuators have been found.

5.4 Sensors

In general, a sensor is a device that immediately sends an electrical signal in response to an external factor and is part of a larger system working together with actuators, signal processors, or detectors [50]. Among the various types of use nowadays sensors and materials for their fabrication TPE-A is also taken into account. Mainly, due to the easily doping of TPE-A as well as good processability especially toward thin films. One of the first described sensors which uses TPE-A is a vapor sensor which is part of an interdigital capacitor in an electronic nose for wine fermentation process monitoring [51]. This consists of a transducer in which the electrode is made of polymer-coated glass. PEBA was used as a polymer layer because of its high sensitivity to aroma and because of its flexibility and chemical resistance. A second example is a moisture sensor for detecting humidity by monitoring the resistance of the material [52]. According to the presented invention, the sensor consists of two layers of hygroscopic polymer doped with conductive particles and a layer of pure polymer between them. When the humidity level increases, the sensitive polymer with conductive particles absorbs water, which causes a reduction in contact between particles. A sensor measures the changes in resistance in the material when humidity is more than 60 %. Three grades of Pebax® were tested as a polymer because of its high sensitivity. In dry conditions, the level of resistance is between 4×10^3 and 2×10^3 Ω, but after moisture absorption, the resistance increases to 10^4–10^6 Ω. Such detected resistance values are required for sensing composites (over 10^4 Ω) even under low moisture levels [53].

The most sophisticated humidity sensor was presented by Liang et al. [54] In this case, Pebax® was also used, but it was mixed with 44.4 wt% content of $LiCl{\cdot}H_2O$ resulting in a highly sensitive material. The synthesis procedure includes mixing of ionic salt and PEBA in isopropanol and then electrospinning the solution into fabrics with thin nanofibers. When amount of $LiCl{\cdot}H_2O$ is increased from 16.8 to 44.4 wt%, the measured resistance decreases from 10^8 to 10^5 Ω for 60 % of humidity.

5.5 Textiles Composites

Due to its thermoplastic properties, all types of TPE can easily be processed by common methods used for common thermoplastic polymers. Dependent on the applications, almost all shapes can be formed from them, for instance, thin films (membranes) or fibers. Nowadays, it is common to decrease the total weight of structural materials by decreasing the total mass of components. Promising candidates for lightweight structures are fibers and fabrics, called textile products. Such structures can be formed from doped and non-doped TPE-A. Similarly, the addition of TPE-A into a thermoplastic matrix can improve processability as TPE-A has a high level of flexibility. From a technical point of view, making a polymer blend is better than adding fillers if the aim is to achieve antistatic behavior. Obviously, if the main target is EMI shielding, the priority is to obtain great electrical conductivity. However, it should be emphasized that the addition of nanofillers hampers the processing of the material into the final product due to the extreme increase in the viscosity of polymer [55, 56].

TPE-A-based composites and nanocomposites can be processed into fibers by extrusion and drawing or various types of spinning. Hence, it is possible to obtain fibers with various types of electrical properties [57, 58]. Antistatic or conductive fibers are perfect precursors for spinning, knitting, or weaving techniques resulting in fabrics, mats, or even carpets [59, 60]. Conductive textiles belong to a relatively new group of materials which are intensively studied nowadays. They are mainly produced from polyesters or PAs coated by metallic powders with different areal weight. There is nothing to prevent textiles being produced from TPE-A or its blends. For instance, antistatic fabrics with satisfactory dissipation properties are easily produced from PS mixed with antistatic Pebax®MV 1074. Due to the outstanding properties of TPE-A, the resulting textile products are lighter and more flexible. As previously mentioned, fabrics for the electronic industry are easily created by mixing antistatic grades of PEBA with thermoplastic polymer. On the one hand, non-woven fabrics with high areal weight are suitable for filtration processes. The same non-wovens with areal weight less than 15 g/m^2 are interesting for the surface reinforcement of electronic device components [61]. On the other hand, woven fabrics fabricated from stretchable fibers are a promising material for woven electrodes [62]. For instance, they could replace the knitted metallized textiles in display technologies as shown in Fig. 11. As use of computer controllers increases in many areas of life, the development of conductive textiles made of flexible polymers should be considered.

Conductive textiles are a promising candidate to protect electronic parts of devices where a small amount of material is required or the total mass should be decreased. If the conductive properties are high enough, they will control static discharge and EMI shielding and monitor flame resistance [63]. One of the newest trends in the creation of conductive composites is the addition of multi-walled carbon nanotubes or graphene instead of metallic particles. On the one hand, this is due to the difference in the amount of fillers required to reach the desired electrical

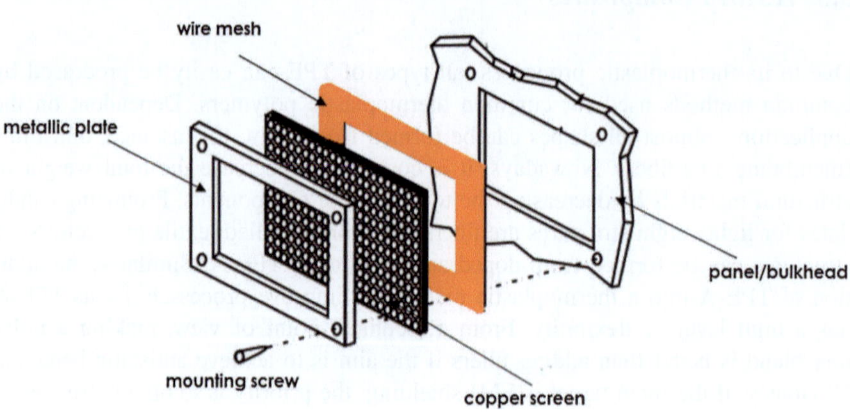

Fig. 11 Metallized textiles used to wires discernible

conductivity. On the other hand, the metal sheets used to shield electronic equipment are heavy and undergo thermal expansion. Undoubtedly, a great advantage is the ease with which various sizes and shapes can be formed from composites based on TPE-A. Therefore, a new idea involves developing micro- or nanotextiles with conductive properties. Depending on the level of conductivity achieved, they can reduce reflection of radar signals, protect from electrostatic discharge, or be used in pressure sensors (piezoresistive fabrics) [64]. To the best of our knowledge, there is lack of research papers describing textile products fabricated from TPE-A with conductive fillers.

6 Conclusions

The suitability of TPE-A as a material for electronic applications is analyzed in this chapter. Until now, it has mainly been used as insulation for jacketing due to its high electrical resistivity, great dielectric constant, and its outstanding flexibility across a broad range of temperatures. TPE-A behaves much better than TPU or TPE-E at low temperatures and does not lose its properties even at 300 °C. Therefore, TPE-A is applied as high-resistance insulation in, for instance, electronic circuits. Moreover, owing to its abrasion resistance and mechanical stability, it is suitable as an insulator in hazardous environments. Commercially available antistatic grades of TPE-A are added to thermoplastic polymers in order to decrease electrical resistivity and improve material processing. Doping TPE-A with conductive particles results in a reduction of electrical resistance which makes it popular as EMI shielding structures, electrodes in actuators, and sensors. These applications are described in this chapter.

TPE-A is still the smallest and the most expensive TPE group with a global production of just 1 %. Based on market analysis and study of the literature, TPE-A has clear advantages in membranes for filtration and separation and also medical devices. In recent years, some papers described mixing TPE-A with conductive fillers, but most of them did not take electrical properties into account. To sum up, the diversity of TPE-A types makes this polymer interesting for the electronic sector, but it is still not often used. Hopefully, in the near future, more work will be done to exploit the potential of composites and nanocomposites based on TPE-A.

References

1. Drobny JG (2014) Handbook of thermoplastic elastomers. Elsevier, Amsterdam
2. Bonart R (1979) Thermoplastic elastomers. Polymer (Guildf) 20:1389–1403. doi: -10.1016/0032-3861(79)90280-5
3. Evaluation P, States U, Report P, The T (2006) PERP program—ethanol new report alert, pp 9–12

4. Fakirov S (2006) Handbook of condensation thermoplastic elastomers. doi:10.1002/3527606610
5. Liu L, Chakma A, Feng X (2004) A novel method of preparing ultrathin poly(ether block amide) membranes. J Membr Sci 235:43–52
6. Ahmadpour E, Shamsabadi AA, Behbahani RM, Aghajani M, Kargari A (2014) Study of CO_2 separation with PVC/Pebax composite membrane. J Nat Gas Sci Eng 21:518–523
7. Armstrong S, Freeman B, Hiltner A, Baer E (2012) Gas permeability of melt-processed poly(ether block amide) copolymers and the effects of orientation. Polymer (Guildf) 53:1383–1392. doi:10.1016/j.polymer.2012.01.037
8. Armstrong SR (2013) Novel applications of Co-extruded multilayer polymeric films
9. Kamal T, Park SY, Choi MC, Chang YW, Chuang WT, Jeng US (2012) An in-situ simultaneous SAXS and WAXS survey of PEBAX nanocomposites reinforced with organoclay and POSS during uniaxial deformation. Polym (United Kingdom) 53:3360–3367
10. Shanks R, Kong I (2012) Thermoplastic elastomers. doi:10.5772/2038
11. Hoffendahl C, Fontaine G, Bourbigot S (2013) Flame retardancy of bio-based polyether-block-amide polymer (PEBAX). Polym Degrad Stab 98:1247–1255
12. De Lannoy CF, Jassby D, Davis DD, Wiesner MR (2012) A highly electrically conductive polymer-multiwalled carbon nanotube nanocomposite membrane. J Memb Sci 415-416:718–724
13. Visakh PM, Thomas S, Chandra AK, Mathew AP (2013) Advances in elastomers I: blends and interpenetrating networks. Springer, Berlin
14. Flesher JR (1986) Pebax® polyether block amide—a new family of engineering thermoplastic elastomers. Atochem Inc., Birdsboro, PA
15. Kucera LR, Brei MR, Storey RF (2013) Synthesis and characterization of polyisobutylene-b-polyamide multi-block copolymer thermoplastic elastomers. Polym (United Kingdom) 54:3796–3805
16. Rabani G, Rosair GM, Kraft A (2003) Low-temperature route to thermoplastic polyamide elastomers, pp 1449–1460
17. Boulares A, Tessier M, Mare E (2000) Synthesis and characterization of poly (copolyethers-block-polyamides) II: characterization and properties of the multiblock copolymers. Water 41:3561–3580
18. Castaldo L, Maglio G, Palumbo R (1978) Synthesis of polyamide-polyether block copolymers. J Polym Sci Polym Lett Ed 16:643–645
19. Acevedo M, Fradet A (1993) Study of bulk chain coupling reactions II: reaction between bisoxazolones and amine-terminated polyether, synthesis of polyether-block-polyamides. J Polym Sci, Part A: Polym Chem 31:1579–1588
20. Sheth JP, Xu J, Wilkes GL (2002) Solid state structure-property behavior of semicrystalline poly (ether-block-amide) PEBAX? Thermoplastic elastomers. Polymer (Guildf) 44:743–756
21. Drobny JG (2012) Polymers for electricity and electronics: materials, properties, and applications. Polym Electr Electron Mater Prop Appl. doi:10.1002/9781118160121
22. Ghosh S, Khastgir D, Bhowmick A (1998) Influence of block molecular weight on the dielectric properties of segmented polyamides. Polym Polym Compos 6:323–330
23. Amin S, Amin M (2011) Thermoplastic elastomeric (TPE) materials and their use in outdoor electrical insulation. Rev Adv Mat Sci 29:15–30
24. Additives P (2009) Wire and cable compounds for extreme end-use conditions. Plast Addit Compd 11:32–35
25. Njuguna J, Pielichowski K, Leszczyńska A (2011) Recent advances in elastomeric nanocomposites. Adv Struc Mat. doi:10.1007/978-3-642-15787-5
26. Pinfa (Cefic) (2009) Innovative flame retardants in E&E applications. Non-halogenated phosphorus, inorganic and nitrogen flame retardants. Innovation 34
27. Choi MC, Jung JY, Yeom HS, Chang YW (2013) Mechanical, thermal, barrier, and rheological properties of poly(ether-block-amide) elastomer/organoclay nanocomposite prepared by melt blending. Polym Eng Sci 53:982–991

28. Lara-estévez JCI, Antônio L, De Almeida S, Schulte K, Bucio E (2012) PEBAX TM-silanized Al_2O_3 composite. Synth Charact 2012:63–69
29. Topp K, Haase H, Degen C, Illing G, Mahltig B (2014) Coatings with metallic effect pigments for antimicrobial and conductive coating of textiles with electromagnetic shielding properties. J Coat Technol Res 11:943–957
30. Buckwalter DJ, Dennis JM, Long TE (2015) Amide-containing segmented copolymers. Prog Polym Sci 45:1–22. doi:10.1016/j.progpolymsci.2014.11.003
31. Wang J, Bao L, Zhao H, Lei J (2012) Preparation and characterization of permanently antistatic packaging composites composed of high impact polystyrene and ion-conductive polyamide elastomer. Compos Sci Technol 72:976–981
32. Sridhar S, Smitha B, Suryamurali R, Aminabhavi TM (2008) Synthesis, characterization and gas permeability of an activated carbon-loaded PEBAX 2533 membrane. Des Monomers Polym 11:17–27
33. Utracki LA, Wilkie CA (eds) (2014) Polymer blends handbook. Springer, Berlin
34. Utracki LA (2013) Commercial polymer blends. Springer, Berlin
35. Pang H, Xu L, Yan DX, Li ZM (2014) Conductive polymer composites with segregated structures. Prog Polym Sci 39:1908–1933
36. Strumpel R, Glatz-Reichenbach J (1999) Feature article conducting polymer composites. J Electroceram 3:329–346
37. McNally T, Potschke P (2011) Polymer–carbon nanotube composites: preparation, properties and applications. ISBN: 978-1-84569-761-7
38. Foulger SH (1998) Electrical properties of composites in the vicinity of the percolation threshold. J Appl Polym Sci 72:1573–1582
39. Surya Murali R, Sridhar S, Sankarshana T, Ravikumar YVL (2010) Gas permeation behavior of PEBAX-1657 nanocomposite membrane incorporated with multiwalled carbon nanotubes. Indian Eng Chem Resour 49:6530–6538
40. Yen HW, Chen ZH, Yang IK (2012) Use of the composite membrane of poly(ether-block-amide) and carbon nanotubes (CNTs) in a pervaporation system incorporated with fermentation for butanol production by clostridium acetobutylicum. Bioresour Technol 109:105–109
41. Yu B, Cong H, Li Z, Tang J, Zhao XS (2013) Pebax-1657 nanocomposite membranes incorporated with nanoparticles/colloids/carbon nanotubes for CO_2/N_2 and CO_2/H_2 separation. J Appl Polym Sci 130:n/a–n/a. doi:10.1002/app.39500
42. Wang S, Liu Y, Huang S, Wu H, Li Y, Tian Z, Jiang Z (2014) Pebax-PEG-MWCNT hybrid membranes with enhanced CO_2 capture properties. J Memb Sci 460:62–70
43. Bae WS, Kwon OJ, Kim BC, Chae DW (2012) Effects of multi-walled carbon nanotubes on rheological and physical properties of polyamide-based thermoplastic elastomers. Korea Austria Rheol J 24:221–227
44. Gonçalves V, Brandão L, Mendes A (2014) Development of porous polymer pressure sensors incorporating graphene platelets. Polym Test 37:129–137
45. Ravati S, Favis BD (2011) 3D porous polymeric conductive material prepared using LbL deposition. Polymer (Guildf) 52:718–731
46. Swann JMG, Topham PD (2010) Design and application of nanoscale actuators using block-copolymers. Polymers (Basel) 2:454–469
47. O'Halloran A, O'Malley F, McHugh P (2008) A review on dielectric elastomer actuators, technology, applications, and challenges. J Appl Phys 104:1–10
48. Shankar R, Ghosh TK, Spontak RJ (2007) Dielectric elastomers as next-generation polymeric actuators. Soft Matter 3:1116
49. Kim B, Park YD, Min K, Lee JH, Hwang SS, Hong SM, Kim BH, Kim SO, Koo CM (2011) Electric actuation of nanostructured thermoplastic elastomer gels with ultralarge electrostriction coefficients. Adv Funct Mater 21:3242–3249
50. Fraden J (2010) Handbook of modern sensors: physics, designs, and applications, 2nd ed. Am J Phys. doi:10.1119/1.18801

51. Igreja R, Marat-Mendes JN, Dias CJ (2002) Dielectric characterization of PEBA and PDMS for capacitive interdigital vapour sensors. In: Proceedings 11th International Symposium on Electrets (pp 1–4)
52. Kobyashi N (2002) United States Patent, US 6343295 B1
53. Feller JF, Langevin D, Marais S (2004) Influence of processing conditions on sensitivity of conductive polymer composites to organic solvent vapours. Synth Met 144:81–88
54. Liang S, He X, Wang F, Geng W, Fu X, Ren J, Jiang X (2015) Highly sensitive humidity sensors based on LiCl–Pebax 2533 composite nanofibers via electrospinning. Sens Actuators B Chem 208:363–368
55. Spitalsky Z, Tasis D, Papagelis K, Galiotis C (2010) Carbon nanotube-polymer composites: chemistry, processing, mechanical and electrical properties. Prog Polym Sci 35:357–401
56. Haider S, Khan Y, Almasry W a, Haider A (2011) Thermoplastic nanocomposites and their processing techniques
57. Gopalan AI, Marquis FDS, Min BG, Chae HG, Minus ML, Kumar S, National K, Engineering F (2009) Polymer/carbon nanotube composite fibers—an overview: composites. doi:10.1002/pc.20058
58. Liu Y, Kumar S (2014) Polymer/carbon nanotube nano composite fibers—applied materials and interfaces, a review
59. Latko P, Kozera R, Salinier A, Boczkowska A (2013) Non-woven veils manufactured from polyamides doped with carbon nanotubes. Text Fibres Eastern Eur 6:45–49
60. Lowell J, McIntyre JE (1978) Antistatic fibres in fabrics and carpets. J Electrostat 4:267–282
61. Cerovic DD, Asanovic KA, Maletic SB, Dojcilovic JR (2013) Comparative study of the electrical and structural properties of woven fabrics. Compos Part B Eng 49:65–70
62. Ma R, Lee J, Choi D, Moon H, Baik S (2014) Knitted fabrics made from highly conductive stretchable fibers. Nano Lett 14:1944–1951
63. Wallace GG, Campbell TE, Innis PC (2007) Putting function into fashion: organic conducting polymer fibres and textiles. Fibers Polym 8:135–142
64. Brzeziński S, Rybicki T, Karbownik I, Malinowska G, Śledzińska K (2012) Textile materials for electromagnetic field shielding made with the use of nano- and micro-technology. Cent Eur J Phys 10(5):1190–1196. doi:10.2478/s11534-012-0094-z

Electronic Applications of Polyacrylic Rubber and Its Composites

Elnaz Esmizadeh, Ghasem Naderi and Ali Vahidifar

Abstract In this chapter, we are going to look at the applications of polyacrylic rubber (ACM) and its composites in electrical industry. The brief introduction to ACM rubbers, its structure, and properties is also included in this chapter. The electrical applications of ACM rubbers are given in two sections: (i) the electrical applications of pristine ACM rubbers; (ii) the electrical applications of ACM composites. In their pristine state, ACM rubbers are widely used as electrical insulators because the atoms in the rubber chain are covalently linked. Furthermore, ACM rubbers are categorized as a dielectric electroactive polymer (EAP), showing the unique electromechanical properties. The most significant characteristic for ACM to our EAPs such as silicone is its ability to undergo large deformation (maximum actuation strains of more than 380 %) and also withstand high elongations, more than 800 %. The most famous application of ACM rubbers as EAP is introduced as "natural muscle" for use in advanced robotics and smart prosthetics. In addition, other applications such as actuators and sensors in optical switches, motors, generators, loudspeakers, as well as haptic and variable-stiffness devices are discussed. This chapter provides the idea of development of composite ACM rubbers which causes the unique versatility of ACM rubbers to manifest extensively. For example, incorporation of carbon black and multiwall carbon nanotube as a conductive filler is discussed as a convenient method to form an interconnected carrier path in insulating ACM rubbers. The conductive ACM composite materials can be widely utilized in electromagnetic (EMI) shielding materials. Micrometer-sized polarizable particles, e.g., piezoelectric particles, can be added to ACM rubbers to make electrorheological (ER) materials. ACM rubbers as ER materials have several applications, such as actuator, MEMS devices, and control systems.

E. Esmizadeh · G. Naderi (✉)
Faculty of Polymer Processing, Iran Polymer and Petrochemical Institute,
P.O. Box 14965/115, Tehran, Iran
e-mail: G.Naderi@ippi.ac.ir

A. Vahidifar
Department of Chemical Engineering, Isfahan University of Technology,
8415683111 Isfahan, Iran

D. Ponnamma et al. (eds.), *Flexible and Stretchable Electronic Composites*,
Springer Series on Polymer and Composite Materials,
DOI 10.1007/978-3-319-23663-6_6

Keywords Polyacrylic rubber · Electrical applications · Dielectric elastomers

1 Introduction

Polyacrylic rubber (ACM) is a copolymer of ethyl, or other, acrylate backbone (95–99 %) and a small amount of a cure site monomer (1–5 %) such as vinyl chloroacetate monomer which facilitates vulcanization process [1, 2] (Fig. 1). ACM is rubbery polymer which contains polar groups to impart hydrocarbon resistance but do not contain reactive carbon-to-carbon double bonds. ACM as a special-purpose rubber illustrates good resistance to mineral oils, oxygen, ozone, and aging, and also performs well at high temperatures. The operating temperature

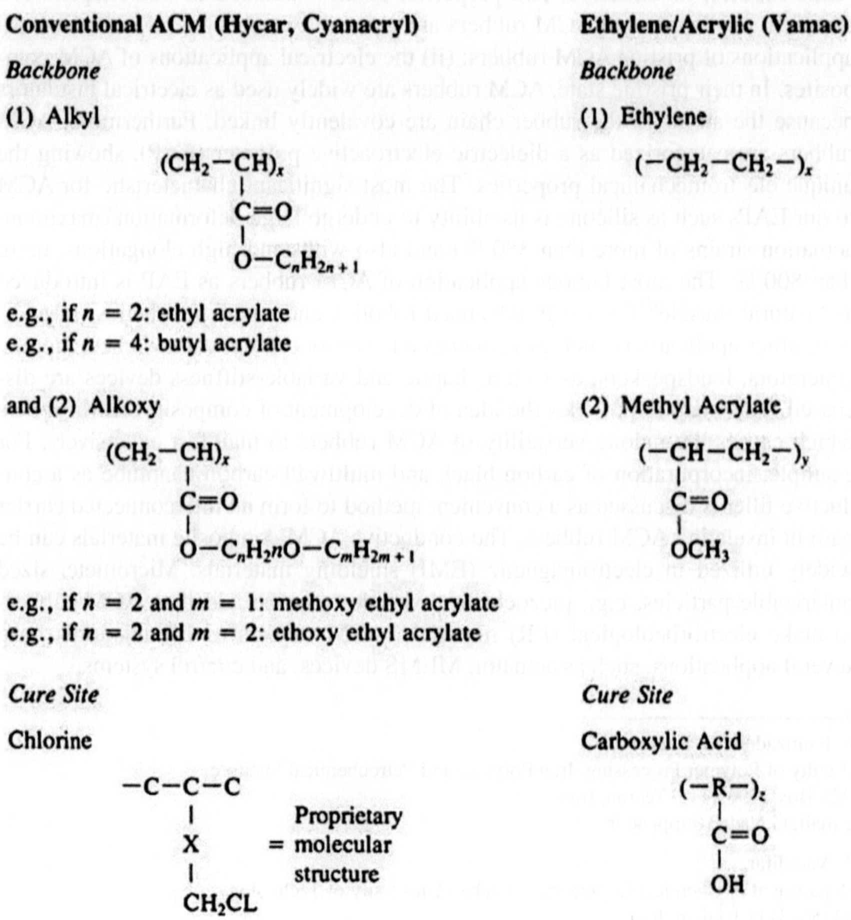

Fig. 1 Major components of ACM rubbers [3]. Copyright 1999. Reproduced with permission from Springer

range for ACM rubbers lies between −40 and +150 °C (−40 and +300 °F), while in some fluids, the upper limit may be extended to 175 °C (345 °F). For this reason, ACM rubbers are used for shaft seals in automotive applications [1, 2]. The water compatibility and cold flexibility of ACM rubbers are better than NBR. These rubbers are frequently used as the alternative material to silicone [2].

ACM rubbers are cured using a system which is typically based on a curative, an accelerator, an activator, and a retarder. An example formulization is as follows: (i) curative, 2,5-dimercapto-1,3,4-thiadiazole at approximately 1 phr; (ii) accelerator, tetrabutylthiuram disulfide at approximately 2.5 phr; (iii) activator, such as lead stearate at approximately 1 phr; and (iv) retarder, such as zinc stearate at approximately 1 phr [1].

ACM rubbers are highly resistant to petroleum-based lubricating oils, transmission fluids, and greases, including the sulfur-bearing and extreme-pressure gear lubricants. Butyl acrylate and ethylene-/acrylic-based polymers are suitable for the applications that enquire high-oil-swell characteristics. ACM rubbers are not recommended for servicing in aromatic hydrocarbons, e.g., gasoline. The most of conventional ACM rubbers are not suitable for continuous use in hot water or steam, or in water-soluble materials such as methanol or ethylene glycol. But the ethylene/acrylic type is recommended for such a service [2].

2 History

Unmodified ACM rubbers were developed in the early 1940s by the BF Goodrich Company. The first commercial products of ACM rubbers were manufactured and marketed by BF Goodrich Chemical in 1948. Since then, considerable advances have been accomplished in the low-temperature capability and cure technology of ACM rubbers by the BF Goodrich Chemical Group and subsequent other suppliers. The most famous manufacturer of ACM rubbers are listed in Table 1 [3].

Table 1 World supplier of ACM rubber [3]

Company	Country	Trade name
BF Goodrich Chemical Group	USA	Hycar
American Cyanamid Company	USA	Cyanacryl
E. I. DuPont DeNemours & Co., Inc.	USA	Vamac
Enichem	Italy	Europrene AR
Nippon Zeon Co., Ltd. (NZ)	Japan	Nipol
TOA Paint Co., Ltd.	Japan	Acron
Nippon Oil Seal-NOK	Japan	Noxtite
Japan Synthetic Rubber Co., Ltd.	Japan	JSR AR

Copyright 1999. Reproduced with permission from Springer

3 Electrical Application of Pristine ACM Rubbers

3.1 ACM Rubbers as the Sealing Rubber for Electric Motors

Rubber seal for extreme temperature application is the primary application of ACM rubbers. NSK Inc. designed the non-contact V-shaped ACM seal for the higher speed, higher temperature application environment such as electric motors. The schematic of the designed non-contact V-shaped ACM seal is given in Fig. 2. The patented V-shaped seal gave an excellent sealing performance without friction for increased bearing life and reduced load on the motor. The V-shaped seal performed better sealing capability than a shield. The non-contact lip decreased the drag in the bearing. It is very important in small electric motors where power loss is critical. Furthermore, the speed capability of V-shaped seals is comparable to the shielded bearing. The ACM V-shaped seals are applicable in the temperature range of −40 to 170 °C [4].

In automobile industry, ethylene acrylic rubber (AEM) is commonly used as an inner tube of the two-ply rubber hose for automotive fuel line, which is less resistant to sour gasoline but is superior in cold resistance [5].

3.2 ACM Rubbers as the Dielectric Electroactive Polymers

In the last decades, electroactive polymers (EAPs) are introduced as the new high-performance actuator materials. EAPs exhibit a significant change in size or shape when stimulated by an electric field. Conventionally, several classes of materials including piezoelectric polymer, carbon nanotubes, and piezoelectric single-crystal ceramics have been considered as the suitable materials for actuators. Unlike these rigid and often fragile inorganic counterparts, EAPs are of particular interest because of exhibiting large dimensional changes as well as their low cost,

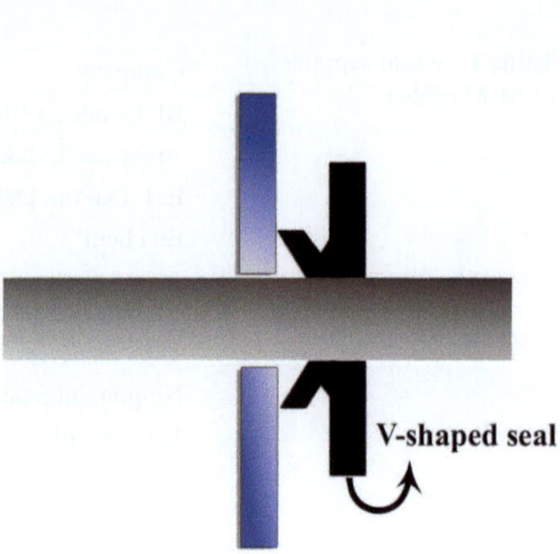

Fig. 2 A typical V-shaped seal (non-contact) applicable in low torque

lightweight, flexibility, toughness, and shape processability [6–8]. In some cases, the electroresponsive properties of EAPs impart lifelike movements of a biological muscle, which explains why EAPs are often referred to "artificial muscles" [7].

A typical electrical actuator based on the dielectric EAPs is made of an EAP film, e.g., ACM rubber or silicone, which is coated on both side with compliant electrode material [6]. When a voltage difference is applied on the top and bottom electrode (external electrical field), the ACM film shrinks in thickness and expands in area due to the electrostatic forces (Fig. 3). The resultant deformation of the ACM film originates from the electrostatic forces of the free charges on the two compliant electrodes [6]. This in-plane expansion is exploited to generate motion or work as an actuator. The performance of an ACM actuator/generator is critically depends on the electrical and mechanical properties of the polymer. In the other words, the macroscopic permittivity of the ACM rubber as well as on its modulus of elasticity determines its performance [9].

The effective actuation pressure, p, produced by the electrodes on the ACM film as a function of the applied voltage, V, can be given with Eq. (1).

$$p = \varepsilon_r \varepsilon_0 E^2 = \varepsilon_r \varepsilon_0 (V/t)^2 \tag{1}$$

where ε_r and ε_0 are the relative dielectric constant of the polymer and the permittivity of free space (8.85×10^{-12} F/m); E is the applied electric field (V/m); V is the applied voltage; and t is the film thickness [10]. Pelrine et al. showed that for low strain deformations (<20 %), the thickness strain, s_z, can be calculated by Eq. (2).

$$s_z = -p/Y = -\varepsilon_r \varepsilon_0 E^2 \Big/ Y \tag{2}$$

where Y is the modulus of elasticity of the polymer film. In this equation, the ACM rubber is assumed to be an ideal rubber with Poisson's ratio of 0.5. The effect of the electrical and mechanical properties, on actuation performance of ACM rubber, can be simply observed in the equation. Equation (2) is not applicable for high strain deformations, $s_z > 20\,\%$, because Y is not a constant parameter and it is generally depends on the strain value [6, 11]. The elastic strain energy per unit

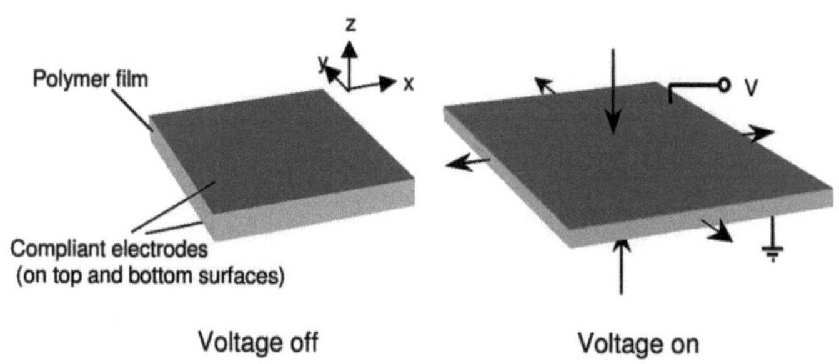

Fig. 3 The dielectric EAP film in an electrical field [10]. Copyright 2002. Reproduced with permission from SPIE

volume (u_e, elastic strain energy density) is the most common parameter for comparing output capacities of actuator materials in low strains. u_e can be calculated by Eq. (3).

$$u_e = 1/2ps_z = 1/2Ys_z^2 \tag{3}$$

.

For high strain values, another parameter is introduced to measure the performance of the actuator materials as the electromechanical energy density, *e*. This parameter shows the amount of electrical energy which was converted to mechanical energy per unit volume of the actuator material in a cycle and can be written as Eq. (4).

$$e = -p\ln(1+s_z) = -Ys_z\ln(1+s_z) \tag{4}$$

Expanding the logarithm above, $u_e = 1/2e$ is obtained in small strain values [6]. Typical thickness or planar strain response of EAP actuators to the applied electric field for a film with no external load is shown in Fig. 4 [10].

As mentioned before, in an electrical actuator based on ACM rubber, the elastomeric film is sandwiched between two compliant electrodes. Common electrodes for EAPs include carbon black, conducting grease, powder, glue or rubber, patterned metal, liquid, and ion-implanted metal. The appropriate electrode material should be highly conductive, compliant in at least one direction, as well as uniform to prevent local breakdown [12]. Goulbourne et al. conducted an experimental evaluation to find the most compliant electrode material for ACM rubber at large strains. The ideal electrode should not affect the stress–strain behavior of the ACM rubber during actuation. Figure 5 shows the micrograph and the surface of the ACM rubber covered with different type of electrode material in a pressurized diaphragm configuration. Gaps in the electrode surface are observed for graphite powder, and graphite spray electrodes indicate that they are not good compliant

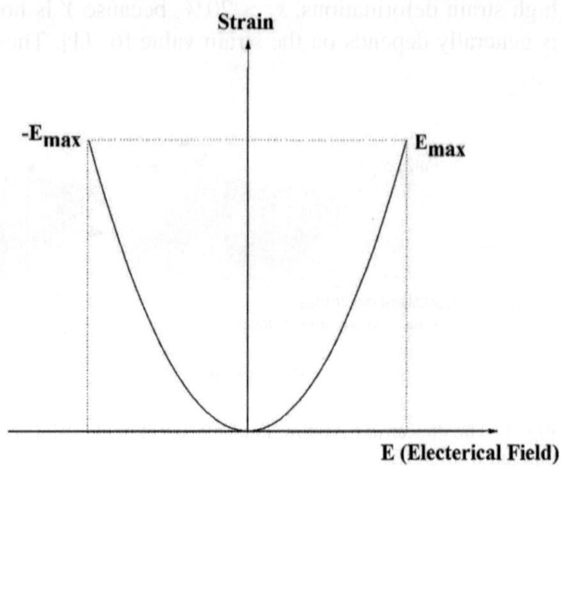

Fig. 4 Simple principle of operation for dielectric elastomer actuators in electrical field [10]. Copyright 2002. Reproduced with permission from SPIE

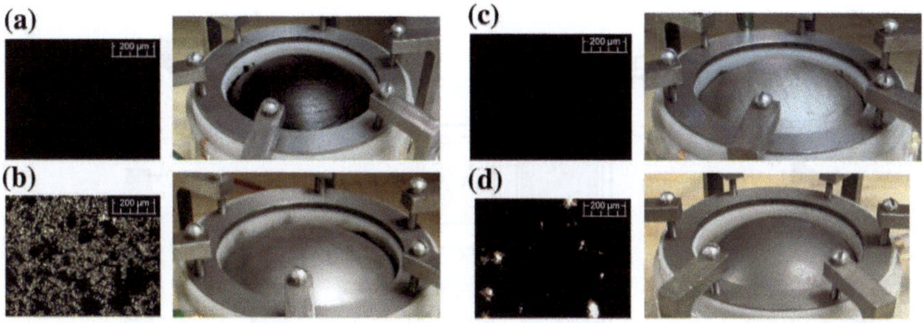

Fig. 5 Electrode surface and the micrographs taken from ACM rubber covered with different electrodes **a** carbon grease, **b** graphite powder, **c** silver grease, and **d** graphite spray [12]. Copyright 2007. Reproduced with permission from SPIE

electrodes for ACM rubber. It can be concluded from the micrographs that carbon grease and silver grease are the best compliant electrodes illustrating better performance with smooth surface [12].

A polymer film sandwiched between two electrodes as the basic element of an actuator (Fig. 3) can be incorporated into a wide variety of configuration. Some configurations are analogous to conventional piezoelectric configurations such as tube form or extender one. Other configurations take advantage of the unique capabilities and flexibility of EAPs. For example, they can be rolled into a scroll, supported by a frame, or laminated to a flexible substrate to produce bending [13]. Some various configurations for EAP-based actuators are given in Fig. 6. The most efficient configuration should be selected depending on the application and the essential dimensions of the process. For example, both directions of deformation are employed in the diaphragm configuration. The framed configuration is designed when two-dimensional process is essential. Other configurations, such as the spider and bowtie, couple both planar directions into a single linear direction [10]. Unimorph and bimorph configurations are potentially applicable for low-force robotic elements such as grippers. High bending angles and high strains can be achieved using ACM rubber as a dielectric film of the actuator. Rolled configuration is suitable for robotic legs and fingers as well as high-force grippers [13].

Conventional dielectric elastomers, silicones, demonstrated actuated strains of 32–41 % at actuation pressure of 0.1–2 MPa with specific elastic energy densities up to 0.038–0.15 J/cm^3 [11]. Nowadays, the commercially available ACM rubber, VHB 4910 (3M), is the most commonly used material to produce muscle-like actuators. It is due to ACM's large electromechanical strains, as well as its ability to produce high actuation pressure and extraordinarily high-specific elastic energy density [9]. The combination of high energy density, good efficiency, and high actuation strain is unique and suggests ACM rubber as a promising material for wide range of applications. Pelrine et al. proposed the pre-straining as an effective approach to further improve the actuation performance of the EAPs. This means that high amount of pre-strain that can be produced by stretching and holding the

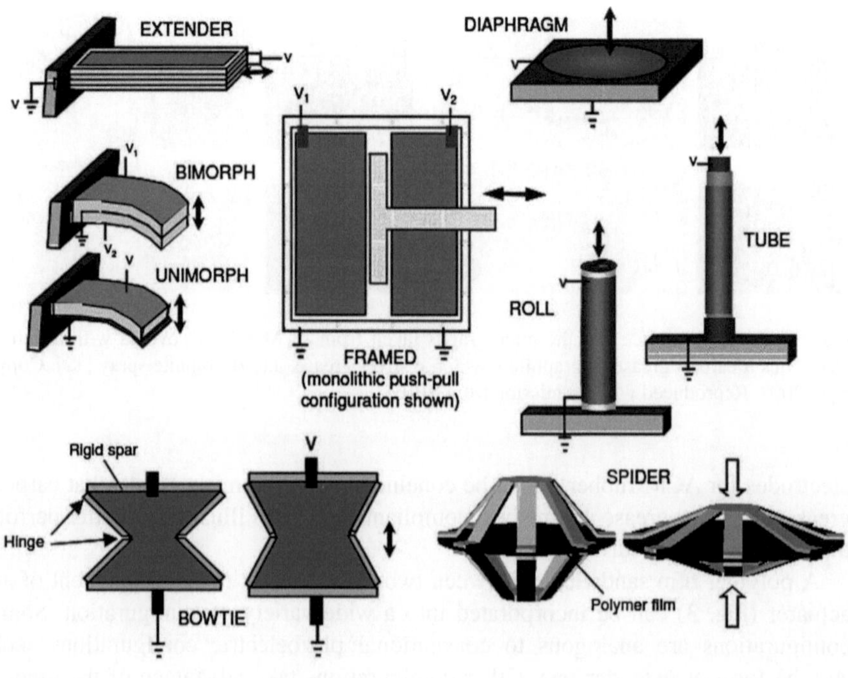

Fig. 6 Various configurations for dielectric elastomer actuators [10]. Copyright 2002. Reproduced with permission from SPIE

EAP film under tension can increase the maximum pressure and strain of them. The circular and linear strain test was accomplished on silicon and ACM rubber as illustrated in Fig. 7. In electrical field, the strain response of the EAP film was measured by optically measuring the change in area [14].

The results obtained for silicone and ACM rubber in both circular and linear strain test are given in Table 2 [6]. Actuated strains up to 215 and 158 % were observed for ACM rubbers which were pre-strained biaxially and uniaxially, respectively [6]. ACM rubber showed excellent overall performance including higher actuated thickness and area strain as well as high elastic strain energy density. It can be concluded that ACM rubber appears more attractive than its competitive elastomer (silicone) [14].

When the applied voltage is raised very much in VHB 4910 film, the mechanical instability is reached. This mechanical instability can be seen in the form of buckling or vertical wrinkles in Fig. 7b. When the electrical-induced stress exceeds the elastic stress needed to strain the ACM film, the film starts to wrinkle with ripples in the width direction of the sample. The voltage at which the wrinkling happens is determined as the maximum voltage can be applied to the ACM film [15]. The maximum electric field that the ACM film can withstand is known as the electrical breakdown. It is critical to determine the limitations on the applied voltage before application of the ACM film. Because the electric

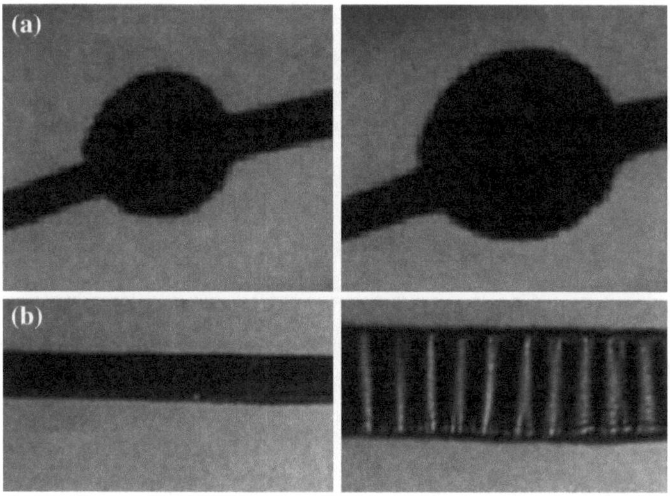

Fig. 7 **a** The circular strain test and **b** the linear strain test of the EAP film [14]. Copyright 2000. Reproduced with permission from SPIE

Table 2 The effect of pre-strain on circular and linear strain test results of silicone and acrylic rubber [14]

Material	Pre-strain value x, y (%, %)	Pre-strain mode	Actuated thickness strain (%)	Actuated area strain (%)	Electrical field strength (MV/m)	Effective compressive stress (MPa)	Elastic strain energy density (MJ/m^3)
VHB 4910 acrylic	540, 75	Linear	68	215	239	2.4	1.36
	300, 300	Circular	61	158	412	7.2	3.4
	15, 15	Circular	29	40	55	0.13	0.022
HS3 silicon	280, 0	Linear	54	117	128	0.4	0.16
	68, 68	Circular	48	93	110	0.3	0.098
	14, 14	Circular	41	69	72	0.13	0.034
CF 19-2186 silicone	100, 0	Linear	39	63	181	0.8	0.2
	45, 45	Circular	39	64	350	3.0	0.75
	15, 15	Circular	25	33	160	0.6	0.091

Copyright 2000. Reproduced with permission from SPIE

breakdown is fatal and when an actuator "burns," it is beyond repair [15]. When the electrical filed is very high (beyond electrical breakdown), a catastrophic avalanche spark will occur. In ACM rubber, electrical breakdown event is fatal for two main reasons: (i) The environment around the spark becomes conductive and (ii) the cracks in the spark can propagate very easy causing the whole film to rip. The improvement in the breakdown electrical field and voltage of ACM rubber with

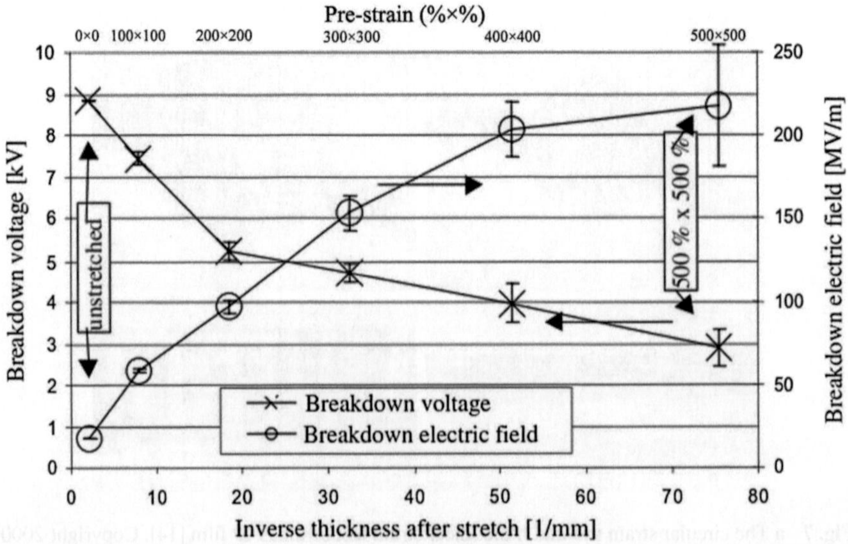

Fig. 8 The effect of isotropic pre-strain on breakdown voltage and electrical field of ACM rubber [15]. Copyright 2001. Reproduced with permission from SPIE

pre-straining is proved by Kofod et al. [15]. For the experiment, the ACM film was inserted between the clamps and the applied voltage was increased gradually until electrical breakdown occurred. The experimental data obtained for ACM rubber in different pre-strain values are given in Fig. 8. For example, the film with an isotropic pre-strain of 500 % × 500 % has increased (1 + 500 %) × (1 + 500 %) = 36 times in area. The trend of the data in Fig. 8 showed that the increasing the pre-strain value of ACM rubber from 0 to 500 % increased the electrical breakdown strength from 18 to 218 MV/m [16]. The increase in breakdown electrical field with increasing the pre-strain value is consistent with what is reported for the ACM rubber by Shankar et al. [8]. The observed phenomenon can be related to the orientation of the polymer chains in the plane of the ACM film caused by the pre-strain [16].

To eliminate the requirement of high pre-strain, Zhang et al. proposed two other approaches to enhance the actuation performance of VHB 4910 ACM rubber. In the first approach, interpenetrating polymer networks (IPNs) based on poly (trimethylolpropane trimethacrylate) were formed in the original acrylic network. In the second approach, plasticizers were employed to reduce the glass transition temperature (T_g), broaden the operating temperature range, decrease the elastic modulus (hence decrease the electric field required for actuation), and reduce the viscoelastic loss (hence increase response speed). The effect of two different plasticizing additives, bis 2-ethylhexyl phthalate (DOP) and dibutoxyethoxyethyl formal (DBEF), on actuation performance of ACM rubber was investigated by Zhang et al. [17].

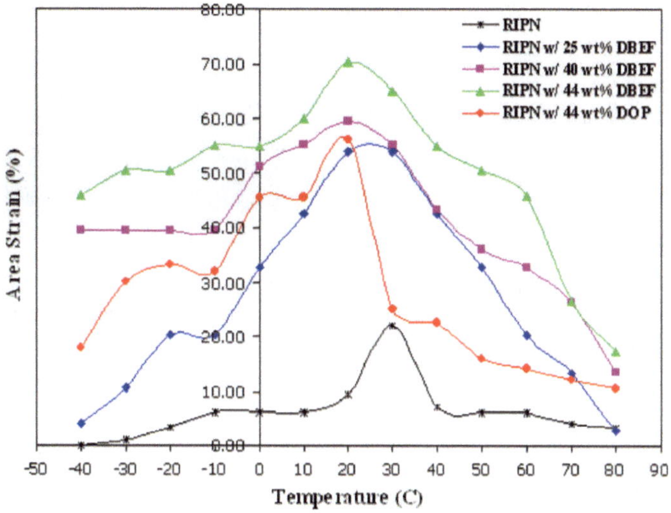

Fig. 9 The effect of different plasticizer on functional temperature range of relaxed ACM rubber IPNs (RIPN) at a constant voltage of 5 kV [17]. Copyright 2010. Reproduced with permission from John Wiley & Sons

Broadening of functional temperature range of ACM rubber IPNs in the presence of plasticizer can be seen in Fig. 9. All ACM rubber samples showed a peak of area strain at around 20–30 °C. A decrease in strain was observed for all the samples as the temperature deviates from 20 to 30 °C. In particular, un-plasticized ACM rubber IPNs showed a steep reduction at either temperature extreme with almost no response to electric field. It can be seen from the figure that higher DBEF concentration resulted in more pronounced deformation at whole range of temperature. Furthermore, at the same plasticizer content (44 wt%), DBEF was the more effective plasticizing agent compared to DOP. The DBEF increased the deformation of ACM rubber IPNs in the entire temperature range, while the DOP was not so effective at high temperatures $T > 30$ °C [17].

The actuation behavior of ACM rubber IPNs plasticized with 40 wt% of DBEF under electric voltage of 6.5 kV is shown in Fig. 10. An area strain of 215 % is achieved for ACM rubber IPNs for the mentioned condition (Fig. 10b).

According to the data given in Table 2, the major shortcoming of ACM rubber comparing with other common EAP, silicone, is its low speed of response due to viscoelasticity. It means that the actuated area strain significantly decreases as frequency increases [17]. For example, the actuated area strain of un-plasticized ACM rubber IPN decreases to half the original strain at 2 Hz. To overcome this disadvantage, some chemical and processing modifications have been proposed [18]. For instance, plasticization of ACM rubber IPNs was reported to be an effective approach to increase the response speed of them to electrical field (Fig. 11).

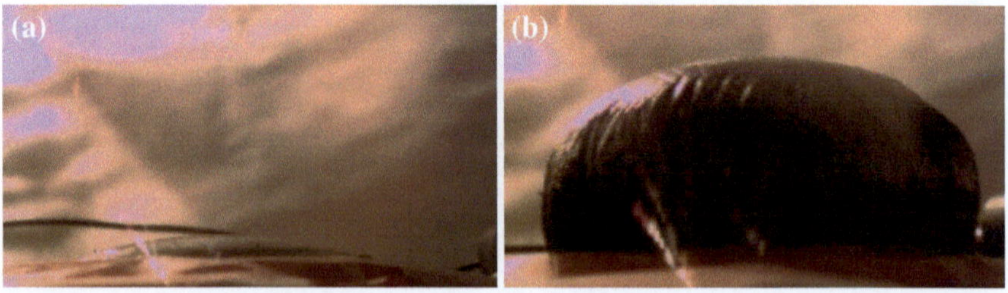

Fig. 10 Actuation of ACM rubber IPN plasticized with 40 wt% DBEF at −40 C, **a** voltage off, **b** voltage on, 6.5 kV (215 % areal strain) [17]. Copyright 2010. Reproduced with permission from John Wiley & Sons

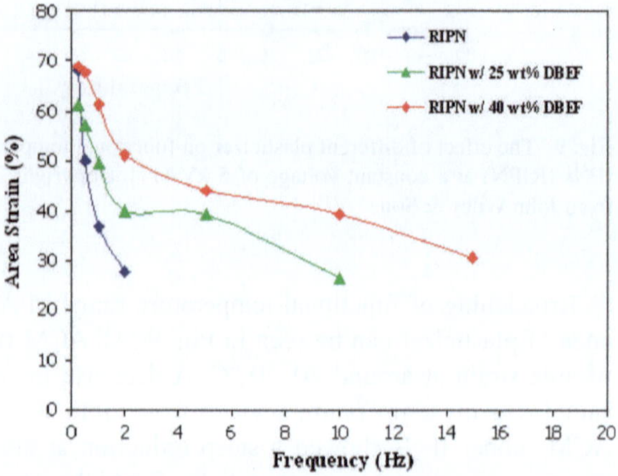

Fig. 11 Effect of plasticizer on the response speed of ACM rubber IPNs [17]. Copyright 2010. Reproduced with permission from John Wiley & Sons

With addition of DBEF plasticizer, the decrease in strain becomes more moderate at high frequencies. For the ACM rubber IPNs plasticized by the 40 wt% of DBEF, the decrease to half original strain was happened at 15 Hz [17].

The presented advantageous characteristics of ACM rubbers over other EAPs as well as inexpensiveness, good resilience, and low swelling in water offer unique performance and application capabilities of the ACM rubbers. A few common and also potential transducer and actuator applications of poly ACM rubber are highlighted in the following.

3.2.1 Artificial Muscles for Robots

Due to the fact that the ACM rubber-based actuators show a stress–strain behavior similar to that of natural muscle, these actuators are often called "artificial muscle." Table 3 serves a basic qualitative comparison of candidate artificial muscle technologies including dielectric EAPs with natural human muscle [10]. ACM

Table 3 Comparison of candidate artificial muscle technologies including dielectric EAPs with natural muscle [10]

Actuator type	Maximum strain (%)	Maximum pressure (MPa)	Specific elastic energy density (J/g)	Specific elastic energy (J/cm^3)	Maximum efficiency (%)	Relative speed (full cycle)
Dielectric elastomer						
ACM rubber	380	7.2	3.4	3.4	60–80	Medium
Silicone rubber	63	3.0	0.75	0.75	90	Fast
Electrostrictive polymer						
PVDF-TrFE[a]	4.3	4.3	0.49	0.92	~80	Fast
Electromagnetic						
Voice coil	50	0.10	0.003	0.025	>90	Fast
Piezoelectrics						
PZT[b] ceramic	0.2	110	0.013	0.10	90	Fast
PZN-PT[c] single crystal	1.7	131	0.13	1.0	90	Fast
PVDF polymer	0.1	4.8	0.0013	0.0024	~80	Fast
Shape memory alloy						
TiNi[d]	>5	>200	>15	>100	<10	Slow
Shape memory polymer						
PU[e]	100	4	2	2	<10	Slow
Electrochemo-mechanical conducting polymer						
Polyaniline	10	450	23	23	<5%	Slow
Mechano-chemical polymer/gels						
Polyelectrolyte	>40	0.3	0.06	0.06	30	Slow
Magnetostrictive						
Terfenol-D	0.2	70	0.0027	0.025	60	Fast
Natural human muscle	>40	0.35	0.07	0.07	–	Medium

[a]Polyvinylidene fluoride-trifluoroethylene
[b]Lead zirconate titantate
[c]$Pb(Zn_{1/3}Nb_{2/3})O_3–PbTiO_3$
[d]Metal alloy of nickel and titanium, also known as nitinol
[e]Polyurethane

rubbers have demonstrated an estimated 3.4 J/cm^3 specific energy density, which is incomparably greater than that of piezoelectric materials [19]. It is encouraging enough to suggest that ACM rubbers can be used competitively in energy-efficient actuators [13]. As it can be seen in Table 2, the most significant characteristic for ACM rubbers are their ability to undergo large deformation comparing to the other counterparts. ACM rubbers have achieved maximum actuation strains of more than 380 %. According to the table, no other material comes close to this strain in the applied electrical field. Furthermore, they can also withstand high elongations, more than 800 % [10].

The similarities between ACM rubber as the artificial muscle and natural muscle are shown in Fig. 12 in more details. Strain/stress and network of natural muscle as a function of frequency were gathered from data on oscillatory contractions. The increase in the frequency of the contractions yields the network output decrease (10-fold for a 100-fold increase in frequency). The decrease in network values results in a decrease in stress and strain of the natural muscle. As it can be seen in the figure, the network and strain/stress behavior of the ACM rubber as the artificial muscles are similar to those of natural muscle [20].

Besides conventional actuation, strain/stress behavior, natural muscle has a number of functions that must be considered in designing an elastomer actuator for artificial muscle application. For example, in the case of locomotion and other biological motions, viscoelastic behavior of the natural muscle helps organisms achieve energy-efficient locomotion. These viscoelastic properties include the compliance, elastic modulus, and viscoelastic damping of natural muscle. Figure 13 illustrates a comparison of the viscoelastic properties of ACM rubber as

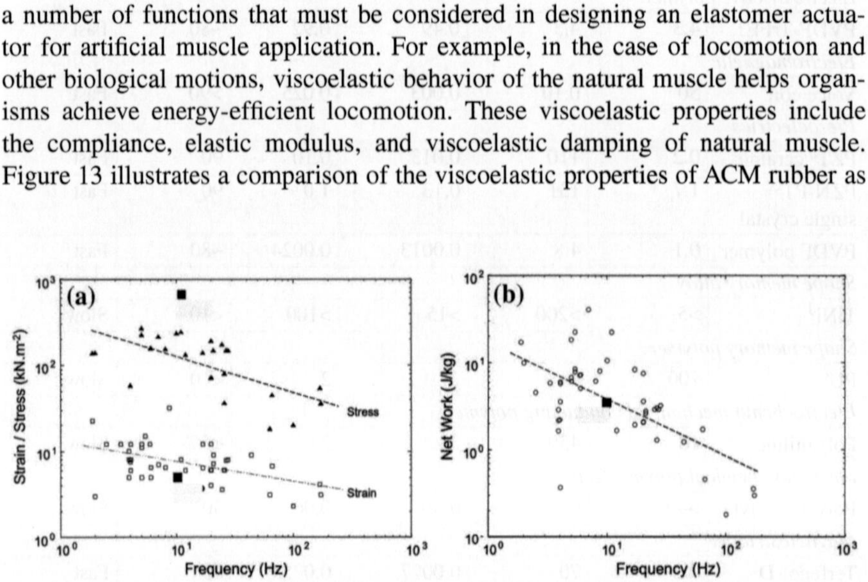

Fig. 12 Comparison of the behavior of natural muscle and artificial muscle (ACM rubber) as the function of frequency: **a** stress/strain of natural muscle, artificial muscle; **b** network of natural muscle and artificial muscle [20]. Copyright 2004. Reproduced with permission from SPIE

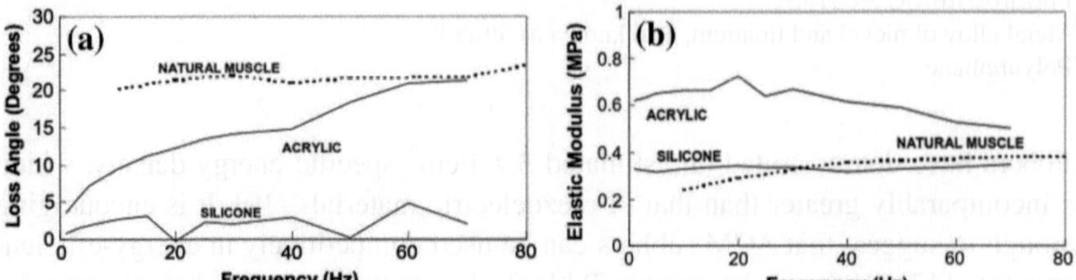

Fig. 13 Comparison of viscoelastic properties of natural muscle (cockroach leg muscle) and artificial muscle (ACM rubber) **a** loss angle; **b** storage modulus [20]. Copyright 2004. Reproduced with permission from SPIE

the artificial muscle with those of a cockroach leg muscle as the natural muscle. It can be seen that the viscoelastic damping and elastic modulus of ACM rubber are similar to that of the natural muscle in high frequencies [20].

Considering these similarities, ACM rubber-based actuators were proposed to be suited for application to biologically inspired robots or biomedical applications such as lifelike prosthetic limbs, exoskeletons for the mobility impaired, and even artificial hearts [20]. Examples of ACM rubber actuators used as the artificial muscles are illustrated in Fig. 14. Figure 14a shows a linear actuator made from rolled ACM rubber actuator with approximately 15 g weight. This actuator can produce forces of more than 5 N with actuation strain of about 25 % [10]. Figure 14b shows the application of this linear actuator as an artificial muscle for the arm of a human skeleton. Although the artificial muscle shown here cannot match the performance of a natural muscle, it can be a promising candidate after some developments in future [20]. Double bow-tie actuator based on ACM rubber is illustrated in Fig. 14c. This kind of actuator can be employed in insect-inspired walking robot as shown in Fig. 14b [21].

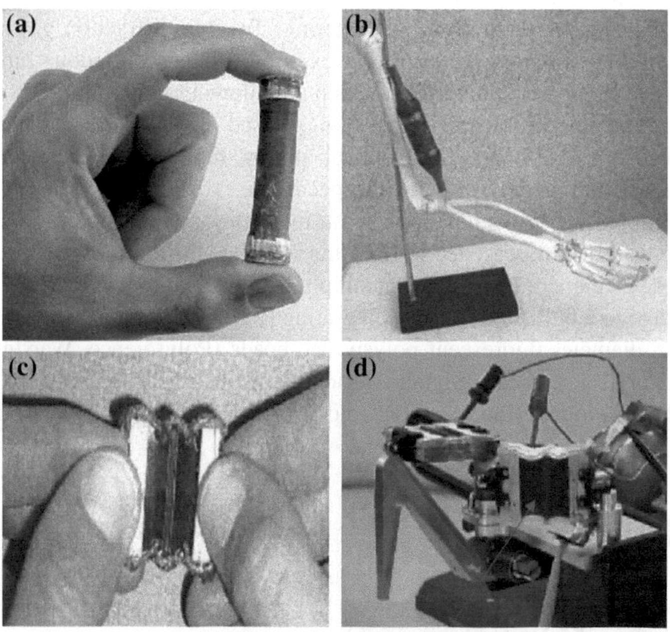

Fig. 14 **a** "Spring roll" artificial muscle actuator [10]. Copyright 2002. Reproduced with permission from SPIE; **b** "spring roll" actuator attached to a model of the arm of a human skeleton; **c** "double bow-tie" artificial muscle actuator [20]. Copyright 2004. Reproduced with permission from SPIE; **d** "Double bow-tie" actuator in insect-inspired walking robot [21]. Copyright 2004. Reproduced with permission from SPIE

3.2.2 Optical Switches, Micro-lenses, and Tunable Transmission Gratings

The framed ACM rubber actuator can be employed for designing the optical switches or micro-lenses due to their controllable variation in optical density or refractive index. These parameters change significantly as the actuator thickness changes with the variation of applied electrical field [7]. ACM rubber actuators are potentially optical switches because the optical density of the area of the ACM rubber film varies during actuation. Figure 15 shows that when the actuation occurs in the ACM rubber film, the optical density changes from non-transparent dark to semitransparent. The change in the optical density is the direct result of stretching expansion of the film [22].

Unique features of the ACM rubber film, large strain capability as well as transparency, are exploited in the design of optical switches or opto-mechanical devices. The resultant apparatus is simple, inexpensive, solid state and contains just one moving part, the ACM rubber film [10].

Aschwanden et al. demonstrated that ACM rubber actuators can be employed to fabricate highly tunable transmission gratings which further can be used for light steering, wavelength separation, and filtering. Tunable diffraction gratings based on ACM rubber actuators are relatively exerting a large continuous tuning range comparing to their stiff counterparts which are allowing only a limited spatial tuning range. The optical transmission gratings based on ACM rubbers operate with high transmission (>90 %), good optical quality, high optical damage threshold (>93 kW/cm), and moderate switching speed. The excellent properties of the ACM rubbers enabled the implementation of an inexpensive beam steering unit in total internal reflection fluorescence (TIRF) microscope. In this microscope, standard epi-fluorescence operation (EPI) can simply be switched to TIRF mode by applying a voltage to a built-in transmission grating [23]. Figure 16a, b show the images of 200-nm-diameter fluorescent polystyrene beads (Polyscience, Warrington, PA) immersed in 1 % agarose (Sigma-Aldrich, St. Louis, MO) recorded in standard EPI mode (no voltage) and in TIRF mode (3.5 kV applied to electrodes of the

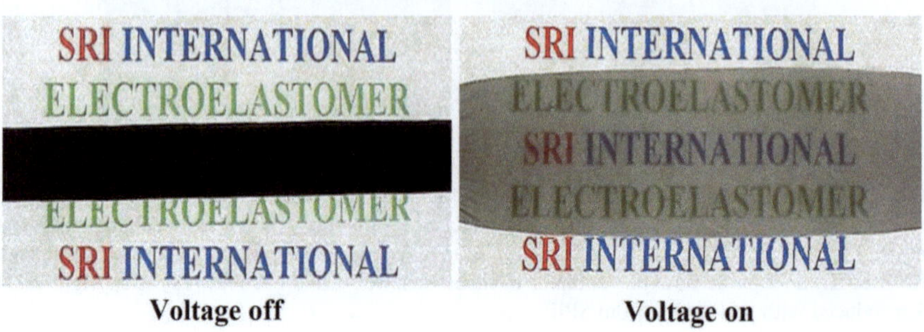

Fig. 15 Variation in optical density of ACM rubber film during actuation; voltage off (*left*) and voltage on (*right*) [22]. Copyright 2004. Reproduced with permission from SPIE

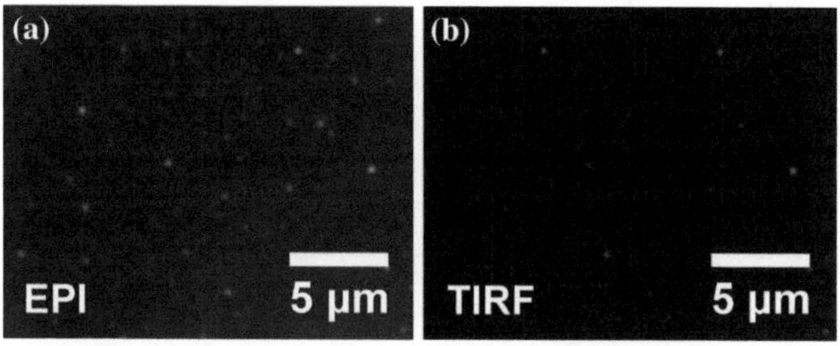

Fig. 16 **a** Micrographs of 200-nm-diameter fluorescent polystyrene beads in EPI mode (no voltage) and **b** TIRF mode (3.5 kV applied to tunable transmission grating) [23]. Copyright 2008. Reproduced with permission from SPIE

ACM rubber transmission grating), respectively. As expected, TIRF image shows much less background fluorescence and less blur. Moreover, the image in EIP mode displayed much less contrast comparing to TIRF mode [23].

3.2.3 Diaphragm Actuators

Diaphragm actuators based on ACM rubber can exploit both directions of planar expansion to produce large out-of-plane deflections and correspondingly large volume displacements. This will cause the diaphragm changes shape from flat to hemispherical in electrical filed [24, 25]. The ACM rubber diaphragm actuators exhibit high efficiency and energy output since they can couple the stresses in two planar directions to drive the load [24]. Diaphragm actuators based on ACM rubber can be used in pumps, valves, surfaces, loud speaker, as well as in areas such as adaptive optics, controllable surface roughness for aerodynamics, and refreshable Braille displays [25]. As shown in Fig. 17, diaphragm actuators are made of a stretching EAP film over an opening in a rigid frame. The diaphragm typically is biased, or pushed up or down when the voltage is applied. Comparing to alternative drive technologies such as piezoelectric materials, EAP diaphragms offer larger displacements [26].

ACM rubber diaphragms are particularly well suited to pumps with some commercial applications including compressors and small cooling systems. In addition, these diaphragms can duplicate biological pumps and may be used in artificial hearts in future. ACM rubber diaphragms could also be used in arrays to produce controllable surface roughness where the actuator need only create a change in appearance or texture such as refreshable Braille displays and touch-based or haptic displays [24, 26]. Surface texturing can also be employed to control air or water flow over the surfaces of airplanes or ships [26]. Nowadays, piezoelectric bending beam actuators are commonly used in Braille displays to raise the Braille dots. Space limitations associated with piezoelectric actuation, and the cost of manufacturing large actuator arrays, limit their use in refreshable displays for many visually disabled people [27].

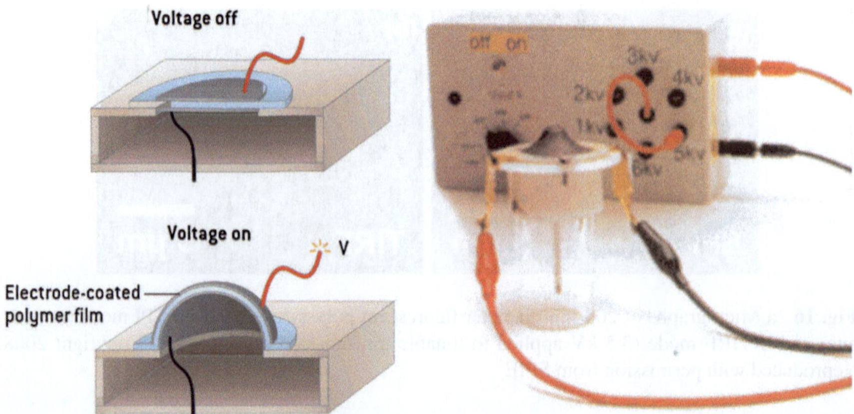

Fig. 17 Diaphragm actuator setup [26]. Copyright 2003. Reproduced with permission from Nature

The design of a proposed refreshable Braille cell based on EAPs and the tip displacement of the cell in electrical field is shown in Fig. 18a. The designed Braille cell basically is a simple tubular dielectric actuator that prolongs when it is subjected to electrical field. The produced maximum axial actuation strain (tip displacement) of the Braille cell based on ACM rubber as a function of applied electric is given in Fig. 18b. It shows that upon exposure to 120 kV/mm electric field, the tip displacement is more than 1 mm. This confirms that the proposed design clearly meets that performance requirement since the minimum tip displacement required for refreshable Braille displays at electric fields beyond ~100 kV/mm is 0.5 mm [28].

An eight-dot Braille cell prototype based on arrays of EAP diaphragm actuators designed by SRI international is given in Fig. 19. Such 2-mm-diameter diaphragm

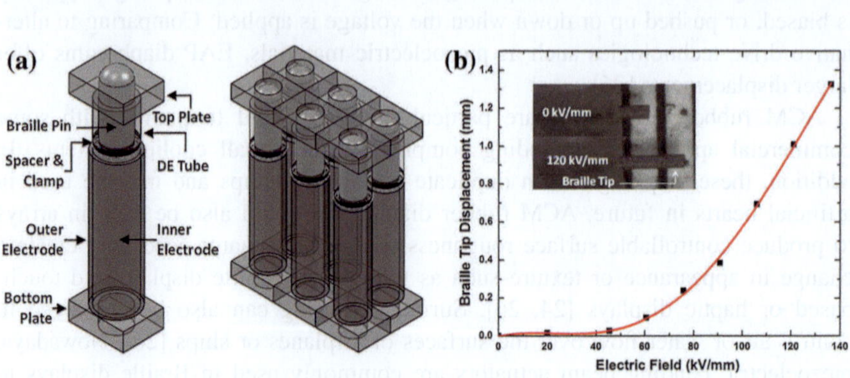

Fig. 18 **a** Schematic diagram of the Braille cell based on EAPs, **b** the tip displacement of the Braille actuator as a function of applied electrical field. (*Insert* the images of the tip at 0 and 120 kV/mm) [28]. Copyright 2012. Reproduced with permission from Elsevier

Fig. 19 The refreshable Braille display prototype based on dielectric elastomer prototype designed by SRI International's [22]. Copyright 2004. Reproduced with permission from SPIE

actuators built with acrylic films can produce pressures of up to 25 kPa that can result in the adequate actuation force on the Braille dot that is needed for easy reading [27].

Expanding and contracting rapidly according to the applied voltage signal, the ACM rubber diaphragm can emit sound yielding in a lightweight, inexpensive loudspeaker. Generally, EAP-based loudspeakers suggest the advantage of scalability to small and large areas, and the ability to conform to both flat and non-flat surfaces [25]. Figure 20 shows a sample of flat surface loudspeakers made of ACM rubber diaphragm actuator. In these speakers, the vibrating medium, the ACM rubber, is both the driver and sound-generating panel [26]. A uniform response over a wide frequency range and good sensitivity and fidelity in the

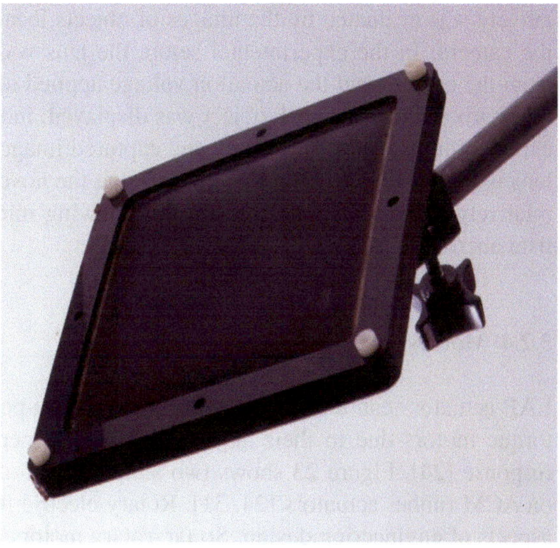

Fig. 20 Flat dielectric elastomer-based loudspeaker [29]. Copyright 2006. Reproduced with permission from SPIE

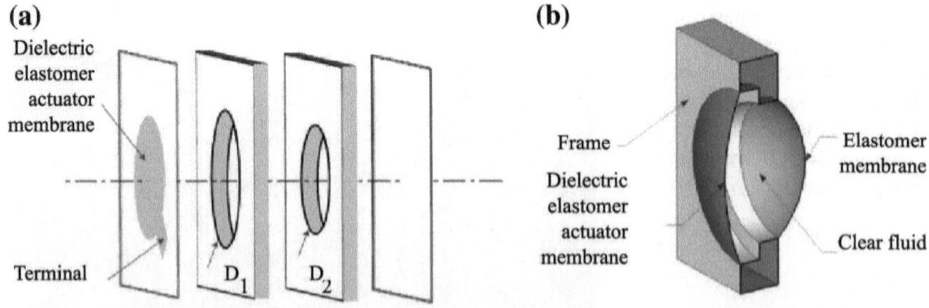

Fig. 21 **a** Schematic construction of the tunable lens: EAP actuator membrane, two frames with diameters of D_1 and D_2, and a passive elastomer membrane, **b** 3-D cross section of the novel lens [30]. Copyright 2013. Reproduced with permission from OSA

upper middle to tweeter ranges were observed for loudspeakers based on ACM rubbers [24].

Recently, EAP-based diaphragm actuators afford the opportunity to employ an electrical signal to alter the shape of a liquid lens without external actuation. Shian et al. presented a focus tunable elastomer-liquid lens system which makes use of an inline, transparent ACM rubber actuator. This adaptive lens provides several advantages over traditional lens assemblies such as compactness, cost, efficiency, and flexibility. The novel focus tunable lens includes four main components: a frame, a passive membrane, a dielectric elastomer actuator membrane, and a clear liquid, as shown in Fig. 21. The focal length variation greater than 100 % and responding time less than one second were observed in this system. Even larger focusing dynamic ranges and higher speed of responding are expected to be achievable selecting appropriate lens dimensions [30].

Focusing capability of the novel focal tunable lens based on ACM rubber qualitatively was evaluated by the images of objects located at different distances from the camera. In the experimental setup, the lens was mounted at a fixed distance from the camera and the actuation voltage applied to ACM film was adjusted until the sharpest image of each object was displayed, indicative of the focal plane. The experimental setup of schematic and captured images with the novel focal tunable lens is shown in Fig. 22. It can be seen that the novel lens is capable of producing relatively sharp images of the objects, showing minimal chromatic and astigmatism aberrations [30].

3.2.4 Motors

EAP actuators can also be utilized to make high-power density and high-specific torque motors due to their large energy output per cycle, and the fast speed of response [24]. Figure 23 shows two simple proof-of-concept rotary motors based on ACM rubber actuators [24, 31]. Rotary electric motors are very practical in all aspects of engineering design. So far, rotary motors are fabricated from hard, stiff,

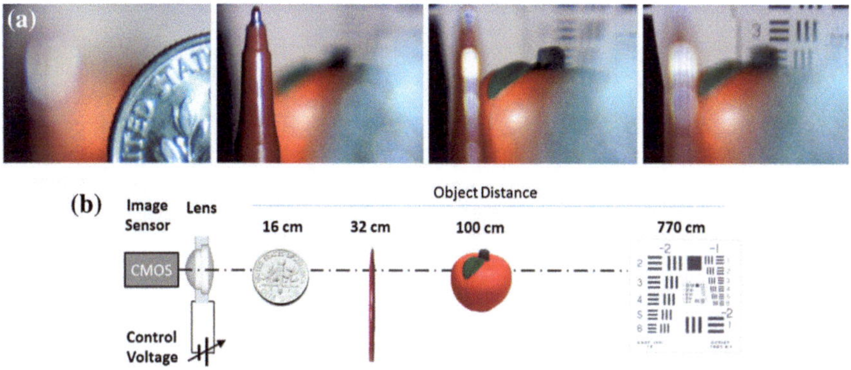

Fig. 22 **a** Photographic images at various focal lengths in order to check the optical performance of the lens. **b** A schematic of the measurement setup used for assess focusing capability of novel focal tunable lens [30]. Copyright 2013. Reproduced with permission from OSA

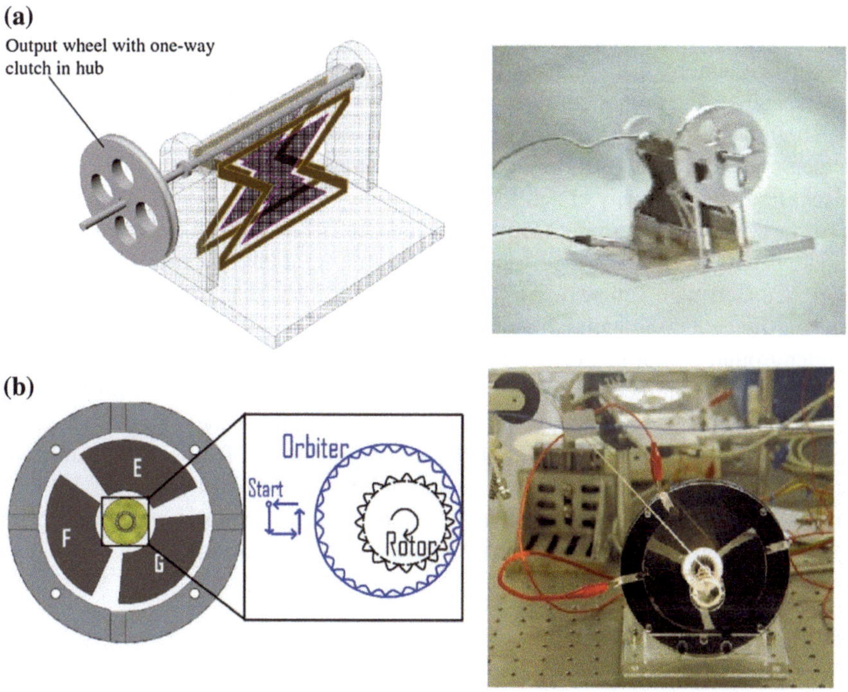

Fig. 23 Two simple rotary motors based on ACM rubber actuators: **a** two bow-tie rotary motor [10]. Copyright 2002. Reproduced with permission from SPIE. **b** A three-phase (120°/sector) rotary motor [31]. Copyright 2009. Reproduced with permission from OSA

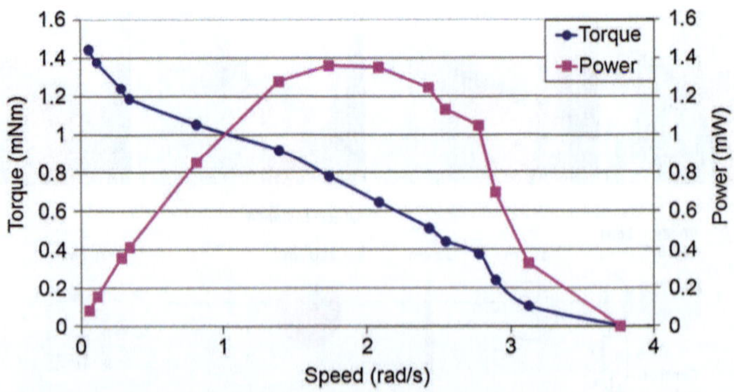

Fig. 24 Torque and power versus speed for a 3-phase rotary motor based on ACM rubber actuator [31]. Copyright 2009. Reproduced with permission from OSA

and dense materials such as metals and typically activated by a magnetic field. Recently, fabricating the rotary motors based on EAPs clearly shows good promise of light, flexible, and low-cost motors for the future [31]. Unlike their electromagnetic counterparts, which are inductive in nature, the EAP-based rotary motors are capacitive and do not consume much power when stalled [24].

The first example of ACM-based rotary motors (Fig. 23a) is a pair of bow-tie actuators that operate 180° out of phase to oscillate a shaft. The oscillatory motion is rectified by a one-way clutch to produce rotary motion at speeds up to 650 rpm [24]. In the second example (Fig. 23b), three electrode sectors (120°/sector) with square segments of ACM tape were attached to a rigid circular frame. The sectors were actuated with square wave voltages up to 2.5 kV that support the "orbiter" gear in the center. Sequential actuation of electrode membrane sectors moves the orbiter around a rotor in a closed path and causes the motor to rotate [31].

The results obtained by torque and power measurements of the 3-phase motor (Fig. 23b) are given in Fig. 24. As it can be seen from the figure, this motor resulted in a peak speed of 3.8 rad/s and peak power of 1.36 mW. Anderson et al. showed that these values can be substantially improved by using more EAP phases. They expect that further improvement can be achieved by reducing the frame mass and using a multilayer EAP actuators configuration [31].

3.2.5 Generators

In generator mode, EAPs can produce electrical power from mechanical input. In this case as the shape of the elastomer changes by an external force, the effective capacitance of the device also alters and, with the appropriate electronics, electrical energy generates. Comparing to other generator technologies, the energy density of EAP-based generators is high that leads to the lighter generators. Nowadays, heel-strike generator (Fig. 25) is proposed as a good way to generate

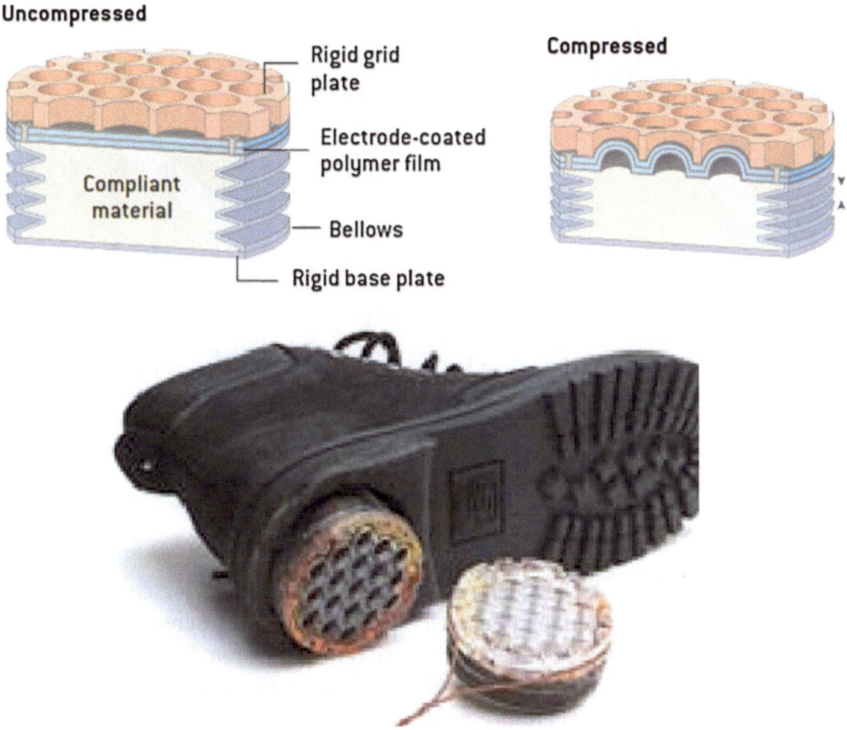

Fig. 25 Heel-strike generator based on acrylic rubber located in a boot heel [26]. Copyright 2003. Reproduced with permission from Nature

portable electrical power. In this approach, the compression energy of a shoe heel is captured when the shoe heel strikes the ground during walking. The deformation of an array of multilayer diaphragms applied by compression of the heel employed to generate electrical energy in heel-strike generator [25]. It was shown that the heel-strike generator can produce a maximum of 0.8 J per cycle [24]. With further developments, it is expected that heel-strike generator will be able to generate about a watt during normal walking. That energy will be able to provide enough electricity to power a cellular phone, on-shoe devices such as lights, health and performance monitors, or navigation devices [24, 25].

3.2.6 Sensor

A sensor is a material or device which is capable of monitoring changes in a specific system parameter without altering that [7]. Based on the unique properties of ACM rubber mentioned before, especially high strain capability, they can be considered as a suitable material to fabricate sensors. Various configurations for ACM-based sensors including thin tubes, flat strips, and large-area sheets are

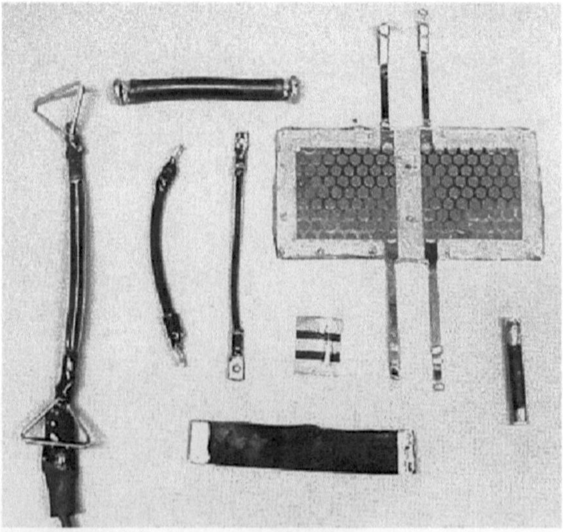

Fig. 26 Some sensors based on ACM rubber [10]. Copyright 2002. Reproduced with permission from SPIE

given in Fig. 26. These sensors can replace expensive and heavy conventional sensors such as potentiometers and encoders. The ACM rubber sensor in its fiber or ribbon form can be woven into textiles, providing position feedback of human motion. Considering sensor mode for EAPs, it is usually not important to maximize energy conversion. Therefore, selection of an appropriate material is based on other criteria such as maximum strain, environmental survivability, and cost [24].

When ACM rubber is used as an actuator, it needs to be controlled or sensed with a sensor that classically done with an external one. Developing smart and self-sensing actuators can reduce the cost, complexity, volume, and weight of their parent systems. To develop the self-sensing actuators, some features are required such as tolerance to very high voltages, geometric changes, temperature and humidity changes, hysteretic effects, and creep [32].

Goulbourne et al. fabricated a self-sensing fiber-reinforced cylindrical construct (McKibben actuator). They placed a cylindrical ACM rubber sensor in direct contact with the inner surface of the actuator which facilitates in situ monitoring of actuator strains and loads. The deformation of the actuator and ACM rubber sensor results in a change in the electrical signal that can be read from the surfaces of the electrodes that cover the elastomer. First, ACM rubber experiences a change in geometry due to an applied mechanical load or pressure exerted by actuator. Then, this mechanical input in the ACM rubber sensor is converted into an electrical signal. Finally, the signal can be detected by the capacitance change at the compliant electrode surface [12]. The accuracy of ACM rubber sensor was checked with the data obtained from a non-contact laser displacement sensor. The experiment setup is shown in Fig. 27a. As shown in the figure, the laser sensor is used to measure the displacement of the pole of the membrane during actuation. The comparison between the peak stretch values obtained from the ACM rubber sensor and

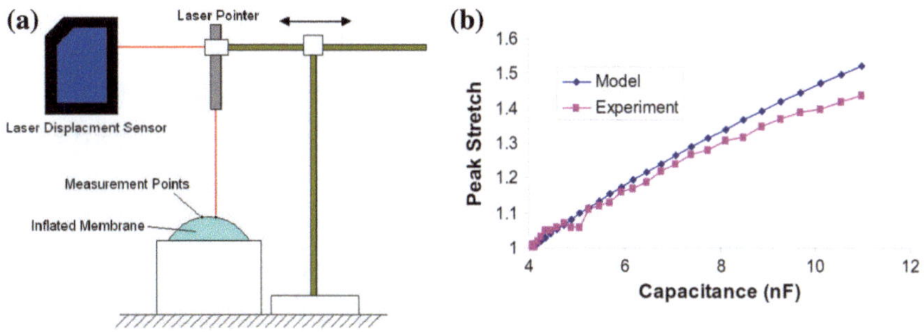

Fig. 27 **a** The experiment setup to measure and validate of peak stretch of ACM rubber sensor using a non-contact laser displacement sensor; **b** comparison between the stretch values calculated from capacitance measurements with ACM rubber sensor and the data measured by the laser sensor [12]. Copyright 2007. Reproduced with permission from SPIE

non-contact laser displacement sensor is illustrated in Fig. 27b. It can be observed that the ACM rubber sensor results obtained using the capacitance as input are in good agreement with the laser sensor readings especially at small strains. Some deviation from the laser sensor data is observed at large strains [12].

3.2.7 Variable-Stiffness Devices

Up to here, the three main application domains of EAPs (highlighting ACM rubber) including actuator, generator, and sensor are discussed. This section, however, discusses a fourth, potentially broad-based, application domain, namely variable-stiffness or variable-damping devices. Natural muscles demonstrate highly tailored stiffness to optimize their function on different conditions. Seeking for a material to mimic natural behavior determines the key role of variable-stiffness and variable-damping behavior in biomimetic robot performance [33].

Due to the fact that EAPs possess excellent ability to transduce electrical and mechanical energies, it is expected that they can function effectively for variable-stiffness and variable-damping devices. As discussed before, when the electrical stress expands the area of the EAPs, it means that they operate in actuator mode and convert the electrical energy to mechanical one. In the other side, when the EAPs are contracted in area by an external force, they operate in generator mode and convert mechanical energy to electrical one. The desired goal for a variable-stiffness device is not to generate a net mechanical or electrical energy output, but rather to control the stiffness [33]. Dastoor et al. presented the design of a variable-stiffness device based on ACM rubber actuators. Due to the fact that the stiffness of a single diaphragm is limited to the stiffness of the actuator material, multilayer diaphragms were proposed to scale the stiffness. In this design, multiple diaphragms are stacked (Fig. 28), providing voltage-controlled stiffness without high damping losses. The change in stiffness of an ACM rubber in a

Fig. 28 Multiunit stack designed for variable-stiffness device based on ACM rubber

stiffness-variable test setup showed that increasing the voltage from 0 to 6 kV, the stiffness of the designed device decreases from 102 to 15 N/mm [34].

Many biomedical applications can be considered for variable-stiffness devices based on EAPs such as external prosthetic and orthotic devices, especially for the ankle or leg, and tuning consumer devices for user comfort or preference. Compared with the conventional variable-stiffness smart materials, EAPs offer significant potential advantages include simplicity, broad dynamic range, ability to reach zero stiffness, low cost, low weight, shock tolerance, simple drive circuitry, and design flexibility [33].

4 Electrical Application of ACM Rubbers Composites

4.1 Conductive ACM Rubbers as Electromagnetic Interference (EMI) Shielding Materials

Nowadays, there is an increasing demand in industrial and military sectors for lightweight electromagnetic interference (EMI) shielding materials due to the rapid increase in electromagnetic radiation sources such as cell phones, computers, power lines, and telecommunication equipments in our surroundings [35]. EMI shielding material must exert high capability to reflect and absorb electromagnetic radiation at high frequencies, acting as a shield against the penetration of the radiation through it. This implies that the EMI shield should be an electrically conducting material [35].

EMI shielding mechanism in a conductive plate can be shown schematically in Fig. 29. When an incident electromagnetic wave with electric filed intensity of E_0 penetrates through a shield with a thickness of t, reflection, transmission, and absorption of the wave can take place. The reflected and transmitted electric filed intensities are shown with E_R and E_1, respectively, in the figure.

The EMI shielding efficiency (SE) or EMI-SE of a material can be defined as the ratio of the incident power (P_0) to the transmitted power (P_1) of an electromagnetic wave. SE parameter can be expressed as a function of the logarithm of the ratio of the incident and transmitted electric filed intensity according to Eq. (5).

$$\mathrm{SE(dB)} = 10\log = \left(\frac{P_0}{P_1}\right) = 20\log\left(\frac{E_0}{E_1}\right) \tag{5}$$

where dB = decibels.

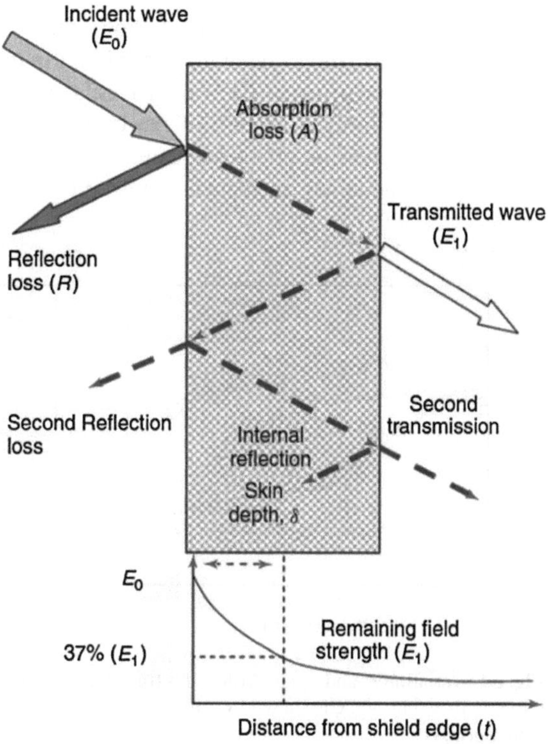

Fig. 29 Schematic representation of EMI shielding in a conductive plate [35]. Copyright 2012. Reproduced with permission from John Wiley & Sons

Conventional EMI shielding materials made from metals are heavy and susceptible to degradation on exposure to corrosive environments. Nowadays, conductive polymer composites are introduced as the novel and promising materials for EMI application. In general, polymers including ACM rubbers have very low EMI-SE due to their insulating nature and transparency to the electromagnetic radiations. Therefore, the incorporation of conductive fillers into the polymer matrix improves EMI-SE owing to improved AC conductivity [36]. The AC conductivity (σ_{ac}) and EMI-SE can be related with Eq. (6).

$$\mathrm{SE} = 20\log\left(1 + \sigma_{ac}\frac{dZ_0}{2}\right) = 20\log\left(1 + \frac{Z_0}{2R_s}\right) \tag{6}$$

where $R_s = \frac{1}{(\sigma_{ac}d)}$, Z_0 and d are the surface resistivity, the wave impedance in free space (377 Ω), and the thickness of the sample, respectively [37].

Electrical properties of EMI shield based on high-structured conductive carbon black (CCB)–filled ethylene ACM rubber (AEM) vulcanizates were studied by Sahoo et al. [38]. In addition to electrical properties, they observed that the bound rubber, crosslink density, tensile strength, and swelling of the AEM vulcanizates increased with the increase in CCB loading. The dielectric permittivity (ε') is the parameter which is proportional to the capacitance and measures the alignment of dipoles. The ε' of AEM, as the ability of a material to store the electric potential energy under the influence of an external electric field, was measured and reported in Fig. 30a. The figure shows that with increasing CCB concentration up to 30 wt%, the ε' value increases in all frequency range. The increase in ε' of AEM in the presence of CCB filler is related to the enhancement of interfacial polarization. A further increase of the CCB concentration does not produce the proportional increasing effect suggesting the saturation of the CCB incorporation. This may be due to any accidental inclusion and inherent imperfection of polymer

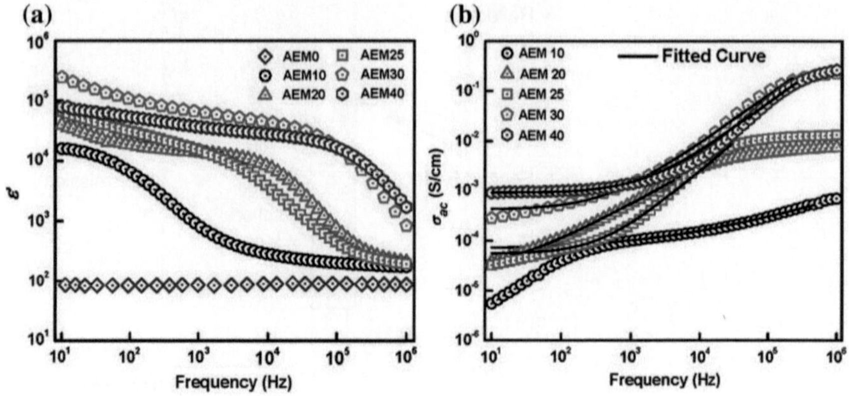

Fig. 30 Variation ε' and σ_{ac} of AEM with frequency considering CCB filler loading (AEM elastomer with × phr of CCB shown by AEMX) [38]. Copyright 2012. Reproduced with permission from Springer

composites such as inhomogeneous dispersion. Furthermore, the sudden reduction in ε' can be related to the restricted motion of polymer chain at higher filler loading [39]. The decrease in ε' on increasing the frequency is well observed in most of the dielectric materials [40]. At the high-frequency range, the dipoles are not be able to orient themselves in the direction of applied field easily. This is due to the less time that the dipoles have to orient themselves in the direction of alternating field. The effect of CCB loading on the electrical AC conductivity of AEM vulcanizates in different frequencies is given in Fig. 30b. Irrespective of the CCB content, σ_{ac} value of the AEM vulcanizates is significantly increased with increase in frequency. Also, σ_{ac} is enhanced with increase in CCB loading in vulcanizates [38].

The variation of WMI-SE with CCB content for AEM vulcanizates at two frequencies, 8 and 12 GHz, is shown in Fig. 31. Increased EMI-SE of AEM vulcanizates with increasing CCB loading shows the direct dependency of EMISE on AC conductivity [37, 38]. The results proved that the CCB-filled AEM vulcanizates can be a good candidate as the EMI shielding material in various electronic equipments.

In order to design electronic systems which are electromagnetically compatible with their environment, Federal Communications Commission (FCC) has different sets of regulations for different types of digital devices [41]. Non-domestic, commercial, militirial, industrial, or business devices connected to low-voltage supply are classified as Class A digital devices. Domestic devices that are marketed for use in residential environment are classified as Class B digital devices. Generally, the regulation used for Class B devices is more stringent than those of Class A ones [41]. In order to meet FCC requirements, SE greater than about 40 dB should be adequate for Class B devices [37].

Recently, the successful preparation of electrically conductive nanocomposites based on AEM and multiwalled carbon nanotube (MWCNT) has been also reported by Sahoo et al. Uniform dispersion of MWCNT in AEM matrix was

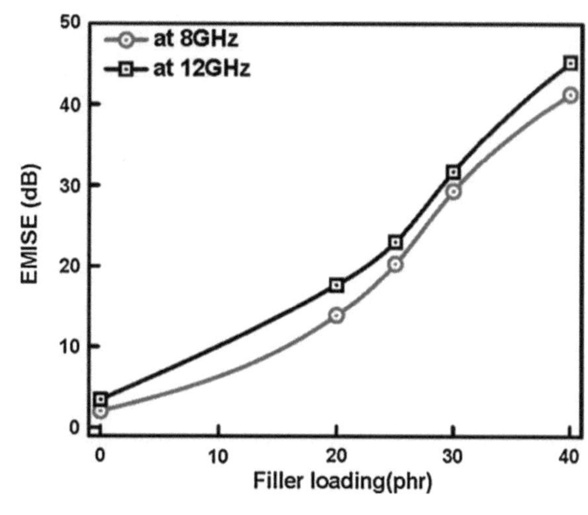

Fig. 31 Variation of EMI shielding efficiency of AEM with CCB loading [38]. Copyright 2012. Reproduced with permission from Springer

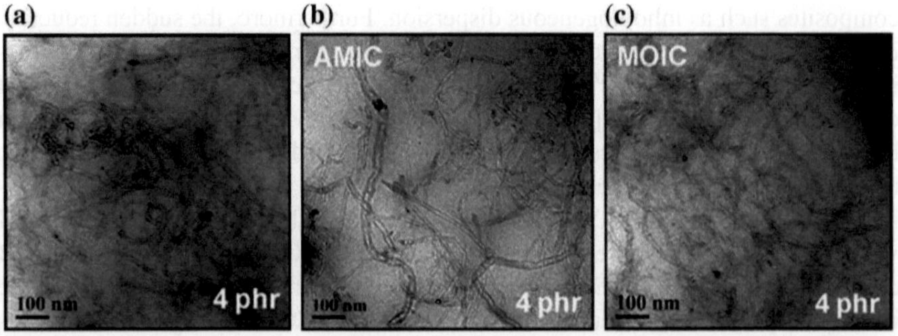

Fig. 32 High-resolution TEM microphotographs of **a** AEM with 4 phr of 4MWNT, **b** in the presence of AMIC, and **c** in the presence of MOIC [42]. Copyright 2015. Reproduced with permission from John Wiley & Sons

achieved in the presence of ionic liquids such as 1-methyl-3-octylimidazolium chloride (MOIC) and 1-allyl-3-methyl imidazolium chloride (AMIC) confirmed by the high-resolution transmission electron microscopic (HRTEM) microphotographs (Fig. 32). The clear MWNT agglomerates are observed in the TEM of the AEM/MWCNT nanocomposites with 4 phr of ethanol dispersed MWNT (Fig. 32a), while in Fig. 32b and c, the individual MWCNTs are clearly observed showing well dispersion of them in the AEM matrix. This is due to the presence of ionic liquids such as MOIC and AMIC along with the ethanol [42].

The dependency of ε' of AEM/MWNT nanocomposites to frequency and MWCNT content is shown in Fig. 33a. With increasing the MWNT content, the ε' of AEM/MWCNT nanocomposites increases and the increase is more significant at lower frequency. The behavior is similar to what was observed for AEM/CCB composites, before [41]. The higher value of ε' is related to proper orientation of dipoles formed by the MWCNT phase in the AEM matrix along the direction of applied electric field. The effect of ionic liquids, MOIC and AMIC in ε' of AEM/MWCNT nanocomposite with 40phr MWCNT, is shown in Fig. 33b. A slight increase in ε' of AEM/MWCNT nanocomposite is observed for the composites prepared in the presence of ionic liquids that is related to the pronounced polarization of the dipoles in the system as well as uniform dispersion of MWCNT in the presence of the ionic liquids [42].

The EMI shielding effectiveness of the AEM/MWNT nanocomposites is illustrated in Fig. 34. As mentioned before for reflection of the electromagnetic radiation, the shield materials should have mobile charge carriers. The number of charge carriers increases with increasing the MWCNT loading in the AEM nanocomposites. Hence, higher the electrical conductivity, higher is the EMISE of the shield materials [42]. Comparing the EMISE results obtained for AEM/CCB composites (Fig. 31) and AEM/MWCNT nanocomposites, Fig. 34b shows that MWCNT is more effective in increasing shielding efficiency, at the same filler concentration. This means that multiple reflections of the electromagnetic ray

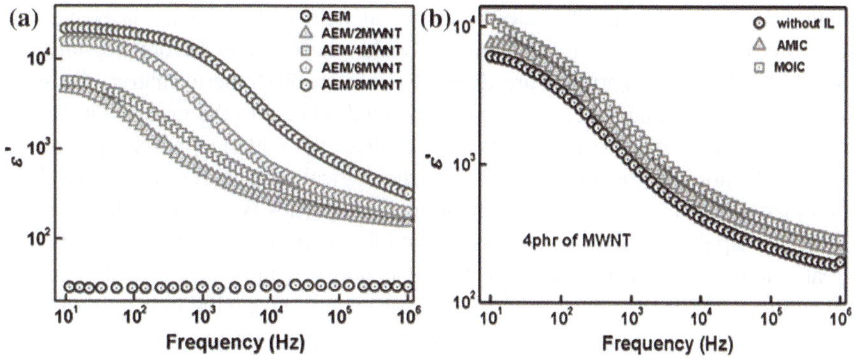

Fig. 33 The effect of **a** MWNT content and **b** different ionic liquids on dielectric permittivity (ε') of AEM [42]. Copyright 2015. Reproduced with permission from John Wiley & Sons

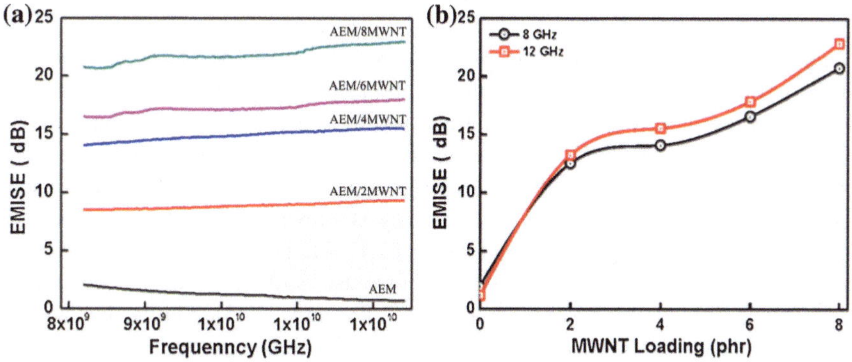

Fig. 34 Variation of EMI shielding efficiency with MWNT loading in different frequencies [42]. Copyright 2015. Reproduced with permission from John Wiley & Sons

occur in the AEM/MWCNT nanocomposites. The phenomenon is related to the nanometric dimension and large surface area of MWCNT causing to be more effective in the enhancement of EMI-SE [43].

4.2 *Electrorheological (ER) Material Based on ACMs*

Electrorheological (ER) materials are typical suspensions composed of micrometer-sized polarizable conductive particles and a non-conductive matrix like ACM rubbers. The rheological properties of ER systems such as viscosity, yield stress, and shear modulus are dramatically changed under applied AC or DC electric fields. Upon applying an external electrical field, ER materials behave like Bingham plastic materials, while without the external electrical field their behavior is Newtonian. It means that ER materials perform the yield phenomenon before beginning to flow [44]. When the particles are dispersed into an elastomeric matrix, the solid-like nature of the matrix keeps the particles fixed relative to the matrix without any movement. But when the electric field is applied, the particle–particle interactions occur and cause significant changes in the rheological properties of the materials. In the other words, the change in the rheological properties is related to the induction of dipole–dipole interactions between suspending particles [45]. ER materials have great potential toward several applications, such as actuators, microelectromechanical systems (MEMS), drive control systems, automotive vibration controls, and robotics control systems [44, 45]. In order to be suitable for these applications, ER materials require high-electromechanical coupling and wide ranges of permittivity and dielectric losses [46].

4.2.1 ER Material Based on ACM/Piezoelectric Particles

Tangboriboon et al. reported the preparation of a novel ER material by mixing lead zirconate titanate (PZT) with the solution of ACM rubber. The piezoelectric PZT particles, at very small volume fractions, were demonstrated to be useful as a dipole moment generating filler to store additional elastic energy within the ACM rubber matrix [45, 47]. Variation of storage modulus (G') and dielectric constant of ACM/PZT ER materials versus frequency is shown in Fig. 35. It can be seen that G' increases linearly with increasing PZT volume fraction, with or without an electric field. In addition, the twofold increase is observed in the modulus as the electric field strength varied from 0 to 2 kV mm^{-1} [45]. Similar behavior in electrical properties of ACM was observed with addition of alumina particles [46].

ER elastomers are considered as the viscoelastic damping materials due to the fact that their physical properties can be instantaneously and reversibly controlled with the application of an electric field. So, ER elastomers can be employed in

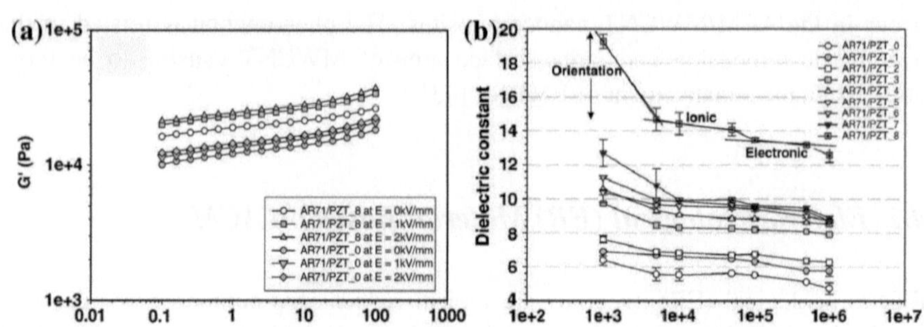

Fig. 35 **a** The storage modulus of ACM/PZT ER materials versus frequency at two different electric field strengths, 0 and 2 kV/mm [45]. Copyright 2009. Reproduced with permission from Elsevier

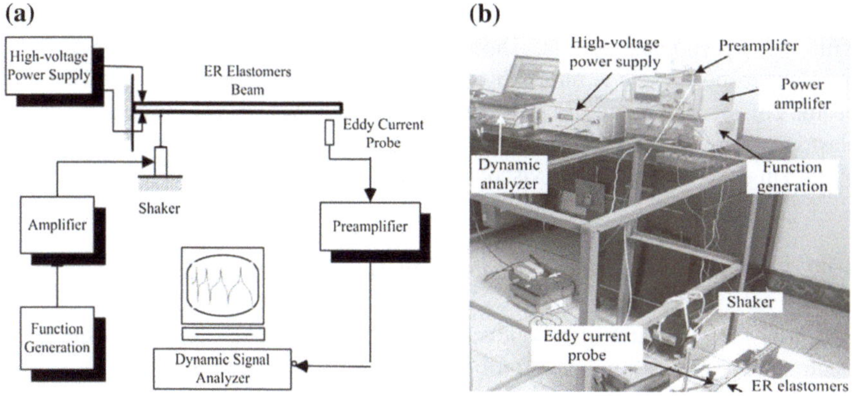

Fig. 36 **a** Schematic diagram of experimental setup for ER elastomer beam testing, **b** photograph of experimental setup [48]. Copyright 2011. Reproduced with permission from IOP science

numerous potential engineering applications, such as shock absorbers, clutch/brake systems, valves, and adaptive structures [48]. Wei et al. showed the application of ER elastomers for the vibration control of flexible beams. A three-layer sandwich beam was considered for the test, in which an ER elastomer layer is restricted by two elastic layers [48]. The schematic configuration and the photograph of the vibration testing setup are given in Fig. 36. The electrical field required during the test is provided by a high-voltage power supply as shown in the figure. A non-contact eddy current probe was employed to measure the vibration displacement of the ER elastomer beam. The measured data were sent to the dynamic signal analyzer for further process [48].

The comparison of theoretical and experimental vibration responses gathered in two different electric fields, $E = 0$ and 1 $\mathrm{kVmm^{-1}}$, is given in Fig. 37. Without

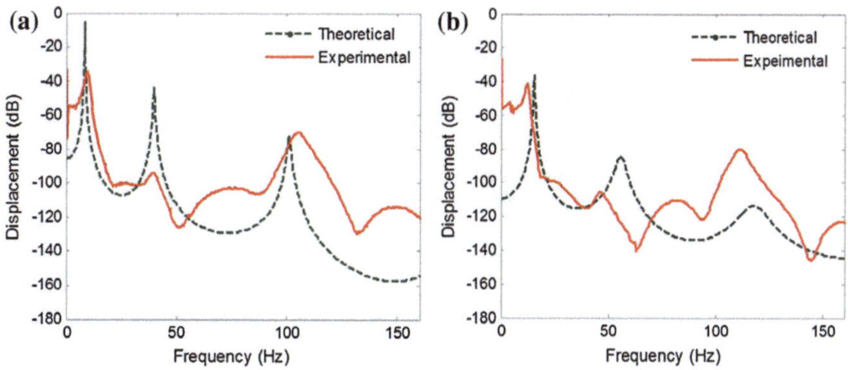

Fig. 37 Comparison of theoretical and experimental frequency responses of vibration of ER elastomers at different electric fields **a** $E = 0$ kV $\mathrm{mm^{-1}}$, **b** $E = 1$ kV $\mathrm{mm^{-1}}$ [48]. Copyright 2011. Reproduced with permission from IOP science

any external electric field, the theoretical resonance frequencies agree acceptably with the experimental results. The deviation of the theoretical response results from experimental one is more obvious when the electric field was applied to the beam. This can be related to the fact that the exact viscoelastic properties of ER elastomers are not definitely known yet. Therefore, this could be improved by optimizing and improving the viscoelastic model of ER elastomers in future [48].

4.2.2 ER Material Based on ACM/Conductive Polymer Particles

Recently, incorporation of a conductive polymer into EAPs to form a blend for electroactive application has been of interest. Kunanuruksapong et al. fabricated the blends of poly(*p*-phenylene) polymer particles (PPPs)/ACM rubbers (Nipol AR71) in order to tailor the electromechanical properties of ACM toward electroactive applications. The microstructure of the synthesized PPT and dispersion morphology of PPT particles within AR71 matrix was shown in Fig. 38 and micrographs obtained using a scanning electron microscope (SEM). The SEM micrograph showed the quite irregular in shape of synthesized PPPs with average diameter of approximately 46 μm. Figure 38b showed that PPPs are dispersed uniformly and randomly within the AR71 matrix [49].

The effect of PPP vol. content on storage modulus of AR71 with and without electrical field is illustrated in Fig. 39. The storage modulus of the blend in electrical field strengths of 0 and 2 kV/mm is shown with G'_0 and G'_2, respectively. With increasing the vol. content of PPP, the G'_0 and G'_2 of the blend increased gradually. The observed increase in storage modulus can be correlated to the presence of the PPPs acting as the filler. The increase trend of G'_0 is linear up to 20 vol.%, while G'_2 is linear up to 40 vol.%. The storage modulus response of the AR71/PPP blends ($\Delta G'_2$) is calculated by $\Delta G'_2 = G'_2 - G'_0$. It can be seen from Fig. 39 that the $\Delta G'_2$ reaches to its maximum value, 107,770 Pa, at 30 vol.% of PPP [49].

(a) (b)

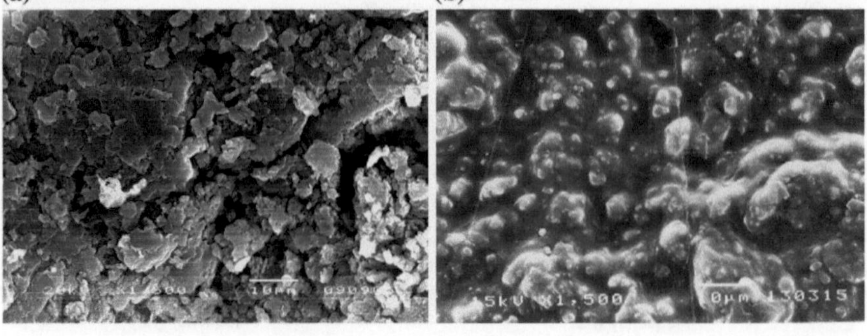

Fig. 38 The morphology **a** synthesized PPP and **b** AR71/PPP blend with 30 vol.% PPP at magnification of 1500 [49]. Copyright 2007. Reproduced with permission from Elsevier

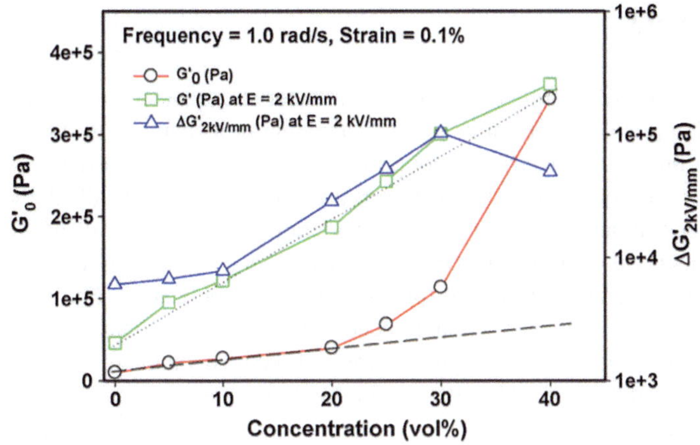

Fig. 39 Effect of vol. concentration of PPP on rheological properties of AR71/PPP blend: the storage modulus at $E = 0$ kV/mm (G'_0), the storage modulus at $E = 2$ kV/mm (G'_2), and the storage modulus responses ($\Delta G'_2$) [49]. Copyright 2007. Reproduced with permission from Elsevier

The effect of vol. concentration of PPP on other electrorheological properties of AR71/PPP blend, including loss modulus without electric field (G''_0), loss modulus at electric field of 2 kV/mm (G''_2), loss modulus response ($\Delta G''_2 = G''_2 - G''_0$), sensitivity value of the storage modulus ($\Delta G'_2/G'_0$), sensitivity value of the loss modulus ($\Delta G''_2/G''_0$), and the electrical conductivity (σ), is tabulated in Table 4. It can be observed that the incorporation of PPPs into AR71 also leads to increase in loss modulus of the blend with and without electrical field. It can be related to the induced free volume with addition of PPPs and the fact that the interface of the PPPs and the AR71 matrix is poor. The sensitivity value of storage modulus is maximum, 0.971, at 30 vol.% of PPP. Furthermore, introduction of PPPs increases the electrical conductivity (σ) of the AR71 ACM rubber. This is due to the higher σ value of PPPs, 5.28×10^{-6} S/cm, compared to that of the ACM matrix (2.35×10^{-12} S/cm) [49].

Table 4 Rheological properties of PPP/AR71 polymer blends with different vol.% of PPT [49]

Vol.% of PPP	G'_0 (Pa)	G'_2 (Pa)	G''_0 (Pa)	G''_2 (Pa)	$\Delta G'_2$ (Pa)	$\Delta G''_2$ (Pa)	$\Delta G'_2/G'_0$	$\Delta G''_2/G''_0$	σ (S/cm)
0	9959	16,105	1124	2206	6146	1082	0.617	0.962	2.35E−12
5	21,375	26,478	2707	2897	5103	190	0.238	0.065	3.53E−12
10	27,250	34,979	2565	3918	7729	1353	0.306	0.527	8.73E−12
20	39,859	68,158	5316	9798	28,299	4482	0.710	0.843	1.03E−11
25	68,353	120,490	10,825	18,503	52,137	7678	0.763	0.415	4.28E−11
30	115,420	218,750	15,111	28,564	107,770	13,453	0.971	0.890	8.95E−11
40	343,330	396,960	42,644	50,312	49,800	7668	0.143	0.179	1.01E−09

Copyright 2007. Reproduced with permission from Elsevier
The test was accomplished at frequency = 1 rad/s, strain 0.1 %
The electrical conductivity of PPP is 5.28×10^{-6} S/cm

5 Summary

ACM rubbers have a number of unique characteristics that suggest a wide range of electrical application both in its pristine state and in its composite state. ACM rubber as an inherent dielectric elastomer can be employed as transducers, actuators, generators, sensors, and variable impedance devices. Indeed, all of the applications described here are under active consideration of the ability of ACM rubber to change electrical energy to mechanical one and vice versa. Conductive ACM rubber composites can be employed for EMI shielding application. For that case, high electrical conductivity is all that is needed which can be obtained by the introduction of conductive filler to ACM rubber. Electrorheological material based on ACM rubber and polarizable filler can be used in some electrical applications such as vibration control devices.

Many more applications of ACM rubbers are bound to emerge as the technology continues to mature. There are still many opportunities for significant improvements in the basic electrical applications of ACM rubber that can further affect the suitability of this elastomer to many other applications.

References

1. Whelan T (1994) Polymer technology dictionary. Springer Science & Business Media, London
2. Demarco RD (1979) New generation polyacrylate elastomers. Rubber Chem Technol 52(1):173–186
3. Seil DA, Wolf FR (1999) Nitrile and polyacrylic rubbers. In: Morton M (ed) Rubber technology, 2nd edn. Springer, Netherlands
4. www.nskamericas.com (2015) NSK Seals and Shields Protect Bearing investment. NSK Americas
5. Satoh S, Suzuki T (1982) Rubber hose for automotive fuel line. USA Patent US4330017 A
6. Pelrine R, Kornbluh R, Pei Q, Joseph J (2000) High-speed electrically actuated elastomers with strain greater than 100 %. Science 287(5454):836–839
7. Shankar R, Ghosh TK, Spontak RJ (2007) Dielectric elastomers as next-generation polymeric actuators. Soft Matter 3(9):1116–1129
8. Shankar R, Ghosh TK, Spontak RJ (2009) Mechanical and actuation behavior of electroactive nanostructured polymers. Sens Actuators A 151(1):46–52
9. Wissler M, Mazza E (2007) Mechanical behavior of an acrylic elastomer used in dielectric elastomer actuators. Sens Actuators A 134(2):494–504
10. Kornbluh RD, Pelrine R, Pei Q, Heydt R, Stanford S, Oh S, Eckerle J (2002) Electroelastomers: applications of dielectric elastomer transducers for actuation, generation, and smart structures. In: Smart structures and materials: industrial and commercial applications of smart structures technologies, San Diego, CA. SPIE, pp 254–270
11. Kornbluh RD, Pelrine R, Joseph J, Heydt R, Pei Q, Chiba S (1999) High-field electrostriction of elastomeric polymer dielectrics for actuation. In: Smart structures and materials: electroactive polymer actuators and devices (EAPAD), Newport Beach, CA. SPIE, pp 149–161
12. Goulbourne N, Son S, Fox J (2007) Self-sensing McKibben actuators using dielectric elastomer sensors. In: Smart structures and materials: electroactive polymer actuators and devices (EAPAD), San Diego, CA. SPIE, pp 652414–652426

13. Kornbluh RD, Pelrine R, Joseph J, Heydt R, Pei Q, Chiba S (1999) High-field electrostriction of elastomeric polymer dielectrics for actuation. In: Smart structures and materials: electroactive polymer actuators and devices (EAPAD), Newport Beach, CA SPIE, pp 149–161
14. Kornbluh RD, Pelrine R, Pei Q, Oh S, Joseph J (2000) Ultrahigh strain response of field-actuated elastomeric polymers. In: Smart structures and materials: electroactive polymer actuators and devices (EAPAD), Newport Beach, CA. SPIE, pp 51–64
15. Kofod G, Kornbluh RD, Pelrine R, Sommer-Larsen P (2001) Actuation response of polyacrylate dielectric elastomers. In: Smart structures and materials: electroactive polymer actuators and devices, Newport Beach, CA, USA. SIPE, pp 141–147
16. Kofod G, Sommer-Larsen P, Kornbluh R, Pelrine R (2003) Actuation response of polyacrylate dielectric elastomers. J Intell Mater Syst Struct 14(12):787–793
17. Zhang H, Düring L, Kovacs G, Yuan W, Niu X, Pei Q (2010) Interpenetrating polymer networks based on acrylic elastomers and plasticizers with improved actuation temperature range. Polym Int 59(3):384–390
18. Carpi F, De Rossi D, Kornbluh R, Pelrine RE, Sommer-Larsen P (2011) Dielectric elastomers as electromechanical transducers: fundamentals, materials, devices, models and applications of an emerging electroactive polymer technology. Elsevier Science, Oxford
19. Pelrine R, Kornbluh RD, Eckerle J, Jeuck P, Oh S, Pei Q, Stanford S (2001) Dielectric elastomers: generator mode fundamentals and applications. In: Smart structures and materials: electroactive polymer actuators and devices (EAPAD), Newport Beach, CA, USA SPIE, pp 148–156
20. Bar-Cohen Y (2004) Electroactive polymer (EAP) actuators as artificial muscles: reality, potential, and challenges. SPIE, Washington
21. Pelrine R, Kornbluh RD, Pei Q, Stanford S, Oh S, Eckerle J, Full RJ, Rosenthal MA, Meijer K (2002) Dielectric elastomer artificial muscle actuators: toward biomimetic motion. In: Smart structures and materials: electroactive polymer actuators and devices (EAPAD), San Diego, CA. SPIE, pp 126–137
22. Kornbluh RD, Pelrine R, Prahlad H, Heydt R (2004) Electroactive polymers: an emerging technology for MEMS. In: MEMS/MOEMS components and their applications, San Jose, CA. SPIE, pp 13–27
23. Aschwanden M, Niederer D, Stemmer A (2008) Tunable transmission grating based on dielectric elastomer actuators. In: Electroactive polymer actuators and devices (EAPAD), San Diego, California. SPIE, p 69271R-69271R-69212
24. Kornbluh RD, Pelrine R, Pei Q, Heydt R, Stanford S, Oh S, Eckerle J (2002) Electroelastomers: applications of dielectric elastomer transducers for actuation, generation, and smart structures. In: Smart structures and materials: industrial and commercial applications of smart structures technologies, San Diego, CA. SPIE, pp 254–270
25. O'Halloran A, O'Malley F, McHugh P (2008) A review on dielectric elastomer actuators, technology, applications, and challenges. J Appl Phys 104(7):071101
26. Ashley S (2003) Artificial muscles. Sci Am 289(4):52–59
27. Kornbluh RD, Pelrine R, Prahlad H, Heydt R (2004) Electroactive polymers: an emerging technology for MEMS. In: MEMS/MOEMS components and their applications, San Jose, CA. SPIE, pp 13–27
28. Chakraborti P, Toprakci HK, Yang P, Di Spigna N, Franzon P, Ghosh T (2012) A compact dielectric elastomer tubular actuator for refreshable Braille displays. Sens Actuators A 179:151–157
29. Heydt R, Kornbluh R, Eckerle J, Pelrine R (2006) Sound radiation properties of dielectric elastomer electroactive polymer loudspeakers. In: Smart structures and materials: electroactive polymer actuators and devices (EAPAD), San Diego, CA. SPIE, pp 61681M-61681M–61688
30. Shian S, Diebold RM, Clarke DR (2013) Tunable lenses using transparent dielectric elastomer actuators. Opt Express 21(7):8669–8676
31. Anderson IA, Calius EP, Gisby T, Hale T, McKay T, O'Brien B, Walbran S (2009) A dielectric elastomer actuator thin membrane rotary motor. In: Smart structures and materials: electroactive polymer actuators and devices (EAPAD). SPIE, pp 72871H-1–10

32. O'Brien B, Thode J, Anderson I, Calius E, Haemmerle E, Xie S (2007) Integrated extension sensor based on resistance and voltage measurement for a dielectric elastomer. In: Smart structures and materials: electroactive polymer actuators and devices (EAPAD), San Diego, California. SPIE, pp 652415-1–11
33. Pelrine R, Kornbluh R (2009) Variable-Stiffness–Mode Dielectric Elastomer Devices. Adv Sci Technol 61:192–201
34. Dastoor S, Cutkosky M (2012) Design of dielectric electroactive polymers for a compact and scalable variable stiffness device. In: IEEE international conference on robotics and automation (ICRA), Saint Paul, MN. IEEE, pp 3745–3750
35. Tjong SC (2012) Polymer composites with carbonaceous nanofillers: properties and applications. Wiley, Weinheim
36. Bhadra S, Singha NK, Khastgir D (2009) Dielectric properties and EMI shielding efficiency of polyaniline and ethylene 1-octene based semi-conducting composites. Curr Appl Phys 9(2):396–403
37. Colaneri NF, Schacklette LW (1992) EMI shielding measurements of conductive polymer blends. Instrum Meas 41(2):291–297
38. Sahoo BP, Naskar K, Tripathy DK (2012) Conductive carbon black-filled ethylene acrylic elastomer vulcanizates: physico-mechanical, thermal, and electrical properties. J Mater Sci 47(5):2421–2433
39. Yuan Q, Wu D (2010) Low percolation threshold and high conductivity in carbon black filled polyethylene and polypropylene composites. J Appl Polym Sci 115(6):3527–3534
40. Sahoo B, Naskar K, Choudhary R, Sabharwal S, Tripathy D (2012) Dielectric relaxation behavior of conducting carbon black reinforced ethylene acrylic elastomer vulcanizates. J Appl Polym Sci 124(1):678–688
41. Weston D (2001) Electromagnetic compatibility: principles and applications, 2nd edn (revised and expanded). Taylor & Francis, Boca Raton
42. Sahoo BP, Naskar K, Tripathy DK (2015) Multiwalled carbon nanotube-filled ethylene acrylic elastomer nanocomposites: influence of ionic liquids on the mechanical, dynamic mechanical, and dielectric properties. Polym Compos. doi:10.1002/pc.23451 (in press)
43. Hajibaba A, Naderi G, Esmizadeh E, Ghoreishy MHR (2014) Morphology and dynamic-mechanical properties of PVC/NBR blends reinforced with two types of nanoparticles. J Compos Mater 48(2):131–141
44. Yanju L, Hejun D, Dianfu W (2001) ER fluid based on inorganic/polymer blend particles and its adaptive viscoelastic properties. Colloids Surf A 189(1–3):203–210
45. Tangboriboon N, Sirivat A, Kunanuruksapong R, Wongkasemjit S (2009) Electrorheological properties of novel piezoelectric lead zirconate titanate Pb (Zr 0.5, Ti 0.5) O 3-acrylic rubber composites. Mater Sci Eng, C 29(6):1913–1918
46. Tangboriboon N, Wongpinthong P, Sirivat A, Kunanuruksapong R (2011) Electroactive alumina particles embedded in an acrylic elastomer. Polym Compos 32(1):44–51
47. Tangboriboon N, Sirivat A, Wongkasemjit S (2008) Electrorheology and characterization of acrylic rubber and lead titanate composite materials. Appl Organomet Chem 22(5):262–269. doi:10.1002/aoc.1388
48. Wei K, Bai Q, Meng G, Ye L (2011) Vibration characteristics of electrorheological elastomer sandwich beams. Smart Mater Struct 20(5):055012
49. Kunanuruksapong R, Sirivat A (2007) Poly(p-phenylene) and acrylic elastomer blends for electroactive application. Mater Sci Eng A 454–455:453–460

Electronic Applications of Polydimethylsiloxane and Its Composites

Kishor Kumar Sadasivuni, Deepalekshmi Ponnamma, John-John Cabibihan and Mariam Al-Ali AlMa'adeed

Abstract Stretchable or elastic electronics is based on building electronic circuits or devices on stretchable substrates and its potential applications include skin sensors for robotics, wearable devices, and flesh-like biodevices. As the physically and chemically stable silicone rubber, polydimethylsiloxane (PDMS) has a unique flexibility with a shear elastic modulus 250 kPa and lowest glass transition temperature among other polymers; it is widely used to fabricate stretchable electronic materials. In addition, PDMS also possesses clean room processability, low curing temperature, high flexibility, possibility to change its functional groups, high compressibility, and very low change in elasticity versus temperature and time. This chapter addresses the applicability of patterned PDMS nanocomposites in electronics along with its advantages and limitations. Major areas such as sensors, actuators, light-emitting diodes, piezoelectrics, dielectrics, electromagnetic radiation shields, transistors, and supercapacitors are covered.

Keywords PDMS · Sensor · Piezoelectric · Actuator · Supercapacitor · LED · Dielectrics

1 Introduction

The potential applications of stretchable electronics range from robotic sensory skins and wearable communication devices to bio-integrated devices [1, 2]. Flexible composite materials are often made by replacing silicon or polymeric

K.K. Sadasivuni (✉) · D. Ponnamma · M.A.-A. AlMa'adeed
Center for Advanced Materials, Qatar University, P.O. Box 2713, Doha, Qatar
e-mail: kishorkumars@qu.edu.qa

M.A.-A. AlMa'adeed
Materials Science and Technology Program, Qatar University, P.O. Box 2713, Doha, Qatar

J.-J. Cabibihan
Mechanical and Industrial Engineering Department, Qatar University, P.O. Box 2713, Doha, Qatar

D. Ponnamma et al. (eds.), *Flexible and Stretchable Electronic Composites*,
Springer Series on Polymer and Composite Materials,
DOI 10.1007/978-3-319-23663-6_7

substrate materials by elastomers. A large number of investigations are going on in this particular area to develop advanced materials for stretchable and wearable electronic devices. However, there are huge numbers of challenges to be rectified in this field. Simultaneous achievement of excellent mechanical robustness and electronic performance at the same time possessing electrical conductivity of metals under high strain values is one among them. By alleviating the challenges, stretchability of materials are achieved by following three important strategies—(i) by varying the geometry of the inorganic conductors using thin gold films [3], wavy silicon nanoribbons [4] rigid semiconductor nanowires, and nanomembranes [5–7], or by patterning spring-shaped metal traces [8, 9]; (ii) by making the substrate porous or engineering microfluidic channels and coating it with a metal layer [10] or filling it with liquid metal alloys [11, 12]; and (iii) by stretchable interconnects strategy, i.e., by dispersing [13], casting [14], reducing [15], or implanting [16] nano- or microconductive particles such as metal wires, conducting polymer, carbon nanotube, or graphene films [17–21] in soft elastic materials such as poly(dimethylsiloxane) (PDMS) or polyurethane (PU) by various processes including vacuum evaporation, photolithographic patterning, transfer printing, and screen printing [22–27].

While the wavy method accommodates external strain by nonlinear buckling phenomena to improve the stretchability [28, 29], the interconnect strategy uses stretchable electrodes to connect rigid active device islands on elastomeric substrates [30]. However, both these methods are limited to 2D structure and low aspect ratio features and do not offer all the possibilities that printed circuit boards do. Large-scale production of omni-directional, solderable and individually integrated stretchable electronics using standard design still remains a challenge and recent methods of multidirectional writing of conducting materials onto elastomers at low temperatures overcome such limitations [31].

PDMS belongs to silicone elastomer family and is in various forms, thin films, tubes, and rods; coatings on a fiber, inside a capillary column, on a magnetic stir bar, etc.; foams or particles packed into sorbent tube are applied mainly in the areas of sampling of organic compounds from air, water, and soil gas matrices, as well as manufacturing of laboratory-on-a-chip devices. Generally, it is obtained by combining a pre-polymer (vinyl terminated PDMS) and a cross-linker (dimethyl, methyl hydrogen siloxane) in specific ratios and subsequently curing them at specified conditions [32]. On PDMS substrate, the composite is stencil printed or screen printed with a resolution of 150 mm to fabricate large stretchable circuit boards. The conductivity of the printed composite increased with the reduced wire dimensions. The Ag-epoxy was also used to bond commercial electrical components on straight tracks for the construction of electronic circuits which interface soft and hard circuits together [33, 34].

Elastomers realize integrating transparency, flexibility, and functionality together in technology, but it cannot withstand high processing temperature. This drawback has again been rectified by Nomura et al. [35]. High-temperature-processed functional oxides with the flexible, elastomeric elastomer thin-film substrate were developed for electrical conduction, ferroelectrics for memories, piezoelectrics for energy harvesting [36], and semiconductors or dielectrics for high-performance transistors. The transparent conductive indium tin oxide (ITO) was used as the functional layer and the technique enabled large area fabrication and scalability by microtectonics in the corrugated oxide thin film [37, 38]. Furthermore, functional, high-temperature-processed thin films with acid-free processing to PDMS substrates were realized from the high-quality semiconducting zinc oxide (ZnO). The stronger bonding nature of ZnO with the PDMS caused an easy processing with no need for tailored etching methods or dissolving of sacrificial layers [39] unlike transfer printing methods.

This particular chapter aims at investigating the various electronic applications of the PDMS-based composites such as sensing, electromagnetic radiation shielding, microwave absorption, and actuation.

2 Structure and Properties

PDMS consists of a flexible Si–O backbone and a repeating $Si(CH_3)_2O$ unit, of which the latter decides the molecular weight and thus most of the viscoelastic properties of the material [40]. Generally, the molecular weight of PDMS varies from 10,000 to 60,000 g mol^{-1} its specific gravity from 0.91 to 1.00 and dielectric constant from 2.72 to 2.75 [41]. This non-toxic, highly hydrophobic, and translucent polymer has a very low glass transition temperature of −127 °C [42], shear elastic modulus of 250 kPa [43], and optical transparency down to 300 nm [44]. PDMS swells in most of the hydrophobic solvents, whereas it is least affected in hydrophilic solvents [45]. Based on the application needs, the properties of PDMS can be altered by cross-linking the polymer [46] (e.g., vinyl cross-linking), or by adding fillers (e.g., silicon dioxide) to the polymer network [47].

The wider applications of PDMS in many fields are due to its simple fabrication process, design adaptability, optical transparency, controllable elastic property, gas permeability, and its spontaneous and reversible adhesion to commonly used substrates such as polystyrene and glass. The microfluidic systems based on the PDMS have increased importance nowadays especially in cell culture applications [48–50]. However, the chemical properties of PDMS, experimental artifacts introduced by it as well as the causes of metabolic and stress pathways variation in this polymer compared to the conventional systems need more investigation. Since PDMS is a cross-linked polymer of hydrophobic dimethylsiloxane oligomers, its use in culture systems arises two concerns—the possibility of leaching the residual uncross-linked oligomers from the bulk polymer to the culture medium and the chance of sequestration of small hydrophobic molecules from culture media by the porous, hydrophobic network.

In order to study the influence of PDMS network stiffness on the behavior of cells, its mechanical properties must be characterized after curing at different cross-link densities. The mechanical properties of PDMS can be tuned by changing its degree of cross-linking, i.e., the lower the degree of cross-linking, the lower its stiffness. But the soft polymers such as PDMS exhibit a significant toe region at the beginning of the stress–strain curves, which is an artifact caused by the slack and specimen misalignment. This region can be quite large according to the strain range. The calculated stress–strain curves are not linear, especially at the higher strain levels. In addition to this, the lack of material to manufacture the standard specimens and fixing the sample in the grips without local damage and slippage exist as other challenges.

3 Preparation and Characterizations

The base PDMS is a viscous liquid and becomes rigid and flexible upon the addition of the curing agent. In a typical preparation process, PDMS and the curing agent tetraethoxysilane (TEOS) were mixed at a ratio of 10:1 in a Petri dish and kept at room temperature for 1 h and then under vacuum for 12 h to obtain elastomeric PDMS. The flexibility and thus the properties vary depending on the preparation parameters and thickness (Table 1).

Table 1 Mode of preparations of various PDMS composites

Base medium	Solvent	Filler	Mode of preparation
Hydroxyl-terminated PDMS with TEOS	Dichloromethane solution mixing	Silica particles	Sonicated silica for 15 min and then added to PDMS
Hydride-terminated PDMS	Toluene solution mixing	Benzoxazine monomer	After thermal treatment at 150 °C for 2 h, add 10 % hexamethyldisiloxane in toluene and refluxed at 110 °C for 3 h, filtered and dried to form PDMS–diaminohexane–benzoxazine
PDMS	Melt mixing	$BaTiO_3$ multipods at 10–70 parts of polymer by weight	Mixed in an internal Brabender mixer at 45 rpm for 10 min to synthesize PDMS-BaTiO3
PDMS	Xylene wet mixing method	G-ODA	Sonicated G-ODA and then added to the polymer, stirred for 30 min, removed solvent, added TEOS with accelerant (dibutyltindilaurate), and molded to obtain G-ODA/PDMS composites

4 Electronic Applications of PDMS Composites

4.1 Sensors

A large number of sensors based on the elastomers composites find useful applications in flexible and wearable electronics [51–54]. The conducting alkyl-functionalized graphene (G-ODA) was prepared from graphite oxide and uniformly dispersed in PDMS using octadecylamine. The PDMS/G-ODA nanocomposites with an ultra-low percolation threshold (0.63 vol.%) was synthesized. However, remarkably highly positive piezoresistive response ($R/R0 > 400$ under the pressure of 1.2 MPa) with excellent repeatability, small hysteresis, and long-term durability was showed by 1.19 vol.% grapheme elastomer composite. The resistance exponentially increased with the pressure under uniaxial compression and the resistance–pressure curves remain unchanged after 1000 loading–unloading cycles [55].

Plasmonic sensors from gold nanostars-filled PDMS were fabricated by Shiohara et al. The nanostars display tunable optical properties in the visible and near IR, leading to strong electromagnetic field enhancement at their tips, and are uniformly immobilized with high density, as well as mechanical flexibility [56].

For the analysis based on both localized surface plasmon resonance (LSPR) and surface-enhanced Raman scattering (SERS), the PDMS strips were fixed onto the wall of the cuvette, to which solutions with increasing MUA concentrations (from 10^{-8} to 10^{-4} M) were added. The absorbance spectrum in Fig. 1a shows the slight red shift of the plasmon peak of Au nanostars from 840 to 871 nm, with MUA concentration. The shift is linearly fitted for quantitative sensing as given in Fig. 1b, by which the detection limit is observed at 10^{-7} M and saturation of the

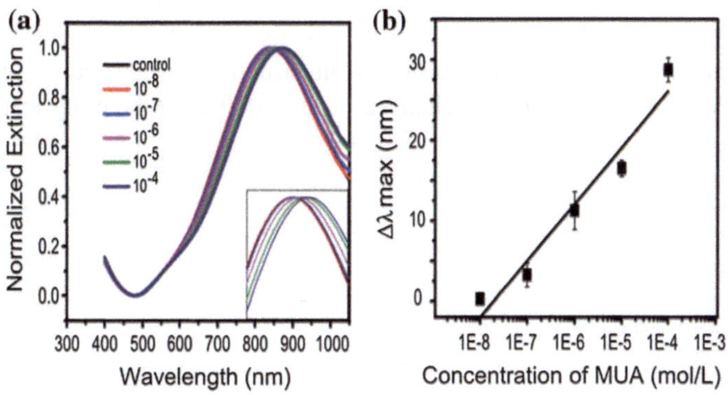

Fig. 1 **a** Extinction spectra of sample on a PDMS film immersed in different concentrations of MUA in ethanol. **b** LSPR shift versus MUA concentration. The *solid line* is a linear fit to the data [56]. Copyright 2014. Reprinted with permission from Royal Society of Chemistry

molecules on the surface of the AuNS at 10^{-4} M. These kinds of flexible sensors find application in detecting pesticides on fruit skin substrates using backside illumination [56].

The superior intrinsic piezoresistivity and the tunneling effect of the MWCNT network were effectively utilized to fabricate patterned MWCNT/PDMS nanocomposite strain sensors by a microelectromechanical system-assisted electrophoretic deposition (EPD) technique [57]. As shown in Fig. 2, the normalized resistance change of the four prepared samples displays similar monotonic increase with strain. At higher *R*0, the strain-normalized resistance change is linear (at strain range 0–2.5 %), with calculated gauge factors of 20 (*R*0-462 kΩ) and 32 (*R*0-380 kΩ). During 2.6–14 %, strain nonlinearity appeared, and upon strain release, the resistances of these two sensors were immediately restored to 420 and 590 kΩ, and after 4 h to 390 and 450 kΩ. This indicates that the sensing process does not permanently deform/damage the microlines. The patternable MWCNT/PDMS sensors possess additional advantages of larger working range (>10 %) and mechanical flexibility compared to the conventional metal foil sensors and find applications in tissues, microfluidic systems, as well as flexible 3D microsystems.

The properties of PDMS such as high flexibility (shear modulus 250 kPa at room temperature) and viscoelasticity [58] are the main reasons for its wider applicability. A multiscale-structured elastomeric electrode is made out of PDMS by utilizing its spontaneous buckle formation with the embedded silver nanowires [59]. This electrode is laminated onto the dielectric layer/bottom electrode template to fabricate pressure sensor.

The stability of the pressure sensor was investigated by performing repeated loading/unloading cycling test repeatedly up to 1500 times. Figure 3a shows the capacitance change at 1.5 kPa which indicates high stability for the pressure sensor. The bending stability of the sensor was also tested by using a homemade bending apparatus (by fixing the multiscale-structured electrode to the bottom plane). The relative capacitance changes of the 5000-cycle bending tested sensor

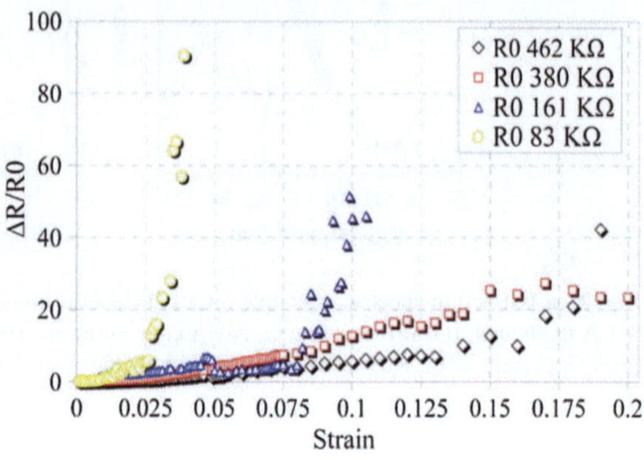

Fig. 2 Strain versus normalized resistance change of the MWCNT/PDMS sensors with different initial resistances (R0) [57]. Copyright 2013. Reprinted with permission from John Wiley & Sons

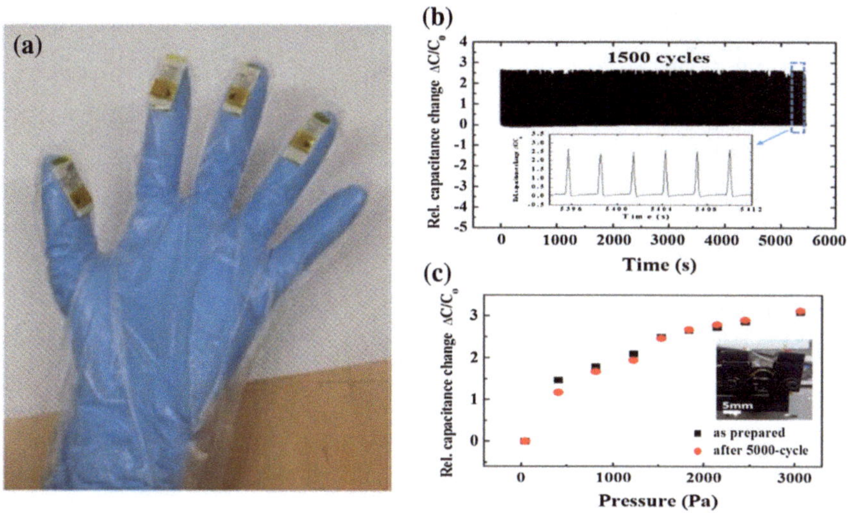

Fig. 3 Fingertip grip pressure sensing device. **a** Photograph of the fingertip grip pressure sensor. Each pressure sensor is attached on the finger tips. Stability of the pressure sensor. **b** Stability of pressure response to the 1500-cycle loading/unloading pressure of 1500 Pa. **c** Bending stability of pressure response after 5000-cycle bending test [59]. Copyright 2015. Reprinted with permission from Royal Society of Chemistry

with 3 mm radius of curvature at each pressure show no appreciable degradation (Fig. 3b), specifying the sensor is robust. The developed pressure sensor measures the pressure of fingertips when an object is grabbed with fingers, thus facilitating its use in wearable electronic skins for rehabilitation treatment of patients and pressure mapping in sports activities or daily activities. Moreover, this highly sensitive and stable capacitive pressure sensor follows low-cost fabrication and easy handling.

The typical amperometric response of a biosensor made by assembling acetylcholinesterase (AChE) on PDMS-poly(diallydimethylammonium) (PDDA)/gold nanoparticles (AuNPs) composite film for organophosphate pesticides is illustrated in Fig. 4. Curve a indicates the behavior of the sensor before inhibition and curves b and c after inhibiting in 10^{-5} g/L and 10^{-9} g/L parathion, respectively, for 10 min [60]. This sensor has utmost importance as it is capable in detecting pesticides and can widely applied to detect adulteration of fruits and vegetables. High-performance UV photodetectors were also fabricated using flexible sensor of ZnO nanowire sheet embedded in PDMS. On exposure to atmosphere, the sensor showed reproducible UV photoresponse and enhanced photoconduction. The reduced response speed was due to the PDMS coating than that of the ZnO nanowire film. Thus, the mechanism for UV photoresponse is understood as the diffusion of oxygen molecules in PDMS [61].

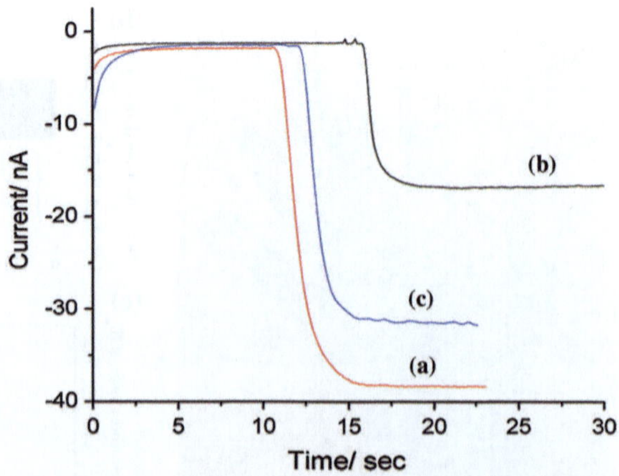

Fig. 4 Amperometric responses of PDMS-PDDA/AuNPs/ChO/AChE-modified electrode before (*a*) and after inhibition in 10^{-5} g/L (*b*) and 10^{-9} g/L (*c*) parathion for 10 min to 0.20 mM acetylcholine in a stirring 0.1 M pH 9.0 borate buffer under 0.7 V [60]. Copyright 2009. Reprinted with permission from American Chemical Society

4.2 Actuators

Actuators respond to applied electric field by body movement and many viscoelastic PDMS-based composites show actuation. Depending on the applied pre-strains, the graphene nanoplatelet (GNP)/PDMS composites showed actuation responses and such large IR-induced reversible and elastic responses were studied with increasing pre-strains by Loomis et al. [62]. At 3–9 % pre-strain (low), the actuators showed reversible expansion, while during 15–40 % pre-strain, the actuators exhibited reversible contraction. For the composites (GNP content of 0.1–5 wt%), a photomechanically induced change in stress of four orders of magnitude was observed upon IR illumination compared to pristine PDMS. An extraordinary optical-to-mechanical energy conversion factor (ηM) of 7–9 $MPaW^{-1}$ for GNP/PDMS composite actuators is reported [62], compared to other forms of nanostructured carbon/PDMS composites, including carbon nanotubes (CNTs), for the same fabrication method.

The photomechanical properties of both single-walled (SWCNT) and multi-walled carbon nanotube (MWCNT)/polymer systems in multilayer were investigated to construct optical actuators. Both CNTs showed similar responses directly related to pre-strain and nanotube alignments. Such composites exhibited easier sample construction, intact polymer, and nanotube properties, and compatibility to MEMS processes [63]. The super-aligned CNT sheets (SACNS) were embedded into PDMS to fabricate a new kind of bending actuator through a two-step method. The coefficient of thermal expansion of the composite was reduced by two orders of magnitude compared to the neat PDMS. The actuator was easily operable and generated exceptionally large bending actuation with controllable motion at very low dc voltage [64]. The performance of the actuator is illustrated by the current voltage (*I*–*V*) curve in Fig. 5a, the linearity of which represents the Ohmic character.

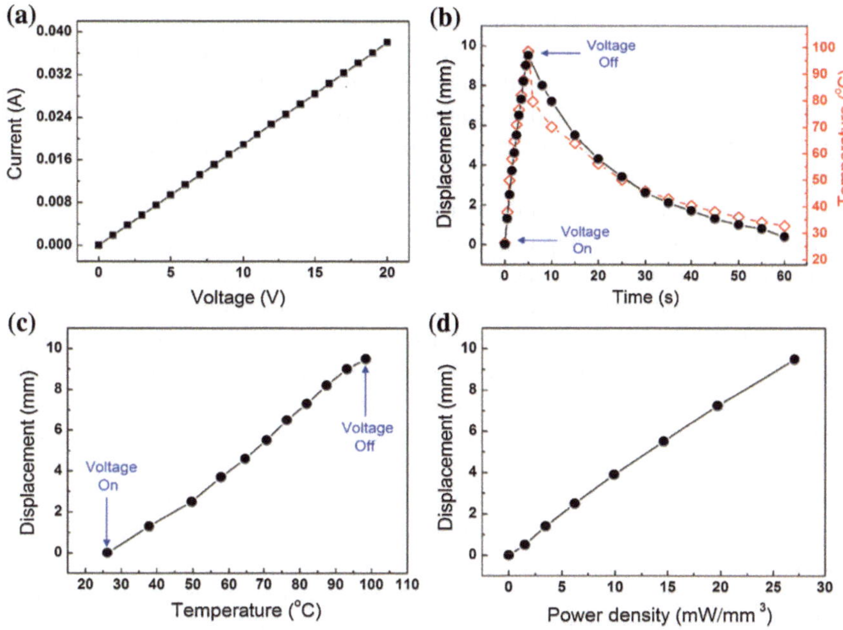

Fig. 5 **a** *I–V* curve, **b** displacements (*black solid cycle*) and temperature (*red open diamond*) variation with an applied dc voltage of 40 V and current of 78.5 mA, **c** temperature dependence of the displacements and **d** power density dependence of the displacements of the SACNS/PDMS composite actuator [64]. Copyright 2011. Reprinted with permission from American Chemical Society

There was about 5 wt% CNT in the CNT-rich layer of the composite and about 0.1 wt% in the entire composite. The continuous and super-aligned structure of the CNT conductive network caused the electrical conductivity to be up to 2280 S/m for the CNT layer and 60 S/m for the entire composite. The conductivity of the CNT layer was two orders of magnitude higher than the reported conductivity for the randomly oriented CNT (5 wt%)/PDMS composites. At 40 V-applied dc, the current became 78.5 mA, and the actuator began to bend immediately. After 5 s, the voltage was cut off and the free-end displacement of the actuator reached up to 9.5 mm. The influence of temperature on actuation displacement was nearly in line. Since the CNTs are perfect black bodies with high thermal conductivity, within the actuator, the electrical energies absorbed by the CNTs are converted to thermal energy and immediately heat up the entire actuator. This temperature enhancement expands the pure PDMS layer more than that of the CNT-rich layer, and this causes exceptional bending actuation of the structure. With temperature, the displacement increased rapidly (Fig. 5c), and once the voltage was cut off, the free-end returned to its initial position in 60 s. From 26 to 98 °C, the maximal displacement became 9.5 mm and the output displacement normalized to the beam length was 0.44 μm. This was much greater than that of the existing bimorph thermal actuators. Figure 5d shows the dependence of electric power density (for 5 s) on displacements where the increasing electric power density make CNTs to convert more thermal energies to heat and resulted in larger displacements. Therefore, the actuator is quite controllable through input energy [64].

The same actuator was used to manipulate a piece of plastic foam as shown in Fig. 6. At 35 V-applied dc, the free-ends of both actuators were getting closer to each other to clamp the small object. After moving to the other Petri dish, voltage was cut off to release the object there [64].

Many laboratory-on-a-chip (LoC) devices, optics, biomedical applications, and electronic paper (e-paper) use the process of electrowetting. PDMS is used for fabricating electrowetting-on-dielectric (EWOD)-based actuators and the influence of electrode-electrolyte-insulator interface layer capacitance at the metal–liquid interface by means of contact angle versus applied voltage experiment at different pH was studied by Sohail et al. [65]. Loomis et al. further studied the response of GNP/PDMS actuators on IR illumination as illustrated in Fig. 7. The upper ends of the three samples were fixed to a rigid plate with the free-ends unrestricted, weight attached and illuminated with weight, respectively, for samples 1, 2, and 3 (Fig. 7a). The IR LED caused in energy transduction to the polymeric chains and thus the actuator contracted.

GNPs absorb light, and the energy is subsequently transduced to the polymeric chains through phonons in the sp2 graphene sheet and this creates entropic elasticity to controllably produce significant amounts of motion in the composite. Along with the macroscopic thermal conductivity of graphene (300 W/m/K), the good dispersion of GNPs within PDMS also cause heat percolation and thus polymeric chain contraction. In Fig. 7b, the thin GNP/PDMS composite strip is held at a fixed pre-strain with a nanopositioning stage mounted in the composite center, and IR light-emitting diodes (LEDs) placed on either side for differential control via both positive and negative stage actuation. When the LED is illuminated, it heats the composite and causes polymeric chain contraction in that region and subsequent stage translation. By dynamically modulating IR intensity, a force balance can be enacted on either side of the stage to maintain it in a desired displacement.

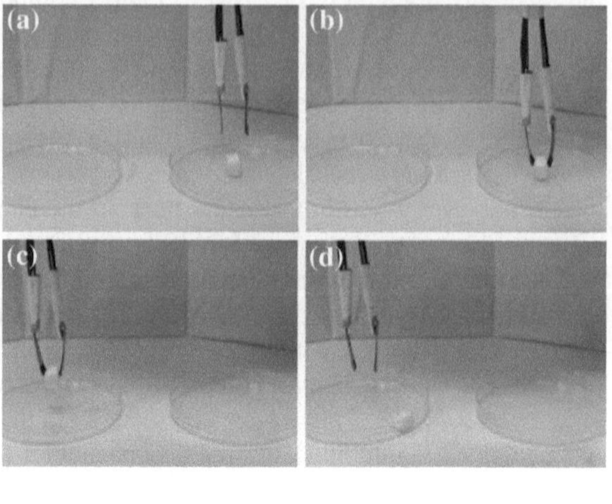

Fig. 6 A demonstration of a gripper manipulating a small object: **a** without, **b** with, **c** with, **d** without an applied dc voltage of 35 V [64]. Copyright 2011. Reprinted with permission from American Chemical Society

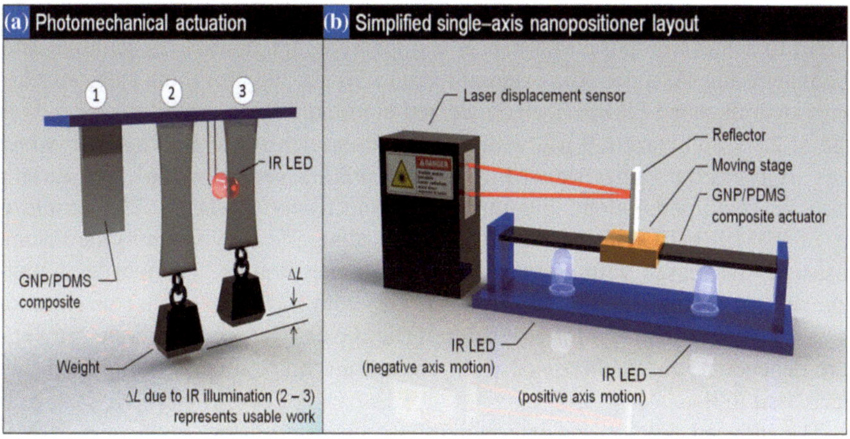

Fig. 7 **a** Photothermal actuation schematic. **b** Simplified single-axis nanopositioner layout. Independently controlled diodes on either side of the stage allow for differential positive or negative axis stage motion [81]. Copyright 2013. Reprinted with permission from Nature, Open Access

However, even when the stage is at a set position, LED intensity is still being continuously tuned via a proportional–integral–derivative (PID) control loop which monitors stage position and compensates for thermal drift. This highly dynamic process achieves nanopositioning by controlling polymeric chain extension/contraction through rapidly modulating thermal energy input. Stage displacement is measured with the aid of high-speed laser displacement sensor and a reflective element placed on the stage. Finally, a two-axis photothermal positioning was achieved with 120 nm resolution, 5 mm/s actuation speed, and 0.03 % maximum actuator efficiency, about 1000 times greater than recently reported for light-driven polymer systems.

In general, the strong covalent sp2 bonds result in efficient heat transfer by lattice vibrations in the GNPs, and allow photothermal actuation. This enables a unique photothermal actuation mechanism in all kinds of nanocarbon/elastomer composites containing sp2 covalent bonds and as the fraction of sp3 bonds increases in the system (as in the case of reduced graphene oxide) the thermal conductivity of the system lowers.

4.3 Piezoelectric Properties

Nanogenerators with high electrical output by cost-effective way of production are attracting wider attention. Such a generator was directly used to lighten 28 commercial LEDs without any energy storage unit or rectification circuit. This was made by filling PDMS with the piezoelectric barium titanate (BT) and MWCNT and thus creating pyramid microstructures. The composition of the film was optimized to achieve output performance, and favorable toughness was achieved after thermal curing. At a periodic external force of 20 Hz, the film (both sides attached with an arch-shape ITO/PET electrode and aluminum film) exhibited a peak voltage of 22 V and current 9 μA with a peak current density of 1.13 $\mu A/cm^2$ which was six times the values showed by the neat PDMS generator [66]. The $BaTiO_3$ particles (multipods) alone also impart piezoelectricity to PDMS. The permittivity of PDMS-$BaTiO_3$ nanocomposites increased significantly whereas the volume resistivity decreased with increase in $BaTiO_3$ concentration. The tensile strength and percent elongation at break also decreased with $BaTiO_3$ particle concentration because of their non-reinforcing nature. The dielectric properties were temperature-dependent, and it increased with temperature up to a certain limit and then decreased [67].

The induced dipoles and interfacial/space charges present within the PDMS-$BaTiO_3$ composites undergo polarization on applying external electric field. In addition, the external stress changes the geometry of the composite and thus the overall charge density of the system (Fig. 8). This creates a change in capacitance/dielectric constant of the composites and the restricted movement of the bound charges causes the current to flow leading to an increase in the conductivity.

Figure 9 shows the effect of external stress on the dc current (I_{dc}), dielectric constant (ε'), and dc conductivity (σ_{dc}) of the the PDMS/$BaTiO_3$ (41.18 wt%). Both dc current and dc conductivity suddenly increased after the initial load of ~74 kPa and thereafter slowly but continuously increased with further increase in pressure. Beyond ~295 kPa, the change in value of the current with applied stress is marginal. The permittivity increase with increase in external pressure at three different frequencies (10 Hz, 1 kHz, and 1 MHz) in a similar fashion is also clear from Fig. 9c.

The detail power generation mechanism is investigated from the piezopotential distributions inside the 10-mm-diameter ZnO hemispheres embedded PDMS composites under the load of 640 mN as in Fig. 10. ZnO material is assumed to be

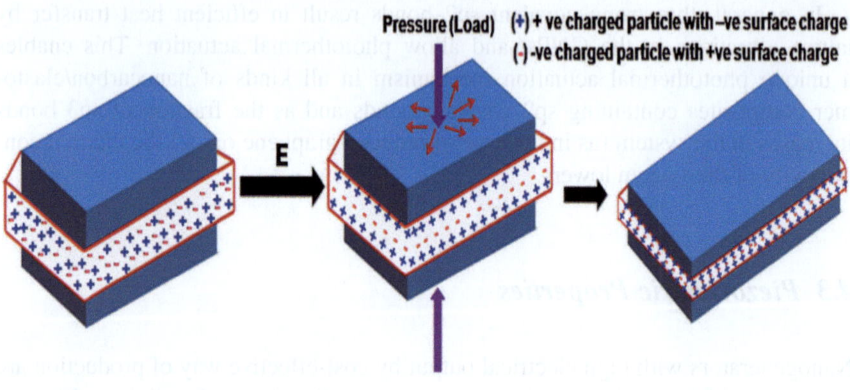

Fig. 8 Schematic representation of positions of the charged species in PDMS-$BaTiO_3$ composites before and after the application of both external electric field and pressure [67]. Copyright 2014. Reprinted with permission from American Chemical Society

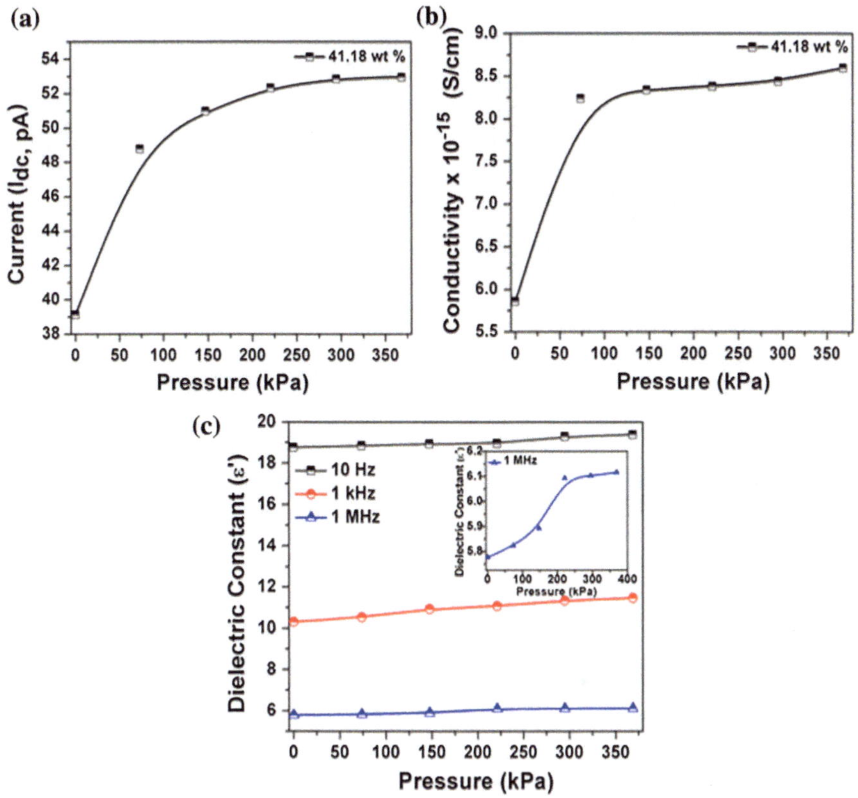

Fig. 9 Effect of pressure on **a** dc current (I_{dc}), **b** dc conductivity (σ_{dc}), and **c** dielectric constant (ε') of composite system filled with 41.18 wt% of $BaTiO_3$ at three different frequencies (10 Hz, 1 kHz, and 1 MHz) [67]. Copyright 2014. Reprinted with permission from American Chemical Society

insulator and its dielectric constant is taken as 8.5446 [67]. Upon convex bending (Fig. 10c), the tensile stress is built up over the entire PDMS matrix along the transverse direction, causing the tensile strain to the ZnO hemispheres also along the same direction. This induces compressive stress on the hemispheres along the vertical direction and promotes the strain distribution along the vertical direction, causing a gradient in the piezoelectric potential. As in Fig. 10d, the tensile strain in the transverse direction under convex bending is $\varepsilon = \Delta x/x$ and the compressive strain (ε_z) in the vertical direction is calculated as $-0.4\ \varepsilon_x$. As the hemisphere diameter increases, the displacement of the hemisphere along the vertical direction also increases and creates larger piezoelectric potential (Fig. 10e). This increase in the displacement may ascribe to the enhancement of the strain confinement effect in larger hemisphere. In addition, the randomly distributed nanoparticles or nanowires in the PDMS matrix do not generate large output power and electrical poling or high applied force is necessary to orient such electric dipoles. The ZnO/PDMS system discussed here gives 2 times higher piezopotential and

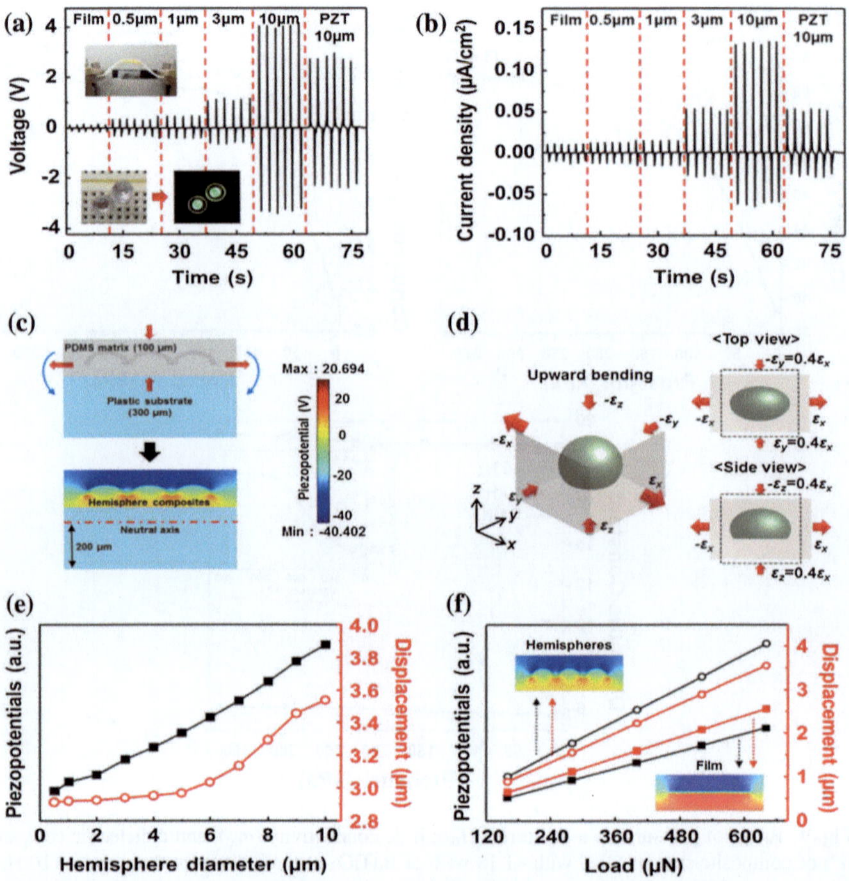

Fig. 10 **a** Output voltage and **b** current density, generated by the ZnO hemisphere-embedded composites as a function of the diameter of the hemispheres from 0.5 to 10 μm and PZT hemispheres under the convex bending strain. **c** Piezopotential and **d** strain distributions of hemispheres in PDMS. **e** Piezopotentials and displacement increase with the hemisphere diameter from 0.5 to 10 μm. **f** Enhancement in the piezopotential and displacement as increase of external force from 160 to 640 μN [68]. Copyright 2015. Reprinted with permission from Elsevier

displacement than that of flat film under the load from 160 to 640 mN (Fig. 10f) and thus proves uniform distribution.

Layer by layer stacking of hemispheres, bending direction, etc. influence the strain sensitivity of the films and high sensitivity is achieved by convex bending due to the strong electric dipole alignment. Such highly ordered and highly stretchable piezoelectric hemisphere composite films are useful in fabricating self-powered nanogenerators as exceptionally sensitive sensor for providing sensitive motion information from a human body. In practice, the films were attached on the wrist and its output voltage/current density provided the information on the wrist motion [68].

4.4 Dielectric Properties and Super Capacitors

Dielectric properties of the composites depend on the size and the surface modification of the reinforcing filler inclusions along with the elastomer properties. Investigation was done on the dielectric performances of the silica nanofillers embedded PDMS as a function of filler nanofiller size, volume fraction, and surface treatment by hexamethyldisiloxane. The broad-spectrum dielectric properties such as dielectric constant, the dielectric loss, and tan δ depended on the size of the nanoinclusions attributed to their high specific surface area and pronounced interfacial interactions. The surface-modified silica improved the dielectric properties of PDMS further due to changes occurred in the surface properties [69]. PDMS-based core shell nanocomposites with high PDMS content (80 wt%) are reported for exhibiting similar interfacial relaxation process of conventional PDMS/silica nanocomposites, whereas composite with lower PDMS content (40 wt%) showed some differences [70].

In another work, stretchable supercapacitors with nonvolatile electrolyte component was developed using an ionic-liquid-based nonvolatile gel in CNT. Such CNT/ion–gel supercapacitors showed high capacitance retention (96.6 %) over 3000 stretch cycles at 20 % strain. High durability against stretch cycles was achieved by introducing microroughness at the interfaces by imprinting the surface microstructure of office paper onto a PDMS substrate. This roughness strengthens the adhesion between different layers and prevents delamination over repeated stretch cycles. Even after 3000 cycles of stretch, the shape of the CV curves was generally maintained except the slope at 0 and 3 V, where the sweep direction changed and became slightly less steep (Fig. 11a). Though variation is observed

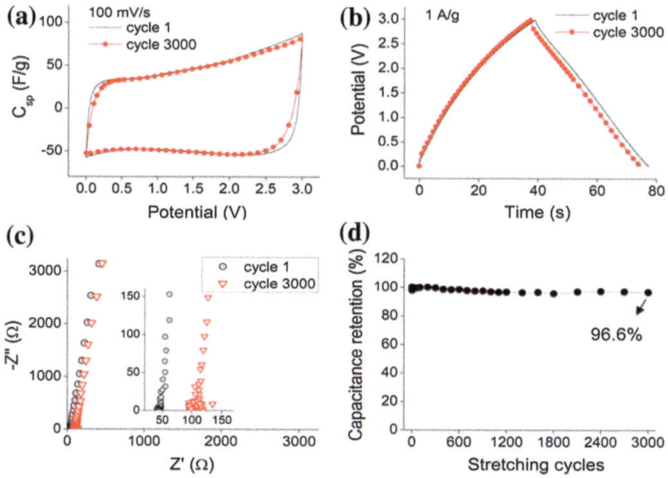

Fig. 11 Supercapacitor characteristics at 3000 stretch cycles: **a** CV curves; **b** galvanostatic charge–discharge curves; **c** Nyquist plots measured at the 1st and 3000th stretch cycles; **d** capacitance retention over 3000 stretch cycles [71]. Copyright 2014. Reprinted with permission from American Chemical Society

in the galvanostatic charge–discharge curves and Nyquist plots, the general shapes of the curves are maintained, indicating negligible change in the capacitance (Fig. 11b, c). In addition, the charge–discharge curves at various current densities confirm high rate capability of the supercapacitors. Thus, the ion–gel increased the operating voltage window (3 V) and hence the energy density, thus facilitating flexible and/or stretchable energy storage devices with high durability [71].

In addition to flexibility, transparency was also achieved by Yuksel et al. in their PDMS/SWCNT solid-state supercapacitor. For this, the symmetric electrodes made up of binder-free SWCNT thin films were deposited on PDMS substrates by vacuum filtration followed by a stamping method. The devices assembled using a gel electrolyte showed optical transmittance of 82 %, and a specific capacitance of 22.2 F/g for 0.02 mg of SWCNTs [72]. The galvanostatic charge/discharge profile for the supercapacitors with respect to SWCNT loading at 1.25 A/g current density is shown in Fig. 12a. The small internal resistance drop observed was caused by equivalent series resistance including electrode, electrolyte, and contact resistance between the electrode and the electrolyte. With SWCNT loading, the specific capacitance significantly increased due to the improved overall charge storage capacity of the composites. The porous SWCNT films allow the gel electrolyte access to the active area. The charge/discharge characteristics at various current densities (Fig. 12b) show similar curve shapes, indicating the dependence of capacitance on

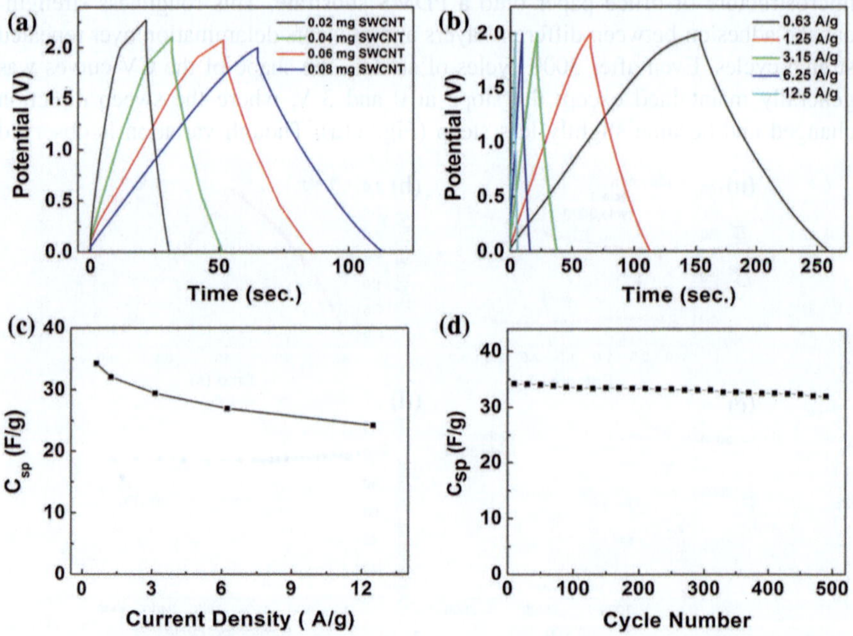

Fig. 12 **a** Galvanostatic charge–discharge profile for supercapacitors at 1.25 A/g and **b** at different current densities. **c** Specific capacitance change with respect to potential scan rate (lines are for visual aid). **d** Cycle performance of flexible supercapacitors at a current density of 1.25 A/g [72]. Copyright 2014. Reprinted with permission from American Chemical Society

the applied current. The power density became 41.5 kW kg^{-1} with good capacity retention (94 %) upon cycling over 500 times (Fig. 12d). The transparency and flexibility find specific applications in the field of displays and touch screens.

Chen et al. also achieved the transparency and flexibility with wrinkled graphene (WG) sheet-filled PDMS materials in which the WG acts both as current collector and electrode materials. The final device showed a high transparency (57 % at 550 nm) and can be stretched up to 40 % strain without obvious performance change over hundreds of stretching cycles.

The authors investigated the stretchability of the supercapacitors at different strains by measuring their CV and galvanostatic charging–discharging characteristics. The supercapacitors were fixed on a mechanically stretching machine and stretched up to 40 % strain (Fig. 13a, b) and the effect of bending was checked (Fig. 13c, d). Both planar graphene (PG) and WG sheets used were with a growth time of 2 min and their electrochemical performance (Fig. 13e) as well as their specific capacitance (Fig. 13f) was almost unchanged at 40 % strain. Over hundreds of cycles of stretching also these supercapacitors showed outstanding stability (Fig. 13g, h). Similar stable performance was also observed when these supercapacitors were bent to a large bending radius.

By using highly aligned CNT sheets of excellent optical transmittance and mechanical stretchability as both the current collector and active electrode, high-performance transparent and stretchable all-solid supercapacitors with a good stability were developed [73]. With two layers of CNT, the composite showed a transmittance of 75 % at 550 nm wavelength and specific capacitance of 7.3 F g at 30 % biaxial stretching without any obvious change in electrochemical performance even over hundred stretching cycles.

The effect of single-layer CNT sheet on PDMS (Fig. 14a) imparted high flexibility and easiness to bend/fold without any effect on the device performance (Fig. 14a, b). In other words, a parallelly assembled capacitor with two transparent electrodes made of one layer of CNT sheet attached on PDMS showed 64 % transmittance at 550 nm wavelength (Fig. 14c), whereas a single electrode with two layers of CNT sheets exhibited higher transmittance value of 68 % (Fig. 14d) (an overall transmittance of 78 %) due to the presence of the polymer (PVA) electrolyte and another PDMS layer in the device assembly. However, if the supercapacitor was assembled in crossover configuration, the CNT overlapping get reduced and the corresponding transmittance became 75 % at 550 nm (Fig. 14c), compared to the parallelly assembled device. Good optical stability was also observed for the assembled supercapacitors after 100 stretching cycles. But the transmittance of the parallelly assembled supercapacitors reduced from 64 to 35 % at 550 nm as the CNT layers on each of the two transparent electrodes increased from one to three. For the cross-assembled supercapacitor, 7.3 F g^{-1} of specific cell capacitance was obtained at 0.23 A g^{-1} discharge current (Fig. 14e), higher than that of the parallelly assembled device (4.9 F g^{-1}) under the discharge current of 0.25 A g^{-1}. For a single electrode (cross-assembled), the specific capacitance was about 29.2 F g^{-1} better than the reported values. For both supercapacitors, the specific capacitance decreased with increase in current density (Fig. 14f) attributed to the high series

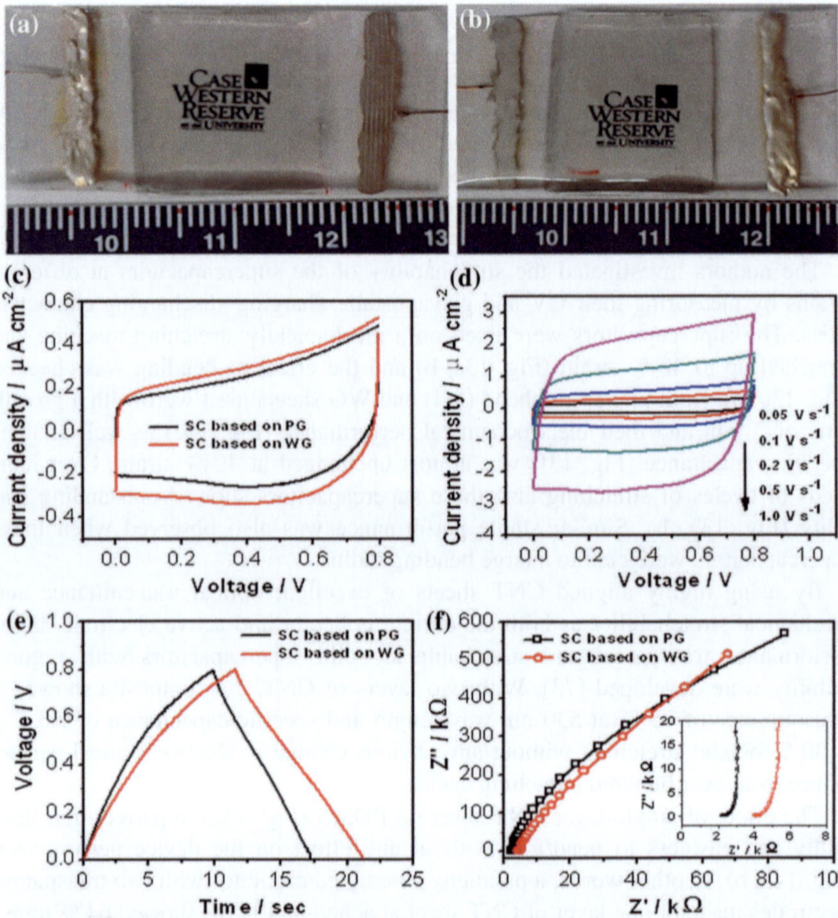

Fig. 13 Digital photographs of the supercapacitors based on the WG (**a**) and PG (**b**). CV curves of the supercapacitors using **c** PG and WG at the scan rate of 0.1 V s^{-1}, **d** WG at different scan rates, **e** galvanostatic charging discharging curves of the supercapacitors at 0.8 μA current, **f** Nyquist plots with a frequency from 10^{-2} to 10^{5} Hz [82]. Copyright 2013. Reprinted with permission from American Chemical Society

resistance of the thin CNT sheet and the inefficient contact among the CNTs. Figure 14g shows the Ragone plots with calculated energy density (E) and power density (P) from which the energy density for the cross- and parallel assembly, respectively, gives 2.4 and 1.4 Wh kg^{-1} at a power density of 0.9 kW kg^{-1}. These values were comparable to the flexible supercapacitor based on In_2O_3 nanowire/carbon nanotube heterogeneous films.

An increase of ten-fold in the dielectric permittivity at 10 Hz is reported for the functionalized graphene sheet (FGS)-filled (2 wt%) PDMS nanocomposite while maintaining low dielectric loss and good mechanical property. Here, the functional groups present on the graphene sheet surface improved the nanofiller/polymer

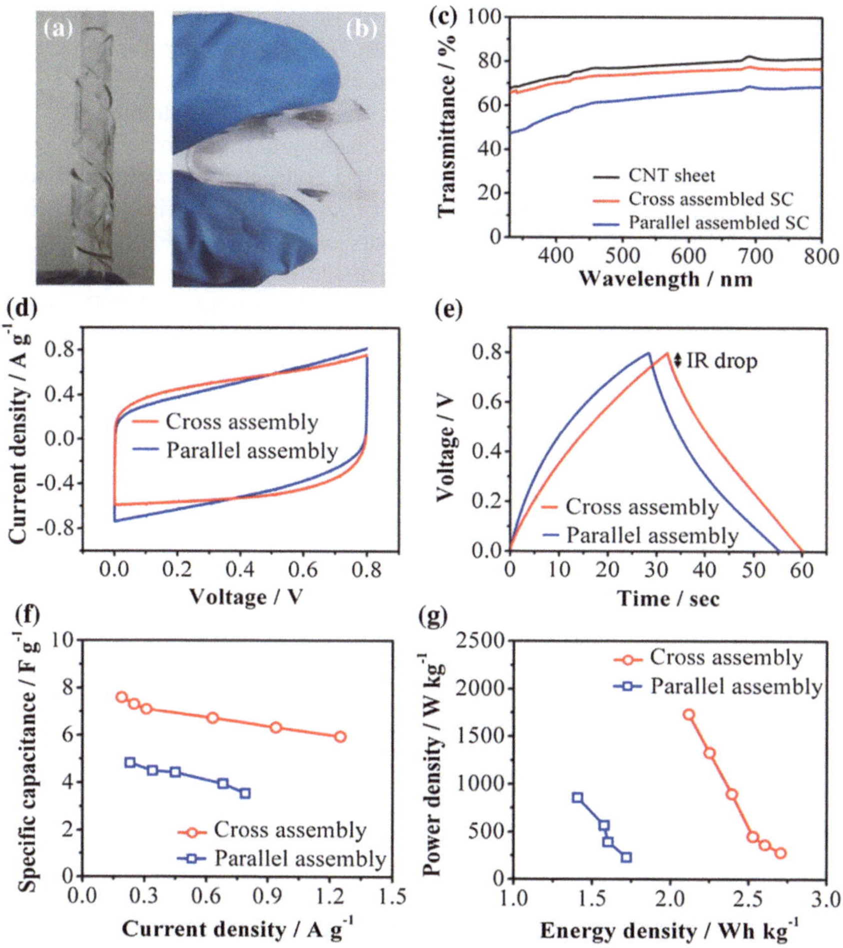

Fig. 14 The optical and electrochemical properties of the CNT/PDMS transparent supercapacitors. A photograph of the supercapacitor **a** wrapped outside along a glass tube. **b** Bent. **c** Transmittance spectra of one-layer CNT sheet on PDMS substrate and its derived supercapacitors with different assembled configurations. **d** CV curves of the supercapacitors at a scan rate of 0.1 V s^{-1}. **e** The galvanostatic charging–discharging curves of the supercapacitors at a constant current density of 0.25 A g^{-1} for the parallel assembly and 0.23 A g^{-1} for the cross-assembly. **f** Dependence of specific capacitance of devices on the discharging current density for both types of devices. **g** Ragone plots of the both types of supercapacitors [73]. Copyright 2014. Reprinted with permission from Nature, Open Access

compatibility at the interface, reducing the polarization process [74]. In another work by Pisanello et al., the magnetic iron oxide nanoparticles (IONPs) were used to fill PDMS to increase its dielectric permittivity in the GHz frequency range [75]. The developed composites are good substrates for radiofrequency applications due to magnetophoretic transport and its efficacy as well as the electromagnetic properties of the nanocomposite which depend on IONPs diameter.

At 8 wt% IONP concentration, the property of aligned samples varied at different diameters $D \approx 15$ nm and $D \approx 29$ nm. The higher the particle diameter, the more pronounced the alignment effect on $\varepsilon' r$ value (Fig. 15). At $D \approx 15$ nm, $\varepsilon' r$ values for random and aligned composites are almost comparable (Fig. 15a), while at $D \approx 29$ nm, $\varepsilon' r$ is higher for aligned nanoparticles, approaching $\varepsilon' r \approx 3$. This indicates the dependency of $\varepsilon' r$ on particle concentration, size distribution, and spatial alignment. The increased value of $\varepsilon' r$ for IONPs with higher diameter ($D \approx 29$ nm) is attributed to their cluster assembly with PDMS, and their interaction compared to the smaller ones. For the random samples with different IONP diameters and for those with magnetically assembled IONPs, almost no differences have been recorded in $\varepsilon' r$, assigned to a thermal demagnetization process. As evident from Fig. 15, average values of both tan $\delta\varepsilon = \varepsilon'' r/\varepsilon' r$ and tan $\delta\mu = \mu'' r/\mu' r$ are well below 0.02 up to 2.5 GHz, indicating the dissipation of less than 2 % of electromagnetic energy passing through it. It is clear from this

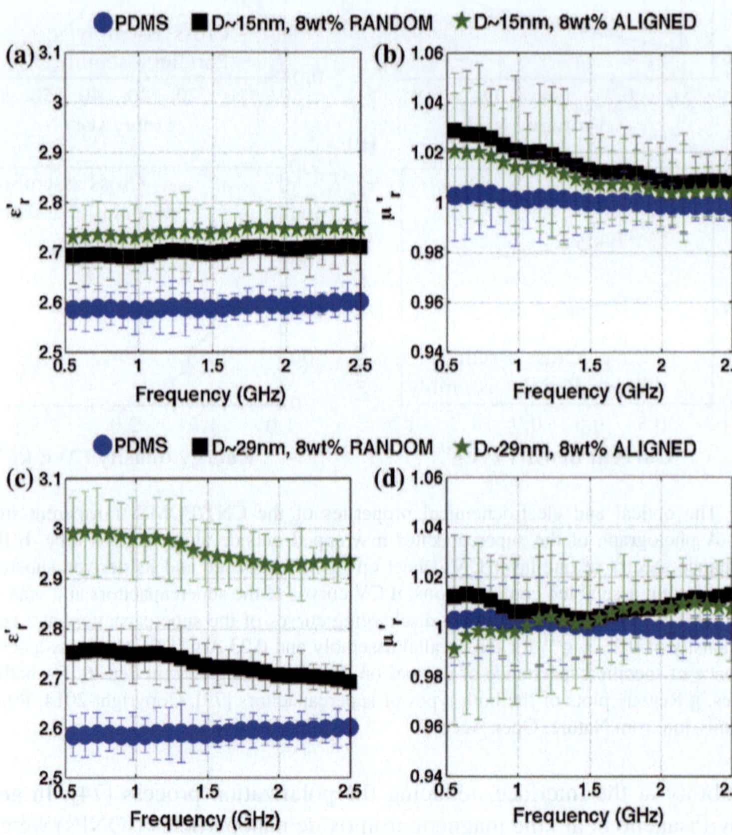

Fig. 15 Radiofrequency characterization of aligned IONP/PDMS. **a**, **c** Dielectric permittivity and **b**, **d** magnetic permeability at 0.5–2.5 GHz for pure PDMS (*blue dots*), random (*black squares*) and aligned (*green stars*) at ~8 wt% NPs concentration for (**a**, **b**) $D \approx 15$ nm and (**c**, **d**) $D \approx 29$ nm [75]. Copyright 2013. Reprinted with permission from American Chemical Society

result that the IONPs do not form conductive paths through the composites and PDMS matrix acts as an electrical insulator between the NPs and prevents them from conducting charges [75]. Due to the lower dielectric loss factors, these kinds of devices are useful for radiofrequency applications, and in particular for antennas' substrates.

4.5 Thermal Conductivity

PDMS is widely used in fabricating light-emitting diodes (LED) and field-effect transistors. An integrated LED device was coated with a PDMS film containing taper holes microstructures arrays (THMA). The electroluminescence spectra, luminous flux, radiant flux, and light extraction efficiency of the device were checked and the experimental results demonstrated enhanced light extraction efficiency up to 23.41–24.43 % under different injection currents for the integrated LED devices by THMA [76].

The integrated LED device (Fig. 16a) consists of 30 LED chips grown by metal organic chemical vapor deposition (MOCVD) reactor, 10 of which are connected in series as one group. All chips are divided into three groups, with parallel configuration. The luminous efficacy of both samples (Fig. 16b) showed decrease with increase in injection current due to an Auger process. When the two samples are compared, B1 showed higher luminous efficacy (about 23.41–24.43 %) at any injection current due to differences of two devices' resistance [76].

Gallium nitride-based green flip-chip light-emitting diodes (FCLEDs) employed with inverted tetrahedron-pyramidal micropatterned PDMS (ITPM PDMS) films also showed enhanced light output power. Soft imprint lithography method was employed to create PDMS micropatterns on sapphire substrates and

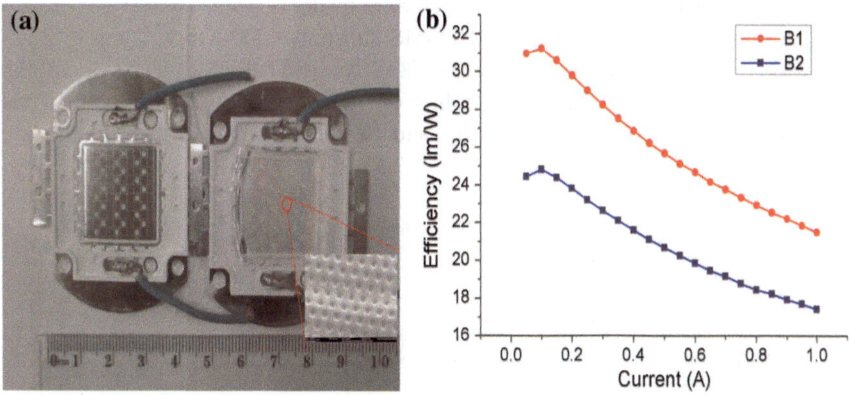

Fig. 16 **a** Samples of integrated LEDs devices: the right is B1 with the THMA, the left B2 without THMA. **b** Comparison of luminous efficacy [76]. Copyright 2015. Reprinted with permission from Elsevier

the laminated ITPM PDMS film enhanced the diffuse transmittance to ~53 % at 525 nm wavelength. When this ITPM PDMS film was introduced on the outer surface of sapphire in FCLEDs, the light output power reached ~11.1 % at 500 mA of injection current compared to the reference FCLED, together with better electroluminescence intensity and far-field radiation pattern [77]. White organic light-emitting device (OLED) with a flexible PDMS film embedded with yellow phosphor is reported through a simple, reproducible, and cost-effective approach. The resultant electroluminescent spectrum showed chromatic coordinates of 0.38 and 0.54 and correlated color temperature of 4200 K for the white OLED. An yield of 10.3 cd/A and the luminous power efficiency of 5.4 lm/W at a luminance of 1000 cd/m^2 along with a desirable Lambertian-like far-field pattern are obtained from the white OLEDs [78].

4.6 Other Electronic Applications

Other electronic applications of PDMS composites include transistors and shielding devices. Using an organic ferroelectric poly(vinylidene-trifluoroethylene) as gate insulator and an organic semiconducting poly(9,9-dioctylfluorene-co-bithiophene) as active channel layers on PDMS, flexible organic ferroelectric nonvolatile memory thin-film transistors (OFMTs) were fabricated. The carrier mobility, on/off ratio, and subthreshold swing of the OFMTs were 5×10^{-2} cm^2 V^{-1} s^{-1}, 7.5×10^3, and 2.5 V/decade, respectively, which did not change when the substrate was bent with 0.6 cm radius of curvature. The memory on/off ratio was maintained to be 20 even after a lapse of 2000 s, and thus, the material can be used to fabricate mechanically stretchable nonvolatile memory devices [79].

Gold electrode pads are plated on both sides of the PVDF composite using e-beam evaporation as illustrated in the scheme in Fig. 17a, and the Au/P(VDF-TrFE)/Au capacitors on PDMS substrates were made.

The photograph of the processed PDMS substrate (300 l m × 2 cm × 2 cm) on which the Al (gate)/P(VDF-TrFE)/F8T2/Au (S/D)/PDMS TFTs were fabricated is given in Fig. 18a. The bending of the substrate was monitored by setting up a proving system (Fig. 18b), and the results for the memory transfer

Fig. 17 **a** Schematic cross-sectional diagram of the fabricated P(VDF-TrFE) capacitors and photograph of **b** PDMS/Au/P(VDF-TrFE)/Au capacitors, **c** capacitor under bending [83]. Copyright 2011. Reprinted with permission from Elsevier

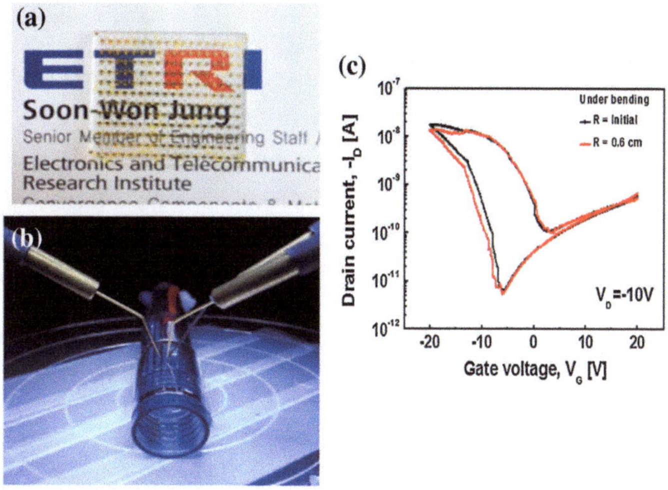

Fig. 18 A typical photograph **a** of a PDMS substrate, **b** electrical evaluation when the processed PDMS substrate was bent, **c** variations of the transfer characteristics and memory behaviors of the fabricated organic memory TFT under bending with radius of 0.6 cm [79]. Copyright 2015. Reprinted with permission from Elsevier

characteristics (Fig. 18c) substantiate the bending durability when the substrate was bent with a radius of curvature of 0.6 cm. Here, the OFMT did not experience marked variations in its device behaviors (the memory window was 1 V at most) and the bending radius could not be reduced to smaller state owing to the limitations of substrate size and machine specification. Thus, cyclic bending tests with large number of bending at smaller radius of curvature are very necessary to be addressed before considering it for practical applications.

Highly flexible transparent and conductive coatings similar to indium tin oxide (ITO) films on glass but with superior conductivity at high bending and durability to scratching were developed by applying annealed silver nanowire network-embedded polyurethane optical adhesive on PET and PDMS substrates [80]. Coatings on PDMS withstand up to 76 % tensile strain and 250 bending cycles of 15 % strain with a negligible increase in electrical resistance. Since the AgNW network is embedded at the surface of the adhesive, these coatings also provide a smooth surface, making them suitable in flexible light-emitting electrochemical cells. These devices continue to emit light even while being bent to radii as low as 1.5 mm and perform as well as unstrained devices after 20 bending cycles of 25 % tensile strain [80] as illustrated in Fig. 19.

The characteristics of unstrained devices and devices subjected to 20 bending cycles of 25 % tensile strain are compared in Fig. 20. The temporal evolution of current and radiance of the devices during ten minutes of operation at 5 V in ambient conditions was recorded, and the external quantum efficiencies (EQE) for the unstrained (Fig. 20a and 20b) and strained devices (Fig. 20c and d) were calculated. Both devices reached maximum EQE by <90 s, followed by a decay

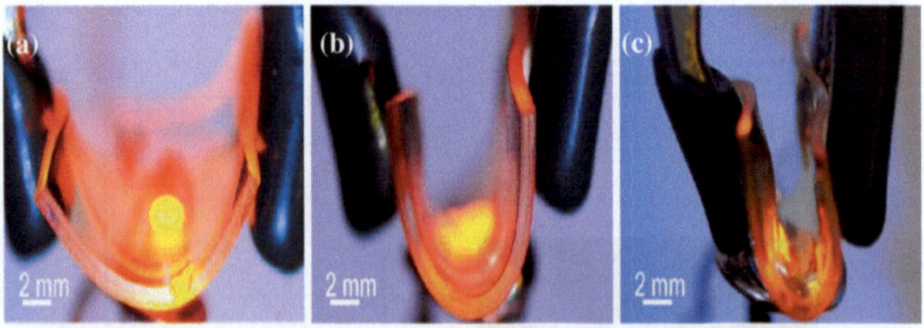

Fig. 19 Photographs of flexible LEECs fabricated with AgNW-OA (14 Ω/sq) transparent anodes on PDMS, bent to radii of **a** 7.0 mm; **b** 3.0 mm; and **c** 1.5 mm [80]. Copyright 2013. Reprinted with permission from American Chemical Society

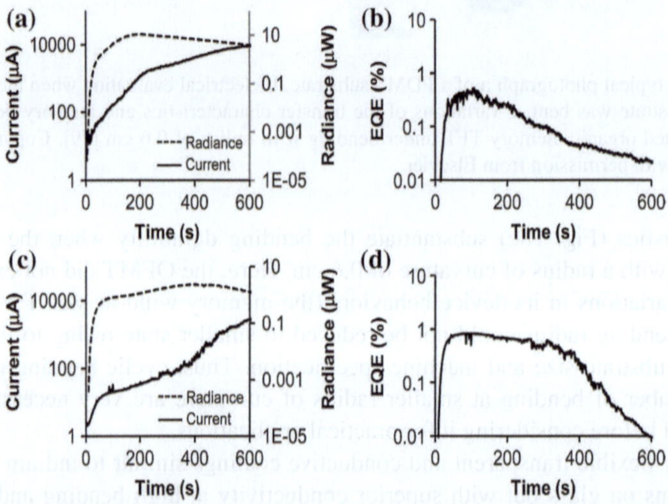

Fig. 20 Characterization of unstrained devices and devices subjected to cycles of tensile strain. Temporal evolution of **a** current (*solid line*) and radiance (*dotted line*) of a typical unstrained device, **b** EQE of a typical unstrained device, **c** current (*solid line*) and radiance (*dotted line*) of a typical device after 20 bending cycles of 25 % tensile strain, **d** EQE of a typical device after 20 bending cycles of 25 % tensile strain [80]. Copyright 2013. Reprinted with permission from American Chemical Society

in radiance over the testing period due to moisture that degrades the ionic transition metal emitter. Maximum EQEs of unstrained (0.57 %) and strained (0.82 %) devices were in line with the reported range of values (0.4–0.9 %) for LEECs. The absence of any negative impact of bending on the device performance indicates that the strain does not damage the flexible LEECs components.

E-skin composed of two layers of SWNTs/PDMS films with the patterned surfaces placed face-to-face, and Ag paste on the edge of both films as source–drain electrodes perform as illustrated in Fig. 21.

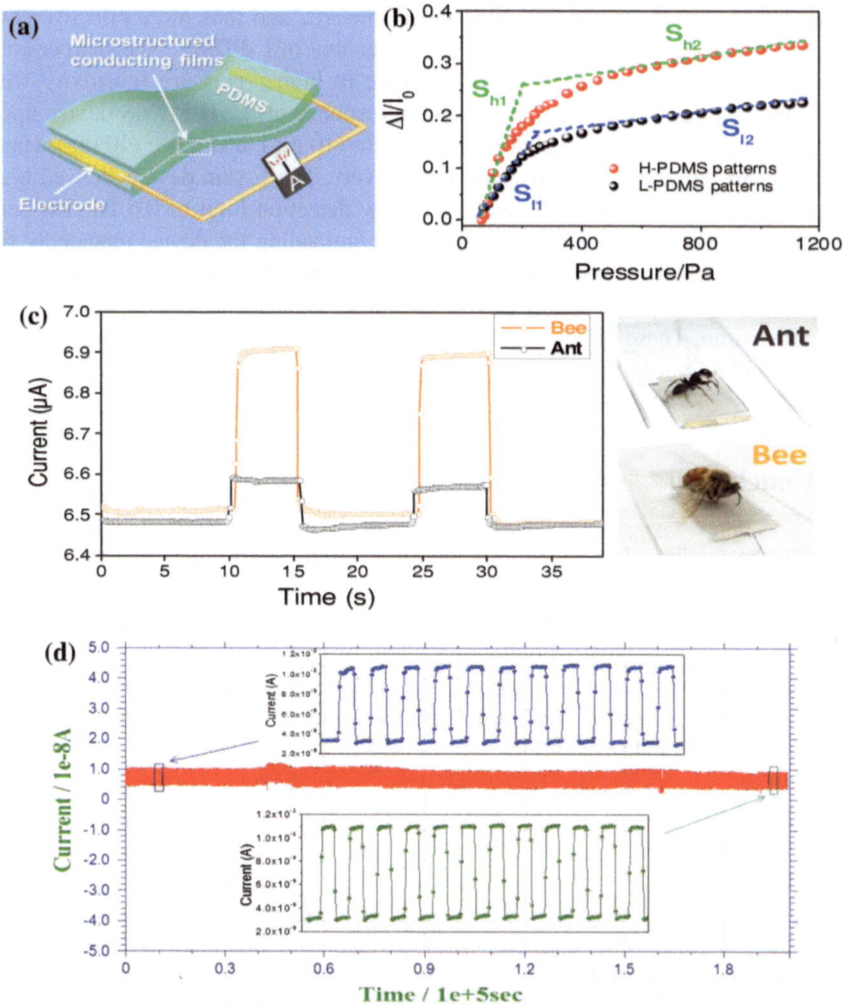

Fig. 21 **a** Schematic of a typical E-skin. **b** Sensitivities of pressure sensors constructed with H-PDMS and L-PDMS real-time *I–t* curves of the E-skin constructed with **c** H-PDMS for detection of a bee (40 mg) and ant (10 mg), **d** H-PDMS for more than 67,500 loading/unloading cycles, at 3 s for each cycle, with an applied pressure of 1 kPa [84]. Copyright 2014. Reprinted with permission from John Wiley & Sons

The sensitivity of the E-skin devices based on L-PDMS and H-PDMS revealed the effect of microstructure density on the sensing performance. Tests were performed by keeping a thin glass slide (17 × 10 mm) over the entire E-skin device to improve its stability. The glass slide imparts a base pressure in addition to the applied pressure. When the pressure is applied (to 1150 Pa), the conductance of E-skin dramatically increased (Fig. 21b) due to the increased contacting sites of top and bottom SWNTs/PDMS films. In H-PDMS-based flexible sensors, the surface contains more microstructures per unit area and thus more effective contact sites leading to a sensitivity of ∼2.3 times that of L-PDMS. The sharper contact edges of H-PDMS films also contribute to higher pressure sensitivity with the same applied force. Fig. 21c shows the response of the H-PDMS device upon loading/unloading the two small insects, an ant (10 mg), and a bee (40 mg) where they impart 0.6 and 2.3 Pa pressure, respectively. The E-skin device showed high sensitivity, fast response time, and a very low detection limit of 0.6 Pa. The sensor stability was further measured by loading/unloading the device repeatedly for more than 67,500 cycles at an applied pressure of 1 kPa, and Fig. 21d demonstrates its high repeatability, stability, durability with no hysteresis and <10 ms response time. These kinds of materials are very significant in the wearable electronics and robotics.

5 Conclusion

PDMS the flexible elastomer forms one of the foundation pillars of flexible electronics. Though the large number of PDMS composites is difficult to introduce in the scenario of a chapter, effort has made to cover most electronic applications of this particular polymer. Conducting fillers of carbon nanotubes, metal oxides, and graphene-based materials are widely addressed. Sensors, actuators, transistors, supercapacitors, and finally the E-skin developed need further investigation to apply them in an industrial scale.

References

1. Someya T, Kato Y, Sekitani T, Lba S, Noguchi Y, Murase Y, Kawaguchi H, Sakurai T, Whitesides GMA (2005) large-area, flexible pressure sensor matrix with organic field-effect transistors for artificial skin applications. Proc Natl Acad Sci USA 102:12321
2. Carta R, Jourand P, Hermans B, Thone J, Brosteaux D, Vervust T, Bossuyt F, Axisa F, Vanfleteren J, Puers R (2009) Design and implementation of advanced systems in a flexible-stretchable technology for biomedical applications. Sens Actuators A 156:79–87
3. Lacour SP, Wagner S, Huang ZY, Suo Z (2003) Stretchable gold conductors on elastomeric substrates. Appl Phys Lett 82:2404–2406
4. Khang DY, Jiang HQ, Huang Y, Rogers JA (2006) A stretchable form of single crystal silicon for high-performance electronics on rubber substrates. Science 311:208–212
5. Kim DH, Ahn JH, Choi WM, Kim HS, Kim TH, Song J, Huang YY, Liu Z, Lu C, Rogers JA (2008) Stretchable and foldable silicon integrated circuits. Science 320:507
6. Rogers JA, Huang Y (2009) A curvy, stretchy future for electronics. Proc Natl Acad Sci USA 106:10875
7. Rogers JA, Someya T, Huang Y (2010) Materials and mechanics for stretchable electronics. Science 327:1603
8. Gray DS, Tien J, Chen CS (2004) High-conductivity elastomeric electronics. Adv Mater 16:393–397
9. Vanfleteren J, Gonzalez M, Bossuyt F, Hsu YY, Vervust T, De Wolf I, Jablonski M (2012) Printed circuit board technology inspired stretchable circuits. MRS Bull 37:254–260

10. Jeong GS, Baek DH, Jung HC, Song JH, Moon JH, Hong SW, Kim IY, Lee SH (2012) Solderable and electroplatable flexible electronic circuit on a porous stretchable elastomer. Nat Commun 3:977
11. Park J, Wang S, Li M, Ahn C, Hyun JK, Kim DS, Kim DK, Rogers JA, Huang Y, Jeon S (2012) Three-dimensional nanonetworks for giant stretchability in dielectrics and conductors. Nat Commun 3:916
12. Zhu S, So JH, Mays R, Desai S, Barnes WR, Pourdeyhimi B, Dickey MD (2013) Ultrastretchable fibers with metallic conductivity using a liquid metal alloy core. Adv Funct Mater 23:2308–2314
13. Niu XZ, Peng SL, Liu LY, Wen WJ, Sheng P (2007) Characterizing and patterning of PDMS-based conducting composites. Adv Mater 19:2682–2686
14. Xu F, Zhu Y (2012) Highly conductive and stretchable silver nanowire conductors. Adv Mater 24:5117–5122
15. Park M, Im J, Shin M, Min Y, Park J, Cho H, Park S, Shim MB, Jeon S, Chung DY, Bae J, Park J, Jeong U, Kim K (2012) Highly stretchable electric circuits from a composite material of silver nanoparticles and elastomeric fibres. Nat Nanotechnol 7:803–809
16. Rosset S, Niklaus M, Dubois P, Shea HR (2009) Metal ion implantation for the fabrication of stretchable electrodes on elastomers. Adv Funct Mater 19:470–478
17. Siegel AC, Bruzewicz DA, Weibel DB, Whitesides GM (2007) Microsolidics: fabrication of three-dimensional metallic microstructures in poly(dimethylsiloxane). Adv Mater 19:727
18. Ponnamma D, Sadasivuni KK, Strankowski M, Guo Q, Thomas S (2013) Synergistic effect of multi walled carbon nanotubes and reduced graphene oxides in natural rubber for sensing application. Soft Matter 9:10343
19. Ponnamma D, Guo Q, Krupa I, Al-Maadeed MAS, Varughese KT, Thomas S, Sadasivuni KK (2015) Graphene and graphitic derivative filled polymer composites as potential sensors. Phys Chem Chem Phys 17:3954–3981
20. Kafy A, Sadasivuni KK, Kim HC, Akther A, Kim J (2015) Designing flexible energy and memory storage materials using cellulose modified graphene oxide nanocomposites. Phys Chem Chem Phys 17:5923–5931
21. Sadasivuni KK, Kafy A, Zhai L, Ko HU, Mun S, Kim J (2015) Transparent and flexible cellulose nanocrystal/reduced graphene oxide film for proximity sensing. Small 11:994–1002
22. Jones J, Lacour SP, Wagner S, Suo Z (2004) Stretchable wavy metal interconnects. J Vac Sci Technol A 22:1723
23. Adrega T, Lacour SP (2010) Stretchable gold conductors embedded in PDMS and patterned by photolithography: fabrication and electromechanical characterization. J Micromech Microeng 20:055025
24. Sadasivuni KK, Ponnamma D, Kim J, Thomas S (eds) (2015) Graphene-based polymer nanocomposites in electronics. Springer, New York
25. Sadasivuni KK, Ponnamma D, Thomas S, Grohens Y (2014) Evolution from graphite to graphene elastomer composites. Prog Polym Sci 39:749–780
26. Loher T, Seckel M, Vieroth R, Dils C, Kallmayer C, Ostmann A, Aschenbrenner R, Reichl H (2009) Stretchable electronic systems: Realization and applications. In: IEEE 11th electronics packing technology Conference, p 893
27. Hosokawa M, Nogi K, Naito M, Yokoyama T (2007) Nanoparticle technology handbook, 1st edn. Elsevier, Oxford
28. Brosteaux D, Axisa F, Gonzalez M, Vanfleteren J (2007) Design and fabrication of elastic interconnections for stretchable electronic circuits. IEEE Electron Device Lett 28:552
29. Kim DH, Song J, Choi WM, Kim HS, Kim RH, Liu Z, Huang YY, Hwang KC, Zhang YW, Rogers JA (2008) Materials and noncoplanar mesh designs for integrated circuits with linear elastic responses to extreme mechanical deformations. Proc Natl Acad Sci USA 105:18675
30. Lacour SP, Jones J, Wagner S, Li T, Suo Z (2005) Stretchable interconnects for elastic electronic surfaces. Proc IEEE 93:1459

31. Deepalekshmi P, Visakh PM, Mathew AP, Chandra AK, Thomas S (2013) Advances in Elastomers: their composites and nanocomposites: state of art, new challenges and opportunities. Advances in elastomers II. Springer, Berlin, pp 1–9
32. Simpson TRE, Parbhoo B, Keddie JL (2003) The dependence of the rate of crosslinking in poly(dimethylsiloxane) on the thickness of coatings. Polymer 44:4829–4838
33. Romeo A, Liu QH, Suo ZG, Lacour SP (2013) Elastomeric substrates with embedded stiff platforms for stretchable electronics. Appl Phys Lett 102:131904
34. Graz IM, Cotton DPJ, Robinson A, Lacour SP (2011) Silicone substrate with in situ strain relief for stretchable thin-film transistors. Appl Phys Lett 98:124101
35. Nomura K, Ohta H, Takagi A, Kamiya T, Hirano M, Hosono H (2004) Room temperature fabrication of transparent flexible thin-film transistors using amorphous oxide semiconductors. Nature 432:488–492
36. Wang ZL, Wu W (2012) Nanotechnology-enabled energy harvesting for self-powered micro-/nanosystems. Angew Chem Int Ed 51:11700–11721
37. Rogers JA, Nuzzo RG (2005) Recent progress in soft lithography. Mater Today 8:50–56
38. Romanowsky MB, Heymann M, Abate AR, Krummel AT, Fraden S, Weitz DA (2010) Functional patterning of PDMS microfluidic devices using integrated chemo-masks. Lab Chip 10:1521–1524
39. Ozgur U, Alivov YI, Liu C, Teke A, Reshchikov M, Dogan S, Avrutin V, Cho SJ, MorkocH AA (2005) comprehensive review of ZnO materials and devices. J Appl Phys 98:041301
40. Deepalekshmi P, Chirayil CJ, Sadasivuni KK, Somasekharan L, Yaragalla S, Abraham J, Thomas S (2013) Special purpose elastomers: synthesis, structure-property relationship, compounding, processing and applications. Advances in elastomers I. Springer, Berlin, pp 47–82
41. Brunchi CE, Filimon A, Cazacu M, Ioan S (2009) Properties of Some poly(siloxane)s for optical applications. High Perform Polym 21:31–47
42. Merkel TC, Bondar VI, Nagai K, Freeman BD, Pinnau IJ (2000) Gas sorption, diffusion, and permeation in poly(dimethylsiloxane). J Polym Sci B 38:415–434
43. Lotters JC, Olthuis W, Veltink PH, Bergveld PJ (1997) The mechanical properties of the rubber elastic polymer polydimethylsiloxane for sensor applications. J Micromech Microeng 7:145–147
44. Dittrich PS, Manz A (2006) Lab-on-a-chip: microfluidics in drug discovery. Nat Rev Drug Discov 5:210–218
45. Lee JN, Park C, Whitesides GM (2003) Solvent compatibility of poly(dimethylsiloxane)-based microfluidic devices. Anal Chem 75:6544–6554
46. Sadasivuni KK, Mohiuddin M, Gao X, Akther A, Mun S, Kim J (2015) Cellulose/PDMS hybrid material for actuating lens. SPIE smart structures and materials + nondestructive evaluation and health monitoring. International Society for Optics and Photonics, pp 94340K–94340K
47. Hanson DE (2004) An explicit polymer and node network model to compute micromechanical properties of silica-filled polydimethylsiloxane. Polymer 45:1055–1062
48. Gross P, Kartalov E, Scherer A, Weiner L (2007) Applications of microfluidics for neuronal studies. J NeurolSci 252:135–143
49. Yeon J, Park J (2007) Microfluidic cell culture systems for cellular analysis. Bio Chip J 1:17–27
50. Meyvantsson I, Beebe D (2008) Cell culture models in microfluidic systems. Annu Rev Anal Chem 1:423–449
51. Sadasivuni KK, Ponnamma D, Kasak P, Krupa I, Al-Maadeed MASA (2014) Designing dual phase sensing materials from polyaniline filled styrene-isopreneestyrene composites. Mater Chem Phys 147(3):1029–1036
52. Kumar SK, Castro M, Pillin I, Feller JF, Thomas S, Grohens Y (2013) Simple technique for the simultaneous determination of solvent diffusion coefficient in polymer by Quantum Resistive Sensors and FT-IR spectroscopy. Polym Adv Technol 24:487–494

53. Sadasivuni KK, Castro M, Saiter A, Delbreilh L, Feller JF, Thomas S, Grohens Y (2013) Development of poly(isobutylene-co-isoprene)/reduced graphene oxide Nanocomposites for barrier, dielectric and sensing applications. Mater Lett 96:109–112
54. Ponnamma D, Sadasivuni KK, Grohens Y, Guo Q, Thomas S (2014) Carbon nanotubes based elastomer composites—an approach towards multifunctional materials. J Mater Chem C 2:8446–8485
55. Hou Y, Wang D, Zhang XM, Zhao H, Zha JW, Dang ZM (2013) Positive piezoresistive behavior of electrically conductive alkyl-functionalized graphene/polydimethylsilicone nanocomposites. J Mater Chem C 1:515
56. Shiohara A, Langer J, Polavarapu L, Liz-Marzan LM (2014) Solution processed polydimethylsiloxane/gold nanostar flexible substrates for plasmonic sensing. Nanoscale 6:9817
57. Xu W, Allen MG (2013) Deformable strain sensors based on patterned MWCNTs/polydimethylsiloxane composites. J Polym Sci, Part B: Polym Phys 51:1505–1512
58. Lottersy C, Olthuis W, Veltink PH, Bergveld P (1997) The mechanical properties of the rubber elastic polymer polydimethylsiloxane for sensor applications. J Micromech Microeng 7:145–147
59. Joo Y, Byun J, Seong N, Ha J, Kim H, Kim S, Kim T, Im H, Kim D, Hong Y (2015) Silver nanowire-embedded PDMS with a multiscale structure for a highly sensitive and robust flexible pressure sensor. Nanoscale 7:6208
60. Zhao W, Yuge P, Juanxu J, Yuanchen A (2009) Selective detection of hypertoxic organophosphates pesticides via PDMS composite based acetylcholinesterase-inhibition biosensor. Environ Sci Technol 43:6724–6729
61. Liu J, Motta N, Lee S (2012) Ultraviolet photodetection of flexible ZnO nanowire sheets in polydimethylsiloxane polymer. Beilstein J Nanotechnol 3:353–359
62. Loomis J, King B, Burkhead T, Xu P, Bessler N, Terentjev E, Panchapakesan B (2012) Graphene-nanoplatelet-based photomechanical actuators. Nanotechnology 23:045501
63. Sadasivuni KK, Kafy A, Zhai L, Ko HU, Mun SC, Kim J (2015) Multifunctional and smart graphene filled polymers as piezoelectrics and actuators. Graphene-based polymer nanocomposites in electronics, p 67
64. Chen L, Liu C, Liu K, Meng C, Hu C, Wang J, Fan S (2011) High-performance, low-voltage, and easy-operable bending actuator based on aligned carbon nanotube/polymer composites. ACS Nano 5(3):1588–1593
65. Sohail S, Das S, Biswas K (2015) Effect of interface layer capacitance on polydimethylsiloxane in electrowetting-on-dielectric actuation. J Exp Phys 2015:426435–426444
66. Xue C, Li J, Zhang Q, Zhang Z, Hai Z, Gao L, Feng R, Tang J, Liu J, Zhang W, Sun D (2015) A novel arch-shape nanogenerator based on piezoelectric and triboelectric mechanism for mechanical energy harvesting. Nanomaterials 5:36–46
67. Nayak S, Chaki TK, Khastgir D (2014) Development of flexible piezoelectric poly(dimethylsiloxane)—$BaTiO_3$ nanocomposites for electrical energy harvesting. Ind Eng Chem Res 53:14982–14992
68. Chun J, Kang NR, Kim JY, Noh MS, Kang CY, Choi D, Kim SW, Wang ZL, Bai JM (2015) Highly anisotropic power generation in piezo electric hemispheres composed stretchable composite film for self-powered motion sensor. Nano Energy 11:1–10
69. Ibrahim IAM, Zikry AAF, Sharaf MA, EMark J, Jacob K, Jasiuk IM, Tannenbaumn R (2012) Dielectric behavior of silica/poly(dimethylsiloxane) nanocomposites. nano size effects. Mater Sci Eng 40: 012011
70. Klonos P, Sulym IY, Borysenko MV, Gun'ko VM, Kripotou S, Kyritsis A, Pissis P (2015) Interfacial interactions and complex segmental dynamics in systems based on silica-polydimethylsiloxane core shell nanoparticles: dielectric and thermal study. Polymer 58:9–21
71. Lee J, Kim W, Kim W (2014) Stretchable carbon nanotube/ion-gel supercapacitors with high durability realized through interfacial microroughness. ACS Appl Mater Interfaces 6:13578–13586

72. Yukse R, Sarioba Z, Cirpan A, Hiralal P, Unalan HE (2014) Transparent and flexible supercapacitors with single walled carbon nanotube thin film electrodes. ACS Appl Mater Interfaces 6:15434–15439
73. Chen T, Peng H, Durstock M, Dai L (2014) High-performance transparent and stretchable all-solid supercapacitors based on highly aligned carbon nanotube sheets. Sci Rep 4
74. Romasanta LJ, Hernández M, López-Manchado MA, Verdejo R (2011) Functionalisedgraphene sheets as effective high dielectric constant fillers. Nanoscale Res Lett 6:508
75. Pisanello F, Paolis RD, Lorenzo D, Guardia P, Nitti S, Monti G, Fragouli D, Athanassiou A, Tarricone L, Manna L, Vittorio MD, Martiradonna L (2013) GHz properties of magnetophoretically aligned iron-oxide nanoparticle doped polymers. ACS Appl Mater Interfaces 5:2908–2914
76. Huang K, Gan Y, Wang Q, Jiang X (2015) Enhanced light extraction efficiency of integrated LEDs devices with the taper holes microstructures arrays. Opt Laser Technol 72:134–138
77. Leem JW, Lee SH, Guan XY, Yu JS (2015) Inverted tetrahedron-pyramidal micropatterned polymer films for boosting light output power in flip-chip light-emitting diodes. Opt Express 23(8):9612
78. Chen CC, Lin HY, Li CH, Wu JH, Tu ZY, Lee LL, Jeng MS, Lin CC, Jou JH, Kuo HC (2014) Enabling Lambertian-like warm white organic light-emitting diodes with a yellow phosphor embedded flexible film. Int J Photoenergy 2014:851371–851377. doi:10.1155/2014/851371
79. Jung SW, Choi JS, Koo JB, Park CW, Na BS, Oh JY, Lim SC, Lee SS, Chu HY, Yoon SM (2015) Flexible nonvolatile organic ferroelectric memory transistors fabricated on polydimethylsiloxane elastomer. Org Electron 16:46–53
80. Miller MS, O'Kane JC, Niec A, Carmichael RS, Carmichael TB (2013) Silver nanowire/optical adhesive coatings as transparent electrodes for flexible electronics. ACS Appl Mater Interfaces 5:10165–10172
81. Loomis J, Fan X, Khosravi F, Xu P, Fletcher M, Cohn RW, Panchapakesan B (2013) Graphene/elastomer composite-based photo-thermal nanopositioners. Sci Rep 3:1900
82. Chen T, Xue Y, Roy AK, Dai L (2013) Transparent and stretchable high-performance supercapacitors based on wrinkled graphene electrodes. ACS Nano 8:1039–1046
83. Yoon SM, Jung SW, Yang S, Park SHK, Yu BG, Ishiwara H (2011) Bending characteristics of ferroelectric poly (vinylidene fluoride trifluoroethylene) capacitors fabricated on flexible polyethylene naphthalate substrate. Curr Appl Phys 11(3):S219–S224
84. Wang X, Gu Y, Xiong Z, Cui Z, Zhang T (2014) Silk-molded flexible, ultrasensitive, and highly stable electronic skin for monitoring human physiological signals. Adv Mater 26:1336–1342

Chlorosulphonated Polyethylene and Its Composites for Electronic Applications

Sayan Ganguly and Narayan Chandra Das

Abstract Chlorosulphonated polyethylene (CSM) occupies an emerging field due to its immense flexibility, excellent heat resistance, excellent oil resistance as well as water resistance, good flame resistance and weathering resistance, and desirable compression set. Generally, this rubber contains 27–45 % chlorine and very nominal sulphur content (approximately 0.8–2.2 %). The enormous improvement in physical, mechanical, and dielectric properties as well as electrical and thermal conductivity of the entire CSM rubber composites is due to the addition of the particulate fillers including conductive fillers. The physico-mechanical behaviour of the composites is quite captivating with respect to other rubber composites in a wide temperature range viz. −80 to 160 °C. It can be used in low-voltage applications in cables where dielectric strength is 500 V/mil with dielectric constant 8^{-10} at 1000 Hz and dissipation factor in the range of 0.05–0.07 at 1000 Hz. CSM is widely practiced in exteriors or outer protective jackets in high-voltage applications due to its outstanding weather-resistant property. Meanwhile, CSM is also widely applied in the fabrication of composites to minimize the effect of radiation pollution emitting uncontrollably from electronic devices. CSM can uptake desirable amount of conducting filler for enrichment or improvement of the conductivity solely to the composites. It is not a hard and fast rule to make the composite highly conducting to inhibit radiation. It is required a conducting pathway to make the composite conducting that can interacts with the electromagnetic wave and reduce its adverse effect. Here, the basic three types of mechanisms of radiation shielding are elucidated to culminate the development of CSM rubber-based electromagnetic interference (EMI)/radiation shielding materials with sufficient flexibility, weather resistance, and chemical resistance features.

S. Ganguly · N.C. Das (✉)
Rubber Technology Center, Indian Institute of Technology, Kharagpur 721302, India
e-mail: ncdas37@gmail.com; ncdas@binghamton.edu

D. Ponnamma et al. (eds.), *Flexible and Stretchable Electronic Composites*,
Springer Series on Polymer and Composite Materials,
DOI 10.1007/978-3-319-23663-6_8

Keywords Chlorosulphonated polyethylene · Conducting filler · Dielectric property · Conductivity · Protective jackets · Electromagnetic interference shielding

Abbreviations

CSM	Chlorosulphonated polyethylene
EMI	Electromagnetic interference
NR	Natural rubber
ASTM	American Society for Testing and Materials
EVA	Ethylene-vinyl acetate
EPR	Ethylene propylene rubber
NBR	Acrylonitrile butadiene rubber
ENR	Epoxidized natural rubber
PE	Polyethylene
LLDPE	Linear low-density polyethylene
HPC	Hydroxypropyl cellulose
AIBN	Azoisobutyronitrile
SBS	Styrene–butadiene–styrene
SNR	Sulphonated natural rubber
C_B	Bulk capacitance
R_B	Bulk resistance
PANI	Polyaniline
CNTs	Carbon nanotubes
SWCNT	Single-wall carbon nanotube
IDE	Interdigitized electrode
HCPE	High chlorinated polyethylene
TEM	Transmission electron microscope
RL	Reflection loss
UV	Ultraviolet
NPPs	Nuclear power plants
EAB	Elongation at break
BACN	Bromoacenaphthylene
MEM	Microelectromechanical
THF	Tetrahydrofuran
NaSS	Sodium 4-styrenesulphonate
HNBR	Hydrogenated nitrile butadiene rubber
U_e	Strain energy density
SE	Shielding effectiveness
dB	Decibel
RFI	Radio-frequency interference
ESD	Electrostatic discharge

IC	Integrated circuit
DBSA	Dodecyl benzene sulphuric acid
CSA	Camphor sulphonic acid

1 Introduction

Chlorosulphonated polyethylene abbreviated as CSM (as per ASTM D1418 rule) is an elastomeric material. The trade names of CSM are KhSPE in the USSR and Hypalon in the USA. CSM is a product of the chemical modification of polyethylene (PE), having a density of 1.11–1.26 g/cm^3, which can be prepared from free radical triggered reaction (low density) of linear PE. CSM is insoluble in aliphatic hydrocarbon and alcohol, slightly soluble in ketones and esters, and readily soluble in aromatic hydrocarbons, such as toluene and xylene, and in chlorinated hydrocarbon. Chlorine content ranges generally from 24 to 43 %, and it is also reported that chlorine content is almost 30 % in case of branched Hypalon 20 and 35 % for linear Hypalon 40. The ability of Hypalon to retain its electrical property after long-term exposure to heat and water is spectacular. These properties make it a commercially viable insulation material in cable sheath compound ingredient in low-voltage applications, particularly in power cable (up to 600 V), control cable, mine trailing cable, locomotive cables, nuclear power station cable, automotive purpose cable, and so on. It is also drawing attention due to easy colourability and unfading after a prolonged time of exposure in atmosphere. It has an important intrinsic property of ozone resistance. There will be no sacrificial lagging due to its colourant incorporation also. Due to almost 30 % halogen content, it is self-extinguishing which leads it to a potential commercially applicable flame-resistant flexible cable coating. The abrasion resistance is superior to natural rubber (NR) and many several elastomers. High flexibility of this rubber in compounded form also marketed this in a dynamic application-related service. The effective chlorine content in the CSM prohibited the termite as well as micro-organism attack towards it, and that is why there will be no such mould growth, mildew, fungus, or bacterial are observed. The extensive wide range of temperature tolerance is another aspect of this material to maintain its position in the commercial sector. The flexibility is also sustained at −40 °C which makes this material a low-temperature applicable material with desired flexibility [1]. Physical and mechanical properties of CSM are summarized in Table 1. Owing to versatility properties, CSM is used for automotive hoses, gas and industrial applications, electric cable jacketing, machine parts, floor tiles, magnetic rubber, rubber-coated cloth, coatings, gaskets, flexible tubes, rolls and linings, and EMI shielding. Special applications for CSM can include escalator handrails, diaphragms and lining for chemical processing equipment, sensors, and actuator. Now a day Hypalon treademark has become a common name for all kind of CSM regardless of manufacturer, for example, Halion CSM from Lianda (distributed by Alternative Rubber & Plastics) and TOSO-CSM® and extos® alkylated chlorosulphonated polyethylene from

Table 1 Physical and mechanical properties of CSM (Hypalon)

Specific gravity	1.08–1.28
Brittle point (°C)	−40 to −62
Dielectric strength (V/mil)	500
Dielectric constant @ 1000 Hz	8^{-10}
Dissipation factor @ 1000 Hz	0.05–0.07
Tensile strength (MPa)	17.236
Elongation (% EB)	430–540
Hardness (shore A)	60
Abrasion resistance	Excellent
Continuous use temperature (CUT)	121 °C
Impact resistance	Good
Compression set (%)	
@ 70 °C	25
@ 100 °C	44
@ 121 °C	16
Diffused sunlight resistance	Excellent
Effect of ageing	None
Heat resistance	good

TOSOH of Japan and a variety of CSM products from Jiangxi Hongrun Chemical Co., Ltd.

Versatile types of conducting filler are developed to enhance the electrical behaviours of CSM. The most common filler practiced was polyaniline (PANI)-based materials for their cost-effectiveness as well as for their easy production, improved thermal stability, and suitable dispersion in the rubber matrix. The present scenario in the total world market in the production of CSM rubber according to the economic report is pictured (Fig. 1).

The USA, China, and Japan showed nearly 80 % of the world consumption of CSM elastomers in 2011 and expected to grow of annual world consumption approximately 1.7–1.8 % during 2011–2017.

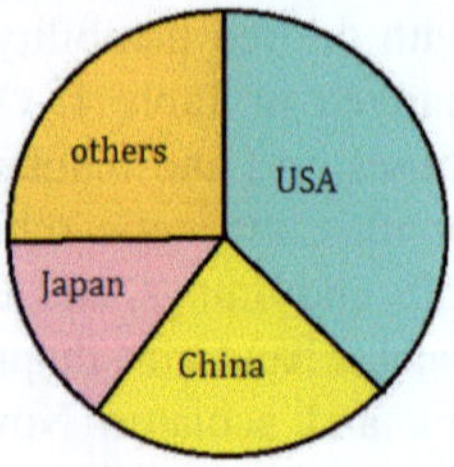

Fig. 1 World consumption of chlorosulphonated polyethylene (2011)

2 Why Chlorosulphonated Polyethylene?

Hypalon is the registered trademark for a series of CSM developed by DuPont Dow Elastomers, USA. Besides several elastomers, CSM engrosses a significant area of elastomer blending. This amorphous, vulcanizable elastomer (Dupont, UK) was formerly applied as solution coating applications. The basic structure of CSM rubber is shown in Fig. 2. The value of z shown in Fig. 2 is usually 17, the typical value of $(x + y)$ is 12 and $z > (x + y)$. There various grades which serve the electrical application fields are jotted down (Table 2) [2].

Generally, when NR is blended with CSM, there is an observation that oil, ozone/oxygen, heat resistance, and elastomers can be improved in a greater degree. It is customary to say that the blend behaviour is potentially dependent on the CSM to other elastomer ratio in the composite blend. Improved oil and thermal ageing resistance were observed when other elastomers were blended with chlorinated polyethylene (CPE). That is why it is used as protective coating or sheath compound materials in cable jacket. Before an era, blends of CSM with NR have been studied by Tanrattanakul and Petchkaew at 2006 [3]. Several elastomer blends are already studied based on CSM due to its potential stability against thermal atmosphere and sunlight. There are a couple of publications that have discussed rubber blends based on CSM, including CSM/EVA [4–6], CSM/EPR [7, 8],

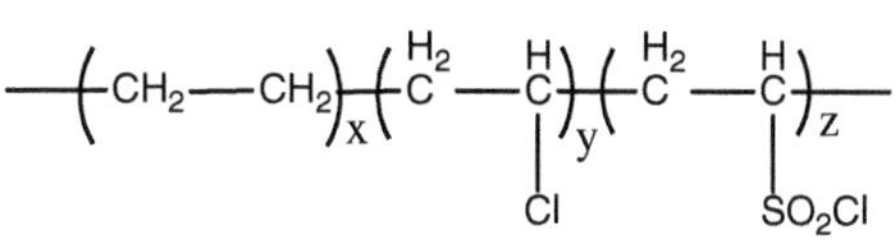

Fig. 2 Structure of chlorosulphonated polyethylene

Table 2 Available grades according to their applications [2]. Copyright 1982. Reproduced with permission from John Wiley & Sons

Quality parameters	Grades							
	20	30	40 soft	40	4085	45	48 soft	48
Chlorine content (%)	29	43	35	35	35	25	43	43
Sulphur content (%)	1.4	1.1	1.0	1.0	1.0	1.0	1.0	1.0
Sp. Gravity	1.14	1.26	1.18	1.18	1.18	1.07	1.27	1.27
Mooney at 100 °C	30	30	45	55	85	40	62	77
Applications	Primary solution coating		General purpose, extruded, calendered, moulded, coloured and black			Hard vulcanizates with moderate loading	Extruded, calendered, and moulded goods with superior oil resistance	
						Unvulcanized compounds with good physical properties		

CSM/NBR [9–12], and CSM/ENR [11–13]. Another aspect for choosing this CSM is that its curing can be tuned inherently in the blend as reported elsewhere, so it is named as "self-vulcanizable rubber blends" [11, 14–17].

2.1 Synthesis Methods of Chlorosulphonated Polyethylene

CSM is synthesized by either solution method, or gas–solid method, or aqueous solution-suspension method. At present, DuPont Dow Elastomers and Jilin petrochemical company produced CSM via solution method in CCl_4 solvent. But the recurring problem for using CCl_4 is its ozone depleting behaviours just like other halo methanes viz. Freons types. There is no other reasonable alteration of CCl_4 solvent to prepare CSM by solution method in the literature. Hence, it is accepted that development in synthetic strategy of CSM is decelerating. Therefore, the development of CSM becomes slow. There are several patents available on preparation of CSM by gas–solid method using PE and a gaseous mixture of chlorine and sulphur dioxide. As an emerging engineering polymeric material, the performance of CSM nearly related to their internal architecture, particularly crystallization process during the production and processing of CSM. In order to enhance the desirable properties of end products, the optimization conditions are adjusted for an industrial process. Wang et al. confronted an aqueous solution-suspension method where a sulphur and chlorine incorporation in LLDPE backbone by means of permutations was done [18]. A typical synthesis scheme of CSM in the aqueous solution-suspension method is shown in Scheme 1.

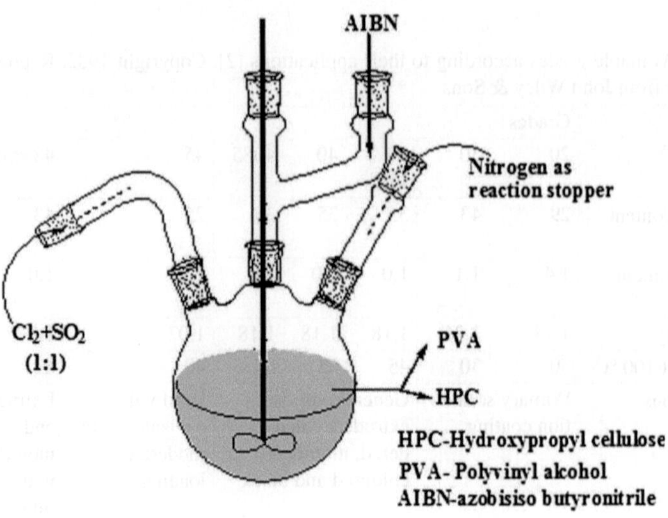

Scheme 1 Synthesis scheme of CSM in aqueous solution-suspension method

$$Cl_2 \longrightarrow 2\,Cl\cdot$$

$$RH + Cl\cdot \longrightarrow R\cdot + HCl$$

$$R\cdot + Cl_2 \longrightarrow RCl + Cl\cdot$$

$$R\cdot + SO_2 \longrightarrow RSO_2\cdot$$

$$RSO_2\cdot + Cl_2 \longrightarrow RSO_2Cl + Cl\cdot$$

Scheme 2 Proposed mechanistic pathway for the solution-suspension method

The aqueous solution-suspension method for CSM preparation is depicted in Scheme 2.

In this work, the CSM with several contents of chlorine and sulphur elements was successfully prepared by an aqueous solution-suspension method. The productive introduction of –Cl and –SO_2Cl groups to the polymers was confirmed by FTIR spectra and also by ion chromatography [18].

Another almost recent method for synthesizing CSM is gas–solid method. The gas–solid method is a novel fashion to prepare CSM compared to the conventional solution procedure which is inexpensive and green process in terms of environment pollution in the absence of solvent. Many patents are available on synthesis of CSM through the gas–solid method using PE and a gaseous mixture of chlorine and sulphur dioxide [19, 21, 22].

Chlorosulphonation of PE by the gas–solid method is not a trouble-free operation. As a highly crystalline feed material, like PE it is very unfortunate to initialize chlorosulfonation reaction under melting temperature of PE. Such type of chlorosulphonation temperature will lead to the aggregation of PE particles, which may forbid the progress of reaction. Noeske et al. discovered a process for chlorosulphonation of PE in a fluidized bed within a temperature range of 40–80 °C [19]. The chlorosulphonated end product prepared with this fluidized bed method is typically not amorphous materials due to sufficient residual inherent crystalline domains already in the feed PE. Later, CSM is also synthesized by the gas–solid method in a tank reactor using CPE and a gaseous mixture of chlorine and dioxide sulphur within a temperature range of 30–50 °C [20]. Another modernized method involves the reaction of amorphous CPE with a gaseous feed mixture of chlorine and sulphur dioxide in a fluidized bed. The yield or the transformation to CSM is much better than using crystalline PE as feed because amorphous nature of CPE. When the crystallinity of amorphous CPE is >15 wt%, and chlorination does not occur in chlorosulphonation, so the reaction can be fulfilled at lower temperature than utilizing PE [23]. The reaction was implemented at a temperature range of 25–100 °C, for a period of more or less 2 h. This process is more acceptable than using crystalline PE as the starting material, but the reaction has not been comprehensively studied. The chemical equation of the chlorosulphonation of CPE is a free radical chain mechanism as shown in Scheme 3.

Scheme 3 Preparation of CSM in gas–solid method

Initiation

$$Cl_2 \longrightarrow 2\dot{Cl}$$

Propagation

$$\dot{Cl} + RH \longrightarrow \dot{R} + HCl$$

$$R + SO_2 \longrightarrow R\dot{S}O_2$$

$$R\dot{S}O_2 + Cl_2 \longrightarrow RSO_2Cl + \dot{Cl}$$

Termination

$$R\dot{S}O_2 + \dot{Cl} \longrightarrow RSO_2Cl$$

Total reaction

$$Cl_2 + SO_2 + RH \longrightarrow RSO_2Cl + HCl$$

3 Filler-Impregnated Chlorosulphonated Polyethylene Composites

A wide range of CSM grades have been used for wire and cable insulation. Older materials include NR, butyl rubber, and styrene–butadiene rubbers (SBR). Younger material community includes crosslinked PE, silicone rubber, ethylene–propylene elastomers, modified or chlorosulphonated polyethylene, and thermoplastic elastomers to overcome the ageing possibility of the compounds. Properties of grandness to electrical insulation exponent include dielectric constant, resistivity, dielectric loss, and dielectric strength. Flame resistance is also important in certain applications.

Sulphonated ionomer moiety in a polymeric material, bearing with a hydrophobic organic backbone chain and a comparatively modest amount of pendant ionic sulphonate groups ($-SO_3^-$) counterbalanced by metal counter ions. A concern in sulphonated ionomer (here CSM) is primarily due to its unique properties, derived from strong associative interactions between ionic groups [24]. In another experiment, the formulation of a quaternary ammonium ionomer of styrene–butadiene–styrene triblock copolymer (SBS) was modernized by a ring opening reaction of epoxidized SBS with triethylamine hydrochloride in the presence of phase transfer catalyst and later on blended with CSM rubber. Thus, the oil-resistant character as well as its electrical application is attributed [25]. Boonsonget al. hypothesized a novel rubbery ionomer of zinc salt of sulphonated NRs (Zn-SNR) and used it as an eminent compatibilizer for the blends of NR and CSM. Establishment of these blending techniques enhances the electrical applicability of CSM rubbers, and in addition, this makes the place in commercialize field [26]. Conductive carbon black (CB) incorporation into the rubber matrix is an effective and easy pathway to make conducting composite for shielding applications. Here, Ensaco 350G, a conducting black, was used in CSM matrix to enhance the conducting property without parting the conductive quality of that composite; 30 phr loading of Ensaco 350G achieves better results without any agglomeration [26]. In this paper, the author revealed that the effect of filler loading is a function of real part of impedance. The complex impedance will be in Eq. 1;

$$Z^* = Z' + jZ'' \tag{1}$$

where Z' and Z'' are real and imaginary parts, respectively. Z' interprets the resistive part of the system and Z'' renders the reactance arising due to the capacitive or inductive nature of the system. Regardless of the filler loading, there is a gradual drop-off of real impedance with increase in frequency, and at higher frequencies (105–106 Hz), the effect is in borderline (Fig. 3). Relaxation phenomenon in crosslinked and reinforced polymer assemblies is basically a hinge on the chemical and physical interactions between the viscoelastic polymer phase and solid filler phase. Crosslinking of the polymers unremarkably enforces intermolecular restraints, which in turn play a crucial role in the segmental dynamics of all polymeric arrangements in the bulk state [27]. Incorporation of fillers such as conductive CB (here used Ensaco 350G) not only results in hydrodynamic interactions but also leads to complex physicochemical interactions between the polymer matrix and the filler surface [26].

Nanda et al. hypothesized another study for the effective filler loading which was supported by Nyquist plots [28]. According to Sluyters-Rehbach and Sluyters, the total impedance of the cell is the series of combinations of resistors, R_B, and capacitors, C_B. Especially in a Nyquist plot of polymer composite system (Fig. 4), the real part of impedance in real axis interprets bulk resistance (R_B) and imaginary axis represents ω_{max} (viz., top of the semicircle) and is given by Eq. 2 [26];

$$\omega\text{max} = \frac{1}{R_B C_B} \tag{2}$$

where C_B is bulk capacitance of the polymer composite.

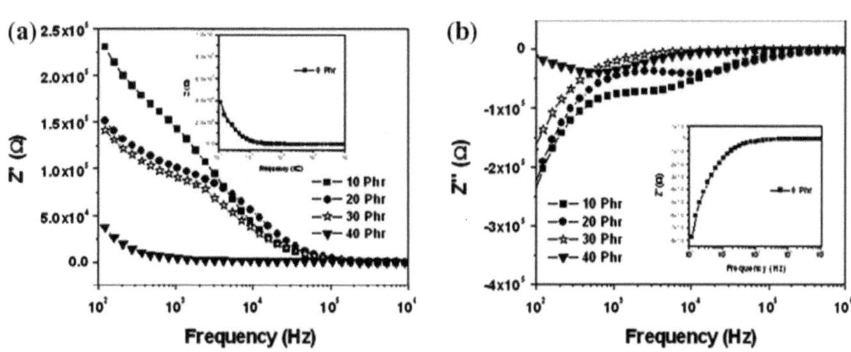

Fig. 3 Effect of filler loading on (**a**) real part of impedance (Z') and (**b**) imaginary part of impedance (Z'') of CSM vulcanizates as a function of frequency at 30 °C [26]. Copyright 2012. Reproduced with permission from John Wiley & Sons

Fig. 4 Nyquist plot (Z' vs. Z'') for different filler loadings in CSM vulcanizates at 30 °C [26]. Copyright 2012. Reproduced with permission from John Wiley & Sons

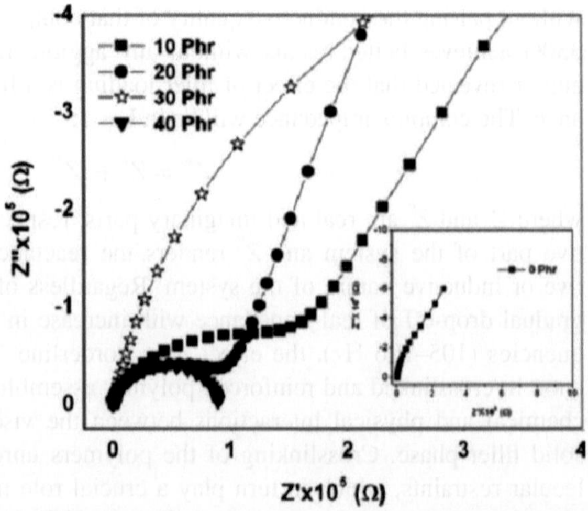

With increase in filler loading, there is a decrease in R_B, which represents increase in conductivity and less lossy response. However, with increase in filler loading, there is a gradual improvement in bulk capacitance (C_B). The gaps in the plot between the C_B aggregates control the electron conduction via non-Ohmic pathway contacts between the C_B agglomerates. The cores of the semicircle plot and the bulk resistance (R_B) values for all experimented filler loading have been elucidated in the plot. The shift in the centre of the semicircle can be used as an evaluation of the breaches in between the agglomerated islands of C_B particulates. It was also discovered that with increase in filler loading, the R_B value is decreasing and core of the semicircle is also decreasing. This may be due to enhancement in conductivity with increase in filler loading, which agreed well with the conductivity plot given in Fig. 5.

Fig. 5 Variation in ac conductivity (σ_{ac}) with frequency for different filler loadings in CSM vulcanizates at 30 °C [26]. Copyright 2012. reproduced with permission from John Wiley & Sons

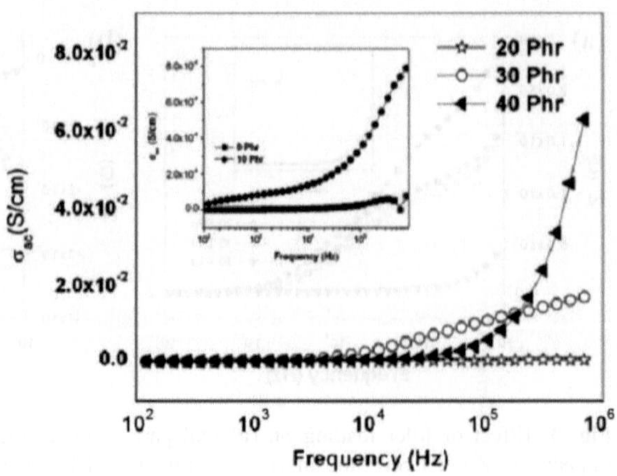

In recent researches, the hybrid fillers act as potential conducting phase in any rubber matrices. Here the main focus is on CSM rubbers where a novel carbonyl iron powder and graphite was used to make more productive conducting composites [29]. The dual behaviour of the composite viz. electric and magnetic sensitivity was also noticed in these types of composites. Such composite absorbent has the maximum magnetic loss for electromagnetic wave at 12.82 GHz and can affect high dielectric loss on the high frequency; hence, this type of composites may serve as a potential shielding material. Marković et al. [30] had considered the effect of gamma radiation on the degradation of blends of NR and CSM [31]. From that standpoint, the CSM-related areas are nurtured by gamma treatment. The effect of gamma irradiation on the acrylonitrile butadiene/chlorosulphonated polyethylene rubber blend (NBR/CSM)-based nanocomposites carrying C_B and silica filler was investigated by Marković et al. also. These nanocomposites containing 20 phr of silica and 30 phr of C_B, respectively, showed high polymer–filler interaction and low filler–filler interaction, so suitable properties are achieved which can drive these CSM-based composites into electrically applicable. Synergism is a predominant effect while 20 phr of silica and 30 phr of conducting C_B-filled NBR/CSM nanocomposites amended the thermal stability, mechanical properties, and conductivity [32].

Emulsion polymerization technique is another process to develop composites. CSM rubber can be enriched with conducting property by incorporation of conducting polymeric fillers with enhanced phase adhesion property. Polymerizing aniline in the presence of CSM rubber where conducting PANI is dispersed phase and CSM rubber solution acts as dispersing medium. The main attraction in this method is that it is low-temperature fabrication technique where thin coherency is maintained. Besides, this conducting sulphur-bearing CSM is used here. Consequently, the composite exposed a discharge capacity of 117.3 $mAhg^{-1}$ and the capacity retention remained at about 78 % after twenty-two cycles, based on the second cycle discharge capacity. Here, PANI officiated as both electrocatalyst and cathode material. At the same time, it improved the conductivity of the active CSM at a molecular level. The results of this study provided a new thought for structure design and development of a potential cathode material for rechargeable magnesium batteries [33]. Among the several nanomaterials, carbon nanotubes (CNTs) have experienced substantial attention due to their unique electronic and extraordinary mechanical properties associated with electrical conductivity [34]. Single-wall carbon nanotubes (SWCNTs) have also a tremendous surface area, as high as approximately 1600 m^2/g [35]. A chemical sensor is fabricated using the CSM rubber nanocomposite where CNT plays a important role in order to construct the mechanically strong as well as electrically potential nanocomposite. He fabricated a simple sensor platform comprising of an interdigitized electrode (IDE) pattern for sensing gas and organic vapours. Purified SWCNTs in the form of a network bedded on the IDE by solution casting function as the sensor material. The electrical conductivity of the SWCNT network changes reproducibly upon exposure to various gases and vapours. Selectivity to specific gases, for example, chlorine and hydrochloric acid vapour, is demonstrated by coating the SWCNTs with CSM [36]. Versatility of CSM rubber in the electrical area is also proved in another experiment. The applicability of polyurethane/CSM rubber blend in such purposes also reported elsewhere [37, 38]. The author developed blends of polyurethane (Vibrathane-SO08 Uniroyal Co., USA) and Hypalon (CSM-350 Du Pont, USA) by three different blending techniques. The ease of processing of the polyurethane rubber was improved as a result of blending with CSM. Experimental data discovered that preheating of the blends, before addition of curatives, enhanced the thermal properties of the blends. The degradation process was delayed along with the retardation in weight loss. The extractability of the single phase by solvent was also restricted on preheating, due to interchain crosslinking reaction. Thus, these typical blends might be used in electrical appliances. Ionic crosslinking efficiency is thoroughly investigated in the presence of silica filler by Owczareket et al. They hypothesized through experimental observation that the CSM rubber composites are formed with core-shell type ZnO-silica engineered particles that paly a novel type crosslinker system [39]. For electrically applicable grades where zinc oxide is enormously used as crosslinker, in this cases zinc oxides not only act as crosslinker but also help to enhance composite strength due to its surprising heat conductivity in the arrested system also [38]. The possible use of zinc oxide and inclinations to substitute CB with precipitated silica modification was achieved to use a new light reinforcing filler with a "core–shell" structure coated with a layer of zinc oxide (ZnO/SiO_2) [39, 40]. CSM is used in rubber engineering due to its prodigious properties resulting from the saturated polymer structure. The vulcanized form of CSM can be used in the production of conveyor belt lining or cable and conductor insulation [41]. Organic acids, thiazoles, dithiocarbamates, or metal oxides are used in CSM crosslinking. The silica containing zinc oxide suggested lower tendency to constitute its own structure in paraffin type oil, which was confirmed by rheological measurements. Zinc oxide doped on the exterior surface of silica particle was a more efficient crosslinking agent as compared to the zinc oxide solely incorporated to the rubber blend in the preparation process. The new filler having a "core–shell"-type architecture enormously amended mechanical properties. The modulus and tensile strength of such vulcanizates were higher in comparison with those with unmodified silica. Silica modified with zinc oxide participated in forming aggregative bonds of the elastomeric network, which was evidenced by high rates of physical relaxation at room temperature. These types of composites revealed the mixture or blend quality as well as their efficiency in the accelerator system. The blend is enriched with well-dispersed silica which can be wholly distributed throughout the rubber matrix [39].

4 Electronic Applications

4.1 Conducting Coating Application

CSM-modified high chlorinated polyethylene (HCPE) anticorrosive coating was canvassed in the paper of Jian, Li et al., and the microstructure and its influence on properties were investigated by transmission electron microscope (TEM) photographs. Results indicated that HCPE particles were implanted in the dispersed phase of CSM rubber, which can improve flexibility of paint film [40]. Meanwhile, it is good for keeping excellent adhesion and toughness of HCPE with low content of CSM in this coating. Besides, the flexibility of CSM-modified HCPE anticorrosion paint is superior to other available paints [41]. The coating applications are also widely exploited in acrylic rubber-based coating application which is also reported elsewhere [42]. Conducting coating applications with sensing behaviour are also achieved by CSM rubber-based nanocomposites. Here, CNT-based gas sensor system was developed for discriminating gases and vapours. The sensor array was framed of nanomaterials, e.g., pristine SWCNTs, and SWCNTs with different metal dopants and CSM rubber coatings. This sensor system was exposed to NO_2, HCN, HCl, Cl_2, acetone, and benzene in parts per million (ppm) concentration levels. The arrangement of the developed nanocomposite data was normalized for eradicating concentration and background noise. The postprocessed arrangement data were then subjected to a principal component analysis and a pattern recognition technique for gas and vapour to distinguish [43]. A novel ceramic filler-impregnated CSM-based composite was developed by Oikonomou et al. where a hexagonal ferrite system was synthesized by sol–gel technique. Ba(Co, $Zn)_2W$-type hexagonal ferrites, prepared by a citrate sol–gel method, have been used as fillers in five layers of CSM composite. Hexagonal ferrites are well long familiar as a class of shielding material with electromagnetic properties (conductivity, permittivity, and permeability) suitable for electromagnetic interference (EMI) prohibition and radar absorbing material (RAM) coatings from centimetre to sub-millimetre wavelengths of the electromagnetic spectrum. Adding an outer layer with carbonyl iron filler, a thin quasi-double-layer microwave absorber has been fabricated that exhibits reflection loss (RL) of 12 dB at 5–12 GHz with a broad minimum of 18 dB at 7.5–9.5 GHz. Layer succession has been optimized for 2 mm total thickness and minimum RL at the X-band as reported by the authors [44].

4.2 Wire and Cable Application

Venturing the discipline of power conveying without power loss is a challenging outcome of several scientific experimentations. Conductor choice depends on ampacity, voltage, mechanical properties, desirable flexibility, enormous dimensional stability, and very acutely cost-effectiveness. The most commonly used metals for conducting core of wire or cables are copper and aluminium possessing solid or stranded winding-type architecture. There are two basic components in a low-voltage cable: the conductor and the electrical insulation (dielectric). There are different types of conductors and insulations. Sometimes, jackets or armouring are added. As voltage increases, other components are added to handle higher electrical stresses. The purpose of this paper is to discuss each component and to

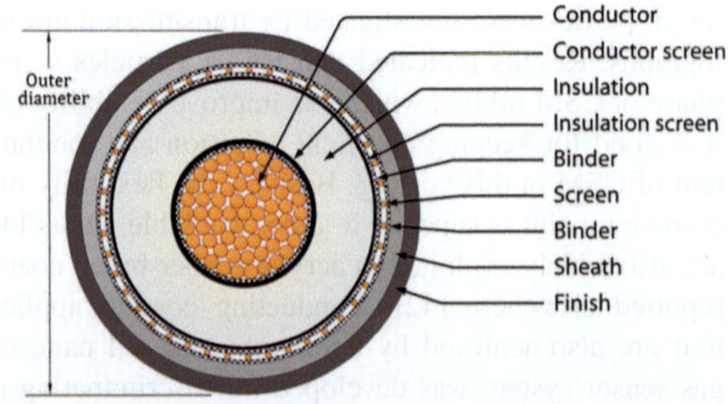

Fig. 6 Typical cable cross section

help an engineer specify the type components required for specific applications. The cable cross-sectional view is elaborated in Fig. 6 as below. It normally consists of two parts with core and outer sheath. Core is the main conductor which carries current, and the outer sheath is the insulation part. Here, the metallic conductor (aluminium, copper) core either in single-core formation or in multicore texture is basically due to the betterment of flexibility. The outer sheath is fabricated by compounded rubber due to their flexibility, weather-resistant nature, insulation behaviour, ease of processing and fabrication, and relatively low cost. The CSM is such a special-purpose one which is normally included in the inner sheath material for its effective adhesion with metal due to its polar moiety (chlorine group) which interacts with the metal as well as outer sheath compound; besides, this CSM rubber is also used in outer sheath also (Scheme 4). The outer sheath should have the effective weather-resistant quality and efficient ultraviolet (UV) resistance. CSM plays an enormous role in such applications due to the absence of unsaturation and chromophore groups (chromophores are liable for UV degradations). The flexibility well balanced with stiffness due to a soft–hard combination of crystalline–amorphous domains inherently present in the CSM rubber makes it an effective outer sheath material.

The atomic energy is most promising for uninterrupted, pollution-free, highly productive source of energy. Countries with forward-thinking atomic/nuclear energy have been investigating methods for assessing the ageing of cables in nuclear power plants (NPPs) and investing on continued research efforts to explore new approaches [45, 46]. Cables used in NPPs must be contrived to last through the plant life on account of the high cost of cable replacement and the complexity of the process. However, cables installed in some speciality areas such as containment buildings may degrade faster than the normal ageing rate, thus impacting cable quality [47, 48]. CSM is synthesized via simultaneous chlorination and chlorosulphonation of polyethylene as feed. Crosslinking can be achieved by using different curing methods (e.g., sulphur, peroxides, and maleimide) to produce commercial generic Hypalon rubber-made cable jackets [49–53]. CSM is a significant and widely used rubber and is mostly used as a sheathing material for electrical cables employed in nuclear power facilities. It is also used in autosupplies, life-saving equipment, and building materials [54, 55]. Jeon, Hwang-Hyun, et al.

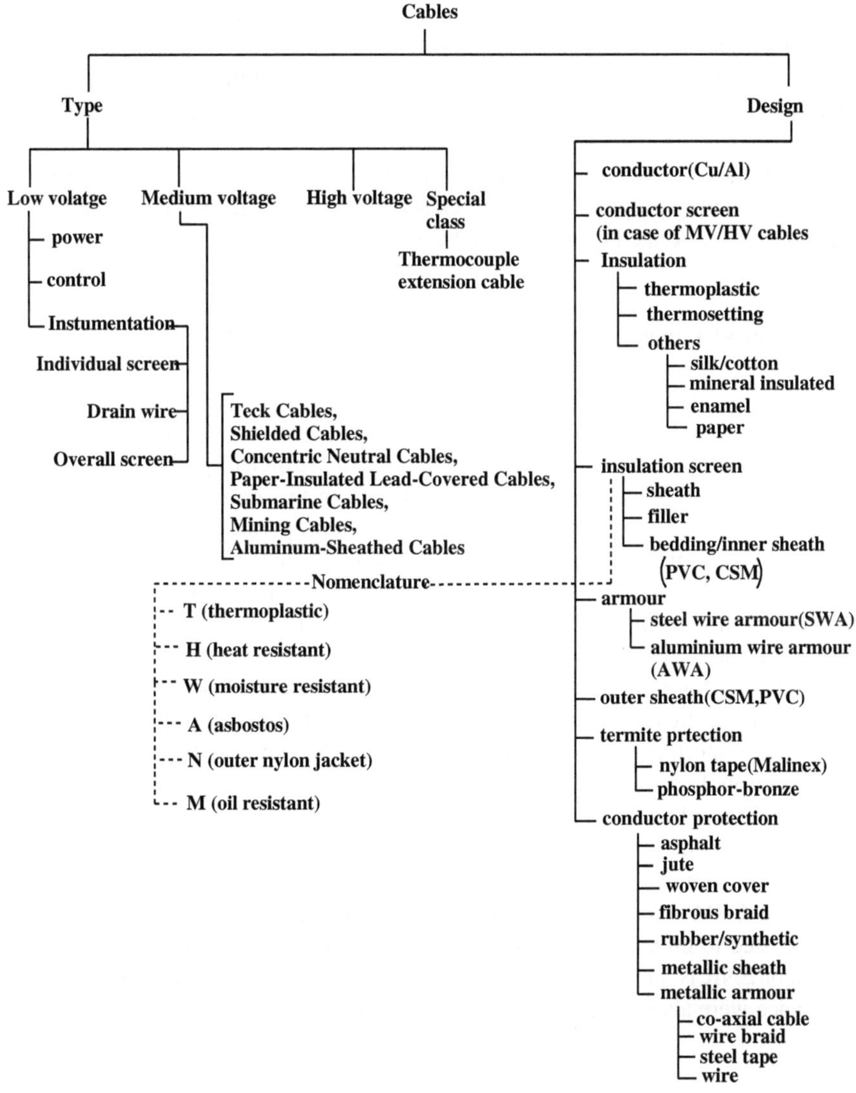

Scheme 4 Cable design overview

measured the dielectric properties of CSM rubber-based cable sheath compound where they revealed that in the sea water atmosphere, the water soaked may disturb the inherent electrical properties of a compounded cable sheath material. He studied dielectric constant of CSM decreases marginally as the number of dried days at room temperature increases. Therefore, dielectric constant cannot be used to determine the degree of deterioration of accelerated thermally aged CSM. The insulation property of cables in NPPs is decreased because of sea water flooding, and the insulation property can be retrieved through freshwater flooding. The volume electrical resistivity, peak binding energy, and microstructural analyses of CSM-based materials suggest appraising the possibility of the ageing state of stimulated thermally aged CSM based on the elongation of break (EAB) and apparent density. It is worth mentioning that the volume electrical resistivity of accelerated thermally aged CSM is lower than that of non-accelerated thermally aged CSM, and the insulating property of non-accelerated thermally aged CSM recovers faster than that of accelerated thermally aged CSM [56].

4.2.1 Special-Purpose Flame-Resistant Cable Application

CSM is a special-purpose elastomer on account of its ideal balance of holdings such as ozone, oxygen, weather, heat, oil, and chemicals resistance with good mechanical properties. Due to the presence of the sufficient polarity of the chlorine group in CSM, it is widely used as blends. In this context, NR/CSM blends are commonly practiced blend which customizes properties such as resistance to ozone, oil, heat, flame, and non-polar chemicals. Grounded on chemical structure, the NR/CSM blend may be an issue in materials' inferior mechanical properties because of the incompatibility originating from the mismatch in polarity [57, 58]. This result was the repetition when higher polarity such as fluoro elastomers is blended with NR [59]. The difference in polarity of the rubbers causes high interfacial energy, which is damaging. It will critically limit mixing at the interface and hence the opportunity for crosslinking between the rubbers. It also causes poor phase morphology, which is characterized by large phase sizes. These problems can be mitigated to an extent by the use of suitable modifiers that minimize the phase separation and increase interfacial adhesion; these include the addition of physical or chemical compatibilizers [60, 61]. The self-crosslink reaction between ENR and CSM improves cure time and automotive fuel resistance. ENR could also be used to improve automotive fuel in the blend system. The optimal concentration of ENR is 5 phr. The good mechanical properties as well as automotive fuel resistance were also achieved when ENR > 5 phr [62].

4.2.2 Shielded Electronic Cable Application

At an early stage, Eisuke Oda et al. developed a cable coating formulation where CSM acts as a rubber sheathing material. In that work, condensed bromoacenaphthylene (Con-BACN) was synthesized and later on incorporated into rubber phase due to its extraordinary radiation shielding efficiency as well as flame resistance property. Con-BACN improves the radiation resistance of polymer matrices. The plausible mechanisms were (i) energy channelized from the polymer matrix to Con-BACN molecules; (ii) radical-scavenging graft reactions of Con-BACN to the polymer backbone; and (iii) energy transfer from methylene linkages to aryl

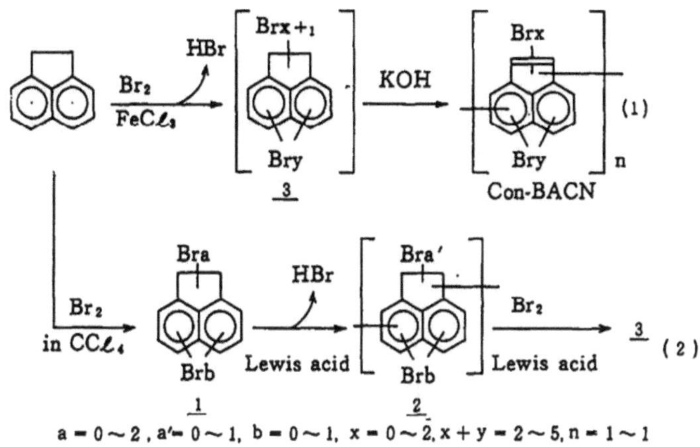

Scheme 5 Synthesis scheme of Con-BACN [63]. Copyright 2012. Reproduced with permission from IEEE

groups of grafted Con-BACN (Scheme 5). The composite showed the radiation resistance of CSM sheet. Con-BACN-bearing sheets show less decrement in elongation accompanying irradiation than do raw sheets; thus, these results indicate that radiation degradation is curbed. The radiation shielding of the cables, with sustained flexibility and the elongation of their insulation and sheath, is excellent at 600 Mrad. The cables can therefore be used not only in the limited wiring of NPPs, but also in any area such as high-level radioactive waste deactivation facilities where flexibility is commanded under exposure to radiation [63].

4.3 CSM Applications in Sensor and Actuator Field

Fabrication of microelectromechanical (MEM) devices in order to apply in the sensor and actuator development is a potential field in recent days. Gas sensors or chemical sensors are attracting magnificent interest on account of their far-flung practical application in industry, environmental supervising, space geographic expedition, medical specialty, and pharmaceuticals. Gas sensors with high sensitivity and selectivity with high confidence levels are demanded for leakage detections of explosive gases viz. hydrogen, and for real-time detections of pathogenic or any sort of lethal gases in industries. There is also a strong demand for the ability to monitor and control our ambient environment, particularly with the raising concern of the globe warming [64]. Typically, there are various introductory criteria for good and efficient gas-sensing systems: (a) high sensitivity and selectivity; (b) fast response time and retrieval time; (c) low functioning temperature and temperature independency; (d) low analyst consumption; and (e) stability in performances. Commonly used gas-sensing materials include vapour-sensitive polymers, semiconductor metal oxides, and other porous structured materials such as porous silicon [65–67].

Organic synthetic polymers or composites based on elastomers are one of the principal materials implemented in gas-sensing systems. Some conducting polymers can behave like semiconductors due to their transient inherent electron cloud generation compounds which display physicochemical characteristics. As a matter of fact, reversible changes in the detection level's conductivity can be detected upon polar material adsorption on the surfaces at ambient temperature [68]. This outcome is considered to be caused by the charge transfer among gas molecules and the polymer [69]. A typical schematic diagram of the chemically functionalized CNTs is shown in Fig. 7 [70].

The perception features of polymer-coated CNTs in nancomposite film to particular gases were also investigated. Typically, CNTs do not have sensing response to all gases and vapours but only the ones with high adsorption energy or that can interact with them. The coating or doping of a material on CNTs' surface may diversify the application range. The sensing character of the same structure but with polymer coatings on CNTs can be achieved. The CSM was dissolved in tetrahydrofuran (THF) solvent as a coating solution for Cl_2 sensing, and hydroxypropyl cellulose (HPC) was dissolved in chloroform as coating to detect HCl. The resistance of polymer-coated SWCNT film changed upon exposure to Cl_2 and HCl, while uncoated pure SWCNTs showed no sensing signal. The sensing response demonstrates a huge potential of using modified CNT gas-sensing materials for a broad range of gases and chemical vapours [71].

In sensing applications as artificial muscles, polymers already adopt a wide area of functional advanced materials. Natural muscles have self-repairing potentiality providing billions of work cycles with greater than 20 % of contractions at a speed of 50 % per second, stresses of approximately 0.35 MPa, and adjustable strength and stiffness [72]. Artificial muscles have been attempted for artificial

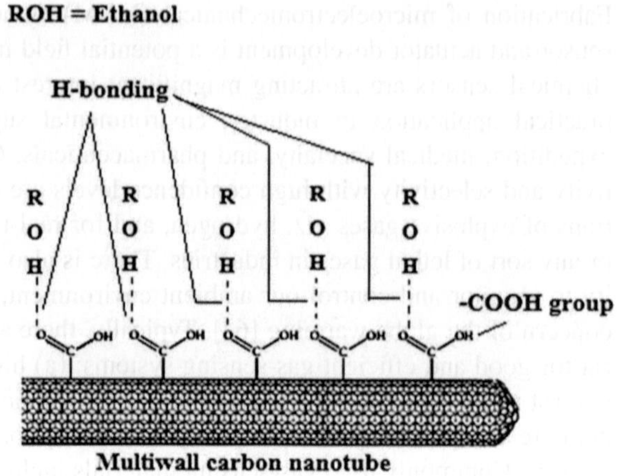

Fig. 7 Proposed mechanism for alcohol vapour detection using functionalized CNT sensors. The COOH groups tend to form hydrogen bonding with the ethanol molecules at room temperature

hearts, artificial limbs, air vehicles, and humanoid robots. Diverse artificial muscles have been investigated for high response rate, large strain, and eminent output power at minimal strain employing their own material characteristics [73]. Among the family of artificial muscles, the dielectric elastomers have typical lineaments of low specific gravity, flexibility, moderate cost reduction, and easy fabrication, which survives it in many attractive applications such as mobile robots, micropumps, microvalves, disc drives, flat panel speakers, and intelligent endoscope. [74–79].

Dielectric elastomer-based actuators have been known for their unique properties of large elongation strain (120–380 %), large stresses (3.2 MPa), high specific elastic energy density (3.4 J/g), high speed of response in 10^{-3} s, and high peak strain rate (34,000 %/s) [79]. Another advantage of dielectric elastomer actuators is the functioning of elastomer actuators can be tailor-made by choosing dissimilar types of elastomers, altering the crosslinking chemistry of polymer chains, adding functional entities, and improving fabrication techniques with ease and versatility in most works. The deformation of elastomers follows with the theories of rubber elasticity and nonlinear viscoelasticity. When an electrical field is applied, the elastomer distortion is tempted primarily by the inherent properties of moduli and dielectric constants of elastomers. Additionally, desirable actuation potentialities are frequently restricted by the dielectric strength (or breakdown voltage) of respective elastomer films. The property processing structure relationship of elastomers particularly below the electrical field and large deformation should be understood on the basis of the fundamental principles of deforming elastomers and practical experience in actuator fabrication. The electrical properties and modulus of the different rubber are shown in Table 3.

Extreme sensitivity to surface adsorption phenomena that make a way for SWCNTs into sensing devices is a very efficient practice. Even so, a practical sensor demands to have selective response in addition to sensitivity. Vichchulada, Pornnipa, et al. have focused on efforts to increase selectivity in SWCNT sensors. This manuscript article presents recent progress towards incorporation of SWCNTs into enhanced sensor and electronic applications. SWCNTs are in nanometre scale with properties that have made them the foci of research for a diversified outcome in electronic applications. In particular, SWCNT gas sensors have experienced significant interest recently. Electrical transducers based on supervising the alteration of conductance or threshold voltage of SWCNT devices upon exposure to an analyte shows the advantage of being simple and low power devices [82].

Solvent-free method to prepare proton-conducting membranes based on commercial elastomers is an effective practice. CSM-based elastomeric blends were studied where the CSM phase used as the matrix materials in the preparation of non-fluorinated proton exchange membranes utilizing a solvent-free route via the in situ reaction of sodium 4-styrenesulphonate (NaSS) as monomer. The morphology of the elastomer/NaSS vulcanizates was deliberated to assess the effect of polarity, viscosity, and saturation degree of the elastomer matrixes. Much better

Table 3 Dielectric and mechanical properties of elastomers

Material	Dielectric constant at 1 kHz	Dielectric los factor at 1 kHz	Young's modulus ($\times 10^6$ pa)	Engineering stress (MPa)	Break stress (MPa)	Ultimate strain (%)
IR, NR [80]	2.68	0.002–0.04	1.3	15.4	30.7	470
CR [80, 81]	6.5–8.1	0.03/0.86	1.6	20.3	22.9	350
BR [80]	–	–	1.3	8.4	18.6	610
Poly(isobutene-co-isoprene) butyl rubber [80]	2.42	0.0054	1	–	17.23	–
Poly(butadiene-co-isoprene) butyl rubber	5.5 (10^6 Hz)	35 (10^6 Hz)		16.2	22.1	440
Poly(butadiene-co-styrene) (SBR, 25 % styrene content) [80, 81]	2.66	0.00009	1.6	17.9	22.1	440
Poly(isobutyl-co-isoprene) (IIR) [80]	2.1–2.4	0.003	–	5.5	15.7	650
CSM [80]	7–10	0.03–0.07	–	–	24.13	–
EPR [80]	3.17–3.34	0.0066–0.0079	–	–	20.68	–
EPDM [80]	3.0–3.5	0.0004 at 60 Hz	2	7.6	18.1	420
Urethane [80]	5–8	0.015–0.09	–	–	20.55	–
Silicone [81]	3.0–3.5	0.001–0.01	–	–	2–10	80.500

dispersion of NaSS was found in CSM rubber and hydrogenated nitrile butadiene rubber (HNBR) matrixes than in the other three types of elastomer matrixes due to fractional polarity in them which is the main driving force of compatibilization. The CSM/NaSS and HNBR/NaSS proton exchange membrane exhibited the proton conductivity as high as ~0.03 S cm^{-1}, and the ratio of proton conductivity to methanol permeability is achieved higher than that of Nafion [83].

4.3.1 Why Elastomers Deform in the Presence of Electric Stimuli?

The response due to the presence of external electrical stimuli helps the elastomers' deformation to a surprising degree. The physics behind this extraordinary behaviour is hypothesized by R. Pelrine [84]. The concept of electronic behaviours explicitly divulges by the electrostatic energy which is already stored in the elastomer systems with an arbitrary amount of film thickness "z" and surface area "A". The electrostatically stored energy can be written as in the Eq. 3,

$$U = \frac{Q^2}{2C} = \frac{Q^2 z}{2\varepsilon_0 \varepsilon_r A} \quad (3)$$

where C, Q, ε_0, and ε_r are the capacitance, electrical charge, free-space permittivity (8.85×10^{-12} F/m), and relative permittivity, respectively. The capacitance is defined in Eq. 4,

$$U = \frac{\varepsilon_0 \varepsilon_r A}{z} \quad (4)$$

From this above equation, the deviation in electrostatic energy can be related to the differential changes in thickness (dt) and area (dA) with a constraint that the total volume is constant (At = constant). Then, the electrostatic pressure generated by the actuator can be derived as below,

$$P = \varepsilon_0 \varepsilon_r E^2 = \varepsilon_0 \varepsilon_r (V/z)^2 \quad (5)$$

where *E and V* are the applied electric field and voltage, respectively.

The electrostatic pressure in previous equation is doubling larger than the pressure in a parallel plate capacitor. This attributes the fact that the energy would switch over with the changes in both the thickness and area of contact of actuator systems.

Actuator performance has been calculated by Eq. 4, and the Hooke's law with Young's modulus (Y) relates the stress (precisely electrostatic pressure) to thickness strain (s_t) as

$$P = -Y s_z \quad (6)$$

where $t = t_0(1 + s_z)$ and t_0 is the initial thickness of the elastomer film. Using the same constraint that the volume of the elastomer is conserved,

$$(1 + S_x)(1 + S_y)(1 + S_z) = 1,$$

$S_x = S_y$ (in-plane strain) can be derived from Eqs. 3 to 4. The through thickness strain S_z can be converted to in-plane strain by solving the quadratic equation for the strain constraint as

$$S_x = (1 + S_z)^{-0.5} - 1. \quad (7)$$

For low-strain materials, the elastic strain energy density (U_e) of actuator materials has been estimated which also shows proportionality to Young's modulus of the material [84]

$$U_e = \frac{1}{2} P S_z = \frac{1}{2} Y S_z^2 \quad (8)$$

In general, for high-strain materials, the in-plane expansion upon the compression phenomena markedly envisioned as the material is compressed, and therefore, the elastic U_e can be obtained by integrating the compressive stress times the varying planar area over the displacement, ensuing in the adopting Eq. 9 [84]

$$U_e = \frac{1}{2} Pln(1 + S_z) \quad (9)$$

4.4 CSM-Based Radiation Shielding Applications

4.4.1 Mechanisms of Electromagnetic Interference (EMI) Shielding

EMI shielding pertains to the reflection and/or adsorption of electromagnetic radiation by means of a material which plays a role in shielding with proficiency against the incursion of the radiation through the shield. Electromagnetic radiation, especially at high frequencies (i.e. radio waves, radiation emanating from cellular phones), tends to interfere with electronics. EMI shielding of electronics as well as radiation source is vehemently required also by the governments around the world. EMI shielding is a slight bit distinguished from the magnetic shielding. The mechanisms of radiation shielding are lucidly stated which were resulted from several hypothesises.

Basically, two major mechanisms are reflected in papers: one is primary mechanism and another one is secondary mechanism. It is also noteworthy that the primary one is related to the reflection phenomena and the latest one is the absorption.

Primary Mechanism of Shielding: Reflection

In case of primary mechanisms, the utmost priority to be a shielding prohibition is that material should be a mobile charge carrier that may come from either electrons or mobile hole carriers. These conducting particulate matters interact with the electromagnetic waves; hence, it is proved that conductivity is prime here but not necessarily very high conductivity. Electrical conductivity is not the essential scientific prerequisite item for shielding. Shielding does not require connectivity, and it is enhanced by connectivity [85]. Metals are the most common materials for EMI shielding. These EMI shielding ismainly by reflectiondue to the inherentfree electrons cloud.But metal sheets arebulky, so metalcoatings, made byeither electroplating,electrode less platingor vacuum depositionare used for radiationshielding [86–88].

Secondary Mechanism of Shielding: Absorption

A secondary mechanism of EMI shielding is usually absorption. Significant absorption of the radiation by the shield is due to the electric and/or magnetic dipoles intrinsically occurred in the shielding materials, which interact with the electromagnetic fields exposed to the radiation. The electric dipoles generally provided by $BaTiO_3$ or other materials having sufficiently high value of the dielectric

Table 4 Electrical conductivity relative to copper (σ_r) and relative magnetic permeability (μ_r) of selected materials [92]

Material	σ_r	μ_r	$\sigma_r\mu_r$	σ_r/μ_r
Silver	1.05	1	1.05	1.05
Copper	1	1	1	1
Gold	0.7	1	0.7	0.7
Brass	0.61	1	0.61	0.61
Bronze	0.18	1	0.18	0.18
Tin	0.15	1	0.15	0.15
Lead	0.08	1	0.08	0.08
Nickel	0.2	100	20	2×10^{-3}
Stainless steel (430)	0.02	500	10	4×10^{-5}
Mu-metal (at 1 kHz)	0.03	20,000	600	1.5×0^{-6}
Superpermalloy (at 1 kHz)	0.03	100,000	3000	3×10^{-7}

constant. The magnetic dipoles may be provided by some sort of magnetic particles viz. Fe_3O_4 or other magnetic materials having a prominent value of the magnetic permeability [86], which may be enhanced by reducing the number of magnetic domain walls through the use of a multilayer magnetic films [89, 90].

The absorption loss is designated by $\sigma_r\mu_r$, whereas the RL is a function of the ratio σ_r/μ_r, where σ_r is the electrical conductivity relative to copper and μ_r is the relative magnetic permeability. Table 4 shows these factors for various materials. Silver, copper, gold, and aluminium are splendid for reflection, due to their high conductivity. Superpermalloy and Mu-metal are excellent for absorption due to bearing of their high magnetic permeability. The RL recedes with evoking frequency, whereas the absorption loss increment is directly related to frequency.

The absorption loss represents the "skin effect" attenuation among the thickness of the material, and it is the function of its relative permeability and conductivity. The attenuation through absorption via a shield of thickness "t" is given by the Eq. 10 [91]

$$A(dB) = 131.4t\sqrt{f\mu_r\sigma_r}. \tag{10}$$

Theory of Multiple Reflections

In addition to reflection and absorption, a multiple reflection, which is basically the consecutive reflections at respective surfaces or interfaces in the shield, is also contributing to shielding. This mechanism needs the presence of a more surface area or interface area in the shield. Any porous substrate or foam-type materials are the relevant examples with large surface area, and conducting particulate materials are also the examples with sufficient surface area in composite-type materials. The loss due to multiple reflections in the respective shields can be neglected when the skin depth is lower than the distance between the reflecting surfaces and interface. The losses, which may be either due to reflection or absorption or multiple reflections, are usually expressed in dB; hence, the algebraic sum of all the losses is the shielding effectiveness (SE, expressed in dB). The absorption loss is directly proportional to the thickness of the shielding material or composite. Electromagnetic radiation at high frequencies pervades exclusively the near surface region of an electrical conductor. This is known as the skin effect. The electric field due to a plane wave penetrating a conductor drops exponentially with enhancement of depth into the conductor. The depth at which the field drops to 1/*e* of the incident value is anticipated as the skin depth (δ), which is expressed in Eq. 11

$$\delta = \frac{1}{\sqrt{\pi f \mu \sigma}} \tag{11}$$

where f = frequency, μ = magnetic permeability = $\mu_0\mu_r$, μ_r = relative magnetic permeability, $\mu_0 = 4\pi \times 10-7$ H/m, and σ = electrical conductivity in $\Omega^{-1}\text{m}^{-1}$.

The losses by multiples reflection are given by Eq. 12 [91]

$$M(dB) = 20 \log\left(1 - e^{-2\frac{t}{\delta}}\right) \tag{12}$$

Multiple reflections and transmissions can broadly dismiss for shield thicknesses which are practically greater than a skin depth. The coefficient is negative; the multiple reflections reinforce the field transmitted to the output of shielding. At low thickness of the material with respect to the thickness of skin, it is accepted that multiple reflections take place on the interface. When the thickness of the material is the identical magnitude with the thickness of skin, reducing electromagnetic fields is less. The term "*M*" multiple reflections is negligible.

4.4.2 Chlorosulphonated Polyethylene Used in Radiation Shielding Applications

Electrically conductive polymer composites have invoked a great deal of scientific and commercial concern. The possible mechanistic pathway of electrical conduction in these composites is the organization of a continuous network of conductive fillers throughout the insulating polymer matrix. Conductive composites have been widely applied in the area of electromagnetic/radio-frequency interference (EMI/RFI) shielding, electrostatic discharge (ESD), conductive adhesives for die attach in electronic packaging applications, and elastomer interconnect devices used for integrated circuit (IC) package fabrication [93]. Electrically conductive elastomer composites, which display variable conductivity in response to varying external loading, are widely used for various electronic applications, such as touch-control switches and strain and pressure sensors for applications in robot hands or artificial limbs, gaskets etc. As per ASTM designation, the working formula for SE is

$$SE = 10 \log \frac{P_1}{P_2} \tag{13}$$

where P_1 is the obtained power with the material present and P_2 is the obtained power without the material present.

EMI shielding is the consequence of reflection (R), absorption (A), and internal multiple reflection (I) of the incident electromagnetic waves in the samples. These three quantities are related to each other by the following expression for SE:

$$SE = 10 \log \frac{P_1}{P_2} = (R + A + I) \tag{14}$$

"I" is negligible when A is over 10 dB or greater than 15 dB. R and A are given by the following equation [94, 95]

$$R = 108 + \log \frac{G}{\mu f} \tag{15}$$

$$A = 1.32t\sqrt{G\mu f} \tag{16}$$

where G is the conductivity of the samples relative to copper, m is the magnetic permeability of the sample relative to vacuum or copper (m is usually 1.0), f is the radiation frequency in MHz, and t is the thickness of the specimen in cm. Here, a microwave absorbing rubber composites was prepared by addition of conductive CB and aluminium powder in NR, ENR, and CSM where the comparative study followed by the selection of rubber system to enhance the SE was investigated. The range of frequencies (8.5–12 GHz) and the effect of fillers on SE of three rubber systems were similar. SE of the rubber increased from lower level (i.e. <1 dB) to 18–28 dB for 50 phr loading of CB or to 30–40 dB for 50 phr loading of CB and 50 phr of aluminium powder. While the conductivity was higher, aluminium powder was less effective than conductive CB because of the lower volume fraction and larger grain size, hence lesser dispersion in the rubber matrix. It is hereby significant that the impregnation of binary filler exhibited higher SE than single filler. SE of the rubber composites can be ranked in the following order, ENR > CSM > NR, and a rank of their mechanical properties, excluded compression set, was CSM > ENR > NR. The SE (dB) is plotted with frequency (Hz) as shown in Fig. 8.

PANI-impregnated CSM is another fruitful development of radiation protective film. PANI and its derivatives or modified PANI have generated enormous involvement among scientists and technologists because of their wide variety of desirable properties and potential technological applications. Outcome of secondary doping of PANI composites, synthesized by in situ emulsion polymerization of aniline in the presence of CSM or SBS and dodecyl benzene sulphuric acid (DBSA) on conductivity, was investigated [97]. PANI can be doped by perchloric acid or an organic sulphonic acid (camphor sulphonic acid, CSA), and the dielectric behaviours are studied. CSA doped samples evidenced better properties which can be used as efficient filler for the development of inorganic–organic hybrid shielding materials which also reported elsewhere [98].

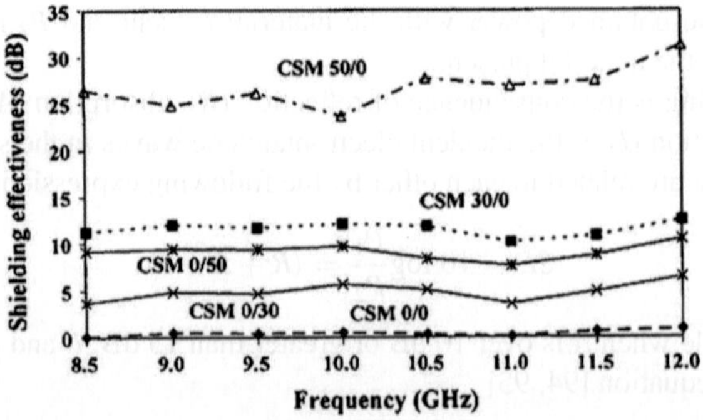

Fig. 8 Effect of filler in CSM matrix to measure SE [96]. Copyright 2007. Reproduced with the permission from John Wiley & Sons

5 Conclusion

The quite promising applicability of chlorosulphonated polyethylene in the electronic application area is a well-practiced topic now in rubber commercialization. The eminent compatibility with the conducting organic as well as inorganic particulate materials can sharpen the exploitation of CSM conducting composites for electronic applications. The flexibility as well as weather resistance makes this material more applicable in external service use. The manufacturing procedures of CSM are elucidated also where the solution-suspension method and the gas phase methods are emphatically explained due to their acceptability for green and better yield. Altering the feed quality from crystalline to amorphous arrangements, the increment of yield and ease of synthesis are also documented. The electronic behaviours such as fabricated flexible electronic devices, MEM, sensors and actuators, and shielding materials where they are loaded with conducting particulate materials are authenticated from the several scientific publications as already documented in this chapter. The wide range of applications are not only in the electronic fields such as electronic coating and cable jacketing either as sheath or inner core material formulations but also in the efficient EMI/RFI/radiation shielding material fabrication and conduction adhesive applications. Sensor and actuator fields make this material with high proficiency in end use. The actuation behaviours are well established by the mathematical models of various scientists, and the deformation functions with the electrical pulse are also studied here. Besides these works, there are wide scopes for graphene oxide-impregnated CSM which is now not practiced enormously for achieving better conducting, flexible, high strength rubber composite comparatively to other commercial rubber composite for radiation shielding applications. The in situ reduction to prepare graphene from the graphene oxide is also an upcoming field to cultivate the radiation pollution inhibition as well as flexible electrosensors and actuators.

References

1. Schweitzer PA (ed) (2004) Encyclopedia of corrosion technology (vol 20, p 138). CRC Press, New York
2. Morrell SH (1982) Rubber technology and manufacture. Blow, CM, Ed 162, p 135
3. Tanrattanakul V, Petchkaew A (2006) Mechanical properties and blend compatibility of natural rubber–chlorosulfonated polyethylene blends. J Appl Polym Sci 99(1):127–140
4. Chowdhury S, Das CK (2000) Structure–shrinkability correlation in polymer blends of ethylene vinyl acetate and chlorosulfonatedpolyethylene. J Appl Polym Sci 78(4):707–715
5. Antony P, Datta S, Bhattacharya A, De SK (2001) Mixed crosslinking in chlorosulfonated polyethylene and its blend with ethylene vinyl acetate copolymer. J Elastomer Plastics 33(3):196–210
6. Chowdhury S, Das CK (2001) Development of heat shrinkable and flame-retardant EVA/CSM blends. Fire Mater 25(5):199–202
7. Gupta S, Chowdhury SR, Mishra JK, Das CK, Patra PK, Tripathy AK, Banerjee MS (2000) Heat-shrinkable polymer blends through interchain crosslinking based on ethylene propylene and chlorosulfonated polyethylene. Mater Lett 46(2):125–130
8. Calmet JF, Carlin F, Nguyen TM, Bousquet S, Quinot P (2002) Irradiation ageing of CSPE/EPR control command electric cables. Correlation between mechanical properties and oxidation. Rad Phys Chem 63(3):235–239
9. Giurginca M, Zaharescu T (2000) Thermal and radiation behaviour of HNBR and CSPE blends. Polymer 41(20):7583–7587
10. Roychoudhury A, De PP (1997) Studies on chemical interactions between chlorosulphonated polyethylene and carboxylated nitrile rubber. J Appl Polym Sci 63(13):1761–1768
11. Mukhopadhyay S, De PP, De SK (1991) A self-vulcanizable and miscible blend system based on hypalon and carboxylated nitrile rubber. J Appl Polym Sci 43(2):347–355
12. Roychoudhury A, De PP, Bhowmick AK, De SK (1992) Self-crosslinkable ternary blend of chlorosulphonated polyethylene, epoxidized natural rubber and carboxylated nitrile rubber. Polymer 33(22):4737–4740
13. Antony P, De SK, van Duin M (2001) Self-crosslinking rubber/rubber and rubber/thermoplastic blends: a review. Rubber Chem Technol 74(3):376–408
14. Roychoudhury A, De PP (1997) Studies on chemical interactions between chlorosulphonated polyethylene and carboxylated nitrile rubber. J Appl Polym Sci 63(13):1761–1768
15. Mukhopadhyay S, Chaki TK, De SK (1990) Self-vulcanizablerubber blend system based on epoxidized natural rubber and hypalon. J Polym Sci Part C Polym Lett 28(1):25–30
16. Roychoudhury A, De PP, Dutta NK, Roychoudhury N, Haidar B, Vidal A (1993) FTIR and NMR studies on crosslinking reaction between chlorosulfonated polyethylene and epoxidized natural rubber. Rubber Chem Technol 66(2):230–241
17. Wang Z, Ni H, Bian Y, Zhang M, Zhang H (2010) The preparation and thermodynamic behaviors of chlorosulfonated polyethylene. J Appl Polym Sci 116(4):2095–2100
18. Noeske H, Roelen O (1959) U.S. Patent 2,889,259 (CA: 54, 14790c)
19. Zhao R, Cheng S, Shun Y, Huang Y (2001) Preparation of chlorosulfonated polyethylene by the gas-solid method. J Appl Polym Sci 81(14):3582–3588
20. Benedirt GM, Hodge IM (1985) E.P. 133,686 (CA: 102, 168178n)
21. Benedirt GM (1984) U.S. Patent 4,452,953 (CA: 101, 39028u)
22. Blanchard RR (1986) U.S. Patent 4,584,351 (CA: 105, 7765y)
23. Bazuin CG, Eisenberg A (1981) Modification of polymer properties through ion incorporation. Ind Eng Chem Prod Res Develop 20(2):271–286
24. Xie HQ, Chen Y, Guan JG, Xie D (2006) Novel method for preparation of quaternary ammonium ionomer from epoxidized styrene–butadiene–styrene triblock copolymer and its use as compatibilizer for blending of styrene–butadiene–styrene and chlorosulfonated polyethylene. J Appl Polym Sci 99(4):1975–1980

25. Boonsong K, Seadan M, Lopattananon N (2008) Compatibilization of natural rubber (NR) and chlorosulfonated polyethylene (CSM) blends with zinc salts of sulfonated natural rubber. Songklanakarin J Sci Technol 30(4):491
26. Nanda M, Tripathy DK (2010) Relaxation behavior of conductive carbon black reinforced chlorosulfonated polyethylene composites. J Appl Polym Sci 116(5):2758–2767
27. Mahapatra SP, Tripathy DK, Lee Y (2012) Electrical response of microcellular EPDM rubber composites: complex dielectric modulus formalism and current–voltage characteristics. Polym Bull 68(7):1965–1976
28. Brug GJ, Van Den Eeden ALG, Sluyters-Rehbach M, Sluyters JH (1984) The analysis of electrode impedances complicated by the presence of a constant phase element. J Electroanal Chem Interfac Electrochem 176(1):275–295
29. Tan Y, Tang J, Deng A, Wu Q, Zhang T, Li H (2013) Magnetic properties and microwave absorption properties of chlorosulfonated polyethylene matrices containing graphite and carbonyl-iron powder. J Mag Magn Mater 326:41–44
30. Markovic G, Veljkovic O, Radovanovic B, Marinovic-Cincovic M, Budinski-Simendic J, Sad N (2008) High energy radiation resistance of composites based on NR/CSM rubber blend: influence of carbon black and wood flour. KGK Kautsch, Gummi, Kunstst 61(7–8):363–367
31. Marković G, Marinović-Cincović MS, Jovanović V, Samaržija-Jovanović S, Budinski-Simendić J (2012) Gamma irradiation aging of NBR/CSM rubber nanocomposites. Comp Part B Eng 43(2):609–615
32. Feng Z, Nuli Y, Yang J (2007) Conductive sulfur-containing material/polyaniline composite for cathode material of rechargeable magnesium batteries. Acta Phys Chim Sin 23(3):327–331
33. Meyyappan M (2004) Carbon nanotubes: science and applications. CRC, Boca Raton, FL
34. Moisan M, Barbeau J, Crevier MC, Pelletier J, Philip N, Saoudi B (2002) Plasma sterilization: methods and mechanisms. Pure Appl Chem 74(3):349–358
35. Li J, Lu Y, Meyyappan M (2006) Nano chemical sensors with polymer-coated carbon nanotubes. Sensors J IEEE 6(5):1047–1051
36. Khatua BB, Das CK, Patra PK, Banerjee MS, Millins W (2000) Speciality polymer blends of polyurethane and chlorosulfonated polyethylene (S-cure). Int J Polym Mater 46(1–2):347–360
37. Dick JS (2014) How to improve rubber compounds: 1800 experimental ideas for problem solving. Carl Hanser Verlag GmbH Co KG, München
38. Cannas C, Mainas M, Musinu A, Piccaluga G (2003) ZnO/SiO_2 nanocomposites obtained by impregnation of mesoporous silica. Comp Sci Technol 63(8):1187–1191
39. Owczarek M, Zaborski M, Stefanowskiego K (2004) Chlorosulfonated polyethylene elastomers containing zinc oxide incorporated on SiO_2. Kautsch Gummi Kunstst 57(5):218–223
40. Jian L, Jun W, Shujun C (1997) Application of chlorosulfonated polyethylene elastomer in coating industry. Paint Coat ind 3:016
41. Ding Y, Qiao H, Zhou M (2002) Study on chlorosulphonated polyethylene (CSM) modified high chlorinated polyethylene (HCPE) anticorrosive coatings. J Chinese Soc Corr Protect 22(4):221–224
42. Lin TIAN (2006) Status and progress on special synthetic rubber in China. China Elastomerics 3:018
43. Lu Y, Partridge C, Meyyappan M, Li J (2006) A carbon nanotube sensor array for sensitive gas discrimination using principal component analysis. J Electroanal Chem 593(1):105–110
44. Oikonomou A, Giannakopoulou T, Litsardakis G (2007) Design, fabrication and characterization of hexagonal ferrite multi-layer microwave absorber. J Magn Mag Mater 316(2):e827–e830
45. U.S. Nuclear Regulatory Commission/Regulatory Guide 1.218 (2012) Condition monitoring techniques for electric cables used in nuclear power plants. Office of Nuclear Regulatory Research, Apr 2012

46. Gillen KT, Assink R, Bernstein R, Celina M (2006) Condition monitoring methods applied to degradation of chlorosulfonated polyethylene cable jacketing materials. Polym Degrad Stab 91(6):1273–1288
47. Assink RA, Gillen KT, Bernstein R, Celina MC (2005) Condition monitoring methods applied to degradation of chlorosulfonated polyethylene cable jacketing materials (No. SAND2005-3251J). Sandia National Laboratories
48. Kim IY, Goo CS, Lee JH, Jin BS, Kang MK, Shin YD (2012) A study of validation of condition monitoring method of NPPs cable through volume electrical resistivity. In International conference on high voltage engineering and application (ICHVE), 2012 (pp 68–71). IEEE
49. Kim KY, Nho YC, Jung SH, Park EH (1999) Development of evaluation technique on ageing degradation of organic polymer in nuclear power plant. Korea Atomic Energy Research Institute, Republic of Korea
50. Li W, Dichiara A, Zha J, Su Z, Bai J (2014) On improvement of mechanical and thermomechanical properties of glass fabric/epoxy composites by incorporating CNT–Al2 O3 hybrids. Compos Sci Technol 103:36–43
51. Lee JH, Kang MK, Jeon JS, Lee SH, Kim IY, Park HS, Shin YD (2014) A study on the properties of CSPE according to accelerated thermal aging years. J Electr Eng Technol 9(2):643–648
52. Jeon HH, Lee JU, Jeon JS, Lee SH, Shin YD, Effects of dried days on properties of seawater and freshwater flooded CSPE in NPPs
53. Gu Z, Song G, Liu W, Gao J, Dou W, Lu P (2010) Preparation and properties of chlorosulfonated polyethylene/organomontmorillonite nanocomposites. J Appl Polym Sci 115(6):3365–3368
54. Chung SK, Lee DC (1996) High voltage engineering. Munundang
55. Gillen KT, Assink RA, Bernstein R (2004) Condition monitoring approaches applied to a polychloroprene cable jacketing material. Polym Degrad Stab 84(3):419–431
56. Jeon HH, Lee JU, Jeon JS, Lee SH, Shin YD, Effects of dried days on properties of seawater and freshwater flooded CSPE in NPPs
57. Tanrattanakul V, Petchkaew A (2006) Mechanical properties and blend compatibility of natural rubber–chlorosulfonated polyethylene blends. J Appl Polym Sci 99(1):127–140
58. Marković G, Radovanović BC, Budinski-Simendić JK, Marinović-Cincović MT (2005) Curing characteristics of chlorosulphonated polyethylene and natural rubber blends. J Serb Chem Soc 70(5):695–703
59. Phiriyawirut M, Luamlam S (2013) Influence of poly (vinyl chloride) on natural rubber/chlorosulfonated polyethylene blends. Open J Org Polym Mater
60. Plochocki AP, Dagli SS, Andrews RD (1990) The interface in binary mixtures of polymers containing a corresponding block copolymer: effects of industrial mixing processes and of coalescence. Polym Eng Sci 30(12):741–752
61. Sirisinha C, Saeoui P, Guaysomboon J (2004) Oil and thermal aging resistance in compatibilized and thermally stabilized chlorinated polyethylene/natural rubber blends. Polymer 45(14):4909–4916
62. Phiriyawirut M, Luamlam S (2013) Improved automotive fuel resistance of natural rubber/chlorosulfonated polyethylene blends by blending epoxidized natural polymer. Open J Org Polym Mater
63. Oda E, Fujimura S, Kubo M, Tsutsumi Y, Seguchi T, Hagiwara M (2012) Development of super-radiation-resistant cable. Int J Radiat Appl Instrument Part C Radiat Phys Chem 22(3):4003304
64. Ren Z, Lan Y, Wang Y (2013) Carbon nanotubes. Aligned carbon nanotubes. Springer, Berlin Heidelberg, pp 7–43
65. Rittersma ZM (2002) Recent achievements in miniaturised humidity sensors—a review of transduction techniques. Sens Actuat A Phys 96(2):196–210
66. Fenner R, Zdankiewicz E (2001) Micromachined water vapor sensors: a review of sensing technologies. Sensors J IEEE 1(4):309–317

67. Traversa E (1995) Ceramic sensors for humidity detection: the state-of-the-art and future developments. Sens Actuat B Chem 23(2):135–156
68. Bartlett PN, Archer PB, Ling-Chung SK (1989) Conducting polymer gas sensors part I: fabrication and characterization. Sens Actuat 19(2):125–140
69. Petty MC, Casalini R (2001) Gas sensing for the 21st century: the case for organic thin films. Eng Sci Educ J 10(3):99–105
70. Sin ML, Chun Tak Chow G, Wong GM, Li W, Leong P, Wong KW (2007) Ultralow-power alcohol vapor sensors using chemically functionalized multiwalled carbon nanotubes. Nanotechnol IEEE Trans 6(5):571–577
71. Philip B, Abraham JK, Chandrasekhar A, Varadan VK (2003) Carbon nanotube/PMMA composite thin films for gas-sensing applications. Smart Mater Struct 12(6):935
72. Ebron VH, Yang Z, Seyer DJ, Kozlov ME, Oh J, Xie H, Baughman RH (2006) Fuel-powered artificial muscles. Science 311(5767):1580–1583
73. Bar-Cohen Y (2002) Electro-active polymers: current capabilities and challenges. In: Proceedings of SPIE, the international society for optical engineering proceedings of SPIE, the international society for optical engineering, vol 4695, pp 1–7
74. Banerjee S, Banerjee S, Tyagi AK (eds) (2012) Functional materials: preparation, processing and applications. Elsevier
75. Pelrine R, Kornbluh R, Pei Q, Joseph J (2000) High-speed electrically actuated elastomers with strain greater than 100 %. Science 287(5454):836–839
76. Yang G, Ren W, Akhras G, Mukherjee B (2006) Transverse strain of silicone dielectric elastomer actuators. J Adv Sci 18(1/2):166–169
77. Choi HR, Jung KM, Koo JC, Nam JD, Lee YK, Cho MS (2005) Electrostatically driven soft polymer actuator based on dielectric elastomer. Key Eng Mater 297:622–627
78. Koo JC, Choi HR, Jung MY, Jung KM, Nam JD, Lee YK (2005) Design and control of three-DOF dielectric polymer actuator. Key Eng Mater 297:665–670
79. Shoa T, Madden JD, Fok CWE, Mirfakhrai T (2009) Rate limits in conducting polymers. Adv Sci Technol 61:26–33
80. Birley AW (1992) Physics of plastics: processing, properties and materials engineering. Hanser Verlag, Munich
81. Lee MJ, Lee NY, Lim JR, Kim JB, Kim M, Baik HK, Kim YS (2006) Antiadhesion surface treatments of molds for high-resolution unconventional lithography. Adv Mater 18(23):3115–3119
82. Vichchulada P, Lipscomb LD, Zhang Q, Lay MD (2009) Incorporation of single-walled carbon nanotubes into functional sensor applications. J Nanosci Nanotechnol 9(4):2189–2200
83. He S, Lin Y, Wei Z, Zhang L, Lin J, Nazarenko S (2015) Solvent-free fabrication of proton-conducting membranes based on commercial elastomers. Polym Adv Technol 26(4):300–307
84. Pelrine R, Kornbluh RD, Eckerle J, Jeuck P, Oh S, Pei Q, Stanford S (2001). Dielectric elastomers: generator mode fundamentals and applications. In: SPIE's 8th annual international symposium on smart structures and materials, international society for optics and photonics, pp 148–156
85. Chung DDL (2001) Electromagnetic interference shielding effectiveness of carbon materials. Carbon 39(2):279–285
86. Sadchikov VV, Prudnikova ZN (1997) Amorphous materials in electromagnetic shields. Steel in Transl 27(4):71–75
87. Shinagawa S, Kumagai Y, Urabe K (1999) Conductive papers containing metallized polyester fibers for electromagnetic interference shielding. J Porous Mater 6(3):185–190
88. Kumar R, Kumar A, Kumar D (1997) RFI/EMI/microwave shielding behaviour of metallized fabric-a theoretical approach. In: Proceedings of the international conference on IEEE, electromagnetic interference and compatibility'97, pp 447–450
89. Grimes CA (1994) EMI shielding characteristics of permalloy multilayer thin films. In: Aerospace applications conference proceedings IEEE, pp 211–221

90. Biter WJ, Jamnicky PJ, Coburn W (1994) Shielding improvement by use of thin multilayer films. In: International sampe electronics conference society for the advancement of material and process engineering, vol 7, p 234
91. Klinkenbusch L (2005) On the shielding effectiveness of enclosures. Electromagn Compat IEEE Trans 47(3):589–601
92. Paul CR (1992) Introduction to electromagnetic compatibility. NASA STI/Recon Technical Report A, vol 93, p 49330
93. Taya M, Kim WJ, Ono K (1998) Piezoresistivity of a short fiber/elastomer matrix composite. Mech Mater 28(1):53–59
94. Chen G, Zhao W (2010) Polymer/graphite nanocomposites. CRC Press, New York 79
95. Das NC, Khastgir D, Chaki TK, Chakraborty A (2000) Electromagnetic interference shielding effectiveness of carbon black and carbon fibre filled EVA and NR based composites. Compos A Appl Sci Manuf 31(10):1069–1081
96. Tanrattanakul V, Bunchuay A (2007) Microwave absorbing rubber composites containing carbon black and aluminum powder. J Appl Polym Sci 105(4):2036–2045
97. Xie HQ, Ma YM, Guo JS (2001) Secondary doping phenomena of two conductive polyaniline composites. Synt Met 123(1):47–52
98. Thomas RM, John H, Mathew KT, Joseph R, George KE (2009) Optimization of preparation conditions on the dielectric properties of polyaniline. J Appl Polym Sci 112(5):2676–2682

Electronic Applications of Styrene–Butadiene Rubber and Its Composites

Ranimol Stephen and Sabu Thomas

Abstract Stretchable materials in electronics industry have attracted tremendous interest because it can maintain high strain. Therefore, soft materials find application in electromagnetic interference (EMI) shielding, piezoelectric materials, actuators, pressure sensors, capacitive sensors and energy storage devices and solar cells. This chapter provides an outlook into the electrical properties and electronic applications of styrene–butadiene rubber (SBR) composites in the presence of various types of conducting fillers.

Keywords SBR composites · Conducting fillers · Electronic applications

1 Introduction

Conductive rubber composites have received great interest over the last decade from various scientific fields as their applications in electromagnetic shielding, flexible electronic devices, thin film transistors, electrodes and sensors are almost indispensible [1–4]. The possibilities and applications of stretchable and flexible materials in the manufacture of electronic devices is tremendous. However, the basic characteristic properties of both natural and synthetic rubbers are its remarkable electrical resistance, elasticity, toughness, impermeability and adhesiveness which makes it a useful material in different fields rather than in electronics industry. Electrical properties in rubber can be achieved only by the formation of

R. Stephen (✉)
Department of Chemistry, St. Joseph's College (Autonomous), Devagiri, Calicut 673 008, Kerala, India
e-mail: ranistephen@gmail.com; ranistephen@yahoo.com

S. Thomas
International and Inter University Center for Nanoscience and Nanotechnology, School of Chemical Sciences, Mahatma Gandhi University, Priyadarshini Hills P.O., Kottayam 686 560, Kerala, India
e-mail: sabuchathukulam@yahoo.co.uk

D. Ponnamma et al. (eds.), *Flexible and Stretchable Electronic Composites*, Springer Series on Polymer and Composite Materials,
DOI 10.1007/978-3-319-23663-6_9

conductive networks in the matrix, which offers an electrical pathway for the conduction of electrons. It is proved that the conducting fillers (CFs) such as carbon black, nanotubes, graphene and metal particles can create an electrical pathway by the formation of filler–filler networks within the rubber matrix.

Styrene–butadiene rubber (SBR) is a synthetic rubber, also known as Buna-S, used in various sectors such as tyre, medical, footwear and automobile industries. It is produced by the copolymerization of butadiene, $CH_2{=}CH{-}CH{=}CH_2$ (75 %), and styrene, $C_6H_5{-}CH{=}CH_2$ (25 %). Pure gum strength of this non-stereospecific, amorphous polymer is low when compared to stereospecific natural rubber (NR). Subsequently, SBR will not crystallize upon stretching. However, white and black fillers-reinforced SBR exhibit excellent strength and abrasion resistance. It is resistant to mineral oils and fats, and solvents such as aliphatic, aromatic and chlorinated hydrocarbons [5].

The primary goal of this chapter is to give an outlook into the properties of SBR and various CF composites and its challenges and potential applications in electronic devices and sensors.

2 Relevance of Stretchable Materials in Electronics

Stretchable and flexible electronic composites are considered to be a revolutionary technology for both academicians and industrialists. Conductive stretchable materials have been a research theme for a long time. However, the recent developments in the fabrication of stretchable electronic materials are remarkable which mainly focused on electronic devices and sensors from elastomeric materials. Very recently, Zhu and co-workers [6] and Huang and co-workers [7] reviewed the applications of stretchable materials for flexible electronics. The main advantage of stretchable electronics is the designing of flexible, light-weight, cost-effective materials, since the matrix material is highly inexpensive compared to conventional substrates. Moreover, the key attraction of flexible electronics is its ability to bent, fold, twist and even form into different shapes without affecting the performance and properties. Flexible electronic devices are more relevant in the manufacture of medical devices for diagnosis and treatment purposes in human bodies. The potential applications include the fabrication of smart clothing, foldable displays, robotic skins, stretchable solar cells and bioinspired devices [8–10].

3 Conductive Styrene–Butadiene Rubber (SBR) Composites

SBR is an insulator, which makes it conductive by incorporating conductive materials such as metals, carbon-based fillers and ceramics. Common conductive fillers used in elastomers include conductive carbon black, carbon nanotubes (CNTs), graphite, graphite fibres, metal particles, metal wires and powders [11–18]. Conductivity of rubber matrix can be enhanced by changing the weight percentage of these conductive fillers, and it is noteworthy that not only the conductivity but physical properties can also be modified. Rubber composites are used for various applications instead of pure rubber, in which fillers act as reinforcing agents irrespective of their conducting or non-conducting nature. Till date, the actual mechanism involved in the reinforcement of filled rubber is not fully explained, but interpreted in terms of the formation of rubber–filler networks. The inclusion of rigid filler in the soft rubber matrix restricts the segmental mobility of the polymer chain. Conductivity of SBR can be varied over a wide range by varying the CF and its dispersion in the matrix.

Properties of composites depend on several factors such as the nature of individual components, size and shape of fillers, morphology of the system and nature of the interface between the phases. Interface interaction between phases plays an important role in the strength of composite in addition to ensuring uniform dispersion of filler in the rubber matrix. In the manufacturing of rubber nanocomposites, to attain the fine dispersion of filler in the matrix is still a challenging problem for the material scientists. Various methods for the fabrication of SBR composite include solution mixing, latex route, melt compounding and two-roll mill mixing. Structure and morphology of the manufactured rubber composites can be explored using different techniques such as X-ray diffraction (XRD), solid-state nuclear magnetic resonance (NMR), Fourier transform infrared (FTIR), scanning electron microscopy (SEM) and transmission electron microscopy (TEM).

3.1 SBR/Metal, Metal Oxides and Ceramics

Conductive SBR composites are frequently manufactured by incorporating metal fillers such as copper, steel, silver, gold and aluminium. These are generally used in the form of powders or flakes. Ceramic files are also used for the fabrication of conductive composite, which includes silicon carbide, boron nitride, aluminium nitride and alumina.

Negri and co-workers [19] have synthesized magnetic nanoparticles, Fe_3O_4 (magnetite) with average size less than 10 nm to ensure better superparamagnetism to avoid large irreversible magnetic effects. These magnetite particles are then covered with conducting Ag^o to generate microparticles of Fe_3O_4@Ag, which exhibits both magnetic and conductive properties. During the preparation of SBR-DEG-Fe_3O_4 films, the formation of pseudo-chains (needles) of inorganic materials is magnetically aligned and is evident in the morphological analysis presented in Fig. 1. Magnetically aligned needle like Fe_3O_4@Ag can be clearly seen in the optical and SEM images. In the centre, the needles are normal to the surface; however, at the edge, it is diagonal to the surface due to the uniformity of the magnetic field in the centre and not in the edges. They have investigated the magneto- and piezoresistivity of nanostructured composite films based on SBR

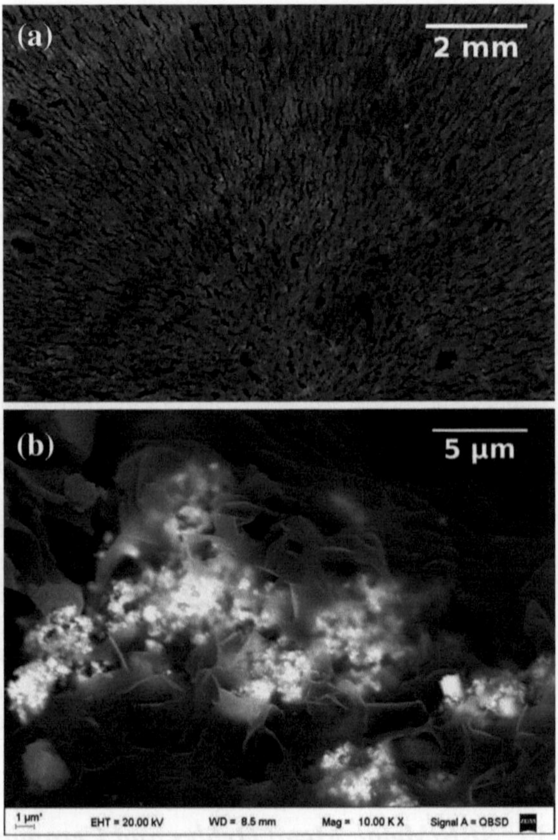

Fig. 1 SBR-DEG—12 % Fe_3O_4@Ag composite films. **a** Optical photograph and **b** SEM image [19]. Copyright 2013. Reprinted with permission from Elsevier Ltd.

filled with electrically CF. Diethylene glycol is added to SBR to get considerable electrical conductivity. The formation of structured SBR-DEG-Fe_3O_4 films with anisotropic properties is the reason behind the conductivity, which decreases the percolation critical point to observe conductivity. These films offer piezoresistive and magneto-resistive properties. Negri and co-workers [19, 20] used Fe_3O_4@Ag filler in various elastomeric matrix due to its simultaneous exhibition of magnetic and electrical properties.

Lee et al. [21] fabricated highly stretchable Ag nanowire–Ag nanoparticles (AgNW–AgNP)-reinforced electrically conductive styrene–butadiene–styrene (SBS) fibre by wet-spinning method and iterative process for Ag precursor absorption and reduction. During stretching AgNW-embedded SBS exhibits high initial electrical conductivity due to the presence of highly conductive fillers. According to them, AgNWs form bridges between AgNPs for electrical conductivity, which will retain up to 100 % strain after that decline of conductivity observed due to the breakdown of networks. SBS–AgNW composite fibre shows strain-sensing behaviour and is incorporated into a smart glove to perceive human motions is shown in Fig. 2.

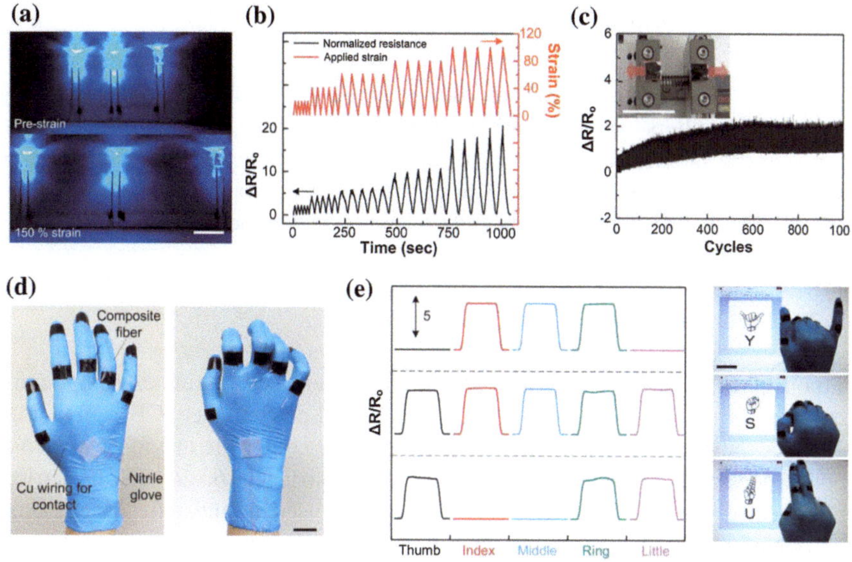

Fig. 2 **a** Images of *blue*-coloured LEDs connected by 0.56 wt% AgNW–AgNP embedded composite fibres. **b** The normalized resistance changes varying the applied strains at 0, 20, 40, 60, 80 and 100 %. **c** Reliability test measured the changes in the normalized resistance of the composite fibre during the stretching and releasing cycle. **d** Photograph of the smart glove attached to the composite fibre on each finger. **e** Motion detection of English letters '*Y*', '*S*' and '*U*' for utilizing sign language glove and photographs of detecting each letter [21]. Copyright 2015. Reprinted with permission from John Wiley & Sons

Titanium carbide (TiC) belongs to the class of interstitial carbides, in which the metal retains its closed packed lattice structure and C atoms occupy the octahedral interstices. TiC is extremely hard refractory materials and good conductors of electricity. Electrical and electromagnetic wave shielding properties of TiC-loaded SBR have been investigated by Sung and El-Tantawy [22]. According to them, SBR–TiC composite is a promising material for applications in self-electrical heating, temperature sensor, time delay switching and electromagnetic wave shielding effectiveness. The electromagnetic wave shielding effectiveness of the composite is due to reflection loss, consisting of the reflections from the material boundaries and surroundings. Reflection loss depends on the electrical conductivity and network structure of the conductive particles in the elastomer.

Measurement of electrical conductivity of filled elastomers gives information about the dispersion of filler in the matrix. Gwaily et al. [23] studied the effect of γ-radiation in the conductivity of Fe and Cu metal particles-filled SBR. In the report, they mentioned that without irradiation, the filled SBR behaves like an insulator, and conductivity, σ is in the range $10^{-12}\ \Omega^{-1}\ \text{cm}^{-1}$. It is due to the weak interfacial interaction of metal–polymer and segregation of filler particles after moulding. Upon γ-irradiation >300 kGy, σ increases significantly due to the formation of filler–filler networks.

3.2 SBR/Carbonaceous Materials

Conducting SBR composites can be prepared by incorporating conducting carbonaceous fillers such as carbon black, carbon fibre, graphite and CNTs. Improvement in mechanical, thermal and electrical properties of composite is strongly influenced by the nature of filler, size, shape, surface area, surface chemistry, distribution within the matrix and processing techniques.

3.2.1 SBR/Carbon Black

Carbon black is one of the major reinforcing filler used in rubber compounds, especially in tyres to impart improved strength and performance. CB is also used as CF for the preparation of conducting elastomeric composite due to its low cost, good processability, availability and corrosion resistant [16, 24, 25]. Electrical properties of conducting carbon black have been used in antistatic, electromagnetic interference (EMI), electroactive pressure sensors and actuators. A special type of highly conductive novel carbon black with superhigh surface area and structure has been synthesized and incorporated in elastomeric matrix, which shows conductivity higher than that of conventional electroconductive carbon [26]. SBR filled with carbon black shows more network chains due to its strong absorbability [27], which thereby reduce the agglomeration of fillers and enhance the conductivity of elastomers. Conductivity of CB depends on three major factors: particle size, structure and surface chemistry. Fillers with high surface area and porosity provide high electrical conductivity to the elastomeric matrix [28]. Podhradska et al. [29] reported on the stability of electrical conductivity during cyclic thermal treatment of SBR/carbon black composites. They observed that the conductivity of SBR/CB composite increased from 1.1×10^{-2} to 3.0×10^{-2} Scm^{-1} when the temperature increased to 85 °C.

Designing of elastomeric materials with conductivity and the maintenance of mechanical performance is a challenging problem for the material scientists. The reason behind this is due to the fact that the lowering of percolation threshold concentration (PTC) is essential for better mechanical properties and good processability. Therefore, filler–filler and polymer–filler network formation is required to obtain elastomeric composite with lower PTC and higher mechanical properties. Due to the high structure of carbon black, the PTC observed is around 5–10 vol.% for conducting rubber composite system [30]. Kawazoe and Ishida [31] proposed a novel mechanism for the network formation of carbon black in an immiscible SBR/NBR blend system. They have demonstrated that the PTC of the conductive rubber composite of binary immiscible blends of SBR/NBR could change by changing the solvent. PTC reduced from 5.7 vol.% for SBR and 8.9 vol.% for NBR to 1.2 vol.% for SBR/NBR (90/10) system. In blend systems, the affinity towards CB is different for individual components, which results in the selective adsorption of filler particles on the polymer matrix. Non-uniform distribution of CB in rubber blend can be explained in terms of the molecular confinement mechanism. PTC can be controlled by the selective adsorption of CB; it reduces the aggregation of filler particles and thus localizing at the interface [32–37]. As the specific surface area of CB increases, PTC is found to be lower. Each CB differs in their electrical conductivity due to the difference in specific surface area as well as PTC. Meissner and co-workers [38] relate the effect of CB concentration and type on the electrical conductivity of SBR composite. Four different grades of CB, ISAF, HAF, FEF and La were used and found that the PTC decreases with increase in specific surface area of filler. They have theoretically correlated the conductivity of composites in terms of percolation theory. Percolation theory predicts the relation between the conductivity of composite, σ, and the volume fraction of filler, X. The equation is,

$$\sigma = \sigma_0 (X - X_c)^s \tag{1}$$

where σ_0 is the conductivity of filler particles, X_c is the volume fraction of the filler at percolation threshold and s is the quantity determining the power of the conductivity increasing above X_c. Unexpectedly, high values of s were found for SBR/CB composite and can be described by the behaviour of a single tunnel junction.

Silica-based fillers, fumed or precipitated silica, are used as reinforcing fillers in tyre industry (green tyres) because they provide better fuel efficiency and low rolling loss when compared to carbon black-filled tyres. Zhao and Wang [39] investigated the antistatic characteristics of solution-polymerized SBR composites filled with hybrid SiO_2/CB. In SSBR/SiO_2 composite, the static accumulation is found to be high, while in the presence of SiO_2/CB in SSBR composite it exhibits antistatic property. This can be attributed in terms of the antistatic nature of CB. According to their studies, the ratio of SiO_2/CB in 35/35 was the percolation threshold of the antistatic property, as the amount of SiO_2 increases the surface and volume resistivity of the composite increases.

He et al. [40] used water-soluble SBR–sodium carboxy methyl cellulose (CMC) mixture as a binder for sulphur cathode, which consists of 60 wt% insulating S active material and 40 wt% conducting carbon black filler. A remarkable change in cyclic performance of sulphur cathode was observed in the presence of SBR-CMC binder than conventional poly (vinylidene fluoride) (PVDF) binder. Binder acts as both adhesion agent and dispersing agent as it accelerates the uniform dispersion of conducting CB and insulating sulphur to ensure a good electrical contact. Capacity of SBR-CMC cathode was found to be 580 $mAhg^{-1}$ after 60 cycles, while that of PVDF cathode is 370 $mAhg^{-1}$.

3.2.2 SBR/Carbon Nanotubes (CNTs)

CNTs have high electrical conductivity along the tube axis. The percolation threshold of polymer/CNT composites is found to be considerably lower when compared to other conventional composites. Therefore, CNTs are considered as one of the ideal conducting material for elastomers. Also, its unique properties such as high mechanical strength, aspect ratio, tubular structure and nanoscale diameter make it a versatile material for the fabrication of stretchable conductive elastomeric matrix. Significant electrical properties of composites of CNTs favour their potential applications in electrostatic dissipation, EMI shielding, super capacitors, chemical sensors and automotive and aviation industries. On incorporating CNTs in elastomers, a continuous network of filler is formed across the matrix and enables the material to a sudden transition from an insulator to a conductor.

The poor dispersion of CNTs in the polymer matrix is one of the challenging problems for material scientists. In order to overcome this facet, Paul and co-workers [41] prepared masterbatches of oxidized MWCNT with SBR using two methods, one by coagulation with acetone and another by aqueous coagulation method followed by mastication in a twin screw extruder, and then analysed the masterbatch rheology to get an insight into the good dispersion and conformation of the MWCNT in the final product. Electrical resistivity decreases with increase in the concentration of MWCNT; however, the change is much less than that expected for percolation networks [42–45]. The tubes are not actually touching each other, but the decrease in resistivity can be explained by electron hopping [46].

Li and Shimizu [47] fabricated a thermoplastic elastomer based on poly(styrene-b-(ethylene-co-butylene)-b-styrene) (SEBS) triblock copolymer and multiwalled carbon nanotubes (MWCNTs) using high-shear processing technique with a screw rotation speed of 1000 rpm. Nanocomposites with 15 wt% of MWCNT show excellent electrical conductivity and stability upon uniaxial deformation with a strain less than 50 %.

Another important factor for determining the electrical properties of composites is the processing method. Nogueira and co-workers [48] reported the electrical conductivity of composites based on MWCNT and styrene–butadiene–styrene block copolymer processed by two methods: melt mixing (extrusion) and solution casting. Composites prepared by solution casting showed higher electrical conductivity than melt mixing. SBS/MWCNT with 6 wt% prepared by melt mixing and the one with 1 wt% prepared by solution casting show the same magnitude of conductivity. Lower PTC is observed for the samples prepared by solution casting method. The reason behind the decrease in electrical conductivity of extruded sample is the decrease in aspect ratio of MWCNT on applying shearing force.

In order to prevent the breaking of CNTs during processing, a novel mixing method rotation–revolution mixer was employed for the preparation of SBR/CNT composite [49]. The obtained results indicated that the new mixing technique reduced the amount of CNT required to provide high electrical conductivity to the SBR matrix than conventional processing methods.

Materials such as CNTs and FG have high aspect ratio, which enable both electrical and mechanical percolation at low filler concentration. Heinrich and co-workers [50] prepared solution styrene–butadiene rubber (S-SBR) composite using graphene nanoplates (GnPs), expanded graphite (EG), MWCNTs and a combination of EG with MWCNT. To get an insight into the specific dispersion and reinforcement of the fillers, they compared the mechanical and electrical

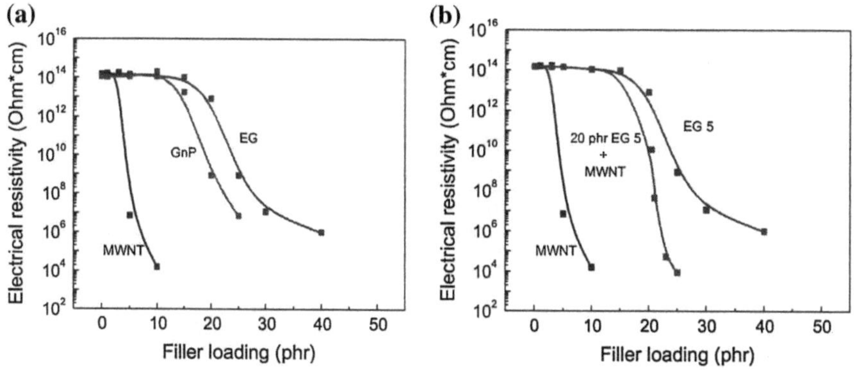

Fig. 3 Percolation of S-SBR filled with **a** GnP, EG, MWCNT and **b** 20 phr EG and various amount of MWCNT [50]. Copyright 2012. Reprinted with permission from Elsevier Ltd.

properties, presented in Fig. 3a. A remarkable drop in resistivity was observed for MWCNT composite at 3 wt%. This is attributed to the high aspect ratio of MWCNT, leading to the formation of filler–filler network. At higher concentration of EG and GnP, particles form networks, resulting in a conducting pathway and the resistivity of the matrix decreased. Mixed filler system exhibits a synergistic effect in the electrical resistivity of composites, shown in Fig. 3b.

Challenges in the design and fabrication of conducting elastomers are the homogeneous dispersion of CFs in the matrix. Graphene sheets have a tendency for restacking in the matrix, and CNTS forms entanglements. These challenges have been addressed by many researchers by using hybrid fillers in elastomers. For example, graphene and MWCNT hybrid fillers, in which MWCNT helps to form bridges between graphene platelets, provide more interface area between the filler and matrix [50, 51].

3.2.3 SBR/Graphene (GN)

Recently, graphene has attracted much attention of material scientists due to the similarities in properties to carbon nanotubes. The production of carbon nanotubes is highly expensive, so material scientists thought about an alternative and cost-effective material for potential applications. Apart from the above reason, graphene is preferred over other fillers due to high surface area, aspect ratio, mechanical properties, thermal and electrical conductivity, EMI shielding ability, flexibility and transparency [52–55]. Graphene is a single carbon layer of graphite. It is a two-dimensional one-atom-thick sp^2-bonded carbon atom densely packed like a honeycomb crystal lattice [56]. In 2010, Andre Geim and Konstantin Novoselov won the Nobel Prize in Physics for their outstanding work in graphene. In 2004, they have identified single layers of graphene and other 2-D crystals [57]. Recently, many researchers reviewed the advances in the properties of graphene-based polymer composites [58–62].

Graphene and functionalized graphene (FG) are extensively used in elastomers as a reinforcing and conducting material [63]. It is a semiconductor with a zero gap and is different from conventional Si semiconductor. Graphene sheets impart percolated pathways for the conduction of electrons in insulators like other CFs such as carbon black, carbon nanotubes and carbon nanofibres. However, the advantage of graphene is its property to enable the insulator to be a conductor at a lower PTC when compared to the PTC of other fillers [64]. It is accounted in terms of the disc-like structure of graphene and can percolate easily in the polymer matrix than rod-like CNTs. Graphene is considered to be a promising candidate to replace Si in microchip electronics and in faster-switching transistor. Modified graphenes can also be used for the fabrication of materials for potential applications by enhancing the interfacial interaction between filler and matrix. Properties of alkylamine (oleylamine and octadecylamine)-modified graphene oxide (GO)–SBR composite were found to be improved significantly [65]. SBR/graphene nanocomposite fabricated by latex compounding method attains antistatic criterion (10^{-6} S/m) at a GN loading of 3 phr, and when the loading is 7 phr, it becomes a good electrical conductor, shown in Fig. 4 [66]. Graphite flakes are converted into graphene oxide (GO) by Hummer's method [67].

Mulhaupt and co-workers [68] reported about the properties of FG and other carbonaceous materials-incorporated SBR nanocomposites prepared by aqueous dispersion blending. They have compared the performance of thermally reduced graphite oxide (TRGO), chemically reduced graphite oxide (CRGO) with that of multilayered graphene, conventional graphite carbon black and CNTs. TRGO-reinforced SBR showed improved mechanical properties and electrical conductivity when compared to other systems. Elastomers show significant improvement in mechanical, electrical and low gas permeability with functionalized graphene sheets (FGS) owing to its high aspect ratio and carbon-based structure [69, 70]. The volume resistivity of SBR (3.43×10^{15} Ω cm) reduced dramatically when compounding with electrically conducting graphene (6000 S/cm) at a vol.% of 16.5 [71].

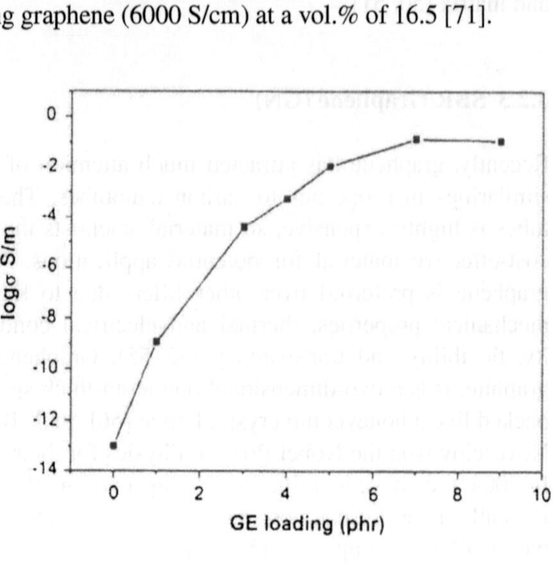

Fig. 4 Electrical conductivity of GE/SBR nanocomposites as a function of GE loading [66]. Copyright 2014. Reprinted with permission from Elsevier Ltd.

Uniform dispersion of graphene in elastomeric matrix is still a challenge. In order to overcome this, hybrid fillers of CB-reduced graphene (RG) were prepared and blended with SBR. Agglomerated CB get adsorbed on the surface of graphene to form microstructured hybrid filler. The role of CB is to block the restacking of RG sheets even after drying. When compared to SBR/RG and SBR/CB, the SBR/CB-RG showed significant improvement in mechanical properties and reduction volume resistivity due to the strong interfacial interaction between RG and SBR [72].

4 Electronic Applications of SBR Composites

Structured elastomer composite with anisotropic physical properties can be used in flexible electronics and in electronic devices. Addition of CFs makes 'Conductive Rubber', which can reduce or eliminate EMI and radio frequency interference (RFI) associated with stereo systems. Depending upon the intended environment and application, one can fine tune the properties of final product by varying the filler. Higher level of shielding could be achieved with the use of highly CF. Conductive elastomers can find applications in military and various industries such as telecom, medical devices and computing. Electrically conductive graphene-based SBR can be used for the manufacture of flexible electronic devices, thin film transistors, electrodes and sensors. In the near future, graphene oxide will be used for the manufacture of stretchable and electrically conductive fabrics by incorporating in elastomers [73].

The electron transport in elastomers can be explained in terms of quantum mechanical tunnelling and percolation theory [74]. If the distance between two conductive particles is large compared to atomic dimension, then the bulk resistivity of insulator predominates. However, if the distance is in the nanometre range <100 A^{o}, electrons can tunnel quantum mechanically between conductive fillers, resulting in less resistivity from the insulator than expected. At lower concentration of filler, electron transport is taking place by quantum mechanical tunnelling. As the filler concentration increases, the CFs come into contact with each other to form more filler–filler networks, leading to increase in conducting pathways. At a particular concentration of filler, the resistivity of the matrix drops dramatically and is called the PTC. Elastomeric composites exhibiting magneto-piezoresistivity find potential applications in sensors, flexible devices and electronic connectors [75–81].

4.1 Electromagnetic Interference (EMI) Shielding

A major concern in electronic industry is the increasing EMI among devices. Therefore, EMI shielding is essential at least for the frequency ranges

50–60 Hz in stereo systems and for 1–100 GHz in strategic applications. Electromagnetic shielding reduces the coupling of radio waves and electromagnetic and electrostatic fields by hindering the field by conducting or magnetic materials. EMI shielding used to protect the electronic devices from radio frequency electromagnetic radiation is also called radio frequency (RF) shielding. Metals such as copper, nickel or aluminium are extensively used for the EMI shielding. Although metals are very effective for EMI shielding, certain factors such as high weight, less corrosion resistance and difficulty in processing lead to the fabrication of rubber composite materials, which can substitute metals in this particular application. Insulating polymers can be converted into conducting or semiconducting polymer composite by the incorporation of CFs. Such materials find application for EMI shielding, microwave absorption and replacing conventional metals [82]. It is essential for medical and laboratory equipments to avoid the interference of other signals. Conductive SBR composite can find application as EMI and RF shielding materials in various electronic equipments.

4.2 Piezoelectric Materials

Conductive SBR composite is coming under the category ‘Flexible and Stretchable’ materials for electronic applications. Piezoelectricity means the accumulation of electrical charge on a certain material on applying a physical stress. As discussed earlier, CFs-reinforced SBR composite becomes a piezoelectric material upon stretching [19]. Piezoelectric behaviour is considered as a reversible process, in the presence of a mechanical strain charge accumulates, similarly, in the presence of an external electric field mechanical strain occurs and charge accumulates. The gripping performance of a tyre can be improved by the incorporation of piezoelectric materials [81]. Due to the flexible nature, rubber materials can generate more energy in response to a mechanical strain. Rubber composite materials can capture energy from movement or any kind of mechanical stress four times higher than the currently available piezoelectric materials. This generated energy could be used for the charging of portable electronic devices such as mobile phone. This technique can be made use of in the charging of medical devices used inside the body (pace maker). The body movements itself can generate enough power to charge the device. Another important area which will find application of flexible piezoelectric materials is shoe soles of soldiers as a source for power. Thin films of piezoelectric materials-incorporated rubber can harness tremendous amount of energy by human body movements; therefore, ‘piezo-rubber’ can find tremendous application in the near future.

4.3 Other Applications

Rubbery materials can be used for the manufacture of medical devices for diagnostic and treatment purposes due to its bendable, curved and stretchable nature. In addition to the above-mentioned applications, SBR composites with CFs are being explored for the manufacture of LEDs, sensors, actuators, smart clothes, artificial muscles, foldable displays, robotic skin (touch sensor) and solar cells. Elastomer-like electrode is fabricated by blending of conductive polyaniline (PANI) and a soft triblock copolymer poly(styrene-co-ethylene-co-styrene) grafted with maleic anhydride (SEBS-g-MA) to be used for artificial muscles [82]. Soft materials are required for smart applications that are to accommodate large actuation strains without increasing the stiffness of the material. Here, SEBS is used to control the stiffness of the material and endow flexibility and stretchability to the material.

5 Concluding Remarks

Stretchable electronics is considered to be a fast-growing revolutionary technology. From the stiff and rigid electronic devices to bendable, curved, twisted and stretchable devices represent the progress in electronic technologies. SBR composites with CFs will benefit significantly in tyre and automobile industries and in the manufacture of medical devices, shoe soles, artificial electrodes and sensors as it can sustain large strain. By incorporating conductive fillers, not only the electrical properties but also the mechanical properties of the matrix get improved. However, one of the biggest challenging problem in the fabrication of rubber composites is the formation of uniform dispersion of filler particles in the matrix. The development of advanced technologies and strategies for optimization and dispersion of fillers for improved properties will be required in the future.

References

1. Chen G, Zhao W (2010) Rubber/graphite nanocomposites, chapter 19. In: Rubber nanocomposites: preparation, properties and applications. Wiley, Singapore
2. Pramanik PK, Khastgir D, Saha TN (1991) Electro magnetic interference shielding by conductive nitrile rubber composites containing carbon fillers. J Elastomers Plast 23(4):345–361
3. Mattson B, Stenberg B (1992) Electrical conductivity of thermo-oxidatively-degraded EPDM rubber. Rubber Chem Technol 65(2):315–328
4. Chen X, Jiang Y, Wu Z, Li D, Yang J (2000) Morphology and gas-sensitive properties of polymer based composite films. Sens Actuators B Chem 66(1):37–39
5. Blow CM, Hepburn C (1982) Rubber technology and manufacture, 2nd edn. Butterworth Scientific, UK

6. Yao S, Zhu Y (2015) Nanomaterial-enabled stretchable conductors: strategies. Mater Devices Adv Mater. doi:10.1002/adma.201404446
7. Cheng T, Zhang Y, Lai WY, Huang W (2015) Stretchable thin- film electrodes for flexible electronics with high deformability and stretchability. Adv Mater. doi:10.1002/adma.201405864
8. Marshall JE, Gallagher S, Terentjev EM, Smoukov SK (2014) Anisotropic colloidal micro muscles from liquid crystal elastomers. J Am Chem Soc 136:474–479
9. Kaltenbrunner M, Sekitani T, Reeder J, Yokota T, Kuribara K, Tokuhara T, Drack M, Schwodiauer R, Graz I, Gogonea SB, Bauer S, Someya T (2013) An ultra-lightweight design for imperceptible plastic electronics. Nature 499:458–463
10. Lipomi DJ, Tee BC-K, Vosgueritchian M, Bao ZN (2011) Stretchable organic solar cells. Adv Mater 23(15):1771–1775
11. Osman HM, Abdel Chani SA, Madkour TM, Mohammed AR (2000) Stress relaxation in carbon black loaded butyl rubber. J Appl Polym Sci 77:1067–1076
12. Chodak I, Omastova M, Pionteck J (2001) Relation between electrical and mechanical properties of conducting polymer composites. J Appl Polym Sci 82:1903–1906
13. Omastova M, Podhradska S, Prokes J, Janigova I, Stejskal J (2003) Thermal ageing of conducting polymeric composites. Polym Degrad Stab 82:251–256
14. Krupa I, Chodak I (2001) Physical properties of thermoplastic/graphite composites. Eur Polym J 37:2159–2168
15. Das NC, Chaki TK, Khastgir D (2002) Effect of processing parameters, applied pressure and temperature on the electrical resistivity of rubber-based conductive composites. Carbon 40:807–816
16. Sichel EK (1982) Carbon black polymer composites. Marcel Dekker, NY, p 214
17. Shioyama H, Akita T (2003) A new route to carbon nanotubes. Carbon 41:179–181
18. Tasis D, Tagmatarchis N, Bianco A, Prato M (2006) Chemistry of carbon nanotubes. Chem Rev 106:1105–1136
19. Mietta JL, Jorge GE, Perez OE, Maeder T, Negri RM (2013) Superparamagnetic anisotropic elastomer connectors exhibiting reversible magneto-piezoresistivity. Sens Actuators A Phys 192:34–41
20. Mietta JL, Ruiz MM, Antonel PS, Perez OE, Butera A, Jorge GE, Negri RM (2012) Anisotropic magnetoresistance and piezoresistivity in structured Fe_3O_4-silver particles in PDMS elastomers at room temperature. Langmuir 28:6985–6996
21. Lee S, Shin S, Lee S, Seo J, Lee J, Son S, Cho HJ, Algadi H, Al-Sayari S, Kim DE, Lee T (2015) Ag nanowire reinforced highly stretchable conductive fibers for wearable electronics. Adv Funct Mater. doi:10.1002/adfm.201500628
22. Sung Y, El-Tantawy F (2002) Novel smart polymeric composites for thermistors and electromagnetic wave shielding effectiveness from TiC loaded styrene-butadiene rubber. Macromol Res 10(6):345–358
23. Gwaily SE, Nasr GM, Badawy MM (2001) Thermal and electrical properties of irradiated styrene butadiene rubber-metal composites. Egypt J Sol 24(2):193–205
24. Dannenberg EM (1978) In encyclopedia of chemical technology, vol 4, 3rd edn. Wiley, NY, p 631
25. Lyon F (1985) Encyclopedia of polymer science and engineering, vol 2, 2nd edn. Wiley, NY, p 623
26. Al-Hartomy OA, Al-Ghamd A, Dishovsky N, Malinova P, El-Tantawy F (2012) Some factors determining the volume resistivity of filled natural rubber-based nanocomposites. Prog Rubber Plast Recycl Technol 28:95–110
27. Luo H, Kluppel M, Schneider H (2004) Study of filled SBR elastomers using NMR and mechanical measurements. Macromolecules 37:8000–8009
28. Morsy RM, Ismaiel MN, Yehia AA (2013) Conductivity studies on acrylonitrile butadiene rubber loaded with different types of carbon blacks. Int J Mater Meth Technol 1(4):22
29. Podhradska S, Prokes J, Omastova M, Chodak I (2009) Stability of electrical properties of carbon black-filled rubbers. J Appl Polym Sci 112:2918–2924

30. Janzen J (1975) On the critical conductive filler loading in antistatic composites. J Appl Phys 46:966
31. Kawazoe M, Ishida H (2008) A new concept for nanoparticle distribution in SBR/NBR blend solution and films via molecular confinement. Macromolecules 41:2931–2937
32. Gubbels F, Jerome R, Vanlathem E, Deltour R, Blacher S, Brouers F (1998) Kinetic and thermodynamic control of the selective localization of carbon black at the interface of immiscible polymer blends. Chem Mater 10:1227–1235
33. Cheah K, Forsyth M, Simon GP (2000) Processing and morphological development of carbon black filled conducting blends using a binary host of poly (styrene co-acrylonitrile) and poly (styrene). J Polym Sci Part B Polym Phys 38:3106–3119
34. Zhang C, Han HF, Yi XS, Asai S, Sumita M (1999) Selective location of the filler and double percolation of Ketjenblack filled high density polyethylene/isotactic polypropylene blends compos. Interfaces 6:227–236
35. Feng JY, Chan CM (1998) Carbon black-filled immiscible blends of poly (vinylidene fluoride) and high density polyethylene: electrical properties and morphology. Polym Eng Sci 38:1649–1657
36. Mamunya YP (1999) Morphology and percolation conductivity of polymer blends containing carbon black. Macromol Sci Phys 38:615–622
37. Diaz R, Diani J, Gilormini P (2014) Physical interpretation of the Mullins softening in a carbon-black filled SBR. Poymer 55:4942–4947
38. Wang L, Zhao S (2010) Study on the structure-mechanical properties relationship and antistatic characteristics of SSBR composites filled with SiO_2/CB. J Appl Polym Sci 118:338–345
39. Karasek L, Meissner B, Asai S, Sumita M (1996) Percolation concept: polymer-filler gel formation, electrical conductivity and dynamic electrical properties of carbon-black-filled rubbers. Polym J 28(2):121–126
40. He M, Yuan LX, Zhang WX, Hu XL, Huang YH (2011) Enhanced cyclability for sulfur cathode achieved by a water-soluble binder. J Phys Chem C 115:15703–15709
41. Peddini SK, Bosnyak CP, Henderson NM, Ellison CJ, Paul DR (2014) Nanocomposites from styrene-butadiene rubber (SBR) and multiwall carbon nanotubes (MWCNT) part 1: morphology and rheology. Polymer 55:258–270
42. Hu GJ, Zhao CG, Zhang SM, Yang MS, Wang ZG (2006) Low percolation thresholds of electrical conductivity and rheology in poly (ethylene terephthalate) through the networks of multi-walled carbon nanotubes. Polymer 47(1):480–488
43. Abdel-Goad M, Potscheke J (2005) Rheological characterization of melt processed polycarbonate-multiwalled carbon nanotube composites. J Nonnewton Fluid 128(1):2–6
44. Alig I, Skipa T, Lellinger D, Potscheke P (2008) Destruction and formation of a carbon nanotube network in polymer melts: rheology and conductivity spectroscopy. Polymer 49(16):3524–3532
45. Das A, Stockelhuber KW, Jurk R, Fritzsche J, Kluppel M, Heinrich G (2009) Coupling activity of ionic liquids between diene elastomers and multi-walled carbon nanotubes. Carbon 47(14):3313–3321
46. McNally T, Ptscheke P, Halley P, Murphy M, Martin D, Bell SEJ et al (2005) Polyethylene multiwalled carbon nanotube composites. Polymer 46(9):8222–8232
47. Li Y, Shimizu H (2009) Toward a stretchable, elastic, and electrically conductive nanocomposite: morphology and properties of poly [styrene-b-(ethylene-co-butylene)-b-styrene]/multiwalled carbon nanotube composites fabricated by high-shear processing. Macromolecules 42:2587–2593
48. Pedroni LG, Araujo JR, Felisberti MI, Nogueira F (2012) Nanocomposites based on MWCNT and styrene–butadiene–styrene block copolymers: effect of the preparation method on dispersion and polymer–filler interactions. Compo Sci Technol 72:1487–1492
49. Tsuchiya K, Sakai A, Nagaoka T, Uchida K, Furukawa T, Yajima H (2011) High electrical performance of carbon nanotubes/rubber composites with low percolation threshold prepared with a rotation–revolution mixing technique. Compos Sci Technol 71:1098–1104

50. Das A, Kasaliwal GR, Jurk R, Boldt R, Fischer D, Stockelhuber KW, Heinrich G (2012) Rubber composites based on graphene nanoplatelets, expanded graphite, carbon nanotubes and their combination: a comparative study. Compo Sci Technol 72:1961–1967
51. Araby S, Saber N, Ma X, Kawashima N, Kang H, Shen H, Zhang L, Xu J, Majewski P, Ma J (2015) Implication of multi-walled carbon nanotubes on polymer/graphene composites. Mater Desig 65:690–699
52. Dreyer RD, Park S, Bielawski CW, Ruoff RS (2010) The chemistry of graphene oxide. Chem Soc Rev 39:228–240
53. Wang G, Yang J, Park J, Gou X, Wang B, Liu H et al (2008) Facile synthesis and characterization of graphene nanosheets. J Phys Chem C 112:8192–8195
54. Li X, Wang X, Zhang L, Lee S, Dai H (2008) Chemically derived, ultrasmooth graphene nanoribbon semiconductors. Science 319:1229–1232
55. Blake P, Brimicombe PD, Nair RR, Booth TJ, Jiang D, Schedin F et al (2008) Graphene-based liquid crystal device. Nano Lett 8:1704–1708
56. www.wikipedia.org/wiki/Graphene
57. Novoselov KS, Geim AK, Morozov SV, Jiang D, Zhang Y, Dubonos SV et al (2004) Electric field effect in atomically thin carbon films. Science 306:666–669
58. Kim H, Abdala AA, Maosko CW (2010) Graphene/polymer nanocomposites. Macromolecules 43:6515–6530
59. Kuilla T, Bhadra S, Yao D, Kim NH, Bose S, Lee JH (2010) Recent advances in graphene based polymer composites. Prog Polym Sci 35:1350–1375
60. Sengupta R, Bhattacharya M, Bandyopadhyay S, Bhowmick AK (2011) A review on the mechanical and electrical properties of graphite and modified graphite reinforced polymer composites. Prog Polym Sci 36:638–670
61. Sadasivuni KK, Ponnamma D, Thomas S, Grohens Y (2014) Evolution from graphite to graphene elastomer composites. Prog Poym Sci 39:749–780
62. Araby S, Meng Q, Zhang L, Zaman I, Majewski P, Ma J (2015) Elastomeric composites based on carbon nanomaterials. Nanotechnology 26:112001
63. Ozbas B, O'Neill CD, Register RA, Aksay IA, Prud'homme RK, Adamson DH (2012) Multifunctional elastomer nanocomposites with functionalized graphene single sheets. J Polym Sci Part B Polym Phy 50:910–916
64. Steurer P, Wissert R, Thomann R, Muelhaupt R (2009) Functionalized graphenes and thermoplastic nanocomposites based upon expanded graphite oxide. Macromol Rapid Commun 30:316–327
65. Liu X, Kuang W, Guo B (2015) Preparation of rubber/graphene oxide composites with in-situ interfacial design. Polymer 56:553–562
66. Xing W, Tang M, Wu J, Huang G, Li H, Lei Z, Fu X, Li H (2014) Multifunctional properties of graphene/rubber nanocomposites fabricated by a modified latex compounding method. Compo Sci Technol 99:67–74
67. Hummers WS, Offeman RE (1958) Preparation of graphitic oxide. J Am Chem 80:1339
68. Schopp S, Thomann R, Ratzsch KF, Kerling S, Altstadt V, Mulhaupt R (2013) Functionalized graphene and carbon materials as components of styrene-butadiene rubber nanocomposites prepared by aqueous dispersion blending. Macromol Mater Eng 299(3):319–329
69. Ozbas B, O'Neill CD, Register RA, Aksay IA, Prud'homme RK, Adamson DH (2012) Multifunctional elastomer nanocomposites with functionalized graphene single sheets. J Polym Sci Part B Polym Phy 50:910–916
70. Song SH, Jeong HK, Kang YG (2010) Preparation and characterization of exfoliated graphite and its styrene butadiene rubber nanocomposites. J Indust Eng Chem 16:1059–1065
71. Araby S, Zhang L, Kuan HC, Dai JB, Majewski P, Ma J (2013) A novel approach to electrically and thermally conductive elastomers using graphene. Polymer 54:3663–3670
72. Zhang H, Wang C, Zhang Y (2015) Preparation and properties of styrene-butadiene rubber nanocomposites blended with carbon black-graphene hybrid filler. J Appl Polym Sci. doi:10.1002/APP.41309

73. Fugetsu B, Sano E, Yu H, Mori K, Tanaka T (2010) Graphene oxide as dyestuffs for the creation of electrically conductive fabrics. Carbon 48:3340–3345
74. Sherman RD, Middelman LM, Jacobs SM (1983) Electron transport processes in conductor-filled polymers. Polym Eng Sci 23(1):36–46
75. Ruiz MM, Marchi MC, Perez OE, Jorge GE, Fascio M, D'Accorso N, Negri RM (2015) Structured elastomeric submillimeter films displaying magneto and piezo resistivity. Polym Phy 53:574–586
76. Ivanekyo D, Toshchevikov VP, Saphiannikova M, Heinrich G (2011) Magneto-sensitive elastomers in a homogeneous magnetic field: a regular rectangular lattice model. Macromol Theory Simul 20:411–424
77. Shahrivar K, De Vicente J (2013) Thermoresponsive polymer-based magneto-rheological (MR) composites as a bridge between MR fluids and MR elastomers. Soft Matter 9:11451–11456
78. Danas K, Kankanala SV, Triantafyllidis N (2012) Experiments and modeling of iron-particle-filled magnetorheological elastomers. J Mech Phys Solids 60:120–138
79. Bica I, Liu YD, Choi HJ (2012) Magnetic field intensity effect on plane electric capacitor characteristics and viscoelasticity of magnetorheological elastomer. Colloid Polym Sci 290:1115–1122
80. Mietta JL, Jorge GE, Negri RM (2014) A flexible strain gauge exhibiting reversible piezoresistivity based on an anisotropic magnetorheological polymer. Smart Mater Struct 23:85026–85038
81. Kirkpatrick S (1973) Percolation and conduction. Rev Mod Phys 45:574
82. Schettini ARA, Khastigir D, Soares BG (2012) Microwave dielectric properties and EMI shielding effectiveness of poly (styrene-b-styrene-butadiene-styrene) copolymer filled with PAni, Dodecylbenzenesulfonic acid and carbon black. Polym Eng Sci 52(9):2041–2048

Electronic Applications of Chloroprene Rubber and Its Composites

Bharat P. Kapgate and Chayan Das

Abstract This chapter is intended to give an overview of electrical conductivity and dielectric properties of chloroprene rubber (CR) composites with reference to their application in the field of electronics. Influence of different types of filler on the dielectric properties of the CR composites, in a wide frequency range and varying temperature condition, is thoroughly discussed. Type, concentration, and state of dispersion of fillers as well as the polar nature of CR are noticed to play a significant role in governing the electrical conductivity of CR composites. Electrical conductivity studies of carbon black (CB) and carbon fiber-based CR composites are discussed in terms of filler concentration, filler dispersion, and processing techniques. The role of ionic liquid-modified multi-walled carbon nanotube (MWCNT) in improving the electrical conductivity of CR composites is emphatically reviewed. The effects of different types of aging on the electrical properties of CR composites are also focused in this chapter. The CR composites discussed herewith could potentially be used in the field of electromagnetic interference (EMI) shielding, microwave absorbance, and conductive adhesive.

Keywords Chloroprene rubber · Dielectric property · Electrical conductivity · Electromagnetic interference shielding · Microwave absorbance

Abbreviations

AC	Alternating current
BMI 1	Butyl -3-methyl imidazolium bis(trifluoromethylsulphonyl) imide
BaHF	Barium hexaferrite (BaFe12O19)
CB	Carbon black

B.P. Kapgate · C. Das (✉)
Department of Chemistry, Visvesvaraya National Institute of Technology,
Nagpur 440010, India
e-mail: chayandas@hotmail.com; chnds@rediffmail.com; chayandas@chm.vnit.ac.in

D. Ponnamma et al. (eds.), *Flexible and Stretchable Electronic Composites*,
Springer Series on Polymer and Composite Materials,
DOI 10.1007/978-3-319-23663-6_10

CR	Chloroprene rubber
CNT	Carbon nanotube
CIP	Carbonyl-iron powder
DBSA	Dodecylbenzenesulfonic acid
DE	Dielectric elastomer
DC	Direct current
EAP	Electroactive polymers
EMI	Electromagnetic interference
FEF	Fast extrusion furnace
GHz	Gigahertz
HAF	High abrasion furnace
MT	Thermal medium
MWCNT	Multi-walled carbon nanotube
PANI	Polyaniline
Phr	Parts by weight per hundred parts of rubber
PTC	Positive temperature coefficient
PVC	Polyvinyl chloride
RAM	Radar-absorbing material
SAF	Super abrasion furnace
SE	Shielding effectiveness
SEM	Scanning electron microscopy
SRF	Semi-reinforcing furnace
Co-TiBaHF	Ti–Co-doped barium hexaferrite

1 Introduction

Electronic devices that operate at high frequencies require materials of high dielectric constant. Nevertheless, combination of adequate mechanical strength and processability of the material with good dielectric properties are also necessary for this purpose which is not easy to achieve in a one-component material [1, 2]. For example, organic polymers can be processed very easily into mechanically strong components, but they possess low dielectric constant [3]. On the other hand, ceramic materials having high dielectric constant are brittle in nature and less compatible with current circuit's integration technologies, and therefore, they offer difficulty in processing. In this context, a lot more attention has been paid to produce hybrid materials with high dielectric constant, such as conducting polymer composites that combine good mechanical properties, thermal stability, processability with adequate electrical properties which would be suitable for different useful electrical and electronic applications in the field of military, industrial, commercial, and consumer products [4, 5].

Electrically conducting polymers are the class of materials that offer several important advantages over metals such as ease of processing, resistance to corrosion, and air tight contact interface, owing to their combined features of electrical and elastic properties, which make them suitable for high-density micro-sized connectors [6]. Extensive work has been carried out to prepare electrically conductive and mechanically strong polymers such as polypropylene, polycarbonate, and epoxies. However, plastics are relatively more rigid and hence, they are not found suitable in many practical application where flexible and elastic materials are required [7, 8]. Electrically conducting rubber composites appear to be very promising, due to their stretchable and deformable properties in combination with sufficient high electric conduction capability which makes them fit for various applications such as electromagnetic interference (EMI) shielding, microwave absorbance, sensors, and electrostatic charge dissipation [9–11]. In addition to this, rubbers having high flexibility, light weight, and ability to absorb mechanical shocks are found suitable for different dynamic applications [12, 13].

Rubbers are generally electrically insulating materials, and hence, they are not much useful for electrical and electronic applications except electrical insulation. However, they can be made electrically conductive material by incorporation of suitable conductive fillers [14, 15]. For this purpose, addition of a critical concentration of filler, beyond percolation threshold, is important to produce conductive rubber composites. A percolation threshold happens when the transport property, i.e., electrical conductivity of the material changes from insulating to conducting [16, 17]. Different factors that can influence the conductivity of rubber-filler system, include chemical and physical nature of rubber as well as size, shape, surface area and dispersion state of filler and processing condition of the composites. A very good dispersion of filler throughout the rubber matrix is necessary to form a conductive network that governs the electrical properties of the composite [18, 19]. The common fillers that have been used for this purpose are carbon black (CB), carbon fibers, multi-walled carbon nanotubes (MWCNTs), graphene, metal oxide, and polyaniline (PANI) to name a few. Thermal expansion of rubber matrix can also play an important role in electrical conduction of the composites [17].

In this chapter, the dielectric and conductive properties of the chloroprene rubber (CR) composites are described with reference to their different potential application in the field of electronics.

1.1 Chloroprene Rubber: Dipole Structure and Electrical Properties

CR, also known as neoprene, is an emulsion polymer of 2-chlorobuta-1-3-diene (Fig. 1a). Several grades of CR have tendency to crystallize upon straining due to presence of C–Cl dipoles that enhance the interchain interaction and promote crystallization (Fig. 1b). Polar nature of CR, due to the presence of chlorine atom, enables it to exhibit many useful features such as good mechanical strength, excellent resistance to ozone, weather, aging and chemical attack, good resistance to oil

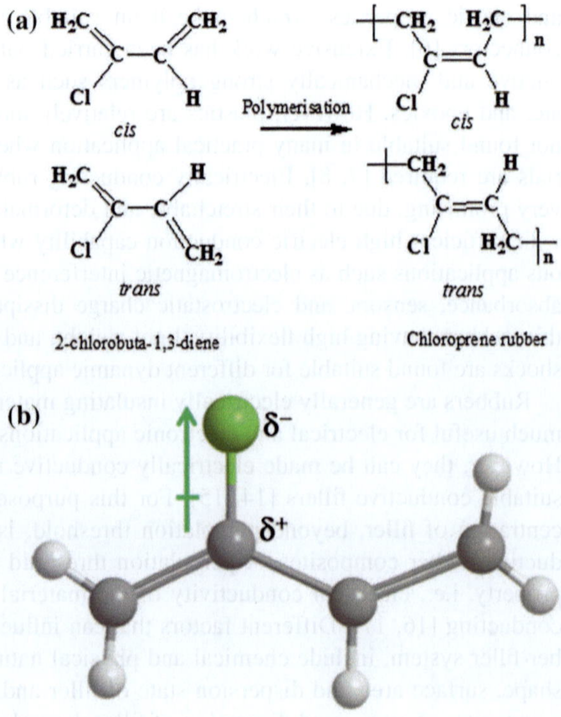

Fig. 1 **a** Chemical structure of chloroprene rubber (CR); **b** dipole in CR

and fuel, adhesive properties, and low flammability. Thus, CR finds useful application in the preparation of electrical wires, cables, and jackets [20, 21]. In addition to this, presence of polar groups in the polymer backbone increases the dielectric constant of the CR matrix. CR, with moderately high permittivity (dielectric constant), offers better microwave absorbance and EMI shielding effectiveness (SE) over other nonpolar elastomers [22]. The dielectric behavior of CR composites filled with channel black, measured by Schneider et al. [23], in the frequency range of 500–10^8 Hz, above room temperature (up to 60 °C), showed high temperature absorption (α absorption). However, additional low temperature absorption (β absorption) was noted when the measurement was carried out in the broader temperature scale (−20 to 100 °C) in the frequency range of 1 c/s–1 Mc/s [24]. Increase in temperature caused appearance of the large α absorption peak in the observable frequency range that moved rapidly toward the high-frequency region. On the other hand, at low temperature, only β absorption peak was observed that moved toward high-frequency region with increase in temperature. It was also observed that α and β absorption peaks appeared at relatively lower temperature and higher frequency side as compared to those of polyvinyl chloride (PVC). This was attributed to more flexible nature of CR over PVC. The chlorine atom in CR, attached to every fourth carbon atom in chloroprene unit, does not cause any steric hindrance. Furthermore, lack of hydrogen atom at doubly bonded carbon atom in

CR imposes less steric hindrance to the immediate single bond for its rotation. This, in turn, favors the movement of rubber chain and makes it more flexible which influences the position of absorption peak with respect to temperature and frequency.

The studies on time and temperature dependence of the electrical conductivities of pure gum and vulcanized CR revealed that the electrical conduction in crystalline CR took place mainly due to the ionic conduction mechanism [25]. For a fresh milled sample, it was observed that the current was independent of time up to 353 K and beyond which current gradients increased with increase in temperature. This was attributed to the mobility of current carrier ions and transition of CR from crystalline phase to amorphous phase (liquid state). During this transition, the impurity ions present in CR caused dipole orientation to occur, which led to slight increase in conductivity that increased with temperature. To have more detailed insight, vulcanized CR was analyzed, and it was observed that the transition region was extended by the effect of vulcanization. This was due to the breaking of C–Cl bonds that released free chloride ions which interacted with ZnO. This resulted in an increase in the concentration of impurity like $ZnCl_2$ which contributed to the increased electrical conductivity.

2 Dielectric Properties of Chloroprene Rubber Composites

In general, all materials have the ability to respond to an applied voltage owing to the presence and movement of electrons, ions, or charged particles within them. In conductive materials, the electrons or ions move over a macroscopic distance. In case of a dielectric material, the charged particles move for a relatively shorter distances [26]. Dielectric materials have the capability of undergoing reversible change in length to a large extent, because of inherent coupling of two processes, i.e., deformation and polarization [27, 28]. Electroactive polymers (EAP) are the class of materials which possess special mechanical and electrical properties. They have the ability to generate mechanical response to an applied electric field. Among these EAP, dielectric elastomer (DE) is the most promising materials for actuator application. The electrochemical transducers (dielectrics) with high energy output, high strains, light weight, low cost, and appreciable damage tolerance capability find many important applications in robotics, automation, and biomedical electronic devices [29, 30].

The basic energy relation of DE transduction is associated with an expansion in area of a film when an electric charge or voltage is applied on it. In that case, the film converts electrical energy to mechanical energy and operates in an actuator mode. If a charged film contracts in area, energy is being converted from mechanical energy to electrical energy and the film operates in a generator mode. The principle mechanism of a DE transducer is illustrated in Fig. 2. A piece of DE is sandwiched between two compliant electrodes provided with a high electric field. The electrodes possess a negligible electrical resistance and mechanical stiffness.

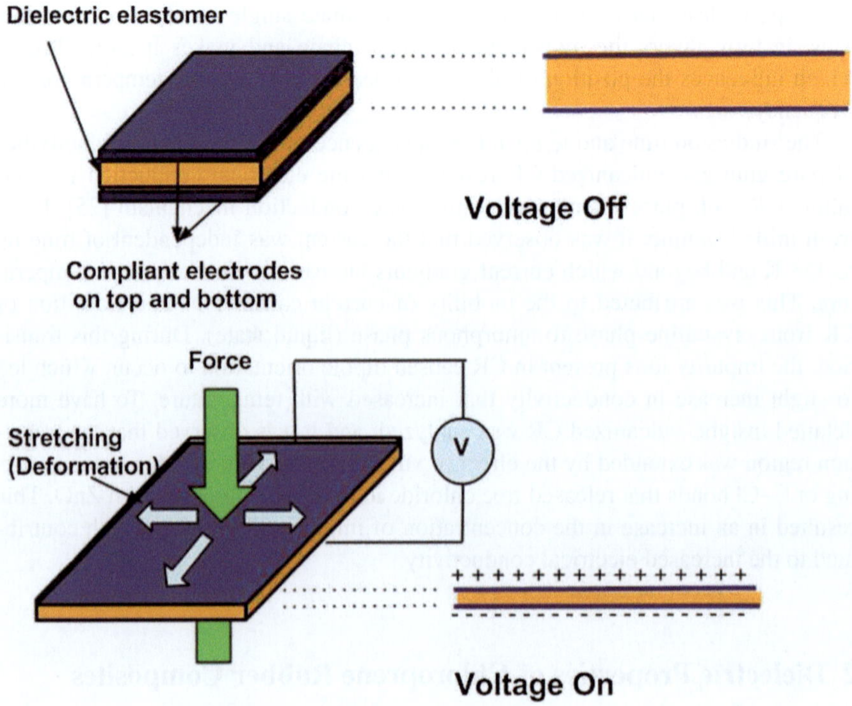

Fig. 2 Principle involved in dielectric elastomer actuator

When a force or voltage is applied to the dielectric specimen, placed between these electrodes, the charge flows through the wire from one electrode to the other. This exerts an electrostatic force between the oppositely charged electrodes which squeezes the elastomer film. The DE film expands in-plane and contracts out of plane simultaneously leading to decrease in the thickness of film. DE can undergo straining even more than 100 %. This super large deformation feature of DE as compared to other electroactive materials makes it useful for novel applications such as preparation of artificial muscles and fabrication of transducers [31].

Rubber composites filled with piezoelectric ceramic particle are found useful for the applications such as probing ultrasonic field in water, blood pressure sensor, guitar pickup, and electric piano pickup. [32]. Banno [32] investigated the dielectric and piezoelectric constants of heterostructure materials composed of ceramic particles dispersed in CR matrix for hydrophone application. Ceramic particles, used in this work, were lead titanate ($PbTiO_3$), $PbTiO_3$ modified with bismuth oxide (Bi_2O_3), lead titanate zirconate ($Pb(Zr,Ti)O_3$), and $Pb(Zr,Ti)O_3$ modified with tungsten trioxide (WO_3). The composites filled with lead titanate particles showed smaller dielectric and larger piezoelectric constant values over composites filled with lead titanate zirconate [32].

The influence of CB on the dielectric properties of CR was studied by Abdel-Nour et al. [33, 34]. Different types of CB viz. thermal medium (MT), semi reinforcing furnace (SRF), high abrasion furnace (HAF), and super abrasion furnace (SAF) were used in this study with varying concentration of CB (up to 60 phr). The dielectric measurements at 1 kHz showed that the dielectric properties (permittivity, ε' and dielectric loss (tan δ)) increased with increase in filler concentration and reached to maximum at a certain concentration. This was attributed to the interfacial polarization that was developed between the CB and rubber chain. It was interesting to note that the type of CB influenced the dielectric properties of the CR composites. For example, the sharp increase in dielectric properties for different CB was found at different concentrations, i.e., at 60 % for MT, 40 % for SRF, and 20 % for both HAF and SAF. The shape factor of different CB was thought to play the key role in this regard. Subsequently, this group carried out more detailed investigation on similar composites to find the influence of type, concentration, and shape factor of different CB on the permittivity and dielectric loss of the CR composite in the wide frequency range (1 kHz–40 MHz), at room temperature. Different absorption regions were observed in the wide frequency range, in the dielectric measurement experiments, which was thought to be originated due to several factors such as change in particle shapes, formation of clusters, charge migration through particles, and mobility of rubber chains.

Al-Hartomy et al. reported the influence of CB (N220) on the dielectric properties of CR composites filled at different filler concentrations (0–100 phr) [35]. The measurement of dielectric permittivity of CR composites in the frequency range of 1–12 GHz showed increase in dielectric permittivity with increase in CB loading. The dielectric permittivity values remained almost constant in the frequency range of 1–7 GHz and thereafter showed a sharp increase with maximum at the highest frequency (12 GHz). This group further explored the effect of Ni and Co powder (10 phr) on the dielectric properties of the CB-filled (0–100 phr) CR composites [36]. The dielectric properties of the composites were studied in terms of relative dielectric permittivity. The values became more than twice in the above frequency range for Ni-containing composites. Also, the addition of varying amount of CB showed considerable effect on the relative dielectric permittivity that increased with increase in filler loading. Above 10 GHz frequency, the increment was noticeable when amount of CB in the composites exceeded 60 phr loading. In contrast, addition of Co showed relatively lower dielectric permittivity values as compared to that of Ni. Nevertheless, the nature of increment in relative dielectric permittivity with increasing frequency and CB concentration was similar for Co-containing composites too. The difference in relative dielectric permittivity for these two types of composites was attributed to the differences in resistivity and conductivity of the Ni and Co particles.

Abd. El. Messieh et al. carried out a dielectric measurement study of the CR composites filled with peroxide (5 phr) and different nitroaniline derivatives (2.5 phr) over a frequency range of 400 Hz–10 MHz [37]. The dielectric permittivity (ε') of the composites increased with the substitution of nitroanilines in the order of ortho > para > meta > *N, N* dinitroaniline. However, the values decreased with the increase in frequency in all cases. The measurements of dielectric loss (ε'') revealed three absorption regions. The first absorption region, around 320 Hz, was due to either DC conductivity or the Maxwell–Wagner effect. The second absorption region was due to the formation of large aggregates of peroxide added to rubber. The third absorption region was explained by Debye losses brought by the dipole rotations due to movement of rubber backbone. The decrease in second and third relaxation was observed with the addition of nitroaniline additives. This was observed in the form of a decrease in relaxation time accompanied by an increase in height of relaxation. This was attributed to the peptizing effect of nitroaniline additives which decreased the glass transition of rubber and retained its flexibility and deformability.

3 Electrical Conductivity of Chloroprene Rubber Composites

3.1 Carbon Black (CB)-Based Electrically Conductive Chloroprene Rubber Composites

The electrical conductivity of CB-filled polymer composites highly depends on the parameters such as particle size, surface area, structure, and dispersion of filler in polymer matrix. In addition to this, thermal expansion of polymer matrix also plays an important role in conduction mechanism [17–19].

Bengston et al. explored the electrical noise measurements, as a new approach to understand the underlying mechanisms of electrical conduction in CB-filled CR [38]. The CR samples, with varying amount of CB, were characterized in terms of their direct current (DC) electrical characteristics and positive temperature coefficient (PTC). There was a gradual increase in the electrical resistance with increase in temperature. Ali and Abo-Hashem [39, 40] studied the percolation characteristics and the electrical conductivity of CB-filled CR composites. Two different systems, viz. crystallisable CR, filled with FEF CB, and non-crystallisable CR, filled with HAF CB, were prepared. The effects of filler concentration and temperature on the electrical conductivity of the composites were investigated. In both the systems, percolation concept was verified by a sharp increase in electrical conductivity beyond 20 phr filler concentration (Fig. 3). Studies on temperature dependence of electrical conductivity in both the system revealed that at low filler loading (below percolation threshold), the conduction mechanism was thermally active, i.e., due to the thermal emission of electrons from and through the carbon clusters. Interestingly, the electrical property of these two systems became different above percolation threshold. In non-crystallisable CR system, the variation in conductivity with temperature was attributed to the breakdown and reformation of CB clusters. However, the inconsistent variation of conductivity with temperature at various loading of CB made it very difficult to draw any specific conclusion. On the contrary, crystallisable CR system showed a decrease in conductivity after percolation threshold (40–50 phr) followed by a sharp increase with increase in

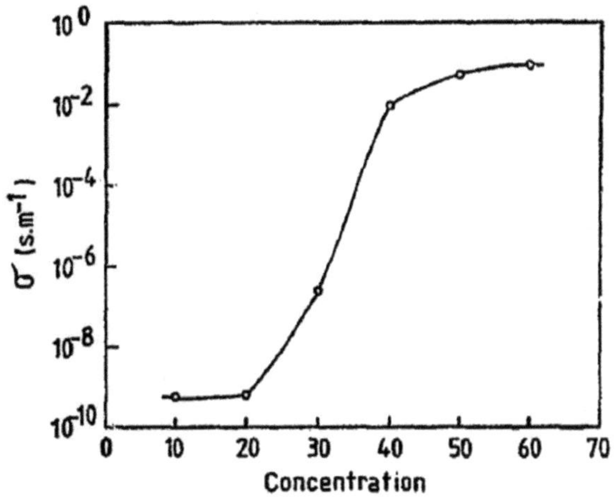

Fig. 3 The concentration dependence of the electrical conductivity (σ) in CB-filled CR composites [39]. Copyright 1997. Reproduced with permission from Elsevier Ltd

temperature. This was attributed to the phase transition of CR from crystalline to amorphous phase at higher temperature that increased the flexibility of rubber chains and favoured easier passage of current carriers.

3.2 *Carbon Fiber-Based Electrically Conductive Chloroprene Rubber Composites*

Other than CB, use of carbon fibers in making conductive CR composite has also been studied. Jana and co-workers [41, 42] studied the effect of carbon fiber on the electrical conductivity of CR composites. The effect of filler concentration and aspect ratio on the volume resistivity at varying temperature was investigated. Filler aspect ratio in the rubber matrix was varied by using different processing techniques. Cement mixing method yielded fibers with higher aspect ratio (~100), whereas mill mixing resulted in fibers with smaller aspect ratio (~25). It was observed that the composites filled with fibers of higher aspect ratio offered higher electrical conductivity than the composites filled with fibers of lower aspect ratio. Thus, critical concentration of fiber was found to be in the range of 5–10 phr for former type, while it was in the range of 20–25 phr for later type (Fig. 4a). This was attributed to the random distribution of former type fiber into rubber matrix (Fig. 4b). Studies on influence of temperature on the electrical conductivity revealed significant increase in volume resistivity and decrease in conductivity of the CR composites with temperature. It was believed that the increase in temperature increased the inter-fiber average distance owing to uneven thermal expansion of CR and carbon fiber. As a result, probability of tunneling of electron or holes became less due to the scattering of carriers in the composite at higher temperature. It was also reported that the conventionally vulcanized composites,

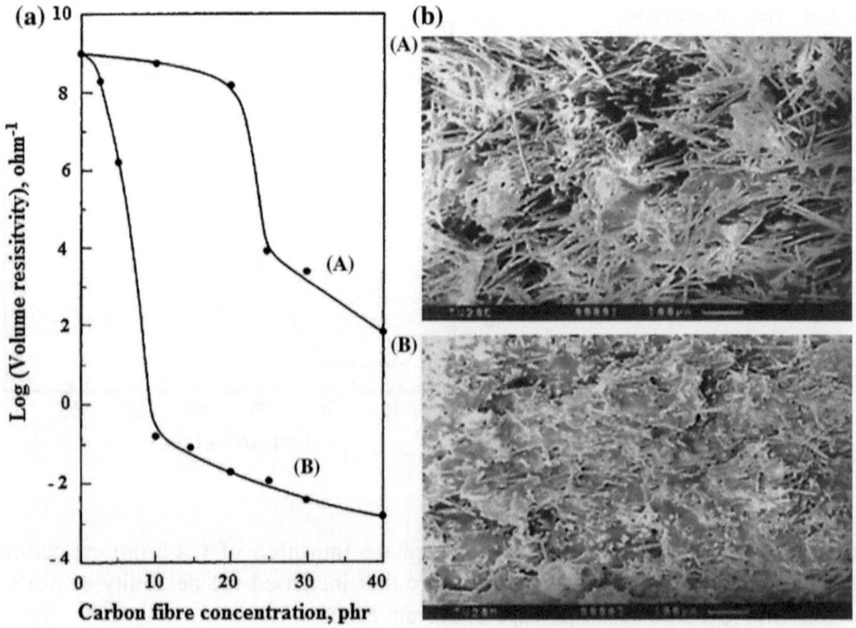

Fig. 4 **a** Effect of carbon fiber concentration (in phr) on volume resistivity of CR-based vulcanizates. **A** Composites prepared by mill mixing method and **B** composites prepared by cement method. **b** SEM photomicrographs of tensile fracture surface of CR-carbon fiber composites: **A** thermovulcanized composite prepared by cement mixing (20 phr) and **B** thermovulcanized composite prepared by mill mixing (20 phr) [41]. Copyright 1992. Reproduced with permission from John Wiley and sons Inc

having more crosslinks, showed higher conductivity over thermovulcanized rubber composites because of the formation of better conductive network in the previous case (Fig. 4b). This group further extended the work where barium ferrite-assisted vulcanization was found to deliver better electrical properties for the carbon fiber-filled CR composites over previously reported conventionally vulcanized CR composites [42].

The DC electrical conductivity of the CR composites filled with PANI and PANI-coated nylon fiber was investigated at varying filler loading [43]. Percolation limit was found to be 100 and 120 phr for PANI and PANI-coated nylon fiber-filled composite, respectively. Since the later type showed relatively better processability and mechanical properties over the other one, it was believed to be more suitable for the electronic devices working at high operating frequencies and requiring high dielectric constant materials with better ease of processing and mechanical strength.

3.3 *Effect of Ionic Liquid on the Electrical Conductivity of MWCNT-Filled Chloroprene Rubber Composites*

The recent interest in the electrically conductive fillers for rubber composites has been extended toward the CNT because of their excellent mechanical and electrical properties. CNTs are highly conductive as electrons can propagate freely through it. The percolation threshold, i.e., critical filler concentration required to obtain the conducting path in the composite very much, depends on the aspect ratio of CNTs, disentanglement of CNT agglomerates, distribution of individual CNTs, and their degree of alignment. Thus, in order to get high conducting CNT-filled elastic materials, a very good dispersion of CNT in the rubber matrix is essential. MWCNTs, in its pure form, have a strong tendency to get agglomerates due to physical force of attraction between them and cannot be separated even by applying high shearing force during mixing with the elastomer matrix. So, several attempts have been made to improve the dispersion of MWCNTs into rubber matrix [44–46].

Heinrich's group [47–50] recently reported an eco-friendly approach for the preparation of MWCNT-filled CR where imidazolium-based ionic liquid, BMI (1-Butyl-3-methyl imidazolium bis(trifluoromethylsulphonyl) imide), was utilized as a modifier of MWCNT (Fig. 5). The exploitation of BMI was found to improve the dispersion of the MWCNT as well as the electrical conductivity of the composites. Moreover, this technique reduced the mixing cycles during the composite preparation. A very high conductivity level of about 0.1 S/cm was observed with low amount (below 3 phr) of MWCNT in CR with more than 500 % of stretchability. BMI favoured very good dispersion of MWCNTs in CR matrix minimizing the agglomeration that was found in case of unmodified MWCNT. This was achieved by the physical interaction (cation-π/π- π) between BMI and MWCNTs. This interaction suppressed the inter-tubular attraction between MWCNTs. It was believed that a secondary network was formed by MWCNTs that enhanced the conductivity. From the measurements of AC and DC conductivity, the percolation threshold of BMI modified MWCNTs was found to be in between 1 and 3 phr. Additionally, the mechanical properties of the composites became much better revealing the efficacy of the ionic liquid in providing conducting and reinforcing networks of the

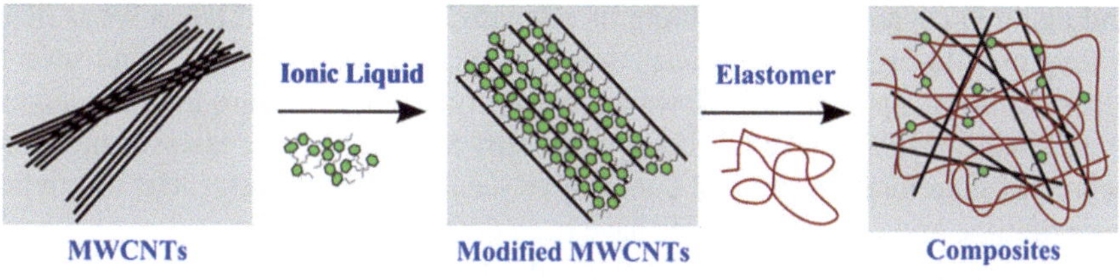

Fig. 5 The preparation of ionic liquid-modified CNT-based CR composite [48]. Copyright 2011. Reproduced with permission from Elsevier Ltd

MWCNT in the composites. The optimum ratio of MWCNT: BMI was found to be 1:5 for reasonably good conductivity and mechanical properties of the composites. This group further carried out electrical resistance measurement of BMI modified MWCNT-filled CR composites as a function of the compression/decompression force. The results showed a hysteresis loop that was formed by the lower and upper curves corresponding to the regimes of the compression and decompression of the sample, respectively. The curves obtained for both compression and decompression modes showed two distinctive regimes associated with the steep increase and gradual decrease in conductivity. This non-monotonic conductivity behavior of the composites was attributed to the alteration in the MWCNT contact separation caused by the compression/decompression force. It was concluded that a high dielectric permittivity of the MWCNT-filled CR composites could be achieved by altering the contact distance between conducting MWCNTs under varying external stress. Role of same ionic liquid on the dielectric relaxation behavior of MWCNT-filled CR composites was also explored in another work. In this study, MWCNT-filled CR composites showed high electrical conductivity due to the transport of electrons along the conductive MWCNT network formed in rubber matrix. From the temperature dependence conductivity measurements, it was evident that the conduction mechanism through the MWCNT network in rubber matrix, at low temperature, was due to the quantum mechanical tunneling while thermal activated hopping of electron conduction became dominant above room temperature. The increased conductivity brought by BMI modified MWCNTs was attributed to their better dispersion and reduced distance between adjacent MWCNTs. Authors believed that the study would be very worthy from the viewpoint of several electrochemical applications such as developing batteries, capacitors, actuators, and sensors.

4 Effect of Aging on Electrical Properties of Chloroprene Rubber Composites

The rubber composites exposed to various service conditions need to retain their elastic property as well as fluid resistance and shock absorption capability to deliver the desired function. The deterioration of rubber products can take place by the action of several factors such as temperature, humidity, radiation, and ozone over long period of time. CR having good resistance to oil, fuel, ozone, weather, and aging is capable of maintaining its strength for long time. That is why CR is widely used in high-voltage electric wire covers, ignition line covers, and several other electronic applications. However, a very long-term exposure to high-temperature conditions can make CR composites stiff and fragile. So, assessment of the degree of deterioration of such composites seems to be very essential to predict their lifetime and a lot of activities have been devoted in this area.

The deterioration behavior of CB-filled CR composite, in an environment of heat and radiation, studied by chemical stress relaxation measurement and vibration fatigue test is reported by Ito et al. [51]. The chemical stress relaxation measurement was carried out from room temperature to 150 °C, and it was observed that cross-linking in the CR dominated over chain scission. Vibration fatigue test of CR samples, carried out under heat and radiation, showed that the CR chains were cross-linked by the effect of heat at elevated temperature while effect of irradiation caused the chain scission due to oxidation. The addition of antioxidant was found to be effective in preventing the chain scission.

Pinho et al. [52] studied the aging effect, by salt spray exposure, on the microwave reflectivity of CR composites filled with various concentrations of CB and carbonyl-iron powder (CIP) in the frequency range 8–16 GHz. The microwave reflectivity measurement, carried out before the exposure to salt spray, showed appreciable microwave absorption for both types of composites filled with CB and CIP though higher CIP content was needed than CB. The exposure of salt spray decreased the microwave-absorbing properties of CIP/CR composites which was due to the formation of iron oxide superficial layer that increased the thickness of the composite and acted as reflective barrier. On the other hand, CB/CR composite showed better resistance to salt spray and it was recommended for naval applications without the need of any anticorrosive agent.

The effects of aging temperature (120–160 °C) and aging time on the dielectric properties of CB (30–50 phr)-filled CR composites, studied in the frequency range of 20 Hz–3 MHz, were reported by Kuwahara et al. [53]. The dielectric loss peaks, observed in the dielectric relaxation curves, indicated two processes. Process I, observed at lower frequency region, was due to the cooperative motion of polychloroprene molecules in CR matrix while process II, observed at high-frequency region, was due to the local motion of polychloroprene molecule. The deterioration of CR, during aging at high temperatures, was attributed to the dehydrochlorination reaction that led to the breaking of C–Cl bond. This decreased the dipole moment density and the relaxation strength. On the other hand, the free carbon atoms that were generated formed linkage with each other and increased the cross-linking density which in turn restricted the motion of the cooperative domains and resulted in the increase in relaxation time. Thus, the results showed that such changes in the molecular structure could alter the dielectric relaxation parameters.

Heinrich and group extended their work on BMI modified MWCNT-filled CR composites, that exhibited high electrical conductivity and improved mechanical properties, to explore the thermal aging resistivity of the composites [54]. From the post-aging results, it was observed that the mechanical properties of the composites were retained as such owing to the antioxidant nature of the ionic liquid which prevented the surface oxidation. The significant outcome of the study was that the modification of the MWCNT by BMI enhanced the electrical conductivity of the composites upon thermal aging which appeared to be very useful for the application where conductivity is required at elevated temperature (Fig. 6).

Effect of self-aging of CR composites, filled with varieties of CB, was studied by exposing them to normal weathering condition for three years [34]. Since no significant change in dielectric properties was observed, it was concluded that the composites, being capable of resisting natural aging, could be used for coating of wires.

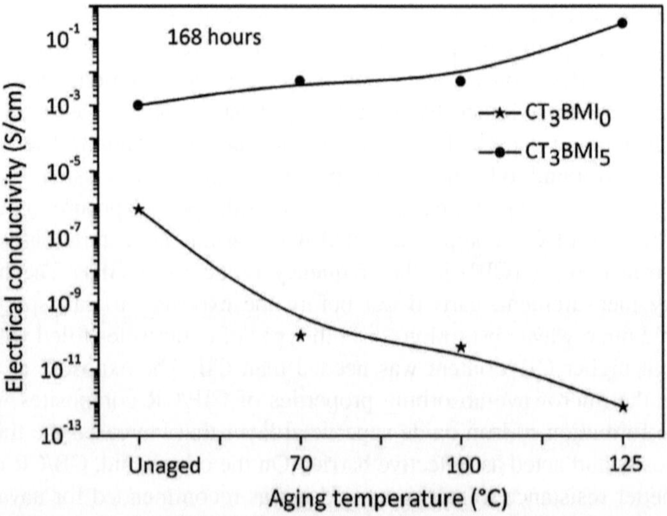

Fig. 6 Electrical conductivity versus aging temperature for unmodified CNT (3 phr) filled CR composite (CT3BMI0) and BMI modified (5 phr) CNT (3 phr) filled CR composite (CT3BMI5) [54]. Copyright 2012. Reproduced with permission from Elsevier Ltd

5 Other Potential Applications of Chloroprene Rubber Composites

5.1 Electromagnetic Interference (EMI) Shielding

EMI shielding of a material refers to its ability to act as a shield against the penetration of an electromagnetic radiation through itself by the phenomenon of reflection, absorption, and internal re-reflection. Electromagnetic radiation of high frequency (radio waves or waves from cellular phones) has a tendency to interfere with the electronic devices. Different sources of radio frequency are increasing day by day with very rapid growth of electronic industries which has led to environmental electronic pollution. In this context, EMI shielding has become a rapidly growing application for both electronic devices and radiation sources [43–58].

The mechanism of EMI shielding is usually associated with the reflection or absorption of electromagnetic radiation by the material (Fig. 7). For the reflection of the radiation by the shield, presence of mobile charge carriers, in the form of electrons or holes, is necessary to interact with the electromagnetic radiation. Although shielding does not require electrical connectivity (as required in electrical conduction), however, shielding power is enhanced by the electrical connectivity. A secondary mechanism of shielding is associated with the absorption, for which presence of electrical or magnetic dipole in the shield is essential that can interact with the electromagnetic field. Other than these, multiple reflections, i.e., re-reflection at the various surfaces or interfaces in the shield also contributes to

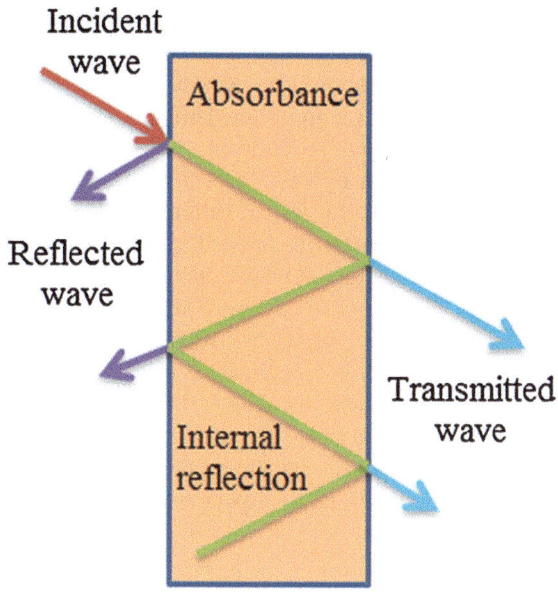

Fig. 7 Mechanism of EMI shielding

shielding. This mechanism essentially requires presence of large surface area or interface area in the shield and the effect becomes insignificant when the distance between the reflecting surfaces or interfaces become relatively large [57, 58].

The SE of a material is defined as the loss of electromagnetic radiation due to the shielding action of the material, i.e., SE = reflected waves + absorbed waves + internally reflected waves. It is expressed by the following equation [59]:

$$\mathrm{SE} = 10 \log_{10} \frac{P_{\mathrm{T}}}{P_{\mathrm{I}}} \tag{1}$$

where P_{T} is the power of the transmitted wave and P_{I} is the power of incident wave.

Although metals have been used for EMI shielding, since long, for their higher conductivity, polymer matrix composites containing conductive filler have become attractive in this regard for their ease of processability, low density, and cost effectiveness along with versatile electrical and microwave properties. In general, polymer matrix is electrically insulating and does not possess shielding effect. Though some electrically conducting polymers are available, they do not show adequate processability or mechanical properties [57, 59]. Nevertheless, addition of suitable electrically conducting fillers, such as CB, graphite, metal fibers, or particulates, imparts EMI shielding effectiveness to polymer composites. However, EMI shielding effectiveness of the composite material not only depends on electrical conductivity of the filler but also depends on their dispersion, aspect ratio, and the sample thickness as well [56–58].

Jana and co-workers [58, 60–62] extensively studied on EMI shielding effectiveness of CR composites. This group thoroughly investigated the effects of carbon fiber concentration, filler aspect ratio, and sample thickness on the shielding effectiveness of CR composites. Different aspect ratios of carbon fiber into rubber matrix are controlled by following two different preparation methods, i.e., conventional two roll mixing (fiber aspect ratio ~25) and cement mixing method (fiber aspect ratio ~100). Moreover, different vulcanization conditions viz. thermovulcanization, barium ferrite-assisted vulcanization, and conventional vulcanization using metal oxide and ethylene thiourea were adopted to prepare the composites. The EMI shielding effectiveness of the composites was measured in both low and high frequency range (100–1000 MHz and 8–12 GHz, respectively) from the viewpoint of electronic application in practical fields. SE of the composites was found to increase with increase in fiber loading for both high and low frequency range. This was attributed to the formation of a good electrically conductive network by the carbon fiber in the rubber matrix. The composites prepared by cement mixing method with larger aspect ratio (~100) showed superior SE than the composites prepared by mill mixing method with smaller aspect ratio (~25). It was also observed that SE was enhanced with increase in sample thickness for fixed fiber concentration and fiber aspect ratio. Moreover, the cross-linking density of the composites significantly influenced the SE which increased in the following order: thermovulcanized composite < conventionally vulcanized composite < barium ferrite vulcanized composite. From the results, it was concluded that at any particular frequency, the SE of the short carbon fiber-filled CR composites strongly depended on the sample thickness and random distribution of the fiber into rubber matrix.

5.2 Microwave Absorbance

Rubber composites, that contain fillers with high magnetic loss, can be used as a microwave absorber and find important application in the area of magnetic field sensors, magnetic data storage, etc. [15, 36]. CR being polar in nature can contribute good microwave properties in the composites designed for microwave application.

Shtarkova et al. reported that chemical structure of rubber could have strong influence on the interaction between the electromagnetic waves and the rubber composite [22]. From the studies on different polar and nonpolar rubber, it was evident that the elastomers such as NBR and CR containing highly polar functional groups exhibited better microwave-absorbing properties over others. This was attributed to the crystallizing ability of the elastomers owing to their polarity. The hybrid combination of fillers containing natural magnetite (having high magnetic losses) and different grades of CB (having high dielectric losses), used in this study, was found to be very effective in expanding the working frequency band and improving the microwave absorbance property of the CR composites.

Saritha Chandran et al. [63] studied the microwave absorption characteristics of CR composites filled with PANI and PANI-coated nylon fiber prepared by conventional mechanical mixing. It was observed that the dielectric permittivity and loss tangent of the composites increased with increase in microwave frequency (in the range of 8–12 GHz) and PANI content in the former type of composite. On the other hand, composites loaded with PANI-coated nylon fiber showed decreased values with increasing fiber loading. Furthermore, permittivity values of the composites loaded with PANI-coated nylon fiber were found lower compared to that of PANI-loaded composite. This was assigned to the low conductivity of the former type composites. The heating coefficient of the composites was found to decrease with increasing frequency and PANI loading. A lower value of heating coefficient is the better indicative of the microwave attenuation. At low and high loading of PANI-coated fiber, heating coefficient became lower, especially at high frequencies. This study revealed that the PANI/CR composite showed very good dielectric properties in the microwave range, whereas PANI-coated nylon fiber-loaded CR composite showed improved mechanical properties with reasonably good dielectric properties. Hence, PANI-coated nylon fiber-loaded composite was regarded as a good choice for the high-frequency heating applications.

The microwave absorption properties of the CB-filled CR composites were studied by Al-Hartomy et al. in terms of coefficient of reflection as a function of frequency [35]. Filler concentration of 60 phr was found to be critical concentration, as coefficient of reflection increased sharply exceeding this concentration. The EMI shielding effectiveness of the CR composites was also determined in this frequency range. The composite filled with 60 phr of CB showed the best shielding effectiveness (SE) compared to those with higher filler content (80 and 100 phr). This was attributed to the complementary effect of coefficient of reflection and coefficient of attenuation which resulted in better SE in a wide frequency range (1–12 GHz) for 60 phr filler loading, in particular. They further studied the influence of hybrid combination of fillers such as CB and conducting nickel (Ni) or cobalt (Co) particles on the EMI shielding effectiveness and microwave properties of the CR composites [36]. The composites showed exceptional EMI SE when filler concentration exceeded 40 phr for CB in combination with 10 phr of Ni or Co particles. Furthermore, microwave-absorbing capacity of the composites was found to depend on the intrinsic properties of the metal (Ni or Co) such as conductivity, complex permittivity, and permeability. It was observed that the Ni-containing composites showed better microwave absorption properties, in the frequency range of 1–10 GHz over Co-containing composite due to higher coefficient of attenuation in former one.

Microwave absorbance property of CR composites has been studied by several groups to investigate their applicability as a radar-absorbing material (RAM) that is used to protect the metallic surface in civil and military applications. CR composites filled with carbonyl-iron powder (CIP) and Ti–Co-doped barium hexaferrite (Co-TiBaHF) have been investigated as microwave absorbers for X-band (8–10 GHz) and Ku-band frequencies (10–16 GHz) [64]. The reflectivity measurement studies revealed better microwave absorption for CIP/CR composites in

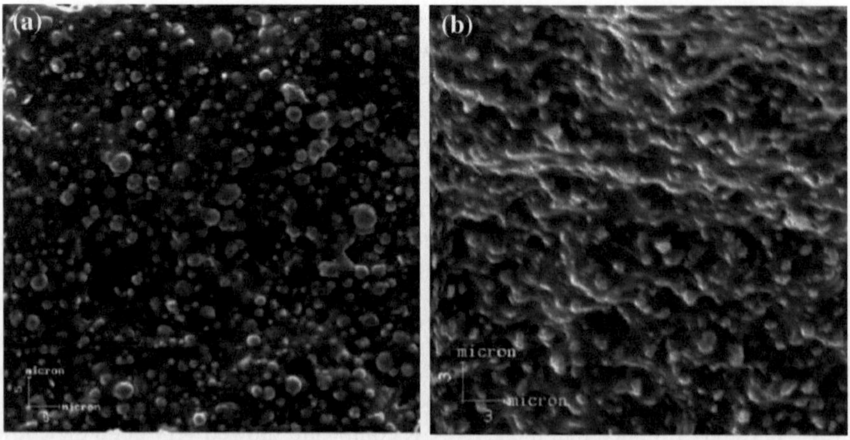

Fig. 8 Scanning electron micrographs of **a** CIP/CR, **b** Co–TiBaHF/CR, (80/20, wt%); magnification = 1500 [64]. Copyright 2002. Reproduced with permission from Elsevier Ltd

the X-band frequency and that for Co-TiBaHF/CR composites in the Ku-band frequency. Such difference in microwave absorbance property for these composites was originated from the different intrinsic nature of the fillers. In the CIP/CR composite, homogenously dispersed CIP (Fig. 8a), possessing highly conductive property, resulted high dielectric loss factor while for the other one with Co-TiBaHF, the magnetic loss increased at higher frequencies [mostly in the frequency range of 12–15 GHz (Fig. 9)].

The effect of reflection and attenuation properties of doubly substituted m-barium hexaferrites (BaHF:$BaFe_{12}O_{19}$) on the microwave absorbance properties of CR composites was studied in terms of complex permeability and permittivity measurement to investigate their use as RAMs [65]. Partial substitution (doping) of Fe^{3+} by Co^{2+}, Ti^{4+}, and Mn^{3+} and that of Ba^{2+} by La^{3+} and Na^{+} was done to prepare varieties of mixed oxides which were then mixed with CR. The microwave absorbance property of the composites was evaluated in the frequency range 8–16 GHz. Doping of metal ions in the hexaferrites increased the complex permeability and reduced the magnetic resonance that enhanced the microwave absorbance properties of the composites. Among all the ferrites, LaNaCoTiMn–BaHF-filled CR composite showed highest microwave absorbance (99.9 %) at 15 GHz. Further studies on the partial substitution of Fe^{3+} by Co^{2+}, Cu^{2+}, and Zn^{2+} in Z-type barium hexaferrite showed more than 90 % microwave absorbance in the frequency range 10–12.5 GHz and 13–16 GHz for CR composites [66]. This was attributed to the high magnetic loss of doped BaHF, because of increase in saturated magnetization that enhanced the magnetic permeability of the composite. Such superparamagnetic properties were achieved by controlling the nanometric size of the ferrite materials (below critical size) that was synthesized by the citrate precursor method. In another study, solgel synthesized cobalt-doped barium hexaferrites was incorporated in CR at high loading (ferrites/CR: 80/20) and the

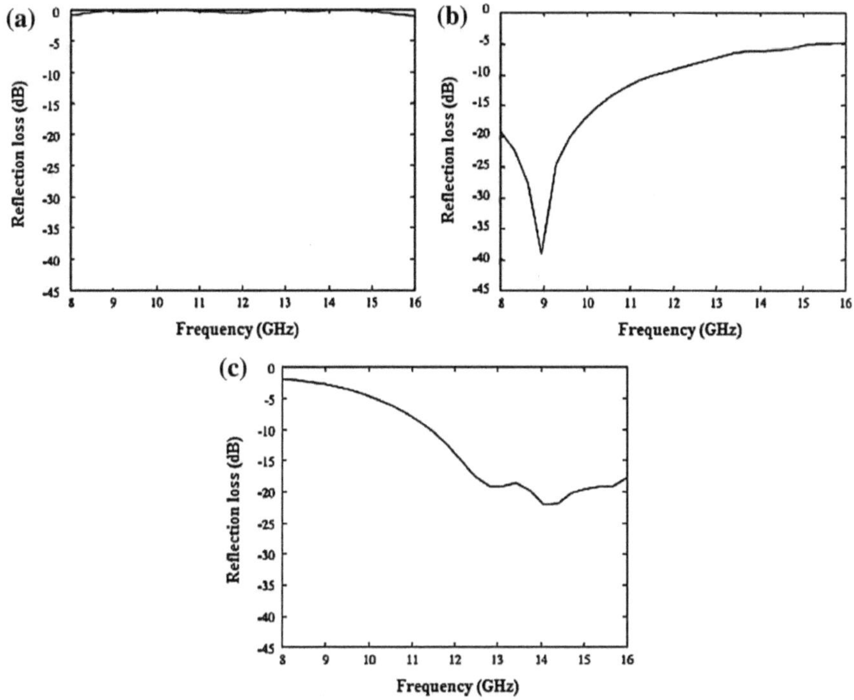

Fig. 9 Reflection loss (dB) as a function of frequency (GHz) for composites with CR (wt%) **a** unfilled composition (0/100), **b** 80/20 of CIP/CR and **c** 80/20 of Co–TiBaHF/CR [64]. Copyright 2002. Reproduced with permission from Elsevier Ltd

prepared nanocomposites were studied as RAM in the frequency range 8–16 GHz [67]. The composites showed more than 90 % microwave absorbance for X/Ku bands. This was attributed to the homogenous dispersion of magnetic hexaferrite nanoparticles even at very high loading (80 %). Three synthetic methods i.e., solgel process, mixed oxide metal approach and citrate precursors method were used for the preparation of substituted BaHF which was proven to be suitable for low temperature synthesis and cost effective as compared to conventional wet-chemical and solid-state synthetic methods. Most importantly, the composites prepared by this method showed better microwave absorbance properties than those prepared by other methods [64–67].

Das et al. studied the microwave absorbance properties of CR composites, as RAMs, filled with doped BaHF and CB [68]. A double-layer RAM was prepared by using different combinations of these fillers in CR. The top layer of RAM was made up of 30 % of BaHF and 10 % of TiO_2-filled CR, whereas bottom layer of RAM was composed of 30 % of BaHF and 10 % of CB filled CR. Each layer of 1 mm thickness were joined together to form a 2-mm-thick sample of RAM. RAM-I with undoped BaHF and RAM-II with doped BaHF were prepared and their microwave absorbance property was studied in the frequency range of 8–12 GHz. The results showed that the maximum reflection loss for RAM-I was −32 dB at 10.64 GHz and that for RAM-II was −29.56 at 11.7 GHz. Moreover, RAM-II showed wide bandwidth of absorption in the X-band frequency range, while RAM-I did not show any consistent reflection loss in the whole frequency range. Better microwave absorbance property of RAM-II was attributed to the reduced particle size of doped BaHF and decreased coercive force.

5.3 Electroresponsive Materials

PANI/CR composite was investigated as an electroresponsive material with reference to studies on electrical conductivity, storage modulus, storage modulus sensitivity, dielectrophoresis force, and deflection angle of the composites [69]. Electrical properties of the CR/PANI blends were found to increase with increase in volume percentage of PANI. This became more significant for nanosized PANI particles that dispersed uniformly throughout the matrix. The storage modulus sensitivities of CR/PANI blends were also found to be very much influenced by the particle size. The dielectrophoresis force and deflection angle, determined from deflection distance, for both the nano- and large PANI particles filled CR composites were measured against increasing electric field strength (Fig. 10). Both dielectrophoresis force and deflection angle increased monotonically for all the specimens as electric field strength increased. CR/PANI blends showed higher dielectrophoresis forces, for both the nano- and large PANI particles, at 0.01 vol% filler content with respect to those at 0.1 vol% filler content or unfilled CR. This was assumed to happen due to suppression of resultant dipole moment generation at high PANI concentration. Similar trend was also observed in the deflection angle measurement of CR/PANI blends.

5.4 Electrically Conductive Chloroprene Rubber-Based Adhesive

About one-third of CR, produced per year, is used as a raw material for the production of adhesives (both solvent-based and water-based). Due to the excellent adhesive properties and scope of formulation at competitive prices, polychloroprene solvent-borne adhesives find dominant position in many market segments in comparison with other adhesive technologies [70–73]. Electrically conductive adhesives are used to paint the inner surface of plastic boxes containing electronic devices. This makes a Faraday Cage saving for the internal components from electromagnetic radiation. Another important application of such adhesive is that in case of a need of a small current, a temperature-sensitive electronic element can be

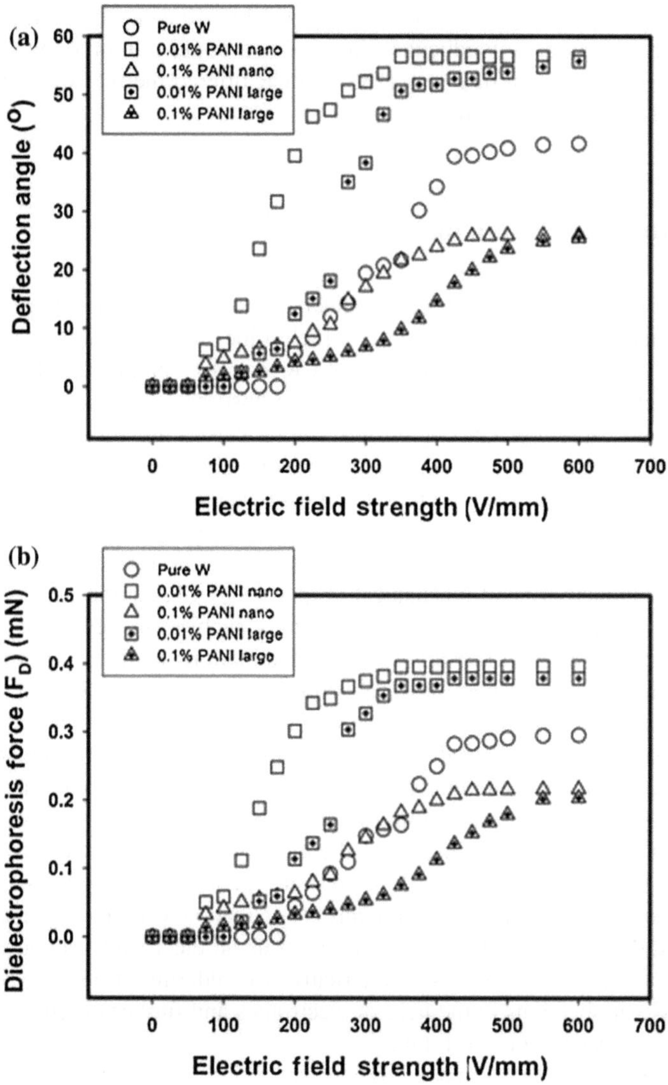

Fig. 10 Electromechanical responses of PANI/CR composites at various DC electric field strengths: **a** deflection angles and **b** dielectrophoresis forces (F_D) [69]. Copyright 2012. Reproduced with permission from John Wiley and Sons

connected to an electrical circuit using an electrically conductive adhesive instead of soldering [74].

Massoumi et al. reported the preparation of conductive PANI–CR blend, with varying amount of dodecylbenzenesulfonic acid (DBSA)-doped PANI, by sonication method [75]. The electrical conductivity of the PANI/CR composites increased with the increase in DBSA content. The adhesive strength of CR was improved with the addition of PANI–DBSA which indicated that PANI–DBSA not only could act as a conductive filler but also could reinforce the CR matrix. Moreover, incorporation of PANI–DBSA in CR offered thermal stability and corrosion resistance to the blend.

6 Summary and Conclusion

Dielectric property and electrical conductivity of CR are mostly governed by its polar nature due to the presence of chlorine atom in its backbone chain. Moderately high dielectric permittivity of CR makes it suitable for different applications such as in EMI shielding, microwave absorbance, preparation of electrical wire, and cable. For CR composites, the properties are very much influenced by several factors such as type, concentration, dispersion, and aspect ratio of filler as well as processing techniques. PANI-coated nylon fiber-filled composites are found suitable for the electronic devices that operate at high frequency. Ionic liquid-modified MWCNT is found to be very effective in improving the electrical conductivity and mechanical properties of CR composite that might be worthy in several electrochemical applications such as in developing batteries, capacitor, actuators, and sensors. Hybrid combinations of fillers are found to be useful in improving the microwave absorption properties of CR composites. CR composites filled with metal ion-doped barium hexaferrite are found promising to be used as RAM. Several studies reveal that incorporation of appropriate filler in optimum concentration can increase the resistance toward different types of aging such as exposure to heat, radiation, salt spray, or self-aging which is essential for a composite to deliver its function in the electronic devices in service condition for longer time. Thus, CB-filled CR composites that show strong resistance to salt spray exposure is recommended for naval application without any anticorrosive agent. PANI–DBSA-based CR could be used as an electrically conductive adhesive, in an electrical circuit, as an alternative to soldering. Thus, CR composites find diverse applications in the field of electronics and further progress in the area is expected to happen in near future.

Acknowledgements Authors wish to thank the editors of this book for the invitation to contribute this chapter.

References

1. Bhattacharya SK, Tummala RR (2000) Next generation integral passives: materials, processes, and integration of resistors and capacitors on PWB substrates. J Mater Sci: Mater Electron 11:253–268
2. Chahal P, Tummala RR, Allen MG, Swaminathan M (1998) A novel integrated decoupling capacitor for MCM-L technology. IEEE Trans Comp Pack Manuf Technol Part B: Adv Packag 21:184–193

3. Brandrup J, Immergut EH (1974) Polymer handbook, 2nd edn. Wiley-Interscience, New York
4. Skotheim T, Elsenbaumer R, Reynolds J (1998) Handbook of conductive polymers. Marcel Dekker, New York
5. Nalwa HS (1997) Handbook of organic conductive molecules and polymers, vol 2. Wiley, New York
6. Tamai T (1982) Electrical properties of conductive elastomer as electrical contact material. Compon Hybrids Manufact Technol IEEE Trans 5:56–61
7. Mather PJ, Thomas KM (1997) Carbon black/high density polyethylene conducting composite materials: part I structural modification of a carbon black by gasification in carbon dioxide and the effect on the electrical and mechanical properties of the composite. J Mater Sci 32:401–407
8. Achour ME, Miane JL, Lahjomri F, El Malhi M, Carmona F (1995) Microwave propagation through carbon black-epoxy resin composites. J Mater Sci Lett 14:1425–1429
9. Makela T, Sten J, Hujanen A, Isotalo H (1999) High frequency polyaniline shields. Synth Met 101:707–717
10. Norman RH (1970) Electrically conducting rubber composites. Elsevier, Oxford
11. Chung D (2000) Materials for electromagnetic interference shielding. J Mater Eng Perform 9:350–354
12. Moniruzzaman M, Winey KI (2006) Polymer nanocomposites containing carbon nanotubes. Macromolecules 39:5194–5205
13. Rizvi R, Cochrane B, Biddiss E, Naguib H (2011) Piezoresistance characterization of poly (dimethyl-siloxane) and poly (ethylene) carbon nanotube composites. Smart Mater Struct 20:094003
14. Bigg DM (1979) Mechanical, thermal, and electrical properties of metal fiber-filled polymer composites. Polym Eng Sci 19:1188–1192
15. Tanrattanakul V, Bunchuay A (2007) Microwave absorbing rubber composites containing carbon black and aluminum powder. J Appl Polym Sci 105:2036–2045
16. Medalia AI (1986) Electrical conduction in carbon black composites. Rubber Chem Technol 59:432–454
17. Sherman RD, Middleman LM, Jacobs SM (1983) Electron transport processes in conductor-filled polymers. Polym Eng Sci 23:36–46
18. Bueche F (1972) Electrical resistivity of conducting particles in an insulating matrix. J Appl Phy 43:4837–4838
19. Bueche F (1973) A new class of switching materials. J Appl Phy 44:532–533
20. De SK, White JR (2001) Rubber technologist's handbook, vol 1. Smithers Rapra Publishing
21. Bhowmick AK, Stephens H (2000) Handbook of elastomers, 2nd edn. CRC Press, Technology and Engineering, Boca Raton
22. Shtarkova R, Dishovsky N (2009) Elastomer-based microwave absorbing materials. J Elastomers Plast 41:163–174
23. Schneider WC, Carter WC, Magat M, Smyth CP (1945) Dielectric dispersion and absorption in neoprene gum and tread stocks. J Am Chem Soc 67:959–963
24. Matsuo M, Ishida Y, Yamafuji K, Takayanagi M, Irie F (1965) Dielectric behavior of polychloroprene. Kolloid-Zeitschrift und Zeitschrift für Polymere 201:89–93
25. Abo-Hashem A (1993) DC conduction in chloroprene rubber (CR). Polym Degrad Stab 40:379–384
26. Baughman RH (1996) Conducting polymer artificial muscles. Synth Metals 78:339–353
27. Suo Z (2010) Theory of dielectric elastomers. Acta Mech Solida Sin 23:549–578
28. Pelrine R, Sommer-Larsen P, Kornbluh RD, Heydt R, Kofod G, Pei Q, Gravesen P (2001) Applications of dielectric elastomer actuators. In: SPIE's 8th annual international symposium on smart structures and materials. International Society for Optics and Photonics, pp 335–349
29. Kornbluh R, Pelrine R, Eckerle J, Joseph J (1998) Electrostrictive polymer artificial muscle actuators. In: Robotics and automation. Proceedings of IEEE international conference, vol 3, pp 2147–2154

30. Heydt R, Kornbluh R, Pelrine R, Mason V (1998) Design and performance of an electrostrictive-polymer-film acoustic actuator. J Sound Vib 215:297–311
31. Carpi F, De Rossi D, Kornbluh R, Pelrine RE, Sommer-Larsen P (2001) (eds.). Dielectric elastomers as electromechanical transducers: Fundamentals, materials, devices, models and applications of an emerging electroactive polymer technology. Elsevier
32. Banno H (1983) Recent developments of piezoelectric ceramic products and composites of synthetic rubber and piezoelectric ceramic particles. Ferroelectrics 50:3–12
33. Lawandy SN, Abd-El-Nour KN (1986) Dielectric properties and stress–strain measurements of chloroprene rubber based on different carbon black fillers. J Appl Polym Sci 31:841–848
34. Hanna FF, Abdel-Nour KN, Abdel-Messieh SL (1992) Dielectric properties of some synthetic rubber mixtures: Part I Neoprene-carbon black mixtures. Polym Degrad Stab 35:49–52
35. Omar Al-Hartomy A, Ahmed Al-Ghamdi A, Falleh Alsolamy J, Dishovsky N, Mihaylov M, Iliev V, El-Tantawy F (2012) Effect of furnace carbon black on the dielectric and microwave properties of chloroprene rubber composites. Kautsch Gummi Kunstst 65:43–48
36. Al-Hartomy OA, Al-Ghamdi A, Dishovsky N, Shtarkova R, Iliev V, El-Tantawy F (2013) Comparison of microwave absorbing properties of chloroprene rubber composites containing carbon black and nickel/cobalt powder. J Elastomers Plast 45:471–485
37. Abd-El-Messieh SL, Younan AF (1996) Dielectric relaxation and mechanical properties of natural and chloroprene rubber with some nitroaniline additives. J Appl Polym Sci 62:805–812
38. Bengtsson P, Klason C, Kubat J, McQueen DH (1989) Electrical noise characteristics of carbon-black-filled chloroprene rubber. J Phys D Appl Phys 22:1736–1741
39. Ali MH, Abo-Hashem A (1997) Percolation concept and the electrical conductivity of carbon black-polymer composites 2: non-crystallisable chloroprene rubber mixed with HAF carbon black. J Mater Process Technol 68:163–167
40. Ali MH, Abo-Hashem A (1997) Percolation concept and the electrical conductivity of carbon black-polymer composites 3: crystallisable chloroprene rubber mixed with FEF carbon black. J Mater Process Technol 68:168–171
41. Jana PB, Chaudhuri S, Pal AK, De SK (1992) Electrical conductivity of short carbon fiber-reinforced polychloroprene rubber and mechanism of conduction. Polym Eng Sci 32:448–456
42. Jana PB, De SK, Chaudhuri S, Pal AK (1992) Electrical conductivity of barium-ferrite-vulcanized polychloroprene filled with short carbon fiber. Rubber Chem Technol 65:7–23
43. Saritha Chandran A, Narayanankutty SK (2008) An elastomeric conducting composite based on polyaniline coated nylon fiber and chloroprene rubber. Eur Polym J 44:2418–2429
44. Kim YA, Hayashi T, Endo M, Gotoh Y, Wada N, Seiyama J (2006) Fabrication of aligned carbon nanotube-filled rubber composite. Scripta Mater 54:31–35
45. Xie XL, Mai YW, Zhou XP (2005) Dispersion and alignment of carbon nanotubes in polymer matrix: a review. Mater Sci Eng R 49:89–112
46. Das A, Stöckelhuber KW, Jurk R, Saphiannikova M, Fritzsche J, Lorenz H, Klüppel M, Heinrich G (2008) Modified and unmodified multiwalled carbon nanotubes in high performance solution-styrene–butadiene and butadiene rubber blends. Polymer 49:5276–5283
47. Subramaniam K, Das A, Steinhauser D, Klüppel M, Heinrich G (2011) Effect of ionic liquid on dielectric, mechanical and dynamic mechanical properties of multi-walled carbon nanotubes/polychloroprene rubber composites. Eur Polym J 47:2234–2243
48. Subramaniam K, Das A, Heinrich G (2011) Development of conducting polychloroprene rubber using imidazolium based ionic liquid modified multi-walled carbon nanotubes. Compos Sci Technol 71:1441–1449
49. Steinhauser D, Subramaniam K, Das A, Heinrich G, Klüppel M (2012) Influence of ionic liquids on the dielectric relaxation behavior of CNT based elastomer nanocomposites. Express Polym Lett 6:927–936
50. Semeriyanov FF, Chervanyov AI, Jurk R, Subramaniam K, König S, Roscher M, Das A, Stöckelhuber KW, Heinrich G (2013) Non-monotonic dependence of the conductivity of

carbon nanotube-filled elastomers subjected to uniaxial compression/decompression. J Appl Phy 113:103706
51. Ito M, Okada S, Kuriyama I (1981) The deterioration of mechanical properties of chloroprene rubber in various conditions. J Mater Sci 16:10–16
52. Pinho MS, Gregori ML, Nunes RCR, Soares BG (2001) Aging effect on the reflectivity measurements of polychloroprene matrices containing carbon black and carbonyl-iron powder. Polym Degrad Stab 73:1–5
53. Kuwahara H, Sudo S, Iijima M, Ohya S (2010) Dielectric properties of thermally degraded chloroprene rubber. Polym Degrad Stab 95:2461–2466
54. Subramaniam K, Das A, Heinrich G (2013) Improved oxidation resistance of conducting polychloroprene composites. Compos Sci Technol 74:14–19
55. Ramasamy SR (1997) A review of EMI shielding and suppression materials. In: Electromagnetic interference and compatibility. Proceedings of the international conference on IEEE pp 459–466
56. Kimmel WD, Gerke DD (1995) Controlling EMI with cable shields. Med Device Diagn Ind 17:112–115
57. Chung DDL (2001) Electromagnetic interference shielding effectiveness of carbon materials. Carbon 39:279–285
58. Jana PB, Mallick AK, De SK (1991) Electromagnetic interference shielding by carbon fiber-filled polychloroprene rubber composites. Composites 22:451–455
59. Joo J, Epstein AJ, MacDiarmid AG (1995) Control of dielectric response of polyanilines: applications to EMI shielding. In: Technical papers of the annual technical conference-society of plastics engineers, pp 1672–1672
60. Jana PB, Mallick K, De SK (1992) Effects of sample thickness and fiber aspect ratio on EMI shielding effectiveness of carbon fiber filled polychloroprene composites in the X-band frequency range. IEEE Trans Electromagn Compat 34:478–481
61. Jana PB, Mallick AK, De SK (1993) Electromagnetic interference shielding effectiveness of short carbon fiber-filled polychloroprene vulcanized by barium ferrite. J Mater Sci 28:2097–2104
62. Jana PB, Mallick AK (1994) Studies on effectiveness of electromagnetic interference shielding in carbon fiber filled polychloroprene composites. J Elastomers Plast 26:58–73
63. Saritha Chandran A, Narayanankutty SK, Mohanan P (2011) Microwave characteristics of polyaniline based short fiber reinforced chloroprene rubber composites. Polym Plast Technol Eng 50:453–458
64. Pinho MS, Gregori ML, Nunes RR, Soares BG (2002) Performance of radar absorbing materials by waveguide measurements for X-and Ku-band frequencies. Eur Polym J 38:2321–2327
65. Lima RC, Pinho MS, Gregori ML, Nunes RR, Ogasawara T (2004) Effect of double substituted m-barium hexaferrites on microwave absorption properties. Mater Sci-Poland 22:245–252
66. Capitaneo JL, Caffarena VDR, Ogasawara T, Pinho MS (2008) Performance of radar absorbing nanocomposites by waveguide measurements. Mater Res 11:319–324
67. Caffarena VDR, Capitaneo JL, Ogasawara T, Pinho MS (2008) Microwave absorption properties of Co, Cu, Zn: substituted hexaferrite polychloroprene nanocomposites. Mater Res 11:335–339
68. Das S, Nayak GC, Sahu SK, Routray PC, Roy AK, Baskey H (2014) Microwave absorption properties of double-layer RADAR absorbing materials based on doped barium Hexaferrite/TiO2/conducting carbon black. J Eng 2014: 1–5. (Article ID 468313, http://dx.doi.org/10.1155/2014/468313)
69. Kunanuruksapong R, Sirivat A (2013) Highly electro responsive polymer blends of polyaniline nanoparticles and chloroprene rubbers. Adv Polym Technol 32:556–571
70. Petrie E M (2000) Handbook of adhesives and sealants. McGraw-Hill Professional

71. Steinfink M (1990) Adhesive materials, Neoprene (polychloroprene)-based solvent and latex adhesives. In Skeist I (ed) Handbook of adhesives. Van Nostrand Reinhold, New York, pp 284–306
72. Pocius AV (2002) Adhesion and adhesives technology: an introduction, 2nd edn. Hanser Gardner Publications
73. Galiatsatos V (1999) Polychloroprene. In: Polymer data handbook. Oxford University Press, pp 375–379
74. Li Y, Wong CP (2006) Recent advances of conductive adhesives as a lead-free alternative in electronic packaging: materials, processing, reliability and applications. Mat Sci Eng R 51:1–35
75. Massoumi B, Farjadbeh F, Mohammadi R, Entezami AA (2013) Synthesis of conductive adhesives based on epoxy resin/nanopolyaniline and chloroprene rubber/nanopolyaniline: characterization of thermal, mechanical and electrical properties. J Compos Mater 47:1185–1195

Electronic Applications of Ethylene Propylene Diene Monomer Rubber and Its Composites

Anjali A. Athawale and Aparna M. Joshi

Abstract Ethylene propylene diene monomer (EPDM), due to its outstanding oxygen, ultraviolet (UV), weather, fatigue and moisture resistance, and good electrical properties, is one of the fast growing elastomer. Besides its general applications such as seals, radiator, and hose, being a flexible polymer and an electrical insulator, it is a suitable candidate for electrical cables, cable terminal protectors, bus bar shrouds, etc., and is advantageous in comparison with glass or ceramic insulators. EPDM can be made electrically conductive by incorporating conductive materials such as conductive carbon black (CCB), metals, conductive polymers, carbon nanotubes, or other nanomaterials. The electrically conductive applications of EPDM include pressure-sensitive switches, actuators, microwave absorption, EMI shielding, conductive gaskets, touch pads, biosensors, and many other fields.

Keywords EPDM · Rubber · Polymer · Insulator · Electronics · Conductor

1 General Aspects of EPDM

1.1 Chemistry and Structure

EPDM is a synthetic rubber consisting of ethylene and propylene where 'M' refers to the saturated backbone in the polymer. It is one of the most important and fastest growing synthetic rubber having both specialty and general-purpose applications.

A.A. Athawale (✉)
Department of Chemistry, University of Pune, Pune 411007, India
e-mail: agbed@chem.unipune.ac.in

A.M. Joshi
K.D. Joshi Rubber Industries Pvt. Ltd., 'A' 82-85, H Block, MIDC, Pimpri, Pune, India
e-mail: aparnamjoshi@rediffmail.com

D. Ponnamma et al. (eds.), *Flexible and Stretchable Electronic Composites*,
Springer Series on Polymer and Composite Materials,
DOI 10.1007/978-3-319-23663-6_11

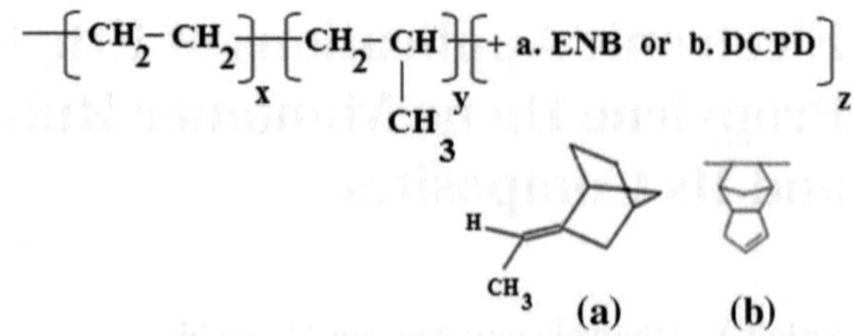

Fig. 1 Ethylene propylene diene monomer with diene as **a** ENB and **b** DCPD

EPDM is an elastic polymer or an elastomer which responds to a large, reversible, and rapid strain to stress. This property differentiates it from the rest of the materials. A broad range of EPDM elastomers can be produced ranging from amorphous and non-crystalline to semi-crystalline structures by varying the polymer composition as well as the way the monomers are combined. The chemical building blocks or monomers in EPDM polymer are similar as for polyethylene (PE) and polypropylene (PP) which combine together randomly. Terpolymerization of a third non-conjugated diene monomer is then carried out by keeping the reactive unsaturation on the side chain [1]. Two most widely used diene termonomers are primarily ethylidene norbornene (ENB) (Fig. 1a) and dicyclopentadiene (DCPD) (Fig. 1b).

1.2 Synthesis, Compounding, and Vulcanization of EPDM

Synthesis of elastomer from simple inexpensive olefins having limited functionality to accommodate vulcanization and to meet the required properties was the scientist's desire. Ziegler et al.'s breakthrough came when they pursued the matter of oligomerization with olefins and metal alkyls. However, as green chemistry as well as new regulations demands the formulations free of hazardous elements, attempts have been made toward developing lead-free EPDM to be used in wire and cable applications. Figure 2 shows the synthesis route of EPDM with Ziegler–Natta catalyst with vanadium as the metal.

EPDM rubbers are manufactured by three commercial processes with the use of catalyst such as solution, slurry, and gas-phase synthesis. Among the three, solution polymerization is the most widely used technique. Further, improvements in the catalyst and the processes have led to increased productivity while maintaining control over the polymer structure. The physical forms of the polymer range from solid to friable bales, pellets, granular forms, and oil blends.

Different materials required for the compounding and processing of EPDM include process aids, antidegradants, tackifiers, pigments, and curatives. The process aids are used to assure the easy incorporation of fillers during mixing with shorter mixing cycles and better dispersion. For the electronic applications at lower voltages, process aids such as esters, amides, and their salts are used. They tend to increase the electrical losses.

Additives such as antioxidants and antiozonants are generally required to maximize the weathering/aging resistance of EPDM for which antidegradants such as

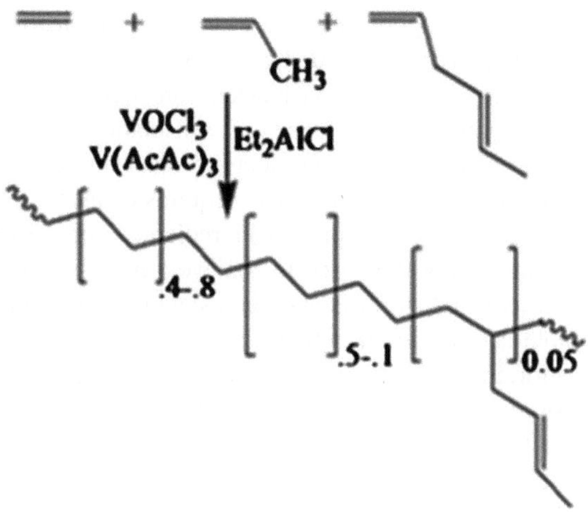

Fig. 2 Synthesis of EPDM [2]. Copyright 2015. Reproduced with permission from John Wiley & Sons. https://s100.copyright.com/CustomerAdmin/PLF.jsp?ref=ab9a5116-2468-45d6-a833-713a45162d2f

amine and phenolic groups are added to the compound. Surface tack is another important property required to improve pliability of the uncured compound. Tackifiers are usually low-molecular weight compounds. Peptizers are chain terminators and reduce the time required to lower the viscosity. Inorganic or organic pigments are another class of materials used in the compounding for improved aesthetics and color coding of products.

Fillers are the essential materials to reinforce the desired properties. The conventional fillers such as carbon black or silica must be used in high loading levels to impart good mechanical reinforcement to the compound. The choice of fillers is substantially narrowed when it comes to the electronic applications of EPDM. The filler should be selected in order to retain the physical properties, electrical losses, electrical stability, breakdown strength, moisture absorption, and increase in the voltage rating. Water-washed hard clay, calcium carbonate or whiting, barium sulfate or barytes, and various types of synthetic silica are the conventional fillers used for low-voltage applications. As the voltage requirements increase, choice of mineral fillers is substantially restricted to untreated and surface-treated calcined clays and to platy structures such as talc which are superior in electrical losses, electrical stability, and moisture absorption [3]. The common mineral fillers include hydrated alumina, precipitated silica, calcium silicate, china clay, and whiting which are used in specific applications. Calcined kaolin is the filler that can be used for insulation [4]. Silane coupling agents improve the filler–matrix interactions [5]. The electrically conductive fillers include CCB, expanded graphite (EG), metal powders, conductive polymers, and nanomaterials.

To achieve best properties in a compound, curing or cross-linking of EPDM is essential. In case of bulky polymer molecules with less sites of unsaturation, peroxide curing agents are best suited, while sulfur curing helps in comparatively simple molecular structures. For low-voltage applications of EPDM components such as insulators, cable jackets, and some molded electrical components, sulfur cure is the better option. For most electrical applications, peroxide cures are the most common curing method. The peroxides such as dicumyl peroxide and bis-(tertiary butylperoxy)-diisopropyl benzene are frequently used in thin-walled low-voltage applications or in semiconductive insulation shields. In both cases, the reaction rate is low. This helps the products to reach the cure temperatures more rapidly. The peroxide curing of elastomers requires coagents to prevent cure inhibition by oxygen or acidic compounding ingredients. Coagents are polyfunctional, unsaturated organic composites used to achieve a more versatile and efficient cure system [6–8]. Elemental sulfur is another specialized coagent. It is frequently used in molded cable accessories in which the improved hot tear strength facilitates demolding. Unfortunately, it also reduces dielectric breakdown strength because of the sulfur cross-links and creates disagreeable odors in the vulcanizate. Phenolic resins have also been used as cross-linking agents [9–11]. Electron beam induced cross-linking of elastomeric materials results in the formation of a three-dimensional network through condensation of the generated macro-radicals [12]. Cure retarders are used to prevent premature curing or scorching of compounds during processing [13, 14].

The functional characteristics of rubber composites depend largely on the mixing process; different mixing facilities are used depending on the type of compound, expected quality of the product, the type of processing method, etc. The commonly used mixing machines include two roll open mixing mill, kneader, internal mixer, and banbury. The two roll open mixing mill is the easiest method in which the mixing process can be inspected visually. The type of mixing in internal mixer is intermeshing. Banbury is used for high-volume applications. It is preferred for easy to mix formulations, sticky materials, multiple step mixing applications, high-viscosity compounds, etc. In this mixer, the rotors are designed for tangential mixing [15].

1.3 Properties of EPDM

EPDM is an inexpensive elastomer with good extendability, low-temperature properties, aging characteristics, and freedom from ozone-induced cracking. In any event, EPDM has found a niche when long-term serviceability is commencing to occupy consumer's minds.

It is a low-cost material with very good low-temperature properties and excellent mechanical strength. The heat aging properties of EPDM are comparatively poor. However, EPDM-based compounds have better wet electrical properties. Being a non-polar elastomer, EPDM has good electrical resistivity and tracking resistance. When it comes to low-voltage application that is up to 35 kV, EPDM supersedes other elastomers and materials [16]. EPDM insulators can maintain the hydrophobicity on the surface even in the presence of pollutants [17–19].

2 General Applications of EPDM

EPDM rubber can be used for different applications in various forms, such as composites or nanocomposite with reinforcing fillers or nanofillers, as blends, coatings etc.

EPDM rubber compound could be a combination of 3–15 different ingredients obtained by mixing with a base polymer. Compound is a substance that consists of chemical bonding between two or more different elements. It has a definite proportion by mass. The bonds between these elements can be broken to get simpler molecules. Science of rubber compounding involves the use of correct ratio of ingredients for fruitful interactions. Composites are engineered materials with two or more constituent materials having significantly different physical and/or chemical properties. The materials comprising a composite remain separate and distinct. They impart the synergic enhancement in the properties of the composite.

Blending of polymers is an attractive way of producing a new material as it does not involve cost and technical uncertainties as that of synthesizing a new one. The blending of elastomers together is to improve the processing and physical properties of cured and uncured elastomers. The physical properties of the cured blends are influenced by vulcanization and filler distribution. Filler distribution in the blend affects the properties of the blend, as it is controlled by the molecular weight of the polymer and filler dispersion in each phase and chemical interaction between the polymer and filler.

The main uses of EPDM are in outdoor and in elevated temperature applications because its saturated structure provides stability under these conditions. Compounding and processing of EPDM leads to cross-linked products with excellent resistance to heat, ozone, oxygen, and weather. The ingredients in the composite play a major role for achieving the desired properties. EPDM being a non-polar elastomer has good chemical resistance towards polar solvents such as salt solutions, alkaline solutions, dilute acids, acetone, alcohols and phosphate ester-based hydraulic fluids. EPDM has poor or limited resistance to aliphatic, aromatic, chlorinated hydrocarbons or petroleum-based fluids, lubricants, and strong oxidizing acids such as nitric acid. However, it is resistant to dilute acids [20, 21].

EPDM elastomer is generally resistant to temperatures up to 121 °C in air and can tolerate higher temperatures with specific compounding. EPDM is often described as a 'preferred' elastomer for nuclear service due to its combined resistance to radiation, chemicals, aging, steam, heat, and other environmental factors [22–25]. Peroxide-cured EPDMs are generally preferred over sulfur-cured types for superior heat aging and radiation resistance properties. However, variations in compounding further influence this behavior.

Versatility in the design of this polymer and its performance have resulted in its broad range of applications in automotive industries, weather-stripping and seals, radiator, garden and appliance hose, tubing, belts, roofing membrane, rubber mechanical goods, plastic impact modification, thermoplastic vulcanizates, and motor oil additive applications. Properly pigmented black and non-black compounds are color stable. It has excellent low-temperature flexibility with glass transition temperature of −60 °C.

EPDM is used in several industrial segments, such as sealing, mechanical and thermal insulation, coatings, tubes, and cushions. In the aerospace sector, EPDM is very important in the development and manufacture of missiles and rockets that demand the use of highly resistant and flexible thermal protections. EPDM is a suitable candidate for these applications due to its compatibility between the materials, compounding techniques, cost of manufacture, aging process, and adhesion properties. Despite these excellent characteristics, EPDM rubber exhibits low values of surface energy, which decreases its adhesion properties to the solid propellant [26–29].

3 Applications of EPDM as Insulator in Electronic Devices

Non-ceramic insulators containing ethylene–propylene rubber (EPR) were first made by different companies in France and USA (1975–1980). In 1976, Rosenthal in Germany and Reliable in USA (1983) presented Silicone rubber (SIR). Ohio Brass produced an alloy consisting of EPDM and SIR in 1986 [30].

EPDM being a non-polar material has good surface and volume resistance, low dielectric permittivity, and high breakdown voltage. EPDM is also a cost-effective material and hence plays a major role as electrical insulator. EPDM insulators are used to support the line conductors to separate them electrically from each other. These composite insulators have substantial advantages such as lightweight, easy installation, and better voltage-withstanding capability compared to inorganic insulators. They exhibit improved contamination performance, resistance to vandalism, high hydrophobicity, and ability to withstand shock loads.

The dielectric constant of EPDM increases with the ENB content as higher number of side chains is available to create greater dipole moments or to store more charges. For example, the dielectric constants of NORDEL IP grades 3670, 4570, 5565, 4520, and 4640, at a frequency of 100 Hz and room temperature, are 1.27, 1.88, 1.92, 1.53, and 1.53, respectively. The dielectric constant of NORDEL IP 5565 is highest as it contains higher percentage of ENB. In addition to this, the dielectric constant of EDPM elastomers increases with increasing molecular weight or decreasing free volume [31].

With the advancement in the technology of large-scale dissipation of electric power through power systems such as substations, distribution, and transmission lines, it became essential that the professionals working in these places were knowledgeable about the insulating materials. In the early days, ceramic and glass materials were used as insulators. However, after 1960s, they were replaced by polymeric insulators due to their excellent insulating properties, ease in designing, manufacturing, and also economic viability. In addition to this, the performance of polymers is good in contaminated environment. They are lightweight, maintenance free, and easy to handle. The basic structure of an insulator consists of a fiberglass core rod covered by weather sheds made up of polymers such as SIR, polytetrafluoroethylene, and EPDM fitted with metal end fittings. Polymers are the commonly used dielectrics; as a result, the commercial and industrial demand for advanced polymers capable of working in harsh environments is constantly increasing [32].

Further, polymer blends are one of the major subjects in polymer industry which are suitable for distribution and transmission class insulations. Their long performance in clean environment has been successful. The purpose of blending is to gain synergistic improvement in properties at both micro- and macro-levels. Batiuk, Herman, and Healy have reported blends of EPDM especially with low-density polyethylene (LDPE) exhibiting superior tensile strength.

Though insulating composites bear several advantages, certain limitations that they pose include chemical changes on the surface due to weathering and dry band arcing together with erosion and tracking, finally leading to their failure as an insulator. It is difficult to detect faulty insulators and evaluate their life expectancy. Suspension type of composite EPDM insulators have been tested by subjecting them to accelerated stress aging as per modified IEC-61109 standards by simulating the inland arid desert's condition. The hydrophobic characteristics have been determined by measuring the contact angle along the surface before and after accelerated aging of the samples. The results show that the composite loses its hydrophobic characteristics in proportion to the intensity of the UV radiations and its rate of recovery is very slow in comparison with SIR. The performance of polymer insulators depends on the selection of materials, design, and its construction. The pollution performance of post-insulators decreases with increasing average diameter [33].

As insulator, the major applications of EPDM include communication cables, magnet wires, power wire, cable apparatus wiring, cable accessories, automotive wiring, control and signal cable and building wire, cable terminal protectors, and bus bar shroud. Due to the flexible characteristics and superior service parameters, EPDM is also widely used as outdoor HV insulator. The flexibility of EPDM has more credentials in the molding process of the complex designs of the insulator components. The complex design of the cable terminal protector can be observed from Fig. 3.

Compared to glass or porcelain insulators, the use of EPDM insulators avoids the unnecessary maintenance such as washing of the insulators and resists deliberate destructions. The ideal composition of an insulator is a resin rod, metallic end fittings, and polymeric weather sheds that protect the resin rod system from the environmental and electric surface discharges due to its water-repellent tendency [35–42]. The use of elastomeric materials in insulation systems of high-voltage apparatuses have increased during the last decades. The apparatuses and components under system voltage operation are expected to remain functional when the over voltage occurs due to unavoidable conditions such as lightning strikes, switching operations, and reversal in polarity in HV dc systems. EPDM insulators are the reliable performers when exposed to such circumstances.

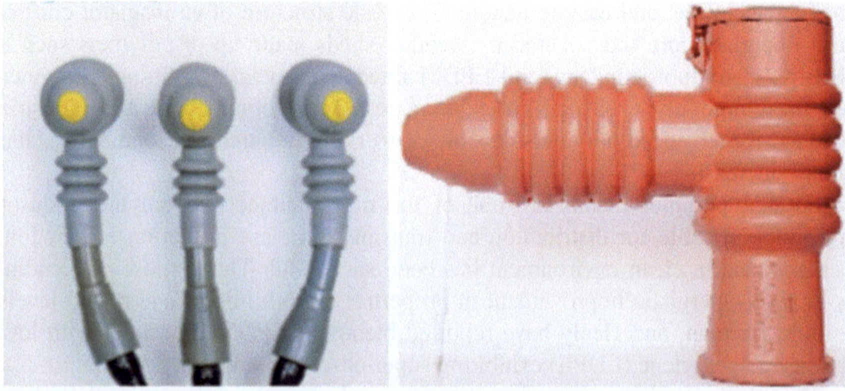

Fig. 3 Cable terminal protector [34]. *Courtesy* K.D. Joshi Rubber Industries

The accumulation and relaxation of electric charges at insulator surface and interface trigger unexpected flashover during testing and operation. The temporal and spatial variations of the surface potentials correlate with the surface and volume resistivity when the surface charge is deposited on EPDM compound surface by impulse positive corona. The rate of potential decay decreases with increasing amplitude of voltage. EPDM polymer is suitable for designing and construction of high-voltage equipment [43, 44]. In comparison with polymeric insulators except silicone insulators, EPDM insulators withstand environmental temperature, pollution, UV radiations, etc., retaining its electrical, chemical, and hydrophobic properties [45]. The degradation process of EPDM is very slow even in the severe pollution conditions such as acid rain and desserts due to its chemically stable structure [46]. EPDM overcomes the leakage current, tracking resistance, and outdoor insulation erosion [47]. Contamination leading to the creepage discharge occurring at the surface is known as tracking. The surface field intensity and the magnitude of current affect the tracking resistance inducing the state of discharge resulting in the loss of its insulating property. The dc power transmission is the preferred mode of power transmission nowadays. Due to greater accumulation of contamination over the insulators under dc voltages, the problem of tracking phenomena is even more severe compared to that with ac voltage. EPDM is an excellent elastomer with its hydrophobic surface properties in the presence of pollutants over the surface of insulators to serve as an outdoor insulator.

Sarathi et al. have studied the behavior of EPDM material under ac and dc voltages, with ammonium chloride/acid rain solution as the contaminant. The tracking time depends on the conductivity and the flow rate of the contaminant. It is a surface degradation process where the tracking time is different for ac and dc voltages [48].

One of the major challenges of polymeric insulators is to minimize the flashover voltage. Its values get affected toward higher risk due to the parameters such as conductivity ($\mu S\ cm^{-1}$) of water droplet, number of droplets, and volume of water droplet (ml). In the field of high-voltage insulators, artificial neural network

(ANN) helps in function approximation, pattern recognition classification, estimation or prediction, etc. [49]. The ANN tool can estimate the level of pollution and flashover voltages and analyze surface tracking on polluted insulators and on hydrophobic polymer insulators [50]. Flashover voltage of hydrophobic polymer insulators has been investigated using a number of different conditions. Composites of EPDM and silicone have been tested for this purpose with silicone content varying from 0 to 100 %.

The prime and most important use of EPDM as insulator is in the wires and cables. Electrical properties that are important for wire and cable applications include resistivity, dielectric strength, dielectric constant, and dielectric loss. For certain applications, fire resistance properties are also important while selecting the material. Voltages up to 150 kV may be required in wire and cable applications; however, each application has its own specific range of voltage. For example, low power transmission cables have low frequencies and high voltages, while telephone cables operate at low voltages and high frequencies. In ultra-high-voltage transmission systems for overhead lines (~1000 kV ac and 800 kV dc) and for submarine lines (~500 and 300 kV), EPDM insulation is becoming prominent.

The construction of basic cable includes three regions. A metallic conductor is encased in a dielectric shield around the conductor system and finally a jacket material for environmental protection. EPDM elastomer due to its environmental stability can be used in many of the components, particularly for insulation and in jackets. It also withstands the temperature of the conductor which increases due to resistive heating when the current passes through the conductor. Electrical properties that are critical to wire and cable applications include resistivity, dielectric strength, dielectric constant, and dielectric loss including fire resistance properties in some applications [51]. EPDM can also accept large amounts of filler and extender oil with no significant prejudice to the final properties. Its good electrical properties make it suitable for application in wires and cables subjected to low to intermediate voltages (up to 35 kV).

As mentioned in the introductory section, some residual ions such as chlorine in the polymer synthesis are susceptible to degradation reactions at elevated temperatures and UV radiations, which cause acceleration in the degradation rate, as well as increased moisture uptake, and have reduced electrical properties. Stabilizing additives are effective in preventing further degradation and yielding reaction products that do not further destabilize the system [52, 53].

In construction wires used in buildings, communication wires, and data transfer and alarm cables, fire or flame retardancy is the most important property to avoid the fire spread through the cables. As a result, the cable materials must possess not only the required electrical properties, but also the appropriate fire resistance as well. EPDM exhibits the electrical resistivity of 10^{15}–$10^{17}\Omega$ cm, dielectric strength of ~35–41 kV mm^{-1}, dielectric constant of 3–3.5 kHz, and dissipation factor of 0.004 kHz [54]. The flame resistance of the composites of EPDM can be improved by using aluminum trihydroxide (ATH). These composites fit the national and international standards to be used for insulation in electrical wires and cables while retaining their property of fire retardancy. Around 170–180 phr of ATH is required in the composition of fire resistant cable [55]. ATH filler also improves the dry band arcing characteristics when exposed to ultraviolet (UV) light [56]. The mechanical properties such as tensile modulus and hardness increase with ATH filler, while elongation at break and tensile strength decrease when ATH (60 phr) is used as filler in the blends of EPDM with other polymers such as polypropylene for cable applications [57].

The ability of SIR insulators is excellent for transferring hydrophobicity, while in operation along with excellent electrical performance under wet and contaminated environments, they are applied widely in polluted condition. The high cost of SIR, poor mechanical properties, and tracking resistance has forced the use of EPDM in HV insulators due to its superior electrical properties, flexibility over a wide temperature range and its resistance to moisture and weather, UV radiation endurance, and hydrophobicity. The blend of EPDM and SIR exhibits excellent electrical and mechanical properties. Contact angle measurements of the blend after aging of the composite determine the surface energy as the blending offers a good degree of protection toward aging of EPDM rubber. EPDM in the blend maintains good mechanical properties such as elongation at break after heat aging [30].

LDPE, EPDM, and their blends are well-known polymers for their use as low-, medium-, and high-voltage insulators. Blends with different weight ratios of LDPE/EPDM prove to be better for mechanical properties such as increasing the tensile strength, modulus, and elongation at break. The properties depend upon the proportion of the blend [58]. Blending PP, EPDM, and thermoplastic elastomer (TPE) combines high-melting temperature and recyclability of PP with the ductility of EPDM. In addition, excellent resistance to degradation and ability to accept high filler loading by EPDM makes TPE suitable for electrical applications with ATH as a flame retarder [55, 57].

Weather sheds for polymer insulators are manufactured from materials such as bisphenol or cycloaliphatic epoxy resins, thermoplastic rubber, EPDM, and silicone elastomers. These materials are compounded with various types of inorganic fillers such as silica and hydrated alumina with concentrations ranging from a few percent to 70 % by weight. Today, the elastomeric materials of EPDM and silicone containing a minimum of 70 % by weight of hydrated alumina that are in use by most of manufacturers are favored for weather sheds with SIR clearly showing the best performance over all other types. Ehsani et al. have studied the mechanical, thermal, dynamic mechanical, and electrical properties of housing (weather shed) materials for outdoor polymeric insulators. The blends of silicone–EPDM show good breakdown voltage strength compared to SIR. Surface and volume resistance of SIR improves with EPDM content. The mechanical properties such as strength, modulus, and elongation at break are improved in the composite [58].

A new technique involving grafting of organofunctional silanes onto the polymer chain has gained importance to attain a homogeneous and maximum mixing of the two different polymers in a blend. A novel copolymer has been synthesized using dicumyl peroxide as initiator. The grafting efficiency of vinyl oxy amino silane (VOS) onto EPDM (EPDM-g-VOS) shows poor mechanical properties due to the flexibility imparted by VOS. However, thermal and dielectric properties are increased due to the introduction of VOS onto EPDM as well as the formation of thermally stable three-dimensional networks through Si–O–Si linkages. Therefore, EPDM-g-VOS could be used as low- and medium-voltage cable insulation for better performance compared to EPDM. Using optimum grafting efficiency through solution-grafting technique, EPDM-g-VOS has been developed in a Haake Rheocord-90, torque rheometer. EPDM-g-VOS can be prepared both by solution polymerization and by melt mixing techniques [59].

The development of polymeric nanocomposites is an interesting area for obtaining materials with high mechanical strength to weight ratio, resistance to vandalism, and high-voltage resistance which make them suitable for application as outdoor insulators. Nanofillers have rendered popularity to rubber nanocomposites over the past decades. These fillers have at least one of their dimensions below 100 nm and a high specific surface area. The specific surface area is responsible for different types of reinforcement in nanocomposites and is manifested even at very low filler loadings (<10 wt%). The rubber nanocomposites are comprised of nanofillers dispersed in rubber matrix. The main families of nanofillers that have become successful commercially are clays and organoclays (OC), carbon nanotubes (CNT), and nanooxides such as nanosilica, TiO_2, polyhedral oligomeric silsesquioxane (POSS), and graphene or graphitic nanofillers. Clays and organoclays are suitable for applications where electrical insulation is required and CNT and graphene are used in conducting applications [60]. Materials with a combination of nanosized organic, inorganic materials, and polymers are expected to give the properties that are synergistic combinations of the individual components with the reinforcing components, such as nanoclay or nanosilica. These are the class of organic, inorganic hybrid materials, where the inorganic components are uniformly distributed in nanometer scale (10–90 nm) within the polymer matrix. The commonly used methods for preparing the nanocomposites are solution blending/melt blending, in situ polymerization and roll milling, and high-shear mixing.

As seen in earlier section, the blend of EPDM and SIR serves the purpose of mechanical strength, tracking resistance and superior temperature stability, excellent UV resistance, better hydrophobicity, etc. However, suitable fillers play a major role in achieving the accurate desired electrical and mechanical properties. The modified inorganic fillers help in this respect. With the increase in the public environmental awareness, research on energy saving and environmentally friendly fillers have become a hot spot for polymer composites in the past two decades. Researchers have used sepiolite, montmorillonite, attapulgite, and kaolin as fillers for polymers for various composite materials with excellent performance. Among them, silica and montmorillonite with layered structure have received a wide interest [61].

Thus, the organically modified montmorillonite (OMMT) clay used in EPDM/SIR composites explores the advantage in high-voltage electrical insulators due to its enhanced mechanical, thermal, and dielectric characteristics. OMMT EPDM/silicone nanocomposites have been prepared by incorporating various phr (1–7) of OMMT onto 50/50 EPDM/SIR blends. Nanocomposites show improved mechanical, thermal, and dielectric properties [62].

4 Applications of EPDM as Conductor in Electronic Devices

4.1 Conductive Composites

By virtue of nature, EPDM is a major candidate for electrically insulating applications. However, inducing electrical conductivity into such a material has enhanced its application horizon. Flexible conductive rubbers have been a subject of interest for a long time and are growing due to their applications in newer areas such as electronic equipment, potential materials for sensors, actuators, supercapacitors, microwave absorbing materials, pressure-sensitive switches, and important strategic materials such as EMI shielding, floor-heating elements, touch-control switches, and conductive coatings for stress cone used in switchgears, applications of semiconducting materials in antistatic rollers for dissipation of static electricity [63], and sensors for vapors and chemicals. EPDM is also a suitable candidate as binder for batteries that binds the active particulate material together and adhere it to the current collector during fabrication of electrodes [64]. Electrical energy storage technologies for stationary applications draw attention to pumped hydroelectric storage, compressed air energy storage, flywheel, capacitor/supercapacitor, and thermal energy storage. The electroactive, flexible, and versatile composites of EPDM with good physicomechanical properties and weather ability serve the purpose for these applications. The piezoelectric characteristics of EPDM, its flexibility, versatility, and low cost make it a promising alternative for fabrication of energy harvesting devices. The most interesting application is a robot arm built with artificial muscles as the mechanical movement of the robot transforms into the electrical energy by piezoelectric energy harvester [65].

These materials need to attain the desired electrical properties as well as adequate mechanical properties. Various rubbers are being widely used for preparation of such composites, e.g., silicone, nitrile, butyl, natural and EPDM rubber. EPDM rubber surpasses other rubbers in these applications because of its mechanical properties and weather aging characteristics. Owing to non-polar and saturated backbone, EPDM exhibits good resistance to heat, oxidation, and polar solvents. However, due to its low crystallinity, EPDM is not self-reinforcing. Generally, there are three main ways to reinforce EPDM to obtain optimal mechanical properties: adding fillers such as carbon black, silica, and clay, blending with crystalline resin (polyethylene, polypropylene, etc.), or changing the cure systems. Adding traditional fillers has become the most significant way to enhance the mechanical properties of EPDM. In addition to traditional reinforcing fillers, there has been a growing interest in several kinds of inorganic fillers which possess excellent electromagnetic properties [66].

Conductive EPDM can be obtained by blending the insulating polymer matrix with conductive fillers. When the percolation threshold of the dispersed particles is achieved, it is possible to obtain this smart material. The conductive fillers or the combination of the conductive fillers that can be used for this purpose are

CCB, metal powders, metal-coated inorganic oxide particles, EG, graphene, carbon fibers, conductive polymers, carbon nanotubes, other nanomaterials, etc. [67]. Various electrically conductive compounds, blends, composites and nanocomposites of EPDM, their preparation techniques, and significance have been discussed in this section.

Metals are conducting due to the presence of metallic bonds in which valence electrons are completely delocalized and form an electron cloud around the metal atoms. In the covalently bonded molecules of saturated carbon compounds, there is no scope of delocalization of the valence electrons; consequently, the electron carrier path is not available. In EPDM molecule, the atoms have covalent bonding; therefore, the movement of electrons over the polymeric chain is restricted. To create the carrier path in the insulating elastomeric matrix, incorporation of conductive additives is the important method. Recent investigations on conducting polymer composites have led to some important revelations relating to attainable electrical conductivity range and trends of change in the conductivity parameter with variations in the filler loading level, nature of the matrix polymer, degree of filler dispersion, and temperature [68].

4.1.1 Composites with Carbon

As the composites of EPDM with carbon are found to be flexible, hence, they have been used for various applications such as EMI shielding, shielding effectiveness, conductive coatings, piezoelectric applications, conductive rollers and other molded components, and pressure-sensitive switches. As carbon provides reinforcement to the composites, they are found to bear superior properties. Improvement in physical and mechanical properties is the result of feasibility of homogeneous dispersion of carbon in the matrix as EPDM can absorb fillers readily. One can obtain composites with good and controllable conductivity by varying the phr concentration of carbon in EPDM.

Among the filled conducting polymer composites, however, those based on the use of different grades of carbon blacks and carbon fibers as filler have been important with reference to their physicomechanical, conducting, and non-corrosive property balance. Low cost, high flexibility, low density, and in particular, specific structures that enable the formation of conductive network inside the polymer matrix at relatively low filler concentration are the added advantages of carbon black. Carbon black is amorphous in nature. Its properties depend upon the particle size, structure, and surface properties. Small particle size and high surface area are the important features of the filler in imparting the electrical conductivity to the composite with smaller loadings. The aggregate size, shape, and the number of particles per aggregate affect the structure of carbon blacks. High structure of the primary aggregate, which is comprised of many prime particles, is important in achieving high electrical conductivity. If the primary aggregates consist of relatively few prime particles, the carbon black is referred to as a low structure black. High structure black has strong attractive forces between their aggregates. This means that the dispersion process should provide more energy to separate them. High structure black tends to produce large aggregates in contact as well as aggregates separated by smaller distance, this results in higher conductivity at the same loading level. Short carbon fibers are better choice in cases where only conductivity is important; it provides high conductivity at lower loading level [69].

Carbon blacks are classified as semi-reinforcement furnace black (SRF), fast extruding furnace black (FEF), high abrasion furnace black (HAF), CCB, and spraying carbon black (SCB). The results of the electrical tests show that the volume and surface resistivity of the EPDM composite is low in the presence of filler having a large surface area that also easily succumbs to dielectric breakdown. The composites show lowest dielectric constant and highest dissipation factor. Hence, CCB with large surface area and high content of sulfur on the surface is suitable for conductive applications; whereas, EPDM filled with other carbon blacks (SRF, SCB, HAF, and FEF) could be useful for insulation applications [70].

The properties of vulcanizates such as physical, mechanical, and DC electrical conductivity show sharp changes on addition of carbon black. When carbon black is loaded above the percolation threshold, it results in more uniform phase morphology. EMI shielding effectiveness generally increases with increase in the carbon black content. Younan et al. have found that EPDM-SRF composites show increase in tensile strength up to 60 phr, while the elongation at break and equilibrium swelling decreases. The permittivity ε' and dielectric loss ε'' increase with increasing carbon content which is abrupt at 60 phr due to the interaction between the rubber and the SRF at higher concentration. Thermal aging up to 7 days shows decrease in ε' and ε'' [71]. The resistivity is also seen to vary significantly as a function of loading level of carbon in case of NBR, EPDM, and their blends (50/50 weight ratio). At low as well as high concentrations of filler, the resistivity is affected negligibly, while in the percolation region (~60 phr), a small increase in filler concentration produces a large increase in conductivity. In the presence of low carbon content, the electrical properties are dominated by the polymer phase between the aggregates with a large distance between the filler particles and the discontinuous conductive path. As the carbon content increases, the average distance between the aggregates decreases forming a discrete chain structure between the small gaps, and at critical concentration, a small addition results in a sharp increase in the conductivity due to the formation of a continuous conductive chain and finally it gains a stable value [72].

Thus, carbon improves the physicomechanical properties of EPDM along with the desired conductivity as per the requirement of applications. However, the thermal stability of the EPDM composites with conductive carbon is comparatively poor; as a result, the concept of using different conductive materials and blends of different polymers has emerged.

The recent development of electrostrictive polymers has generated new opportunities for high-strain sensors and actuators. Recently, investigations related to the use of electrostrictive polymer for energy harvesting, or mechanical to electrical energy conversion, have begun to show its potential for these applications. Strain sensors work on the principle of change in electrical conduction under dynamic and static load. The strain sensitivity of such electrical response depends on various factors such as type of conductive filler, filler concentration, polymer matrix, method of preparation, and mode of application of stress or strain. The electrical conductivity changes significantly when compressive strain and stress are applied on the conductive rubber composites derived from NBR, EPDM, and their 50:50 blends. The resistivity increases during application of the compressive strain which could be due to the destruction of conductive networks formed by the aggregation of conductive particles during compression. When kept under constant strain (compression), there is a reorganization of conductive fillers, which results in some increase in conductivity due to slow relaxation of polymer chains. However, during deloading, there is further breakdown in the conductive networks and hence the conductivity. Conductivity is also affected (increases or decreases) with respect to pressure depending on the viscosity of the matrix. As EPDM is stiff, the relative resistivity (ρ/ρ_0) increases with pressure. Reverse trend is observed for NBR matrix since it is softer, the 50:50 blend exhibits dual behavior. Soft SIR composites also show nature similar to NBR [73].

The electrical conductivity of conductive rubber composites is also seen to vary as a function of applied stress or strain. This property has been very useful for the development of sensors and actuators. Factors such as the type of filler and polymer matrix, the amount of filler added in the composite, method of preparation of composite, and the mode of application of stress and strain affect the electrical response of the conductive rubber composites. The change in electrical conductivity is reported to be due to the change in the internal rearrangement of the conductive particles in the matrix. During the internal arrangement, formation and destruction of the transient structure occurs in the conductive particles present in the matrix together with the change in alignment of the particles. Such materials are characterized by an insulator-to-conductor transition when the concentration of conducting particles reaches a threshold value. This transition has been interpreted within the framework of percolation theory, and the percolation effect is related to the material microstructure. These materials bear the potential to constitute the new generation of sensors or strain gauges [74].

The investigations such as the effect of compression on electrical conductivity and the change in conductivity with time for compressed samples together with the effect of compressive stress on electrical resistivity of conductive rubber have also been reported. The composites are derived from four different matrices such as EPDM, NBR, 50/50 wt% blend of EPDM/NBR, and SIR. Two different loadings of conductive black have been used with 30 phr (percolation limit) and with 60 phr and 20 and 30 phr for SIR. Base polymer has been selected with reference to viscosity. As it has been found from the earlier results, the ease of conductive network formation is strongly dependent on the viscosity of the matrix polymer before vulcanization and the stiffness after vulcanization [75]. Burton and coworkers have carried out extensive work on the change in electrical conductivity and electromechanical properties of conductive black-filled rubber systems. However, they measured the change in electrical conductivity under dynamic conditions when the sample was subjected to repeat flexing.

Biopotential signals are the result of the electrochemical activity of certain cells of the nervous, muscular, or glandular tissue. When one cell is triggered, ion exchange occurs through the membrane of this cell, creating a so-called action potential. A biopotential signal is the action potential from one cell or the average electrical activity of a group of cells, and such signals can be monitored at various locations of the human body. The average activity of the brain cells can be monitored on the scalp, and this signal is called as electroencephalography (EEG). The electric activity of heart cells is monitored by electrocardiography (ECG). To avoid the drawbacks of wet electrodes, various types of dry electrodes have been introduced. Dry electrodes fabricated from flexible conductive polymers are drawing a lot of attention. To obtain optimum properties with reference to material conductivity, hardness, flexibility, and ease of fabrication, various additives have been added in the EPDM matrix and the resulting polymer has been tested as electrode material [76].

The pressure-sensitive EPDM actuators have relatively simple structure, high compliance, and water resistance. They are considered to be very promising for applications such as development of wearable power assist suits, soft handling robots, human support robots, ornamental robots, robot hands of various size say for harvesting fruits, handling animals, etc., since their movement is very smooth similar to a living species. Together with the structure, the working principle of these composites is also simple that consists of rubber structure reinforced with conductive materials or fibers that dissipate charges and have internal chambers in the rubber structure. When pressure is applied to these chamber/chambers in the composites, the rubber undergoes elastic deformation in its structure and thus works as an actuator [77].

The dielectric properties of the composites depend on the volume fraction, size, and shape of conducting fillers and also on other factors such as preparation method, curing method, interface, and interaction between fillers and polymers. It is well established that the effective utilization of filled polymers depends strongly on its ability to disperse the particles homogenously throughout the matrix. Compared to the conventional chemical vulcanizing or curing process using peroxide or sulfur, radiation cross-linking has some advantages. It results in the formation of a three-dimensional network through the union of macro-radicals generated in the system. The process is very fast and clean and requires less energy. The percolation threshold for carbon black in the composites vulcanized by chemical or γ-irradiation dose and by laser beam is different and normally is the lowest with radiation curing method. Dielectric constant (ε') and ac conductivity (σ_{ac}) of the EPDM composite vary with different carbon content. Thus, the desired conductivity can be achieved by selecting both appropriate method of curing and the value of percolation threshold. The high temperatures involved in sulfur or peroxide curing may lead to a variety of uncontrolled side reactions. On the other hand, in radiation curing, the final curing is carried out at ambient temperature under closely controlled conditions, such as radiation dose and dose rate. The type of cross-link formed in this method (–C–C–) gives rise to better mechanical properties such as hot tear strength, abrasion resistance, and superior ozone resistance at higher temperature [78].

The study of the blend of acrylonitrile–butadiene rubber and EPDM rubber (50/50) with different concentrations of HAF carbon black up to 100 phr cured by gamma irradiation indicates that both the degree of loading of the filler and the irradiation dose play an effective role in the conductive properties. Majority of the carbon blacks including HAF have an amorphous quasi-graphite structure with high electron mobility within its hexagonal layer similar to the conducting mechanism of graphite. The electrons flow from one aggregate to another by crossing two barriers such as the aggregate gap width and chemisorbed oxygen complexes existing at the particles or aggregate surfaces acting as insulators. This consumes more energy. The occurrence of the oxygen complexes is diminished to a great extent due to radiation-induced cross-linking and matrix–filler interaction. Thus, the concentration of HAF and the appropriate radiation dose together attain the semiconductive character for the EPDM/NBR blend [79].

Frequency dependence of dielectric constant ε' and ac conductivity of EPDM–HAF black composite increase with phr of HAF (~30 phr and the frequency range 100–105 Hz) and the vulcanization methods. The compound vulcanized by 5 and 10 laser shots shows minimum value of the percolation threshold.

Radiation-induced physicochemical investigations and their impact on the electrical properties of the elastomer blend are studied using FTIR by Deepalaxmi and Rajini [80]. SIR-EPDM blend having 50:50 composition shows electrical changes which reflect in the arc resistance, surface resistivity, and volume resistivity values of the blend after electron beam irradiation (dose = 5 Mrad, 15 Mrad and 25 Mrad).

E-Beam irradiation causes the elastomeric chain cross-linking and chain scission reactions. This results in the formation of new functional groups such as Si–H, Si–CH_3–CH_2, apart from the various functional groups of the component polymers. This indicates that inter molecular cross-linking takes place between SiR and EPDM. The radiation dose of 15 Mrad causes the decreasing effect on the resistivity of the blend due to molecular chain scissions. However, a 73 % increase in volume resistivity has been noticed at 25 Mrad and is attributed to the appearance of new =C–H (alkene, strong) group at 673 cm^{-1} [80].

4.1.2 Composites with Expanded Graphite

The electronic devices demand high protection from microwaves. The microwave absorbing materials have gained importance in communication industry. EPDM is a promising matrix material for this application that is used in conjugation with EG.

Among the various conducting fillers, naturally abundant graphite which possesses good electrical conductivity of about 10^4 S cm^{-1} at ambient temperature has been widely used. In most of the cases, relatively large quantities of graphite are needed to reach the critical percolation value. However, too high concentrations of conductive filler could be detrimental for mechanical properties.

Compatibilized and uncompatibilized EG-loaded EPDM/FKM (50/50, w/w) blends have been prepared and their cure characteristics, filler dispersion, limiting oxygen index, kinetics of thermal degradation, and mechanical and dc electrical properties have been evaluated. Maleic anhydride-grafted EPDM (MA-g-EPDM) has been used as compatibilizer. The dielectric properties have been measured in S (2–4 GHz) band frequency using cavity perturbation technique. The cure time decreases at high loading level due to increase in thermal transition in the presence of EG which promotes vulcanization. Cure characteristics show first order. Tensile properties of composites especially with compatibilizer are enhanced. Addition of EG improves the flame retardancy of the blends and also the ac and dc electrical conductivity due to the formation of conductive networks. The dielectric permittivity and dielectric loss increase with frequency and EG loading [23, 24].

4.1.3 Composites with Conductive Polymers

Among the conductive polymers, polyaniline is one of the most commonly used polymer and it has been evaluated for use in high-voltage engineering for power capacitors. Proposed applications include the use of polyaniline as filler in semiconducting layers as well as in the polymer films used for power capacitors. It has also been shown that polyaniline has nonlinear electrical properties (i.e., nonlinear voltage–current relationship), both when compounded with an insulating matrix and on its own. Nonlinear materials are used as field grading materials in high-voltage engineering and are commonly based on ceramic particles treated in different ways (e.g., by variation of particle size) compounded into a polymer or elastomer matrix. EPDM–polyaniline composites exhibit nonlinear properties, and the conductivity values can be controlled with the addition of carbon black without altering the mechanical strength of the composites [81].

In case of extrinsically conducting polymers, the rate of addition of fillers and the aggregation of particles is critical. The electrical conduction is through interparticle electronic transfer. Large amounts of fillers are required to achieve the conductive phase in the composites. The intrinsically conducting polymers such as polypyrrole when blended with insulating EPDM matrix provide good levels of conductivity. Such combinations are useful in EMI shielding or antistatic protections [82].

4.1.4 Composites with Ceramics

Utilization of ceramic materials as reinforcements in polymer composites is an effective solution to the challenge of developing new polymers for specific sets of properties and applications. With the increase in numerous applications of these materials, more and more knowledge is needed to gain a better understanding of their filler–matrix interaction, which can give them different physical properties. Titanium carbide (TiC) is an important material for high-temperature applications because of its high melting point, hardness, elastic modulus, and electrical conductivity as well as its relatively low coefficient of thermal expansion.

Composites of EPDM/TiC have been prepared as thermistors, with new double negative and positive temperature coefficients of conductivity (NTCC/PTCC). The concentration of TiC has been varied from 0 to 20 phr along with 50 phr HAF. TiC reduces the curing time and accelerates the driving force during the curing process. The electrical properties are also strongly affected by TiC. The dielectric constant increases linearly with temperature. PTCC can be used in antistatic materials, electromagnetic radiation shielding, and electromagnetic interface shielding. The development of NTCC and PTCC polymer composites can be considered as one of the major advances in electronic materials research [73].

Electrical stress grading materials at the insulator–conductor interface for high-voltage applications exhibit nonlinear current–voltage characteristics. Such materials are comprised of an insulating polymer matrix and semiconducting fillers such as ZnO, SiC, or conducting carbon black. One of the challenges in developing field grading materials is the ability to tailor the nonlinear behavior, the low-field conductivity, and the field at which the nonlinear behavior begins. Conductive fillers alone do not provide this tailor ability although mixtures of fillers can. The well-known nonlinear behavior of ZnO and the ability to alter the phases present by the treatment with SnF_2 to tailor the nonlinear behavior of ZnO-filled polymers can be explained on the basis of alteration of the electrical behavior between the heterogeneous particles, which in turn control the nonlinear behavior of the composite. Nonlinear behavior of metal oxide varistors is dominated by a double Schottky barrier mechanism. As such, the nonlinear behavior is controlled by the interface between the particles and the packing of the particles as well as their size. At certain filler concentrations, the resulting nanocomposites exhibit highly nonlinear I-V characteristics. SiC/polymer composites provide further insight into the mechanism leading to nonlinearity in filled polymers. It is also common to use conductive or semiconductive particles in conjunction with ceramic fillers to tailor the electric properties as reported by Wang et al. The ability to tailor the properties makes it possible to use field grading material for both ac and dc as well as for different voltage levels which is advantageous [83].

4.1.5 Composites with Inorganic Conductive Fillers

Apart from traditional fillers, there has been growing interest in inorganic fillers which possess excellent electromagnetic properties. The effect of pH level and surface treatment of samarium oxide (Sm_2O_3), samarium borate ($SmBO_3$), and Sb-doped SnO_2 (ATO) particles on peroxide-cured EPDM rubber composites have been investigated. Coupling agent such as bis-(-3-(triethoxysilyl) propyl) tetra-sulfide (KH845-4) and different level of fillers have significant effect on cure properties of EPDM composites. This is because the coupling agent introduces sulfuric linkages on the surface of the oxide particles which cleave at 170 °C and participate in vulcanization to form C–S linkages. Hence, the cross-link density and mechanical properties of the EPDM composites are enhanced. The pH values of the particles also have a significant effect on the cure and mechanical properties of the composites. The dielectric constant and the dielectric loss of the EPDM composites show an increase with filler loading from 15 to 85 phr followed by decrease at 100 phr. EPDM-ATO composite shows higher dielectric constant compared to other composites at relatively low loading level. Surface and volume resistivities decrease with increasing filler loading in the composites [84].

In another study, Su et al. have studied the cure, mechanical, and electric properties of EPDM composites with 100 phr of Sm_2O_3 treated with different coupling agents such as stearic acid (SA), isopropyl tri(dioctylphosphate) titanate (NDZ102), (KH845-4), and N-b-(aminoethyl)-g-amino propyl methyl dimethoxysilane (SG-Si602). Some functional groups are attached to the filler surface in the presence of the coupling agents. Carboxyl groups retard the EPDM cure, while amino groups, P=O bonds, and S atoms accelerate the cure. Mechanical properties of the composites are enhanced in the presence of amino groups due to the formation of additional rigid C–C linkages, while S atoms boost the mechanical properties of composite by generating flexible S–C linkages. PO bonds probably undergo cleavage during vulcanization and form flexible P–C linkage. Thus, composites with NDZ102- and KH845-4-treated filler exhibit better mechanical properties than those with SG-Si602-treated filler. Composites with KH845-4-treated Sm_2O_3 have the lowest dielectric constant because of minimum polarity of sulfuric atoms. The number of conductive channels and the current density increases with decreasing polarity resulting in low dielectric strength and volume and surface resistivity of the composites [85].

4.1.6 Composites with Steel Fibers

Composites of EPDM with steel fibers have also been reported. They are light in weight since the steel fibers affect the weight negligibly as only small additions (20–30 phr) are enough to reduce the resistivity by 1000 times. Addition of ~20–30 phr (of long as well as short fibers) reduces the resistance of the composite to ~35–40 K Ω from highly insulating electrical range. The steel fibers provide a continuous conducting network in the matrix yielding conductive composite. They can be processed easily, and it is possible to control the conductivity without compromising the physical and mechanical properties. Addition of small amounts of pigments is also permissible by retaining other properties. Important advantage of steel fibers is that they can be recycled. These results indicate that the composite has a well-organized network structure with one component (rubber) being insulating while the other (steel fibers) being conducting, which effectively leads to the induction of conductivity in the matrix. Among the two types of fibers, the SF is expected to form a better network in the matrix and hence exhibits relatively higher conductivity at lower loading levels in comparison with LF. Similar reason holds for the differences observed as a function of concentration [86]. The decreasing trend in the surface resistance can be observed from Fig. 4.

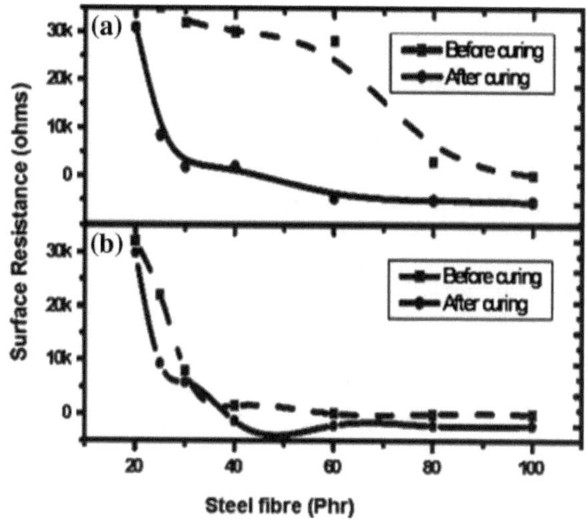

Fig. 4 Effect of **a** LF and **b** SF on electrical conductivity of composites [86]. Copyright 2015. Reproduced with permission from John Wiley & Sons. https://s100.copyright.com/CustomerAdmin/PLF.jsp?ref=14934013-b7fc-4b33-8248-8436a4b5a884

4.2 Conductive Blends

The blending of elastomers together improves the processing and physical properties which are further enhanced by vulcanization and filler distribution. Filler distribution is the important aspect in achieving the properties as it is controlled by the molecular weight of the polymer in each phase and chemical interaction between polymer and filler. Sau et al. have worked on the blends of incompatible elastomer pairs having different polarity which are also very useful for achieving a high degree of conductivity, particularly because of their well-defined interface [87].

Conductive rubber composites from EPDM and acrylonitrile blends have been prepared with acetylene black as the filler. The electrical and mechanical properties of the composites have been investigated. The results indicate that the percolation threshold of filler for achieving high conductivity varies as a function of viscosity of the blends. The conductivity shows an increase with temperature, while the activation energy of conduction is observed to decrease with decrease in filler concentration and percentage of NBR in the blend. The blends exhibit negative coefficient of temperature (NCT) effect in resistivity, and the temperature dependence becomes marginal after percolation limit. The degree of reinforcement in the presence of filler is higher in EPDM and NBR. However, the blends show lower reinforcing potential which can be attributed to the incompatibility of the two constituent polymers in the blend [88].

The service life of rubber is dependent on the behavior of fillers in rubber vulcanizates. Wakil et al. have studied the effect of calcium carbonate, Sillitin N85, and SRF black as filler on EPDM. The cross-link density of the vulcanizates was measured by using ultrasonic and pulse-echo technique. SRF enhances the mechanical properties of the vulcanizate, while Sillitin N85 shows improvement in electrical conductivity after oven aging between 2 and 10 days at 90 °C. The conductivity is basically attributed to the mobility of the free electrons taking place through the interaction between functional filler Sillitin N85 and the EPDM vulcanizate [89].

4.3 Conductive Coatings

A broad range of applications requires an effective conductive coating covering a major segment of antistatic protection such as airplane paints, magnetic tape coatings, spark resistant equipments, and primers for electrostatic spraying. EPDM-conductive filler such as carbon black is used for such applications. The stress cone used in switchgears coated with conductive EPDM filled with carbon black is the example of this application.

4.4 Conductive Foams

In recent years, the users and manufacturers of communication equipment, photographic processing equipment, and electronic devices have already recognized the need to protect their equipment and devices against electrostatic charges, particularly the adverse effects of electrostatic discharge. In order to meet this requirement, a new industry has appeared, called static control industry, and the conductive foam can be used to provide physical protection for electronic components and for electrostatic protection. The foaming conductive polymer is primarily useful for its physical properties, high thermal expansion, its buffering ability, and conductivity. It has been widely used in the detection of static electricity and the elimination of the electrostatic field. Foam composites can be prepared by melt blending and molded foam.

Liu et al. have developed CB/PVC/EPDM foam composites by melt blending and molded foam. The foam with 10 % CB content shows good bubble structure and NTC effect in the heating process. At latter stage of heating, they exhibit PTC phenomena. The DCP content in the composite has great impact on the resistance value and the composite cell formation. The foam plastic has many valuable properties; it has been widely used in industry, agriculture, transportation, military, aerospace industry, and so on. In developed countries, the growth of foam plastic production industry has been rapid and extensive [90].

EPDM structure foams with different apparent densities have been compounded by using different concentrations of foaming agent, azodicarbonamide. The electrical conductivity has been measured under different compression strains (0–60 %). The stress relaxation at different compressive stresses has also been measured. The results show decrease in dielectric permittivity with increasing frequency and foam content which could be due to carbon black dilution. At higher foam concentration, the sample loses its dielectric properties. The conductivity in foams is because of increased mobility of rubber chain molecules. The conductivity shows a sharp decrease with increase in deformation [91].

4.5 Conductive Nanocomposites

The field of nanoscience has blossomed over the last twenty years, and the importance of nanotechnology is increasing as miniaturization has become important in areas such as computing, sensors, biomedical, and many other applications. The discovery of polymer nanocomposites by the Toyota research group has added a new dimension in the field of materials science. In particular, the use of inorganic nanomaterials as fillers in the preparation of polymer/inorganic composites has attracted increasing interest owing to their unique properties and numerous potential applications in the automotive, aerospace, construction, and electronic industries [92].

Elastomer nanocomposites have established a unique position among technologically important materials because of their extensive and potential applications. The dispersion of conductive materials in an elastomer is the most important step during the fabrication of a composite. In general, elastomer nanocomposites can be synthesized in one of the three ways, viz. melt intercalation, solution mixing, and in situ polymerization. These techniques are useful for nanofillers/elastomer composite fabrication as well.

The last few years have seen the extensive use of nanoparticles with large surface area to achieve the required properties at much lower filler loadings. Nanometer-scale particles including spherical particles of silica or titanium dioxide generated in situ by the solgel process are seen to significantly enhance the physical and mechanical properties of rubber matrices. On the other hand, alternative fillers with high aspect ratio such as layered silicates, carbon or clay fibers, and single-wall carbon nanotubes (SWCNTs) or multiwall carbon nanotubes (MWCNTs) often bring about an improvement in mechanical response of the material, along with other interesting properties such as gas barrier, fire resistance, or thermal and electrical conductivities.

The effect of MWCNTs on mechanical and electrical properties of three different sulfur-cured hydrocarbon rubbers, NBR, SBR and EPDM, has been investigated by Bokobza. The composites have been prepared by solution blending using sonication process. The processing conditions have a strong effect on the electrical properties of the composite which are very sensitive to nanotube dispersion in the elastomeric matrix. The potential of CNTs as reinforcing fillers for polymeric matrices has been immediately recognized after their first observation in 1991. CNTs consist of folded graphene layers with cylindrical hexagonal lattice structure. As a result of this unique arrangement, they display exceptional stiffness and strength and remarkable thermal and electrical properties, which make them ideal candidates for the design of advanced materials. These exceptional properties are expected to impart major enhancements in various properties of polymer composites at relatively low filler loadings. The requirement of CNT having diameter of 10 nm, length of 1.5 μm, and surface area of 250–300 $m^2\ g^{-1}$ is less than 10 wt%. The composites can be prepared by solution blending using sonication process [3].

Materials with a combination of nanosized organic, inorganic materials and polymers are expected to show the properties that are synergistic combinations of the individual components with the reinforcing components, such as nanoclay, nanosilica, and nanographite. These are class of organic, inorganic hybrid materials, where the inorganic components are uniformly distributed in nanometer scale (10–90 nm) within the polymer matrix.

EPDM-CNT composites are potential materials to be used as sensors. Microscale change of interelectrical condition in EPDM matrix takes place due to deformation upon exertion of external forces as the contact resistance changes. The composite thus reveals macroscale piezoresistivity. In addition to this, proper alignment of CNT in the matrix also improves mechanical, thermal, and electrical properties [93]. These flexible strain sensors can measure large deformations on flexible structures. They are eligible to develop biomimetic artificial neuron that senses the deformation, pressure, and shear force [94].

Composites of NBR and EPDM have also been prepared by adding short carbon fibers. They are a better choice as conductive filler since they provide higher conductivity at lower loading level. The volume resistivity increase with temperature is probably due to the breakdown of conductive network. It becomes negligible at high loading levels of fillers together with a marginal degree of reinforcement. The resistivity versus temperature plot exhibits electrical set as well as electrical hysteresis during heating–cooling cycle.

Sahoo and Sahoo have synthesized EPDM–MWCNT composites. They have dispersed MWCNTs in xylene by ultrasonication, the concentration of MWCNTs was varied between 0.5 and 5 wt% in the final composite. EPDM solution was slowly poured into the dispersion of MWNTs/xylene. The individual fibrous phases had diameters of about 100 nm which must be due to the coating of EPDM layer on CNT. The nanocomposites exhibited improved mechanical properties. The electrical conductivities of EPDM/c-MWNT increased with the increase of c-MWNT content [95].

Nanocomposites of polymer/graphene have gained importance due to the excellent thermal, mechanical, and electrical properties of graphene. As graphene has a honeycomb structure, it facilitates the design of composites with superior performance at very low loading levels. However, since pristine graphene has a tendency to agglomerate, it has to be chemically oxidized so that it disperses well in the polymer matrix. High-performance nanocomposites exhibiting superior functional and environmental properties have been prepared using EPDM and graphene [96].

5 Conclusions

This chapter presents an overview of the applications of EPDM in electronic industry as electrically insulating as well as conducting material. EPDM is an ideal candidate to be used in the form of compound and blends with other polymers, composites, coatings, and nanocomposites reinforced with conductive fillers for such applications. Conductive carbon, EG, metal fibers, and conductive polymers are the excellent conductive fillers and are feasible substitutes for expensive materials. CNT, carbon fibers, graphene, and other inorganic nanomaterials are the highlights in the electrical applications in electronics. EPDM composites and nanocomposites prove to be promising as dual functional materials for applications; as insulators such as cables, cable terminals, and shrouds; and as conductors in EMI shielding, sensors and actuators, antistatic dissipation, energy harvesting devices, and microwave absorption materials.

References

1. Karpeles R, Grossi AV (2001) EPDM rubber technology. In: Handbook of elastomers, 2nd edn. Marcel Decker, Inc., New York, pp 845–876
2. Cesca S (1975) The chemistry of unsaturated ethylene-propylene-based terpolymers. J Polym Sci Part A: Polym Chem Macromol Rev 10(1):1–231
3. Bokobza L (2012) Enhanced electrical and mechanical properties of multiwall carbon nanotube rubber composites. Polym Adv Technol 23:1543–1549
4. Blodgett RB (1983) Twenty-five years of insulated wire and cable. In: Paper #37 presented at a meeting of the rubber division, ACS, Philadelphia, PA, May 1982. RUBBER CHEM. TECHNOL. 56: 270
5. Mokhothu TH, Luyt AS, Messori M (2014) Reinforcement of EPDM rubber with in situ generated silica particles in the presence of a coupling agent via a sol–gel route. Polym Test 33:97–106
6. Henning SK (2007) The use of coagents in the radical cure of elastomers. In: Proceedings of 56th international wire and cable symposium, Exton, PA, USA, pp 587–593
7. Coran AY (2005) Vulcanisation. In: Mark JE, Eirich FR (ed) The science and technology of rubber, 3rd edn. Elsevier Academic Press, pp 322–3
8. Chokanandsombat Y, Owjinda S, Sirisinha C (2012) Comparison on properties of acrylonitrile styrene butadiene rubber (NSBR) and styrene butadiene rubber (SBR)/nitrile rubber (NBR) blends. KautGummiKunstst 65(11–12):41–46
9. Calleja FJB, Bayer RK, Ezquerra TA (1988) Electrical-conductivity of polyethylene carbon-fiber composites mixed with carbon-black. J Mater Sci 23:1411
10. Genies GM, Boyle A, Lapkowski M, Tsintavis C (1990) Polyaniline: a historical survey. Synth Met 36:139
11. Mattoso LHC (1996) Polyaniline: synthesis, structure and properties. Quim Nova 19:388
12. Faez R, De Paoli MA (2001) A conductive elastomer based on EPDM and Polyaniline: II. Effect of crosslinking method. EurPolym J 37:1139
13. Walker J (2012) Elastomer engineering guide. http://www.jameswalker.biz. Accessed 10 Mar 2014
14. Bhuvaneswari CM, Kakade SD, Gupta M (2006) EPDM as futuristic elastomer for insulation of solid rocket motors. Defense Sci J 56(3):309–320
15. Moribe T (2012) Mitsubishi heavy industries. Tech Rev 49(4):38–43

16. Jl Mead, Tao Z (2002) Insulation materials for wire and cable applications. Rubber Chem Technol 75:701–712
17. Sarathi R, Rao UM, Venkataseshaiah C (2002) Investigation of surface modifications in EPDM rubber due to tracking. Polym Test 21:463–471
18. Vijayalekshmi V, Abdul SSM (2013) Mechanical, thermal and electrical properties of EPDM/silicone blend nanocomposites. Int J Eng Res Appl (IJERA) 3(2): 1177–1180. ISSN: 2248–9622
19. Delor JF, Lacoste J, Barrois ON, Lemaire J (2000) Photo-thermal and natural ageing of (EPDM) rubber used in automotive applications. Influence of carbon black, crosslinking and stabilizing agents. Polym Degrad Stab 67:469–477
20. Su J, Zhang J, Xu Z (2009) Combined effect of pH level and surface treatment of Sm_2O_3, $SmBO_3$ and ATO particles on cure, mechanical and electric properties of EPDM composites. Polym Test 28:419–427
21. Premamoy G, Bibha C, Achintya SK (1996) Thermal and oxidative degradation of PE-EPDM blends vulcanized differently using sulfur accelerator systems. Eur Polym J 32(8):1015
22. Nair AB, Philip K, Joseph R (2013) Effect of expanded graphite on thermal, mechanical and dielectric properties of ethylene–propylene–dieneterpolymer/hexafluoropropylene–vinylidenefluoridedipolymer rubber blends. Eur Polym J 49:247–260
23. Celzard A, Mareche JF, Furdin G, Puricelli S (2000) Electrical conductivity of anisotropic expanded graphite-based monoliths. J Phys D Appl Phys 33:3094–3101
24. Zheng W, Wong SC, Sue HJ (2002) Transport behavior of PMMA/expanded graphite nanocomposites. Polymer 73:6767–6773
25. Saunders DS, Galea SC, Deirmendjian GK (1993) Development of fatigue damage around fastener holes in thick graphite/epoxy composite laminates. Composite 24:309–321
26. Maia JV, Massi M, Sobrinho ASS (2013) Influence of gas and treatment time on the surface modification of EPDM rubber treated at afterglow microwave plasmas J. Appl Surf Sci 285:918–926
27. Chan CM, Ko TM, Hiraoka H (1996) Polymer surface modifications by plasmas and photons. Surf Sci Rep 24:1–54
28. Grythe KF (2006) Surface modification of EPDM rubber by plasma treatment. Langmuir 22:6109–6124
29. Dutra JN, Mello SAC, Massi M, Maciel HS (2006) Surface characterization of synthetic vulcanized EPDM rubber treated with oxygen/argon plasma. In: 41th international symposium on macromolecules, MACRO, Rio de Janeiro 18–21 Feb 2013
30. Ehsani M, Morshedian J, Bakhshandeh R (2004) An investigation of dynamic mechanical, thermal, and electrical properties of housing materials for outdoor polymeric insulators. Eur Polym J 40:2495–2503
31. Canaud C, Nunes CR (2001) EPDM formulations for electric wires and cables. Elastomers Plast 54:56–60
32. Sharma KRET (2001) Polymeric insulators. Technical Article, www.appstate.edu/~clementsjs/surfaceflashover/insulatortesting.pdf
33. ANSI/IEEE Std 4-1978—techniques for high voltage testing, 10.1109/IEEESTD.1995.81695
34. Joshi KD (2014) Rubber components. http://www.kdjoshi.com. Accessed 10 May 2014
35. Ehsani M, Borsi H, Bakhshandeh GR (2004) An investigation of dynamic mechanical, thermal, and electrical properties of housing materials for outdoor polymeric insulators. Eur Polym J 40:2495–2503
36. Tokoro T, Hackam R (2001) Loss and recovery of hydrophobicity and surface energy of HTV silicone rubber. IEEE Trans Dielectr Electr Insulat 8(6):1088–1097
37. Bernstorf RS, Zhao T (1998) Aging tests of polymeric housing materials for non-ceramic insulators. IEEE Electr Insulat Mag 14(2):26–33
38. Hackam R (1999) Outdoor high voltage composite polymeric insulators. IEEE Trans Dielectr Electr Insulat 6(5):557–585

39. Gubanski SM (2003) Modern outdoor insulation—concern and challenges. In: Proceeding of 13th international symposium on high voltage engineering (ISH), Delft, Netherlands
40. Wang X, Schadler LS, Hillborg H, Auletta T (2006) Nonlinear electrical behavior of treated ZnO-EPDM nanocomposites. ieeexplore.ieee.org/ iel5/4105355/ 4105356/ 04105460.pdf
41. Donzel L, Gramespacherz H (2004) Silicone composites for HV applications based on microvaristors. In: 2004 international conference on solid dielectrics
42. Mårtensson E, Nettelblad B, Gäfvert U, Palmqvist L (1998) Electrical properties of field grading materials with silicon carbide and carbon black. In: IEEE international conference on conduction and breakdown in solid dielectrics pp 548–552
43. Mårtensson E, Gäfvert U (2004) A three-dimensional network model describing a nonlinear composite material. J Phys D:Appl Phys 37:112–119
44. Sjöstedt H, Montaño R, Gubanski SM (2009) Charge relaxation on surfaces of polymeric insulating materials for outdoor applications. Mater Sci-Pol 27(4/2):1129–1137
45. Khan Y (2010) Degradation of hydrophobic properties of composite insulators in simulated arid desert environment. Int J Eng Technol IJET-IJENS 10(01): 64–68
46. Deng H, Hackam R (2000) Dielectr Electr Insul IEEE Trans 7(1): 84–94
47. Sarathi R, Rao UM, Venkataseshaiah C (2001) Investigations of surface modifications in ethylene propylene diene monomer (EPDM) rubber due to tracking. Polym Test 21:463–471
48. Yoshimura N, Kumagai S, Du B (1997) Research in Japan on the tracking phenomena of electrical insulating materials. IEEE Electr Insul Mag 13:8–19
49. Haykin S (1999) Neural networks: a comprehensive foundation. Upper Saddle River, Prentice Hall. ISBN 0132733501
50. Ahmad AS, Aljunid ASK (2004) Assessment of ESDD on high-voltage insulators using artificial neural network. Elsevier. Electr Power Syst Res 72(2):131–136
51. Su J, Chen S, Xu Z (2009) Comparison of cure, mechanical, electric properties of EPDM filled with Sm_2O_3 treated by different coupling agents. Polym Test 28:235–242
52. Kang D, Desai K, Lee J, Mead J (2004) Assessing the electrical properties of alternative wire and cable coatings: Metallocene EPDM. Technical report no. 56, pp 1–23
53. Bartnikas R (2000) Characteristics of cable materials. In: Power and communication cables. IEEE Press, New York. doi:10.1109/9780470545546.fmatter
54. Mead JL, Tao Z (2011) Insulation materials for wire and cable applications. Rubber Chem Technol 75:701–712
55. Cheremisinoff NP (ed) (1989) [a]Handbook of polymer science and technology. Marcel Dekker, New York, Chapter 2 p 341
56. Nasrat LS, Hamed M, Ibrahim AA (2014) Effect of ultra violet on dry band arcing behavior of EPDM outdoor insulators. Int J Emerg Technol Adv Eng 4(10): 1–8. ISSN 2250-2459
57. Farzad R, Hassan A, Jawaid M, Piah MAM (2013) Mechanical properties of alumina trihydrate filled polypropylene/ethylene propylene diene monomer composites for cable applications. Sains Malaysiana 42(6):801–810
58. Ehsani M, Zeynali E, Abtahi M, Harati A (2009) LDPE/EPDM blends as electrical insulators with unique surface. Electrical and mechanical properties. Iran Polym J 18(1):37–47
59. Kumar MSC, Alagar M (2002) Development and characterisation of vinyloxyaminosilane grafted ethylene-propylene-dieneterpolymer (EPDM-g-VOS) for engineering applications. Eur Polym J 38:2023–2031
60. Galimberti M, Cipolletti V, Theonis RO, Kumar V (2014) Recent advancements in rubber nanocomposites. Rubber Chem Technol 87(3):417–442
61. Tang Q, Liang J, Zhang F (2015) Effect of coupling agent on surface free energy of organic modified attapulgite (OAT) powders and tensile strength of OAT/ethylene-propylene-diene monomer rubber nanocomposites. Powder Technol 270:92–97
62. Vijayalekshmi V, Majeed ASSM (2013) Mechanical, thermal and electrical properties of EPDM/silicone blend nanocomposites. Int J Eng Res Appl (IJERA) 3(2):1177–1180. ISSN: 2248-9622

63. Sau KP, Chaki TK, Khastgir D (1998) Carbon fibre filled conductive composites based on nitrile rubber (NBR), ethylene propylene diene rubber (EPDM) and their blend. Polymer 39(25):6461–6471
64. Morsy RM, Ismaiel MN, Yehia AA (2013) Conductivity studies on acrylonitrile butadiene rubber loaded with different types of carbon blacks. Int J Mater Methods Technol 1(4): 22–35. ISSN 2327-0322
65. Reinaldo R, João SC (2004) Electroactive polymer (EAP) actuators the artificial muscles—reality, potential and challenges, 2nd edn. SPIE Press, Washington, Yoseph Bar-Cohen, pp p1–p3
66. Ghosh P, Chakrabarti A (2000) Conducting carbon black filled EPDM vulcanizates: assessment of dependence of physical and mechanical properties and conducting character on variation of filler loading. Eur Polym J 36:1043–1054
67. Brosseau C, Boulic F, Le Mest Y, Loaec J, Beroual A (1997) Dielectric and microstructure properties of polymer carbon black composites. J Appl Phys 81:882–891
68. Norman RH (1970) Conductive rubber and plastics. EPC, New York, p 270
69. Sau KP, Chaki TK, Khastgir D (1997) Carbon fibre filled conductive composites based on nitrile rubber (NBR), ethylene propylene diene rubber (EPDM) and their blend. J Mater Sci 32:5717–5724
70. Hoffman W (1998) Rubber technology handbook. Hanser, Munich 431
71. Younan AF, Choneimt AM, Nour AE (1995) Electrical and physical properties of EPDM rubber loaded with semi reinforcing furnace black. Polym Degrad Stab 49:215–222
72. Sau KP, Chaki TK, Khastgir D (1999) Electrical and mechanical properties of conducting carbon black filled composites based on rubber and rubber blends. J Appl Polym Sci 71:887–895
73. Sau KP, Chaki TK, Khastgir D (1999) The effect of compressive strain and stress on electrical conductivity of conductive rubber composites. Rubber Chem Technol 73:310–324
74. Sahoo MK, Sahoo S (2014) Synthesis and characterization of ethylene-propylene diene monomer(EPDM)/MWCNT composites. Adv Polym Sci Technol: Int J 4(1):7–11
75. Kanamori (1986) Structural formation of hybrid siloxane-based polymer monolith in confined spaces. J Sep Sci 27(10–11): 874–886
76. Chen YH, Grundlehner B, Gadeyne S, Boon P, Hoof CV (2014) Soft, comfortable polymer dry electrodes for high quality ECG and EEG recording. Sensors 14: 23758–23780. doi:10.3390/s141223758
77. Suzumori K, Suzuki H (2007) A Bending pneumatic rubber actuator realizing soft-bodied manta swimming robot. In: IEEE international conference on robotics and automation Roma, Italy 10–14 April 2007
78. Shokr FS (2011) Dielectric properties of carbon black loaded EPDM rubber based conductive composites: effect of curing method. J Am Sci 7(9):387–397
79. Magda M, Abou Z (2007) Radiation effect on properties of carbon black filled NBR/EPDM rubber blends. Eur Polym J 43:4415–4422
80. Deepalaxmi R (2014) Property enhancement of SiR-EPDM blend using electron beam irradiation. J Electr Eng Technol 9(3):984–990
81. Sonerud B, Josefsson S, Furuheim K, Frohne C (2013) Nonlinear electrical properties and mechanical strength of EPDM with polyaniline and carbon black filler. In: IEEE international conference on solid dielectrics, Bologna, Italy 30 June–4 July 2013
82. Zoppi RA, De Paoli MA (1996) Chemical preparation of conductive elastomeric blends: polypyrrole/EPDM-II Utilization of matrices containing crosslinking agents, reinforcing fillers and stabilizers. Polymer 37(10):1999–2009
83. Gupta T (1990) Application of zinc oxide varistors. J Am Ceram Soc 73:1817–1840
84. Allen RD (1983) Fundamentals of compounding EPDM for cost/performance. J Elastomer Plast 15(1):19
85. Su J, Chen S, Zhang J, Xu Z (2009) Comparison of cure, mechanical, electric properties of EPDM filled with Sm_2O_3 treated by different coupling agents. Polym Testing 28:235–242

86. Athawale AA, Joshi AM (2011) Studies on electrically conductive composites of ethylene propylene diene monomer rubber and steel fibers. J Appl Polym Sci 120:3036–3041
87. Gessler AM (1969) Evidence for chemical interaction in carbon and polymer associations. Extension of original work on effect of carbon black structure. Rubber Chem Technol 42:850
88. Langley M (1973) Carbon fibres in engineering. Composites 5(2):84
89. Wakil AAE, Megeed AAE (2011) Effect of calcium carbonate, Sillitin N 85 and carbon black fillers on the mechanical and electrical properties of the EPDM. ARPN J of Engg and Appl Sci 6(5):24–29
90. Liu X, Xing H, Zhao L, Chen X (2012) Preparation and electrical properties of CB/PVC/EPDM conductive foam composite. In: 2nd international conference on electronic and mechanical engineering and information technology (EMEIT)
91. Lawindy AMYE (2005) Studies of electrical and physico-mechanical properties of EPDM structure, foams, Egypt. J Solids 28(1):97–107
92. Heggli M (2001) Elastomeric nanocomposites. Research plan. Institute of Polymers, ETH Zurich, Project Review. http://www.mat.ethz.ch
93. Sekitani T, Aida T, Someya T (2002) Carbon nanotubes—the route toward applications. Science 297:787
94. Kim J, Baek WK, Lim KT, Kang I (2010) Flexible strain sensor based on carbon nanotube rubber composites. Nanosensors, biosensors and info-tech sensors and system 76460 N: doi:10.1117/12.847364
95. Smith CS (1954) Piezoresistance effect in germanium and silicon. Phys Rev 94:42
96. Allahbakhsh A, Sharif F (2013) Cure kinetics and chemorheology of EPDM/graphene oxide nanocomposites. Thermochim Acta 563:22–32

Poly(Isobutylene-co-Isoprene) Composites for Flexible Electronic Applications

M.T. Sebastian and J. Chameswary

Abstract Flexible, bendable, and stretchable dielectrics which can cover even curved surfaces are important for applications in electronic control systems, consumer electronics, heart pacemakers, body-worn antenna, etc. The requirements for a material to be used as a flexible dielectric are good mechanical flexibility, low dielectric loss, high thermal conductivity (TC), low coefficient of thermal expansion (CTE), etc. It is very difficult to identify a single material which possesses all these properties simultaneously. There are a number of ceramic materials with high relative permittivity and low dielectric loss that are available but are brittle in nature. Butyl rubber has low dielectric loss with good mechanical flexibility and stretchability, but they have low relative permittivity, low TC, and high CTE. Therefore, the practical applications of an elastomer or a ceramic alone are limited. By integrating the flexibility, stretchability, and low processing temperature of butyl rubber with high relative permittivity and low loss of ceramics, a composite may be formed, which can deliver improved performances. In this review, the preparation, characterization, and properties of butyl rubber composites with several ceramics such as Al_2O_3, $BaTiO_3$ on the microwave dielectric properties, thermal properties, moisture absorption, and mechanical properties are discussed.

Keywords Ceramics · Thermal properties · Dielectrics · Bending

M.T. Sebastian (✉)
Microelectronics and Materials Physics Laboratory, Department of Electrical Engineering, University of Oulu, 90014 Oulu, Finland
e-mail: msebasti@ee.oulu.fi; mailadils@yahoo.com

J. Chameswary
Materials Science and Technology Division, CSIR-NIIST, Trivandrum 695019, India

D. Ponnamma et al. (eds.), *Flexible and Stretchable Electronic Composites*, Springer Series on Polymer and Composite Materials,
DOI 10.1007/978-3-319-23663-6_12

1 Introduction

Microwave dielectric materials play a key role in global society, with a wide range of applications from terrestrial and satellite communication to environmental monitoring via satellites. Recently, an increasing interest has been shown in the development of mechanically flexible electronic devices in various fields such as wireless communication, automotive, biomedical, and aerospace. For many applications, particularly in the biomedical field and body-worn antenna, electronic circuits are to be conformally wrapped around curved surfaces. In such systems, the circuits must be not only flexible but also stretchable. Flexible electronics is an emerging field in electronic industry which enable a wide range of applications such as capacitors for energy storage, wireless communication, electronic packaging, and electromagnetic interference (EMI) shielding. The development of EMI shielding materials is increasing in recent years in order to prevent the electromagnetic radiations emitting from modern electronic devices [1, 2]. Modern electronic industry requires systems that can be fitted into non-planar forms which can be folded and unfolded for packaging or storage [3, 4]. The flexible substrates have many potential applications from neural prosthetics in the medical field to microwave devices in electronics, which provides the advantage of three-dimensional designs of conformal structures [5, 6]. The requirements of a material to be used for flexible substrate applications are low relative permittivity, low loss tangent, low CTE, high TC, good mechanical flexibility, low moisture absorption, etc. Ceramics with optimum relative permittivity and low dielectric loss are available. But their brittle nature precludes them from practical applications in flexible electronics. Low-loss polymers with good flexibility are also available. However, their high CTE and low TC restrict their practical use. Hence, the use of polymers or ceramics alone is limited. The polymer–ceramic composites can deliver better performance than their constituent phases. There are several polymer–ceramic composites such as polytetrafluoroethylene (PTFE), polystyrene, polyether ether ketone (PEEK), and polyethylene with fillers such as silica, alumina, titania, and glasses, and some of these are commercially available. The reader is referred to the recent review by Sebastian and Jantunen [7] on polymer–ceramic composites for details. Conventional polymers filled with ceramic are rigid and cannot be bend or stretched. Stretchability means that the circuits should have the capacity to absorb large levels of strain without degradation in their electronic properties. In this context, the elastomer–ceramic composites can be used for stretchable and bendable electronic applications. However, not much attention was paid on bendable and stretchable materials such as elastomer–ceramic composites. In this chapter, we discuss the recent advances in low-loss butyl rubber–ceramic composites for microwave applications.

1.1 Butyl Rubber

Butyl rubber is a copolymer of 98 % polyisobutylene with 2 % isoprene distributed randomly in the polymer chain as shown in Fig. 1.

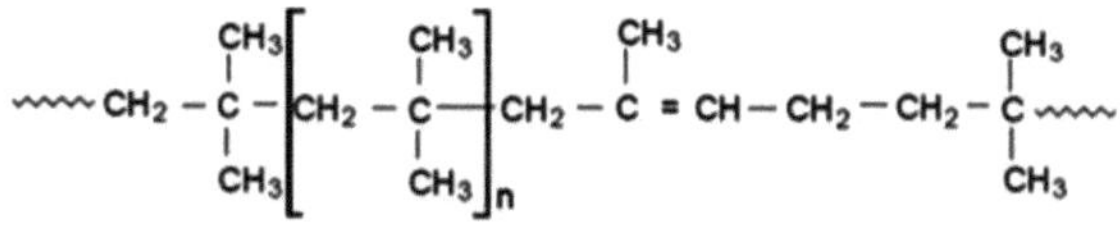

Fig. 1 Butyl rubber

Polymerization of isoprene results in the incorporation of a double bond or unsaturation into the polymer chain. These double bonds serve as cross-linking sites. Vulcanization of the butyl rubber with sulfur results in the formation of a network structure in the form of a cross-linked rubber. As such butyl rubber is a thermoset polymer and once vulcanized, it cannot be reformed into a new shape [8, 9]. Butyl rubber is amorphous and viscoelastic with low Young's modulus and high yield strain as compared to other polymer materials due to its weak intermolecular attractive forces. Butyl rubber exists above the glass transition temperature, and hence, considerable segmental motion is possible. The glass transition temperature of butyl rubber is around −63 °C. At ambient temperatures, rubbers are relatively soft and deformable. The chain segments of elastomers can undergo high local mobility, but the gross mobility of chains is restricted by the introduction of a few cross-links into the structure. The long polymer chains cross-link during curing, and the process is known as vulcanization. The covalent cross-linkages ensure that the elastomer will return to its original configuration when the stress is removed. In the absence of applied stress, [10] molecules of elastomers usually assume coiled shapes. Consequently, elastomers exhibit large elongation (up to about 1000 %) from which they recover rapidly on removal of the applied stress.

Among elastomers, vulcanized butyl rubber has excellent dielectric properties, excellent heat/ozone resistance, high damping properties at ambient temperatures, mechanically flexible, air tight and gas impermeable, low glass transition temperature, good resistance to oxidizing chemicals, good weathering, and heat and chemical resistance. The main use of butyl rubber is in tyre-curing bladders and inner tubes and also used for seals, adhesives, and molded flexible parts [11–13]. These excellent properties of butyl rubber are the result of low levels of unsaturation between the long rubber chain segments. The molecular structure of butyl rubber can be oriented to resist stress, and thus, the mechanical properties are retained over a relatively wide stiffness range since reinforcement is not required for good tensile and tear strength [14].

Materials with varying relative permittivity are required for practical applications in electronic industry. Although butyl rubber is mechanically flexible and has excellent dielectric properties, it cannot be used for flexible electronic applications due to its low ε_r and poor thermal properties. The relative permittivity of butyl rubber can be tailored over a wide range by reinforcing with low-loss ceramics having different range of relative permittivity. For electronic packaging and substrate applications, materials must have a low relative permittivity to minimize capacitive coupling and signal delay along with low loss tangent to reduce signal attenuation. High-permittivity ceramic materials can be used to make butyl rubber composites for flexible dielectric waveguide and capacitor applications.

Table 1 Microwave dielectric properties of ceramics and butyl rubber [23]

Material	Relative permittivity (ε_r)	Loss tangent (tan δ)	CTE (ppm/°C)	TC (W m^{-1} K^{-1})
Al_2O_3	10.1	1×10^{-5}	7.0	30.0
TiO_2	104.0	1×10^{-4}	9.2	11.0
SiO_2	3.8	1×10^{-4}	0.5	1.4
$SrCe_2Ti_5O_{16}$ (SCT)	113.0	1×10^{-4}	1.7	2.9
$SrTiO_3$	290.0	1×10^{-3}	9.4	12.0
$Ba(Zn_{1/3}Ta_{2/3})O_3$ (BZT)	28.0	1×10^{-4}	4.2	3.9
Butyl rubber	2.4	1×10^{-3}	191	0.13
$BaTiO_3$	>1500	1×10^{-2}	5.4	2.6
$Ba_{0.7}Sr_{0.3}TiO_3$	2850	1×10^{-2}	3.0	4.4

In addition, thermal properties of butyl rubber composites can be improved by the addition of ceramics, since ceramics have high TC and low CTE. Elastomer–ceramic composites can combine the advantageous properties of both elastomer and ceramics which can satisfy diverse requirements of present-day electronic industry. Today elastomer–ceramic composites found applications ranging from ultrathin health monitoring to advanced imaging devices. Although extensive researches were done on the microwave dielectric properties of rigid polymer–ceramic composites [7], very little attention was paid on elastomer–ceramic composites. Recently, we have reported the microwave dielectric properties of butyl rubber composites with $SrTiO_3$, $Sr_2Ce_2Ti_5O_{16}$, SiO_2, $Ba(Zn_{1/3}Ta_{2/3})O_3$, $Ba_{0.7}Sr_{0.3}TiO_3$, Al_2O_3, $BaTiO_3$, and TiO_2 composites [15–22]. The microwave dielectric properties of the ceramics used and butyl rubber are given in Table 1.

High-purity alumina has a very low loss tangent of about 10^{-5}. The very high TC (30 W m^{-1} K^{-1}) and low CTE (6–7 ppm/°C) of alumina make it a suitable electronic packaging material [23]. Elastomer–alumina composites are well studied for its mechanical and curing characteristics [24, 25]. Silica is a low-permittivity ceramic with excellent dielectric properties ($\varepsilon_r = 3.8$, tan $\delta \approx 10^{-4}$) [26] and good thermal properties such as TC = 1.4 W m^{-1} K^{-1} and CTE = 0.5 ppm/°C. Silica is important reinforcing filler in elastomers for industrial applications. The effect of silica on dielectric properties of styrene-butadiene rubber was investigated by Hanna et al. [27] in the frequency range of 60–10^8 Hz at room temperature. $Ba(Zn_{1/3}Ta_{2/3})O_3$ is a complex perovskite ceramic with $\varepsilon_r = 28$, tan $\delta \approx 10^{-4}$, and a nearly zero temperature coefficient of resonant frequency ($\tau_f = 1$ ppm/°C). $BaTiO_3$ is a ferroelectric material having relatively high permittivity greater than 1500 and does not resonate in the microwave frequency due to the high dielectric loss factor and has thermal properties such as TC = 2.6 W m^{-1} K^{-1} and CTE = 6 ppm/°C. Butyl rubber is a synthetic elastomer with excellent dielectric properties in the microwave frequencies ($\varepsilon_r = 2.4$, tan $\delta \approx 10^{-3}$), good mechanical flexibility, aging resistance, and weathering resistance [28].

2 Preparation of Butyl Rubber (BR)–Ceramic Composites

The raw materials of alumina, silica, and rutile were procured from Sigma-Aldrich. The $SrCO_3$, $BaCO_3$, and ZnO from Aldrich, Ta_2O_5 from Treibacher Industries, and CeO_2 from Indian Rare Earths Ltd. were used for preparing $Ba(Zn_{1/3}Ta_{2/3})O_3$ (BZT), $Sr_2Ce_2Ti_5O_{15}$ (SCT), $SrTiO_3$, $BaTiO_3$, and $Ba_{0.7}Sr_{0.3}TiO_3$ by conventional solid-state ceramic route. The stoichiometric amount of chemicals were mixed by ball milling and dried. The mixtures were then calcined followed by sintering at appropriate temperatures and ground well. The powders were sieved through a 25-μm sieve and then used for preparing butyl rubber–ceramic composites [15–22]. The nanoalumina (<50 nm particle size), nanorutile (100 nm), and nano-$BaTiO_3$ (100 nm) powders were purchased from Sigma-Aldrich, USA, and were used for preparing butyl rubber–nano composites in the present work. The nanopowders were dried at 100 °C for 24 h before use.

The butyl rubber used for present investigation was IIR grade. The process involving incorporation of ingredients such as activators, accelerators, vulcanizing agent, and fillers into the virgin rubber is known as compounding of rubber [15–22]. The compounding of butyl rubber mix was done by sigma blend method using a kneading machine. The butyl rubber was first masticated through the two counter rotating sigma blades in order to make it soft and more processable. Then, the additives are incorporated one by one as per the order given in Table 2. The additives such as zinc oxide and stearic acid act as activators for vulcanization, tetramethylthiuram disulfide acts as accelerator, and sulfur acts as vulcanizing agent. Finally, appropriate amount of ceramic filler was added. The mixing was done for about 30 min to get uniform composites. Thus, obtained composites were hot pressed at 200 °C for 90 min in suitable dies or molds under a pressure of 2 MPa. After hot pressing, the composites with desired shapes were used for characterization. The physical properties of butyl rubber are given in Table 3. In this review, we describe two representative materials, i.e., butyl rubber composites with alumina and barium titanate. Al_2O_3 is a very low-loss dielectric ceramic, whereas $BaTiO_3$ is ferroelectric ceramic with high relative permittivity and loss factor. Figure 2 shows the photograph of a typical butyl rubber–ceramic composite.

Since alumina is a widely used ceramic packaging material, the effect of filler particle size on dielectric, thermal, and mechanical properties of butyl

Table 2 Formulation of rubber mix [29]

Ingredient	Loading (phr)#
Butyl rubber	100.0
Zinc oxide	5.0
stearic acid	3.0
Tetramethylthiuram disulfide	1.0
Sulfur	0.5

parts per hundred

Table 3 Physical and electrical properties of butyl rubber [22]

Density	0.97 gcm^{-3}
Moisture absorption	0.039 vol.%
Relative permittivity at 5 GHz	2.4
Loss tangent at 5 GHz	$\approx 10^{-3}$
Thermal conductivity	0.13 $W\ m^{-1}\ K^{-1}$
Coefficient of thermal expansion	191 ppm/°C

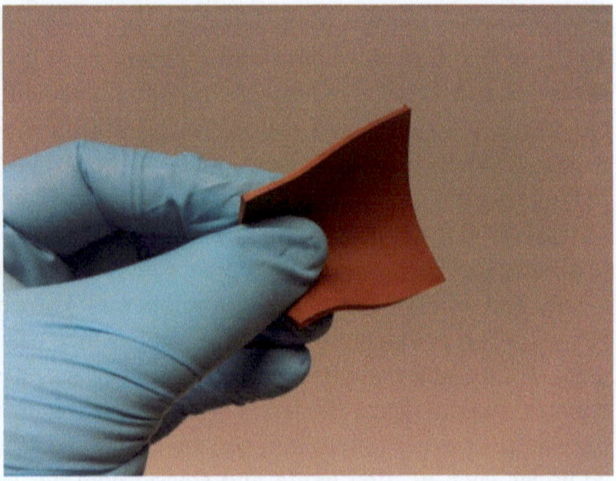

Fig. 2 Photograph of butyl rubber–strontium cerium titanate ceramic composite

rubber–ceramic composites was studied in butyl rubber–alumina composites. The alumina (micron and nano) powders were dried at 100 °C for 24 h before using for composite preparation. The butyl rubber–micron alumina (BR/AL) and butyl rubber–nanoalumina (BR/nAL) composites were prepared as described earlier. The BR/AL composites were prepared with micron alumina loading up to a volume fraction (v_f) 0.42. The nanoalumina has high surface area and large ceramic volume. Hence, the possibility of agglomeration of nanoparticles increases with filler loading. This makes the processing of BR/nAL composites difficult at higher filler loading [30]. Hence, a maximum loading of 0.1 volume fraction of nanoalumina was possible in the case of BR/nAL composites. It was found that the use of coupling agent such as silane treatment although slightly improve the relative permittivity and mechanical properties degrades the microwave dielectric loss factor [22].

3 Dielectric Properties

The dielectric properties of the composites were measured in the radio frequency range by the parallel plate capacitor method employing a LCR meter and using thin pellets of about 1 mm thickness and diameter 11 mm. The dielectric

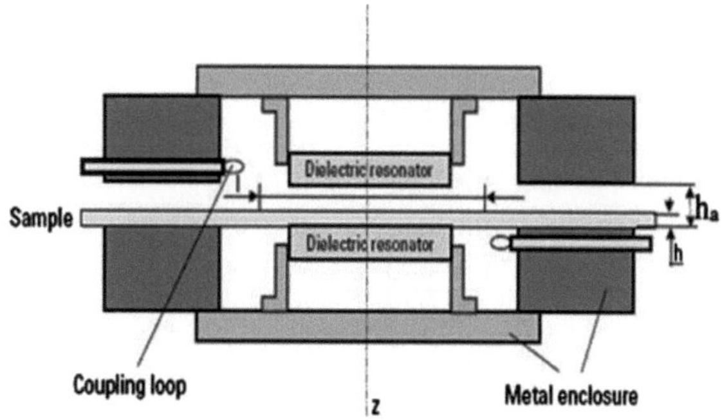

Fig. 3 Schematic representation of SPDR for microwave measurement [23]

properties at microwave frequency range were measured by the split-post dielectric resonator (SPDR) method. The SPDR is an accurate method for measuring the dielectric properties of substrates and thin films at a single frequency in the frequency range of 1–20 GHz [31–33]. In this method, the sample should be in the form of a flat rectangular piece or a sheet. The SPDR uses a particular resonant mode which has a specific resonant frequency depending on the resonator dimensions and the relative permittivity. The resonator mainly consists of two dielectric disks in a metal enclosure. The sample under test is placed in the gap between the two parts of the resonator. Figure 3 schematically shows SPDR fixture for the measurement of the relative permittivity of dielectric and loss tangent of sheet samples. SPDR usually operates with the $TE_{01\delta}$ mode. The relative permittivity of the sample is an iterative solution to the following equation [31]

$$\varepsilon_r = 1 + \frac{f_o - f_s}{hf_o K_\varepsilon(\varepsilon_r, h)} \tag{1}$$

where h is the thickness of the sample under test, f_o is the resonant frequency of empty resonant fixture, f_s is the resonant frequency of the resonant fixture with dielectric sample, and K_ε is a function of ε_r and h and has been evaluated using Rayleigh–Ritz technique [32]. The loss tangent of the sample can be determined by

$$\tan\delta = \left(Q_u^{-1} - Q_{DR}^{-1} - Q_C^{-1}\right)/\rho_{es} \tag{2}$$

In Eqs. 1 and 2

$$p_{es} = h\varepsilon_r K_1(\varepsilon_r, h) \tag{3}$$

$$Q_c = Q_{c0} K_2(\varepsilon_r, h) \tag{4}$$

$$Q_{DR} = Q_{DR0}\frac{f_o}{f_s}\cdot\frac{p_{eDR0}}{p_{eDR}} \tag{5}$$

where p_{es} and p_{eDR} are the electric energy filling factors for the sample and for the split resonator, respectively; p_{eDR0} is the electric energy filling factor of the

Table 4 Dielectric properties at 1 MHz and water absorption of BR/AL and BR/nAL composites [22]

Composite material	Sample designation	Filler in phr# ()$	ε_r (1 MHz)	tan δ (1 MHz)	Water absorption (vol.%)
Butyl rubber–micron alumina composite	BR-0	0 (0.00)	2.44	0.0003	0.039
	BR/AL-1	10 (0.02)	2.48	0.0011	0.045
	BR/AL-2	20 (0.04)	2.54	0.0012	0.045
	BR/AL-3	30 (0.06)	2.65	0.0021	0.056
	BR/AL-4	40 (0.08)	2.73	0.0032	0.057
	BR/AL-5	50 (0.10)	2.78	0.0046	0.065
	BR/AL-6	100 (0.26)	3.01	0.0051	0.067
	BR/AL-7	200 (0.33)	3.80	0.0063	0.071
	BR/AL-8	300 (0.42)	4.61	0.0083	0.082
Butyl rubber–nanoalumina composite	BR/nAL-1	10 (0.02)	2.49	0.0090	0.150
	BR/nAL-2	20 (0.04)	2.55	0.0110	0.220
	BR/nAL-3	30 (0.06)	2.67	0.0160	0.490
	BR/nAL-4	40 (0.08)	2.79	0.0270	0.620
	BR/nAL-5	50 (0.10)	2.81	0.0470	0.700

#Parts per hundred rubber
$The corresponding ceramic volume fraction is given in parenthesis

dielectric split resonator for empty resonant fixture; Q_{c0} is the quality factor depending on metal enclosure losses for empty resonant fixture; Q_{DR0} is the quality factor depending on dielectric losses in dielectric resonators for empty resonant fixture; and Q_u is the unloaded quality factor of the resonant fixture containing the dielectric sample. The values of p_{eDR}, p_{es}, and Q_c for a given resonant structure can be calculated using numerical techniques.

The sample should be flat and must be positioned such that it extends beyond the diameter of two cavity sections. The position of the sample in z-direction is not sensitive to the measurement results. This provides the accuracy of a resonator technique without machining the sample. The uncertainty of the permittivity measurements of a sample of thickness h is $\Delta\varepsilon/\varepsilon = \pm(0.0015 + \Delta h/h)$, and uncertainty in loss tangent measurement is $\Delta(\tan\,\delta) = 2 \times 10^{-5}$. The sample dimensions required for SPDR operating at 5.155 GHz are $50 \times 50 \times 1.8\ \text{mm}^3$.

Table 4 gives the dielectric properties measured at 1 MHz and moisture absorption of BR/AL and BR/nAL composites. The relative permittivity of both composites increases with filler loading. This is due to the high relative permittivity of alumina compared to butyl rubber matrix. From Table 4, it is clear that the composite with nanoalumina has higher relative permittivity than that with micron alumina. The composites filled with nanoparticles have large interfacial area for the same filler loading which promotes interfacial polarization mechanism leading to higher relative permittivity of BR–nanoalumina composites. The loss tangent is the main factor affecting the frequency selectivity of a material and is influenced by many factors such as porosity, microstructure, and defects [23]. The loss tangent of both composites shows same trend as that of relative permittivity.

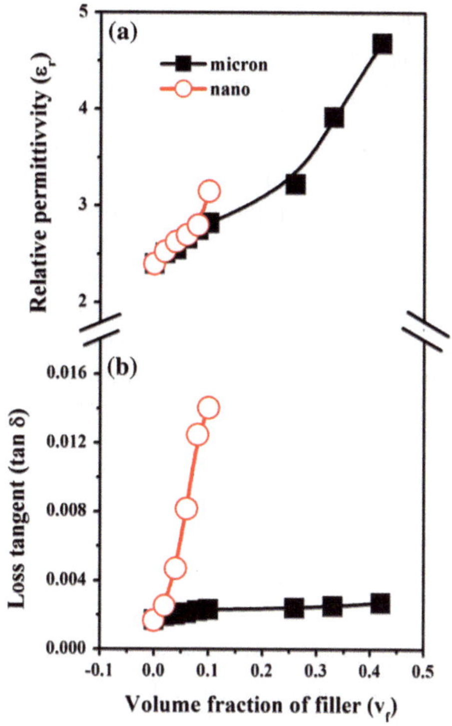

Fig. 4 Variation of **a** relative permittivity and **b** loss tangent of BR/AL and BR/nAL composites at 5 GHz [22]

Figure 4a, b [22] shows the variation of relative permittivity and loss tangent of BR–nanoalumina and BR–micron alumina composites at 5 GHz. As the relative permittivity of alumina is higher than that of butyl rubber, the ε_r of BR/AL composite and BR/nAL composite show an increasing trend with filler content. The nanoalumina-filled butyl rubber composites have higher relative permittivity than micron composites. Similar behavior was observed in PTFE–silica, PTFE–rutile, and PEEK–$SrTiO_3$ composites [34–36]. It was reported that the nanofiller has more polarization at interface region due to its high surface area and also due to its large interface region between the filler and matrix. The nanocomposites also have higher moisture content due to its large surface area. These factors will contribute to the higher relative permittivity of nanocomposites. The relative permittivity of BR/AL composite increases from 2.40 to 4.68 as the micron alumina content increases from 0 to 0.42 v_f and that of BR/nAL composite from 2.40 to 3.15 when the nanoalumina loading increases from 0 to 0.1 v_f. It is also noted that the loss tangent increases with the increase in filler content of both micron and nanosized alumina and a higher loss tangent is exhibited by the nanoalumina-filled composites. The main reason for the higher loss tangent of nanocomposite is due to its high moisture content. A similar observation was reported in PTFE/rutile, epoxy/SiO_2, and PTFE/alumina composites [35, 37, 40]. The loss tangent of BR–micron alumina composite increases from 0.0017 to 0.0027

Table 5 Dielectric properties at 1 MHz and water absorption of BR/BT and BR/nBT composites [20]

Composite material	Sample designation	Filler in phr# ()$	ε_r (1 MHz)	tan δ (1 MHz)	Water absorption (vol.%)
Butyl rubber–micron $BaTiO_3$ composites	BR-0	0 (0.00)	2.44	0.0003	0.039
	BR/BT-1	10 (0.02)	2.56	0.0009	0.040
	BR/BT-2	25 (0.04)	2.74	0.0011	0.045
	BR/BT-3	50 (0.07)	3.14	0.0015	0.046
	BR/BT-4	100 (0.13)	3.88	0.0016	0.048
	BR/BT-5	200 (0.24)	6.06	0.0020	0.057
	BR/BT-6	300 (0.32)	7.94	0.0021	0.068
	BR/BT-7	400 (0.38)	9.18	0.0023	0.120
Butyl rubber–nano $BaTiO_3$ composites	BR/nBT-1	10 (0.02)	2.63	0.0022	0.048
	BR/nBT-2	25 (0.04)	2.86	0.0033	0.052
	BR/nBT-3	50 (0.07)	3.37	0.0077	0.059
	BR/nBT-4	100 (0.13)	4.39	0.0080	0.078
	BR/nBT-5	200 (0.24)	7.02	0.0110	0.091

#Parts per hundred rubber
$The corresponding ceramic volume fraction is given in parenthesis

as the ceramic loading increases from 0 to 0.42 v_f and that of BR–nanoalumina composite increases from 0.0017 to 0.0140 as the nanoalumina content increases from 0 to 0.1 v_f at 5 GHz.

The dielectric properties at 1 MHz and water absorption values of both butyl rubber–micron $BaTiO_3$ and butyl rubber–nano $BaTiO_3$ composites are given in Table 5. The relative permittivity and loss tangent of both BR/BT and BR/nBT composites show an increasing trend with ceramic content since the relative permittivity and loss tangent of $BaTiO_3$ are higher than those of butyl rubber matrix. The relative permittivity and loss tangent of nano-based composites are higher than those of the micron-based composites for the same volume fraction of the filler.

Figure 5a shows the variation of relative permittivity of BR/BT and BR/nBT composites with ceramic content at 5 GHz. The ε_r of both composites shows the same trend as at 1 MHz. The nano-$BaTiO_3$-filled butyl rubber composite shows higher relative permittivity than micron-based composites due to the presence of higher moisture content in the nanocomposite since the relative permittivity of water is high (ε_r ~ 80). The interface region between the filler and matrix is large in the case of nanocomposites which can also contribute to higher relative permittivity [34]. The relative permittivity of BR/BT composite is 7.03 for ceramic loading of 0.24 v_f and that of BR/nBT composite is 8.79 for the same loading of nano-$BaTiO_3$. Although ε_r of $BaTiO_3$ is much higher than that of matrix, the composite cannot attain higher relative permittivity. The relative permittivity of polymer–ferroelectric ceramic composite cannot increase to a very high value even at maximum filler loading [38] since the 0–3 type composite follows an exponential relationship between relative permittivity of composite and the volume fraction

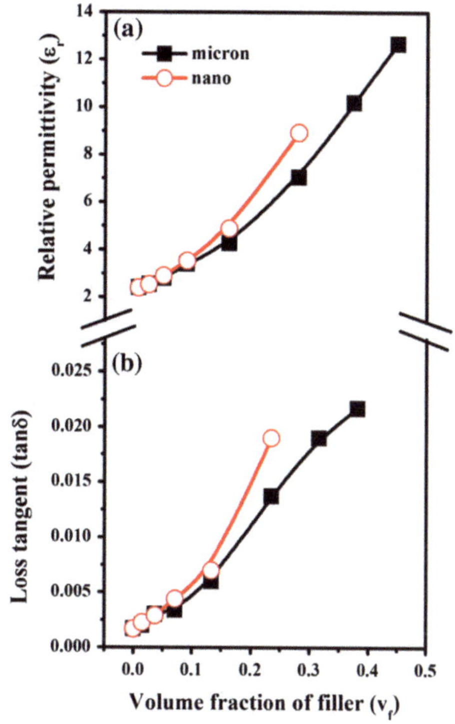

Fig. 5 Variation of **a** relative permittivity and **b** loss tangent of BR/BT and BR/nBT composites at 5 GHz [20]

of the filler. Logarithm of the relative permittivity of such composites ($\varepsilon'_{composite}$) is linearly proportional to the volume fraction of the filler (ϕ_{filler}) with the slope dependent on the dielectric properties of both components:

$$\log \varepsilon'_{composite} = \phi_{filler} \log\left(\varepsilon'_{filler}/\varepsilon'_{polymer}\right) + \log \varepsilon'_{polymer} \tag{6}$$

In general, it is difficult to prepare polymer composite with ceramic loading higher than 0.40 v_f. In order to get a polymer composite with relative permittivity higher than 100 based on typical polymers with relative permittivity of the order of 5, the relative permittivity of the filler must be higher than 9000 [38]. Xie et al. [39] prepared core–shell-structured polymethylmethacrylate–$BaTiO_3$ nanocomposites by in situ atom transfer radical polymerization of methyl methacrylate from the surface of $BaTiO_3$ nanoparticles. For 76.88 wt% of $BaTiO_3$ content, the composite achieved ε_r of 14.6 and tan δ of 0.00372 at 1 kHz [39]. The addition of ceramic fillers gradually increased the relative permittivity. The increase in the relative permittivity depends on the amount of ceramic filler, its relative permittivity, and the frequency of measurement.

The variation of loss tangent of both BR/BT and BR/nBT composites with filler loading is also shown in Fig. 5b. The loss tangent of both composites increases with filler volume fraction since the loss tangent of $BaTiO_3$ is higher than that of butyl rubber. The tan δ of butyl rubber–nanobarium titanate composite is higher than that of butyl rubber–micron barium titanate composites. This is due to the presence of higher moisture content in the nanocomposites, larger surface area, and reactivity of nanoparticles. The presence of large amount of grain boundaries and structural defects like lattice strain can also increase the loss tangent of nanocomposites [40]. The tan δ of micron composite is 0.0140 for a filler loading of 0.24 v_f and that of nanocomposite is 0.0190 for the same filler loading.

3.1 Theoretical Modeling of Relative Permittivity

The prediction of effective relative permittivity of polymer–ceramic composites is very important for engineering applications. In order to understand the physical mechanisms controlling the relative permittivity of a heterogeneous system, the experimental values of ε_r were compared with values predicted using different theoretical models. Several numerical relations have been proposed to predict the relative permittivity of polymer–ceramic composites [41–47]. These models can also be used to determine the extent of filler dispersion and filler/matrix compatibility by comparing the predicted values with the experimental data. The relative permittivity of the composites is influenced not only by the relative permittivity of the individual components but also by other factors such as morphology, porosity, the presence of moisture, impurities, agglomerates, mechanical and chemical coupling between rubber and filler, and dispersion and interaction between the two phases [21, 37, 39, 47–51], and hence, the prediction of relative permittivity of composites is very difficult. The following equations are used to predict the relative permittivity of the rubber–ceramic composites theoretically:

(a) **Lichtenecker's equation**:

The most widely used relation for the prediction of ε_r is Lichtenecker's logarithmic law of mixing. It considers the composite system as randomly oriented spheroids that are uniformly distributed in a continuous matrix [44].

$$\ln \varepsilon_{\mathrm{eff}} = (1 - v_{\mathrm{f}}) \ln \varepsilon_{\mathrm{m}} + v_{\mathrm{f}} \ln \varepsilon_{\mathrm{f}} \tag{7}$$

where $\varepsilon_{\mathrm{eff}}$, ε_{f}, and ε_{m} are the relative permittivity of the composites, filler, and matrix, respectively, and v_{f} is the volume fraction of the filler.

(b) **Maxwell-Garnett equation**:

The Maxwell-Garnett mixing rule was initially used to calculate the effective permittivity of a system where metal particles are encapsulated in an insulating matrix. This mixing rule was modified for polymer–ceramic composites incorporating homogeneous distribution of spherical ceramic particles, and the excitation of dipolar character is considered [45].

$$\frac{\varepsilon_{\mathrm{eff}} - \varepsilon_{\mathrm{m}}}{\varepsilon_{\mathrm{eff}} + 2\varepsilon_{\mathrm{m}}} = v_{\mathrm{f}} \frac{\varepsilon_{\mathrm{f}} - \varepsilon_{\mathrm{m}}}{\varepsilon_{\mathrm{f}} + 2\varepsilon_{\mathrm{m}}} \tag{8}$$

(c) Jayasundere Smith equation:

$$\varepsilon_{\mathrm{eff}} = \frac{\varepsilon_{\mathrm{m}}(1 - v_{\mathrm{f}}) + \varepsilon_{\mathrm{f}} v_{\mathrm{f}}\left[\frac{3\varepsilon_{\mathrm{m}}}{\varepsilon_{\mathrm{f}}+2\varepsilon_{\mathrm{m}}}\right]\left[1 + \left(\frac{3v_{\mathrm{f}}(\varepsilon_{\mathrm{f}}-\varepsilon_{\mathrm{m}})}{\varepsilon_{\mathrm{f}}+2\varepsilon_{\mathrm{m}}}\right)\right]}{1 - v_{\mathrm{f}} + v_{\mathrm{f}}\left[\frac{3\varepsilon_{\mathrm{m}}}{\varepsilon_{\mathrm{f}}+2\varepsilon_{\mathrm{m}}}\right]\left[1 + \left(\frac{3v_{\mathrm{f}}(\varepsilon_{\mathrm{f}}-\varepsilon_{\mathrm{m}})}{\varepsilon_{\mathrm{f}}+2\varepsilon_{\mathrm{m}}}\right)\right]} \tag{9}$$

Jayasundere–Smith equation is a modification of well-known Kerner equation by incorporating the interactions between neighboring spheres. This equation considers composite as a biphase system of dielectric spheres (ε_{f}) dispersed in a continuous medium (ε_{m}) and is valid only when $\varepsilon_{\mathrm{f}} \gg \varepsilon_{\mathrm{m}}$ [46].

(d) **Effective Medium Theory (EMT)**:

In EMT model, composites are treated as an effective medium whose relative permittivity is obtained by averaging over the relative permittivity of the constituents. The basic concept of EMT model is that when a random unit cell (RUC) is embedded in an effective medium, it cannot be detected in the electromagnetic experiment. A RUC is defined as a core of ceramic surrounded by a concentric shell of the polymer [47]. A correction factor '*n*' is used to compensate for the shape of the fillers and is called morphology factor which is related to ceramic particle and can be obtained empirically.

$$\varepsilon_{\mathrm{eff}} = \varepsilon_{\mathrm{m}}\left[1 + \frac{v_{\mathrm{f}}(\varepsilon_{\mathrm{f}} - \varepsilon_{\mathrm{m}})}{\varepsilon_{\mathrm{m}} + n(1 - v_{\mathrm{f}})(\varepsilon_{\mathrm{f}} - \varepsilon_{\mathrm{m}})}\right] \tag{10}$$

where n is the shape factor in EMT model.

The modeling techniques give relative permittivity of a composite system in terms of the relative permittivity of all the constituent components and their volume fractions. Figure 6 compares the experimental and theoretical relative permittivities of BR/AL and BR/nAL composites at 5 GHz. All the models are matching

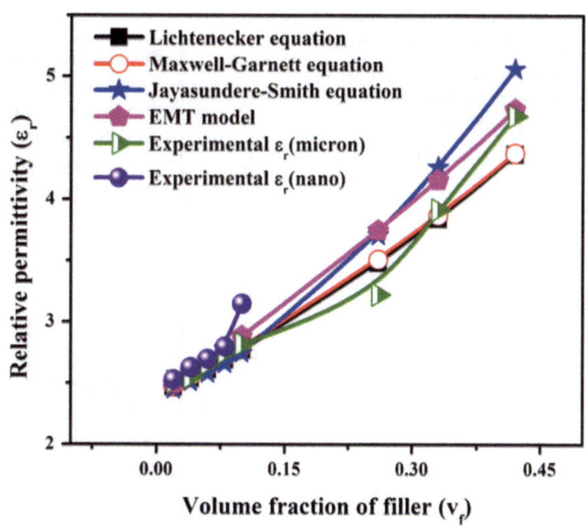

Fig. 6 Comparison of theoretical and experimental relative permittivity of BR/AL and BR/nAL composites at 5 GHz [19]

with the experimental values of ε_r of both BR/AL and BR/nAL composites up to 0.1 v_f. In Maxwell-Garnett model, only the excitation of dipolar character is considered to be important and the correlations between these excitations are not taken into account. The multipolar contributions to the local field are also neglected. It may be noted that these assumptions are valid only in dilute systems [46]. Hence, the Maxwell-Garnett model shows deviation at higher filler content. The most widely used relation for the prediction of ε_r is Lichtenecker's logarithmic law of mixing [44]. The deviation of experimental values of ε_r of both composites from Lichtenecker's equation at higher filler loadings may be due to the lack of consideration of interfacial interaction between the polymer and the filler particles. The EMT model is also in agreement with experimental values of ε_r of both composites at lower filler loadings and shows deviation at higher filler content. The experimental values of ε_r of both composites are matching with those values calculated from Jayasundere–Smith equation and show deviation at higher filler loadings. It is found that EMT and Maxwell-Garnett models are matching more with experimental results. The deviation of all theoretical models at higher filler loading is due to the imperfect dispersion of filler particles in the butyl rubber matrix. The effective permittivity of a composite depends on the various factors such as relative permittivity of individual components in the system, their volume fractions, shape, size, porosity, interphase polarizability, and interphase volume fractions. All these parameters cannot be accounted in a single equation. Hence, the experimental results show deviation from the theoretical values at higher filler content.

Figure 7 compares the experimental and theoretical relative permittivities at 5 GHz for both butyl rubber–micron $BaTiO_3$ and butyl rubber–nano-$BaTiO_3$ composites. The experimental values of ε_r of both BR/BT and BR/nBT composites are in agreement with Lichtenecker's equation and EMT model at low filler loadings. The ε_r of BR/BT composites shows deviation from both models above a filler loading of 0.24 v_f of $BaTiO_3$ content. The relative permittivity of BR/nBT

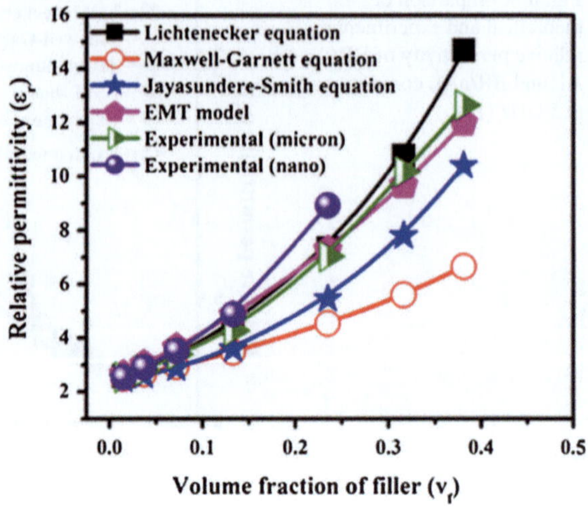

Fig. 7 Comparison between experimental and theoretical relative permittivity of BR/BT and BR/nBT composites at 5 GHz [20]

composites is matching more with Lichtenecker's equation and EMT models. The deviation at higher filler loading may be due to the inhomogenous dispersion of nanofiller particles in the rubber matrix.

3.2 Temperature Coefficient of Relative Permittivity ($\tau_{\varepsilon r}$)

The relative permittivity of a material should be stable within the operational temperature range of electronic devices for practical applications. The temperature coefficient of relative permittivity of the present composites was measured by parallel plate capacitor method. The sample is kept in a chamber, and ε_r is measured from 25 to 75 °C.

Figure 8a, b shows the temperature dependence of relative permittivity of BR/nAL and BR/AL composites, respectively. From the figure, it is clear that all

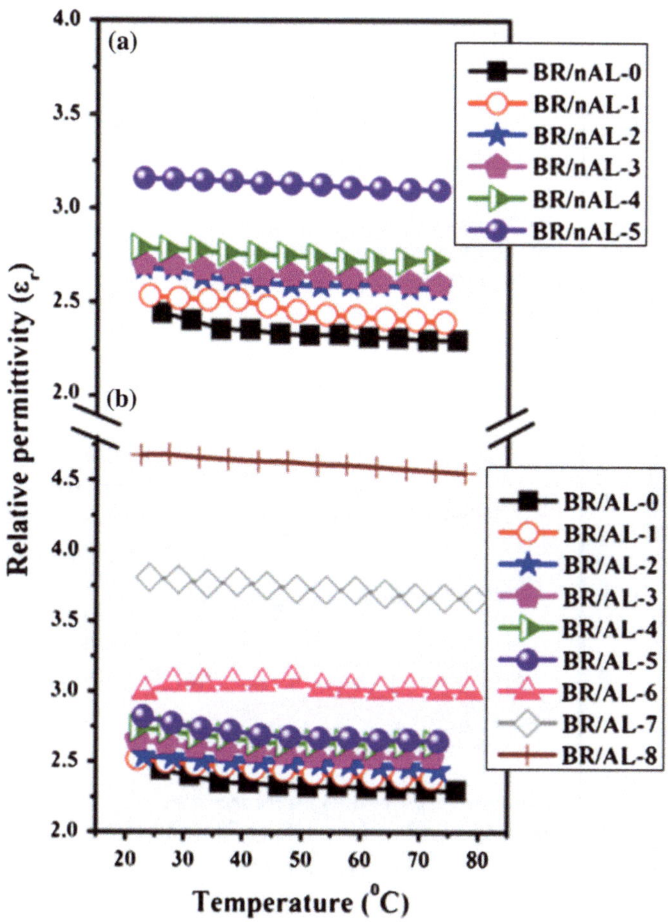

Fig. 8 Variation of relative permittivity of **a** BR/nAL and **b** BR/AL composites with temperature [22]

the composites are almost thermally stable within the measured temperature range. Figure 9a, b shows the temperature variation of relative permittivity of both BR/nBT and BR/BT composites at 1 MHz, respectively. The relative permittivity of all the composites of butyl rubber–micron $BaTiO_3$ and butyl rubber–nano-$BaTiO_3$ is also nearly stable in the measured temperature range. From the Figs. 8 and 9, it is clear the relative permittivity of composites shows a slight decrease with increase in temperature. The small decrease in relative permittivity with temperature may be due to the large difference in thermal expansion coefficient of butyl rubber and the ceramics [52, 53]. The thermal stability of relative permittivity is one of the important properties of the substrate materials that control the overall performance of the device.

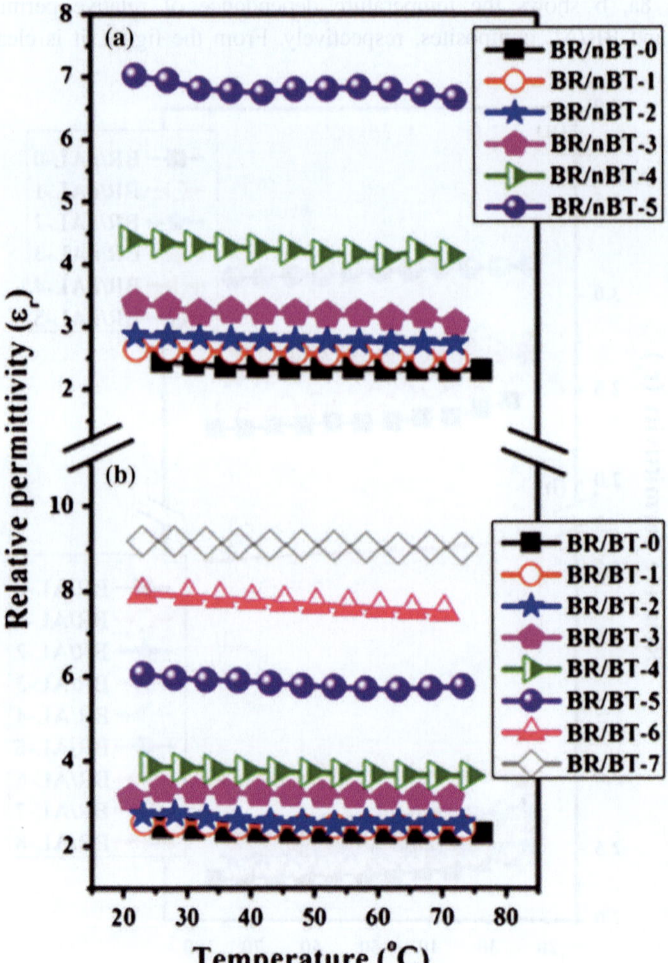

Fig. 9 Variation of relative permittivity of **a** BR/nBT and **b** BR/BT composites with temperature at 1 MHz [20]

3.3 Bending

The bending effect on dielectric properties is important as far as flexible electronic applications are concerned. Bending measurements of the composites were carried out by bending the samples manually in such a way that every part of the sample had undergone 180° bending. The bending cycle was repeated for 125 times, and the microwave dielectric properties were measured by SPDR method after every 25 cycles. Figures 10 and 11 show the variation of dielectric properties of BR/AL and BR/nAL composites with bending. It is clear from Fig. 11a, b that the relative permittivity of all the composites is nearly independent of bending. It is evident from Fig. 10a, b that the loss tangent of the nanocomposites is almost independent of bending and that of micron composite shows slight variation with bending. Figures 12 and 13 show the effect of bending on dielectric properties

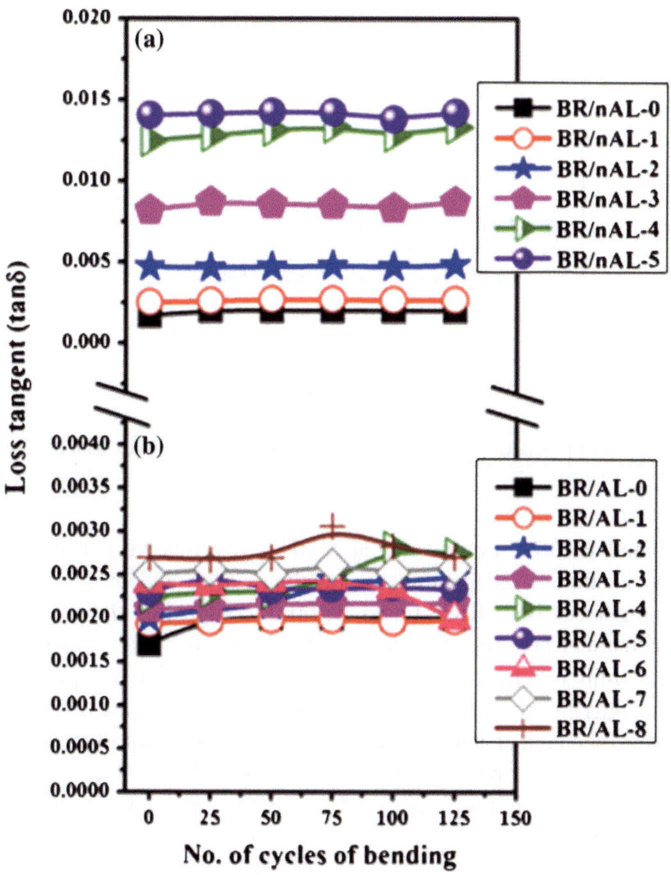

Fig. 10 Variation of relative permittivity of **a** BR/nAL and **b** BR/AL composites with bending [22]

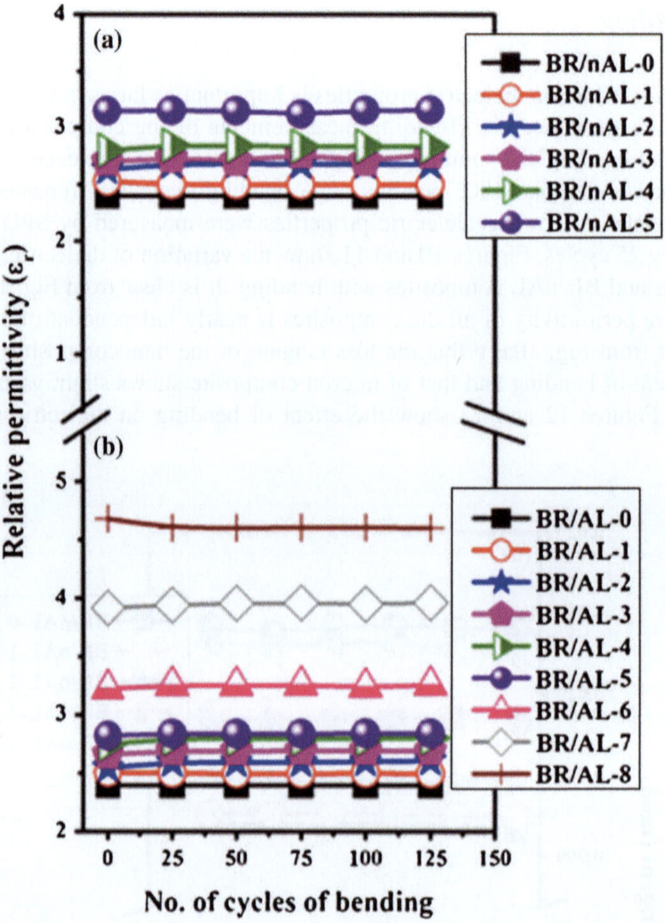

Fig. 11 Variation of loss tangent of **a** BR/nAL and **b** BR/AL composites with bending [22]

of BR/BT and BR/nBT composites at 5 GHz. The relative permittivity of both composites is almost independent of bending as is evident from Fig. 13a, b. The loss tangent of both BR/BT and BR/nBT composites shows a slight variation in mechanical bending, but the variations are marginal. Vrejoiu et al. [54] observed a similar behavior in PFCB (perfluorocyclobutene (poly 1,1,1-triphenyl ethane per-fluorocyclobutyl ether)–$BaTiO_3$ composites. They observed that repeated mechanical bending up to 50 cycles does not change ε_r and tan δ of the composites. The bending of the composites does not considerably affect the microwave dielectric properties, and hence, they are suitable for flexible electronic applications.

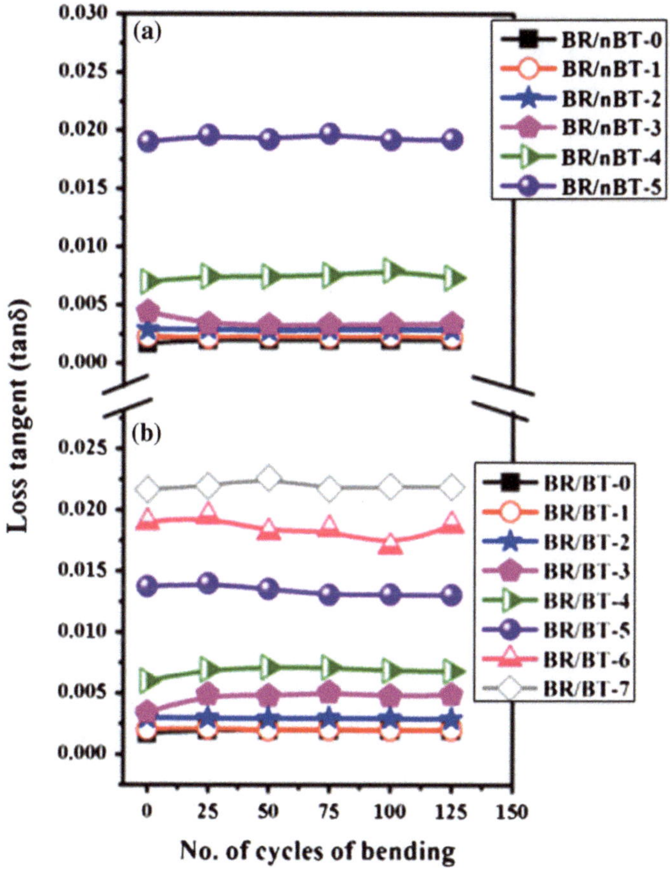

Fig. 12 Variation of relative permittivity of **a** BR/nBT and **b** BR/BT composites with bending [20]

4 Thermal Conductivity

Heat dissipation from integrated circuits is a crucial problem for electronic industry that affects potential miniaturization, speed, and reliability. The thermal conductivity (TC) of polymers is very low, ranging from 0.14 to 0.60 W m^{-1} K^{-1} [55], and can be improved by the addition of ceramic fillers as the TC of ceramic is higher than that of polymers. The TC of the composites depends on the intrinsic thermal conductivities of filler and matrix, shape and size of the filler, and the loading level of filler [56]. The variation of TC of both BR/AL and BR/nAL composites with filler content is shown in Fig. 14. The TC of both composites shows increasing trend with ceramic loading. This is quite expected since the TC of alumina (30 W m^{-1} K^{-1}) is higher than that of butyl

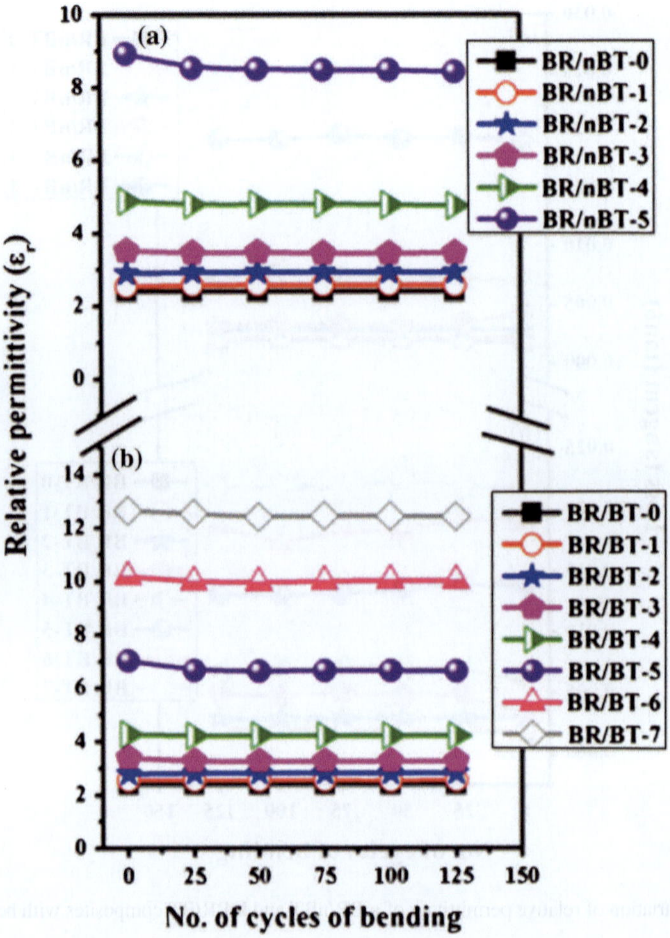

Fig. 13 Variation of loss tangent of **a** BR/nBT and **b** BR/BT composites with bending [20]

rubber matrix (0.13 W m^{-1} K^{-1}). From the inset plot in Fig. 14 [22], it is clear that the TC of nanoalumina-loaded composites is slightly higher than that of micron composite.

Figure 15 shows the variation of TC of BR/BT and BR/nBT composites with ceramic loading. As the ceramic loading increases, the TC of both composite increases since the TC of $BaTiO_3$ (2.6 W m^{-1} K^{-1}) is higher than the butyl rubber (0.13 W m^{-1} K^{-1}). As the filler loading increases, distance between the filler particles decreases and thus filler becomes the main channels for thermal conduction. Hence, TC of both composites can also increase with filler content [56]. From Figs. 14 and 15, it is clear that the nanocomposites have higher TC than that of

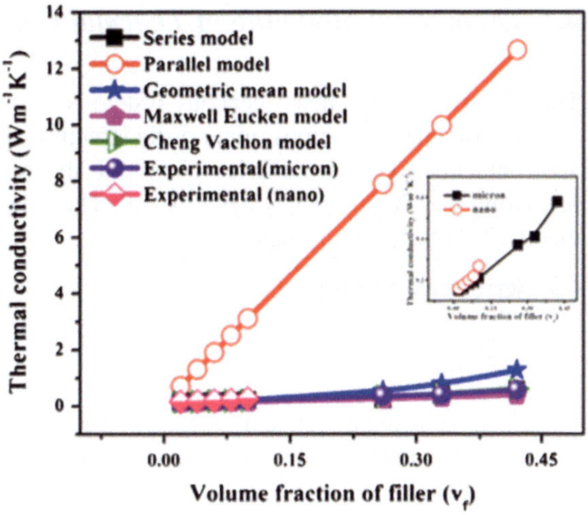

Fig. 14 Variation of thermal conductivity of BR/AL and BR/nAL composites with filler content [22]

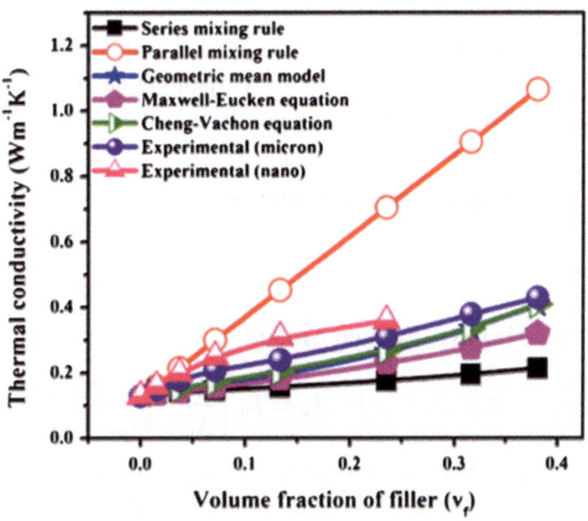

Fig. 15 Variation of thermal conductivity of BR/BT and BR/nBT composites with filler content [20]

micron-based composites. The matrix/filler interface plays a critical role in nanocomposites due to its large surface area. The number of particles increases with decreasing particle size for the same filler content [57, 58]. This will leads to the formation of large number of heat-conductive channels in nanofiller-added composites. Hence, the TC of nano-based composite is higher than micron-based composites. The butyl rubber–micron barium titanate composite with ceramic loading of 0.24 v_f has a TC of 0.31 W m^{-1} K^{-1} and that of butyl rubber–nanobarium titanate composite is 0.36 W m^{-1} K^{-1}.

4.1 Theoretical Modeling of Thermal Conductivity

The effective TC of a heterogeneous system is strongly affected by its TC of individual components, composition, crystal structure, distribution within the medium, and contact between the particles. Numerous theoretical models were proposed for predicting the TC of composites [59, 60]. The following models are used to predict the TC of the composites:

(a) **Series mixing rule**:

$$\frac{1}{k_c} = \frac{v_f}{k_f} + \frac{v_m}{k_m} \tag{11}$$

where k_c is the effective TC of the composite, k_m and k_f are the TC of matrix and filler, respectively, and v_m and v_f are the volume fractions of matrix and filler, respectively.

(b) **Parallel mixing rule**:

$$k_c = v_f k_f + v_m k_m \tag{12}$$

The physical structures assumed in the series and parallel models are of layers of phases aligned either perpendicular or parallel to the heat flow [59]. The series and parallel model of TC gives only lower and upper limits of TC values of composites, respectively.

(c) **Geometric mean model** [60]:

$$k_c = k_f^{v_f} k_m^{1-v_f} \tag{13}$$

(d) **Maxwell-Eucken Model:**

$$k_c = k_m \left[\frac{k_f + 2k_m + 2V_f(k_f - k_m)}{k_f + 2k_m - V_f(k_f - k_m)} \right] \tag{14}$$

The Maxwell model assumes a dispersion of small spheres within a continuous matrix of a different phase, where spheres being far enough apart such that the local distortions to the temperature distributions around each of the spheres do not interfere with their neighbors' temperature distributions [60].

(e) **Cheng–Vachon Model**:
Cheng and Vachon assumed a parabolic distribution of the discontinuous phase in the continuous phase based on Tsao's model [61]. The constants of this parabolic distribution were determined by analysis and presented as a function of the discontinuous phase volume fraction. Thus, the equivalent TC of the two-phase solid mixture was derived in terms of the distribution function and the TC of the constituents.

$$\frac{1}{k_c} = \frac{1}{\sqrt{C(k_f - k_m)[k_m + B(k_f - k_m)]}} \ln \frac{\sqrt{[k_m + B(k_f - k_m)]} + \frac{B}{2}\sqrt{C(k_f - k_m)}}{\sqrt{[k_m + B(k_f - k_m)]} - \frac{B}{2}\sqrt{C(k_f - k_m)}} + \frac{1-B}{k_m} \tag{15}$$

where

$$B = \sqrt{3V_f/2} \quad C = -4\sqrt{2/3V_f} \tag{16}$$

Figure 14 also compares the experimental TC of both BR/AL and BR/nAL composites with theoretical models. The series and parallel model of TC gives only lower and upper limits of TC values of composites, respectively [59]. The experimental TC values of both composites are within the range of series and parallel model. The geometric mean model is in good agreement with experimental TC of both composites up to a volume fraction of 0.1 and has a lower value than predicted TC at higher micron alumina loading. This may be due to the agglomeration of filler particles at higher filler loadings. The models Cheng–Vachon and Maxwell-Eucken are matching with experimental TC of both composites. Figure 15 also shows the comparison between experimental and theoretical TC of $BaTiO_3$-based composites. The series and parallel mixing rule of both composites shows the same trend as that of alumina-based composites and does not match with experimental profiles. All the other theoretical models are matching with TC of BR/nBT composites at low filler loading and show deviation at higher nano-$BaTiO_3$ content. The TC of butyl rubber–micron $BaTiO_3$ composites is in agreement with Cheng–Vachon and geometric mean models. Maxwell-Eucken equation holds good at low filler loading and deviates at higher micron $BaTiO_3$ content [62].

5 Mechanical Properties

The mechanical flexibility is the prime requirement for flexible electronic applications. Tensile test is useful to evaluate mechanical properties of materials in which the sample is pulled to failure in a relatively short period of time. The sample is elongated at a constant rate, and the load required to produce a given elongation is measured as a dependent variable. A stress–strain curve may be plotted from the results of a tension test, and from the plot, the toughness of the material can be assessed [63, 64]. The stress–strain properties of the composites were measured using a universal testing machine (Hounsfield, H5K-S UTM, Redhill, U.K.) with a rate of grip separation of 500 mm/min. Tensile tests were conducted using dumb-bell-shaped samples of width 4 mm and thickness in the range 1.5–2 mm.

Figure 16 shows the stress–strain curve of BR/AL and BR/nAL composites. The mechanical properties of a particulate composite depend on the strength of the adhesive bond between the different phases, the type of dispersion, and the amount of particle agglomeration. From the figure, it is clear that the stress needed for ceramic-filled butyl rubber composite is greater than that of unfilled sample. Among BR-0, BR/AL-5, and BR/nAL-5, the stress required for BR/nAL-5 composite is high. This may be due to more homogeneous dispersion of nanoparticles in the rubber matrix. Chee et al. [65] reported that at lower filler loading, the nanoalumina particle orients along the direction of stress and this would reinforce and increase the stiffness of the

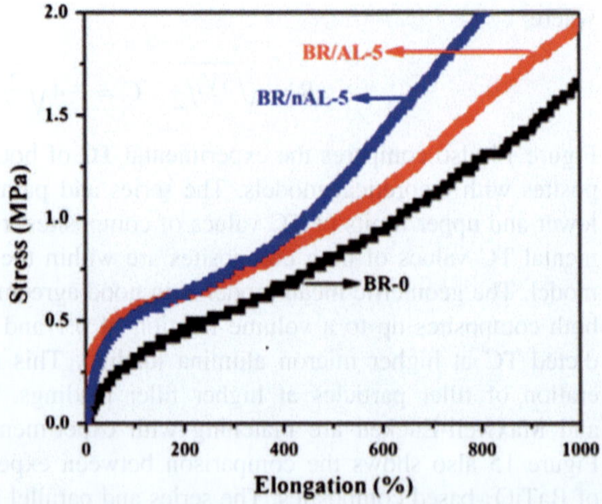

Fig. 16 Stress–strain curves of BR/AL and BR/nAL composites [19]

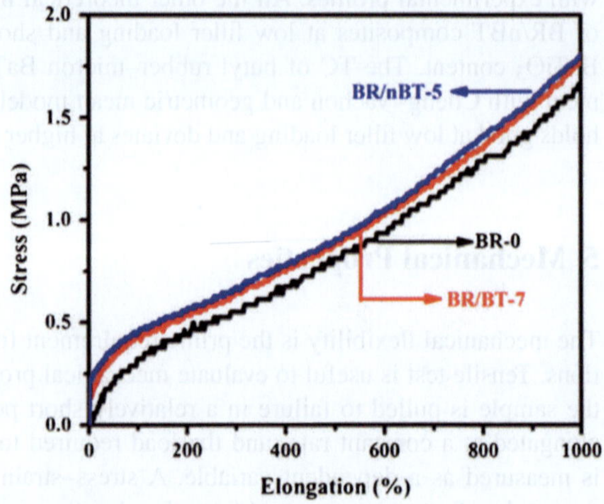

Fig. 17 Stress–strain curves of BR/BT and BR/nBT composites [20]

nanocomposite. Figure 17 shows the stress–strain curves of BR/BT-7 and BR/nBT-5. As the filler loading increases, the stress needed for elongation and also stiffness of the composite increases [66]. This figure shows that the nanocomposite is slightly stiffer than micron composite. Both the composites were not broken up to an elongation of 1000 %, which indicates the excellent mechanical flexibility of the composites.

6 Moisture Absorption of Composites

The moisture content is an important parameter for materials used for practical applications. Absorption of moisture from the working atmosphere will degrade the dielectric properties since water is a polar molecule. The moisture absorption characteristics of the composites were measured using the samples with dimensions 50 mm × 50 mm × 2 mm. The samples were weighed accurately and immersed in distilled water for 24 h. The samples were then taken out and again weighed after removing the excess water from the surface. The volume % of water absorption was then calculated using the relation,

$$\text{Volume \% water absorption} = \frac{(W_f - W_i)/\rho_w}{(W_f - W_i)/\rho_w + W_i/\rho_c} \times 100 \quad (17)$$

where W_i and W_f are the initial and final weights of the sample, and ρ_w and ρ_c are the densities of distilled water and composite, respectively.

It is evident from Tables 4 and 5 that as the filler content increases, the volume % (vol.%) of water content increases for both composites since the ceramic is hydrophilic in nature. Compared to micron composite, nanocomposites have a high tendency to absorb moisture due to the large surface area of the nanoalumina or barium titanate, and also, higher nanofiller loading leads to the formation of pores in the composites due to the agglomeration of the particles. The loss tangent and moisture absorption of butyl rubber–nanoceramic composites are much higher as compared to the micron composites. The dielectric properties of the composites are affected by the presence of moisture content. The absorbed moisture interacts with polymer matrix and also with the filler–matrix interface [67] and thus affects the dielectric properties of composites.

7 Coefficient of Thermal Expansion

Figures 18 and 19 show the variation of coefficient of thermal expansion (CTE) of butyl rubber–alumina and barium titanate composites with filler content. The CTE of both the composites decreases with ceramic loading since

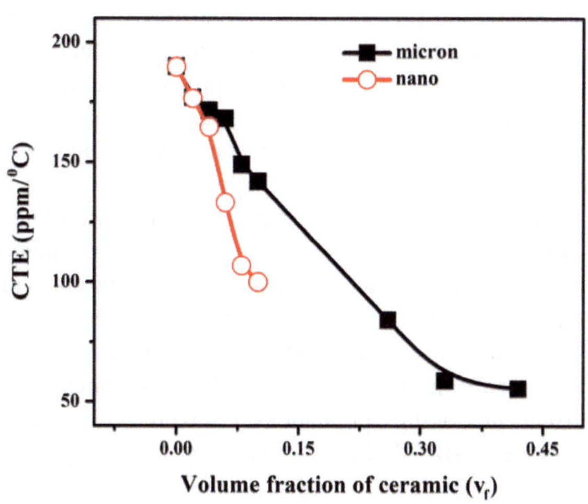

Fig. 18 Variation of CTE of BR/AL and BR/nAL composites with filler content [19]

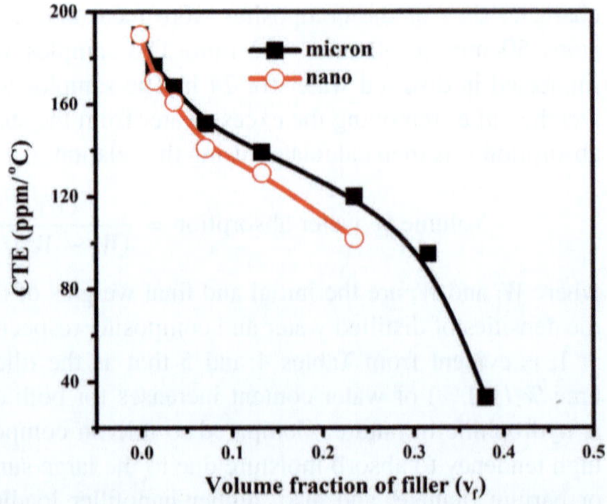

Fig. 19 Variation of CTE of BR/BT and BR/nBT composites [20]

the CTE value of the ceramics is smaller than that of rubber matrix, and also, the mobility of loose molecular bonds in the polymer chains is restrained by the ceramic loading [68, 69]. The BR–nanocomposites show much lower CTE value compared to that of micron composites. The physical cross-linking points are more in the case of nanoparticles because of its high specific surface area, and this will increase the mechanical interaction between the filler and rubber matrix [70]. Hence, the nanocomposites have lower CTE than that of micron composites.

8 Conclusions

The butyl rubber–ceramic composites can be prepared by sigma mixing followed by hot pressing. The composites were found to be flexible, bendable, and stretchable. The microwave dielectric and thermal properties are improved with ceramic loading. The composite preparation is difficult for ceramic loading higher than 40 vol.%. The nanopowder can be loaded to a much lower level. The relative permittivity essentially depends on the rubber matrix, and one can get a relative permittivity of about 15 even with maximum loading with high-permittivity ceramics. The TC of the composites increased and CTE decreased with ceramic content. The water absorption increases with increase in ceramic loading and is higher for nanoceramic loading. The composite with nanoceramic filler has inferior microwave dielectric properties. Table 6 gives the microwave dielectric properties of several butyl rubber–ceramic composites. The BR/AL and BR/BT composites are found to be suitable candidates for microwave substrate and electronic packaging applications. More recently, butyl rubber–single-walled carbon nanotube (BR-SWCNT) composites were prepared by a solution mixing process

Table 6 Microwave dielectric, thermal, and physical properties of several butyl rubber-–ceramic composites [22]

Filler	v_f of filler	ε_r	tan δ	TC (W m^{-1} K^{-1})	CTE (ppm/°C)	Water absorption (vol.%)
SiO_2	0.40	3.09	0.0045	0.56	80	0.09
Al_2O_3	0.40	4.68	0.0027	0.58	55	0.08
$Ba(Zn_{1/3}Ta_{2/3})O_3$	0.32	5.72	0.0025	0.35	86	0.05
TiO_2	0.40	12.5	0.0027	0.72	108	0.08
$Sr_2Ce_2Ti_5O_{15}$	0.43	11.0	0.0018	0.49	30	0.07
$SrTiO_3$	0.42	13.2	0.0028	0.55	26	0.08
$BaTiO_3$	0.38	12.7	0.0200	0.43	33	0.12
$Ba_{0.7}Sr_{0.3}TiO_3$	0.39	13.1	0.0090	0.41	29	0.06

and reported to be useful EMI shielding applications. The BR-SWCNT composite shows enhanced relative permittivity of 14 with a loss tangent of 0.07 at 15 GHz for 8 phr SWCNT [71]. The EMI shielding effect increased with SWCNT content.

References

1. Gargama H, Thakur AK, Chaturvedi SK (2015) Polyvinylidene fluoride/nickel composite materials for charge storing, electromagnetic interference absorption, and shielding applications. J Appl Phys 117: 224903-1-9
2. Sudeep PM, Vinayasree S, Mohanan P, Ajayan PM, Narayanan TN, Anantharaman MR (2015) Fluorinated graphene oxide for enhanced S and X-band microwave absorption. Appl Phys Lett 106:221603-1–221603-5
3. Siegel AC, Phillips ST, Dickey MD, Lu N, Suo Z, Whitesides GM (2010) Foldable printed circuit boards on paper substrates. Adv Funct Mater 20:28–35
4. Kim DH, Rogers JA (2008) Stretchable electronics: materials, strategies and devices. Adv Mater 20:4887–4892
5. Kramer RK, Majidi C, Sahai R, Wood RJ (2011) IEEE/RSJ international conference on intelligent robots and systems, San Francisco, CA, USA, 25–30 September 1919–1926
6. Seol YG, Noh HY, Lee SS, Ahn JH, Lee NE (2008) Mechanically flexible low-leakage multilayer gate dielectrics for flexible organic thin film transistor. Appl Phys Lett 93:013305-1–013305-3
7. Sebastian MT, Jantunen H (2010) Polymer-ceramic composites of 0-3 connectivity for circuits in electronics: a review. Int J Appl Ceram Technol 7:415–434
8. Thomas RM, Sparks WJ (1944) Mixed olefinic polymerization process and product. U.S. patent, 2, 356, 128
9. http://en.wikipedia.org/wiki/Butyl_rubber
10. Ebewele RO (2000) Polymer science and technology. CRC Press, Florida
11. http://en.wikipedia.org/wiki/Elastomer
12. Billmeyer FW (1962) Text book of polymer science. Interscience Publishers, Wiley, New York, London
13. Duffy J, Wilson GJ (1993) Synthesis of butyl rubber by cationic polymerization. In: Ullman's encyclopedia of industrial chemistry, 5th edn. Elsevier, Amsterdam
14. General properties of elastomers, www.elbex-us.com

15. Thomas D, Chameswary J, Sebastian MT (2011) Mechanically flexible butyl rubber-$SrTiO_3$ composites for microwave applications. Int J Appl Ceram Technol 8:1099–1107
16. Chameswary J, Thomas D, Subodh G, Harshan S, Philip J Sebastian MT (2012) Microwave dielectric properties of flexible butyl rubber-strontium cerium titanate composites. J Appl Polym Sci 124:3426–3433
17. Chameswary J, Sebastian MT (2013) Effect of $Ba(Zn_{1/3}Ta_{2/3})O_3$ and SiO_2 ceramic fillers on the microwave dielectric properties of butyl rubber composites. J Mater Sci: Mater Electron 24:4351–4360
18. Chameswary J, Sebastian MT (2013) Butyl rubber-$Ba_{0.7}Sr_{0.3}TiO_3$ composites for flexible microwave electronic applications. Ceram Int 39:2795–2802
19. Chameswary J, Namitha LK, Brahmakumar M, Sebastian MT (2014) Material characterization and microwave substrate applications of alumina filled butyl rubber composites. Int J Appl Ceram Technol 11:919–926
20. Chameswary J, Sebastian MT (2015) Preparation and Properties of $BaTiO_3$ filled butyl rubber composites flexible electronic circuit applications. J Mater Sci: Mater Electron. doi:10.1007/s10854-015-2879-5
21. Chameswary J, Sebastian MT (2014) Development of butyl rubber-rutile composites for flexible microwave substrate applications. Ceram Int 40:7439–7448
22. Janardhanan C (2015) Butyl rubber-ceramic composites for flexible electronic applications. Ph.D thesis, Cochin University of Science & Technology
23. Sebastian MT (2008) Dielectric materials for wireless communication. Elsevier, Amsterdam
24. Mohamad N, Muchtar A, Ghazali MJ, Dahlan HM, Azhari CH (2009) Epoxidised natural rubber—alumina nanoparticle composites (ENRAN): effect of filler loading on the tensile properties. Solid State Sci Technol 17:133–143
25. Mohamad N, Muchtar A, Ghazali MJ, Mohd DH, Azhari CH (2008) The effect of filler on epoxidised natural rubber-alumina nanoparticles composites. Eur J Sci Res 24:538–547
26. Li L, Fang Y, Xiao Q, Wu YJ, Wang N, Chen XM (2014) Microwave dielectric properties of fused silica prepared by different approaches. Int J Appl Ceram Technol 11:193–199
27. Hanna FF, Yehiab AA, About-Bakr AF (1973) Dielectric properties of polar rubber mixtures. Br Polym J 5:49–53
28. Hakim IK, Bishai AM, Saad AI (1988) Dielectric properties of butyl rubber mixtures at 10^6–10^{10} Hz. J Appl Polym Sci 35:1123–1125
29. Barron H (1949) Modern synthetic rubbers. Chapman & Hall Ltd., London
30. Hu T, Juuti J, Jantunen H, Vikkman T (2007) Dielectric properties of BST/polymer composites. J Eur Ceram Soc 27:3997–4001
31. Krupka J, Gregonry AP, Kochard OC, Clarke RN, Riddle B, Baker-Jarvis J (2001) Uncertainity of complex permittivity measurement by split post dielectric resonator techniques'. J Eur Ceram Soc 21:2673–2676
32. Krupka J, Gabelich S, Derzakowski K, Pierce BM (1999) Comparison of split post dielectric resonator and ferrite disc resonator techniques for microwave permittivity measurements of polycrystalline ytterium iron garnet. Meas Sci Technol 10:1004–1008
33. Krupka J (2003) Precise measurement of complex permittivity of dielectric materials at microwave frequencies. Mater Chem Phys 79:195–198
34. Murali KP, Rajesh S, Om Prakash A, Kulkarni R, Ratheesh R (2009) Preparation and properties of Silica filled PTFE Flexible laminates for microwave circuit applications. Compos A 40:1179–1185
35. Rajesh S, Nisa VS, Murali KP, Ratheesh R (2009) Microwave dielectric properties of PTFE/rutile nanocomposites. J Alloy Compd 477:677–682
36. Nisa VS, Rajesh S, Murali KP, Priyadarsini V, Potty SN, Ratheesh R (2008) Preparation, characterization and dielectric properties of temperature stable $SrTiO_3$/PEEK composites for microwave substrate applications. Compos Sci Technol 68:106–112
37. Todd MG, Shi FG (2002) Validation of a novel dielectric constant simulation model and the determination of its physical parameters. Microelectron J 33:627–632

38. Popielarz R, Chiang CK (2007) Polymer composites with dielectric constant comparable to that of barium titanate ceramics. Mater Sci Eng B 139:48–54
39. Xie L, Huang X, Wu C, Jiang P (2011) Core shell structured poly(Polymethyl methalycrate) nanocomposites prepared by insitu atom transfer radical polymerization: a route to high dielectric constant materials with inherent low loss of the base polymer. J Mater Chem 21:5897–5906
40. Murali KP, Rajesh S, Nijesh KJ, Ratheesh R (2010) The effect of particle size on the microwave dielectric properties of alumina filled PTFE substrates. Int J Appl Ceram Technol 7:475–481
41. Sareni B, Krahenubuh L, Beroul A, Brosseau C (1997) Effective dielectric constant of random composite material. J Appl Phys 81:2375–2383
42. Sihvola AH (1988) Effective permittivity of dielectric mixtures. IEEE Trans Geosci Remote Sens 26:420–429
43. Wakino K (1993) A new equation for predicting the dielectric constant of a mixture. J Am Ceram Soc 76:2588–2594
44. Goncharenko AV, Lozovski VZ, Venger EF (2000) Lichtenecker's equation: applicability and limitations. Optics Commun 174:19–32
45. Claro F, Rojas R (1991) Correlation and multipolar effects in the dielectric response of particulate matter: an iterative mean-field theory. Phys Rev B 43:6369–6375
46. Jayasundere N, Smith BV (1993) Dielectric constant for binary piezoelectric 0-3 composites. J Appl Phys 73:2462–2466
47. Rao Y, Qu J, Marinis T, Wong CP (2000) A precise numerical prediction of effective relative permittivity for polymer ceramic composites based on effective medium theory. IEEE Trans Compon Packag Technol 23:680–683
48. Teirikangas M, Juuti J, Hu T, Jantunen H (2009) Extrinsic influences of the polymer matrix on electrical properties of high frequency composites. Ferroelectrics 387:70–76
49. Xiang F, Wang H, Yao X (2006) Preparation and dielectric properties of bismuth based dielectric/PTFE microwave composite. J Eur Ceram Soc 26:1999–2002
50. Murugaraj P, Mainwaring D, Mora-Huertas N (2005) Dielectric enhancement in polymer-nanoparticle composites through interphase polarizability. J Appl Phys 98:054304-1–054304-6
51. Valant M, Suvorov D (2003) Microstructural phenomena in low firing ceramics. Mater Chem Phys 79:104–110
52. Xie SH, Zhu BK, Wei XZ, Xu ZK, Xu YY (2005) Polyimide/$BaTiO_3$ composites with controllable dielectric properties. Compos A Appl Sci Manuf 36:1152–1157
53. Berger MA, Mc Cullough RL (1985) Characterization and analysis of the electrical properties of a metal-filled polymer. Compos Sci Technol 22:81–106
54. Vrejoiu I, Pedarnig JD, Dinescu M, Gogonea SB, Bäuerle D (2002) Flexible ceramic-polymer composite films with temperature-insensitive and tunable dielectric permittivity. Appl Phys A 74:407–409
55. Tavman IH (1998) Effective thermal conductivity of isotropic polymer composites. Int Commun Heat Mass Transfer 25:723–732
56. Kemaloglu S, Ozkoc G, Avtac A (2010) Properties of thermally conductive micro and nano size boron nitride reinforced silicon rubber composites. Thermochim Acta 499:40–47
57. Zhou W, Qi S, Li H, Shao S (2007) Study on insulating thermal conductive BN/HDPE composite. Thermochim Acta 452:36–42
58. Ling W, Gu A, Liang G, Yuan L (2010) New composites with high thermal conductivity and low dielectric constant for microelectronic packaging. Polym Compos 31:307–313
59. Richerson DW (2006) Modern ceramic engineering properties. In: Processing and use in design. Taylor and Francis, CRC Press, London
60. Progelhof RC, Throne JL, Ruetsch RR (1976) Methods for predicting thermal conductivity of composite systems: a review. Polym Eng Sci 16:615–625
61. Tsao GTN (1961) Thermal conductivity of two-phase material. Ind Eng Chem 53:395–397

62. Jaguaribe EF, Beasley DE (1984) Modeling of the effective thermal conductivity and diffusivity of a packed bed with stagnant fluid. Int J Heat Mass Transf 27:399–407
63. Menard KP (1999) Dynamic mechanical analysis: a practical introduction. CRC Press, Boca Raton
64. Hayden HW, Moffatt WG, Wulff J (1984) The structure and properties of materials, vol III. In: Mechanical behaviour, Wiley Eastern Ltd., New Delhi
65. Chee CY, Song NL, Abdullah LC, Choong TSY, Ibrahim A, Chantara TR (2012) Characterization of mechanical properties: low-density polyethylene nanocomposite using nanoalumina particle as filler. J Nanomater 2012(215978):1–6
66. Ismail H, Sam ST, Mohd Noor AF, Bakar AA (2007) Properties of ferrite-filled natural rubber composites. Polym Plast Technol Eng 46:641–650
67. Zhao H, Li RKY (2008) Effect of water absorption on the mechanical and dielectric properties of nano-alumina filled epoxy nanocomposites. Compos A Appl Sci Manuf 39:602–611
68. Li TL, Hsu SLC (2010) Enhanced thermal conductivity of polyimide films via a hybrid of micro- and nano-sized boron nitride. J Phys Chem B 114:6825–6829
69. Goyal RK, Jadhav P, Tiwari AN (2011) Preparation and properties of new polyphenylene sulfide/AlN composites for electronic packaging. J Electron Mater 40:1377–1383
70. Zhou W, Qi S, Tu C, Zhao H, Wang C, Kou J (2007) Effect of the particle size of Al_2O_3 on the properties of filled heat-conductive silicone rubber. J Appl Polym Sci 104:1312–1318
71. Joseph N, Janardhanan C, Sebastian MT (2014) Electromagnetic interference shielding properties of butyl rubber single walled carbon nanotube composites. Compos Sci Technol 101:139–144

Nanomaterials-Embedded Liquid Crystal Elastomers in Electronics Devices Application

Md Mohiuddin and Tran Thanh Tung

Abstract Liquid crystal elastomers (LCEs) are soft, elastic, durable, and lightweight materials that offer different physical phenomena with the presence of external stimuli. Cross-linked natural three-dimensional structures of liquid-crystalline polymer chains (nematic, cholesteric, or smectic) form the elastic network of LCEs. The hybridization by dispersing nanoparticles (NPs) or nanotubes (NTs) in LCEs matrix leads to change in its physical as well as chemical properties. This chapter focuses on recent developments of NPs- and NTs-embedded LCEs that has provided many opportunities for a wide range of applications of electronic devices including actuators, tactile display, soft robotics, and micro-technology. LCEs network with reinforced nanomaterials provide better electrical, optical, thermal, and mechanical properties resulting enhancement of performance in its targeted applications. Initially, brief synthesis processes of zero- to one-dimensional nanomaterials-embedded LCEs composites will be discussed by explaining the core mechanism followed by physical properties. Subsequently, details of devices application such as soft robotics, artificial heliotropism, refreshable braille displays, reversible shape-changing artificial muscles, and shape memory will be discussed.

Keywords Liquid crystal elastomers · Nanocomposite · Nanomaterials · Nanoparticles · Nanotubes · Actuators

M. Mohiuddin (✉)
Bangladesh University of Engineering and Technology, Dhaka 1000, Bangladesh
e-mail: mohipalash@gmail.com

T.T. Tung
School of Chemical Engineering, The University of Adelaide, Adelaide, SA 5005, Australia

D. Ponnamma et al. (eds.), *Flexible and Stretchable Electronic Composites*,
Springer Series on Polymer and Composite Materials,
DOI 10.1007/978-3-319-23663-6_13

1 Introduction

Soft materials of liquid crystal elastomers (LCEs) are shape-changing polymers formed by flexible cross-linked polymer chains composed of *mesogens* (a rigid anisotropic liquid crystalline) units, capable of significant shape change in response to external stimuli (light, temperature, electric field, magnetic field, etc.) without the need for external mechanical manipulation. LCEs are of considerable interest due to combining the anisotropic properties of liquid crystals with the entropy elasticity of polymer networks, and give rise to various anisotropic properties such as the spontaneous macroscopic shape change at the phase transition which is associated with property changes. LCEs complex structure, unique characteristics, and class of physical behaviors including but not limited to lightweight, transparency, environmental friendliness, elasticity, durability, high force density, and mechanical strength are getting focus for research and advance electronics application. LCEs can be used for artificial muscles [1–3], shape memory [4, 5], soft elasticity [6], thermomechanical actuation [7–9], motors [10], microgrippers in microsystems [11], microvalves [12], tunable lasing media [13], tactile display [14], etc. However, their slow response to external stimuli is generally due to the low energy-transfer ability and thermal conductivity, thereby limiting its potential in some other fields. So that challenge possible overcome by introducing nanoparticles (NPs), nanowires (NWs), or carbon nanotubes (CNTs) as fillers into LCEs matrix for the formation of nanocomposite materials, which is due to enhancing the response to an external stimulus.

It has been reported that the response of a polymer to an external stimulus can be modified and improved when incorporated with different nanomaterials that impart new physical response. Zero-dimensional (0D) NPs, one-dimensional (1D) NWs, and nanotubes (NTs) are possible categories of incorporated nanomaterials in LCEs, and many studies have been conducted on both types specially focusing CNTs and iron-oxide-based LCEs nanocomposites and related application. CNT-LCE nanocomposites are responsive to light as CNTs absorb light thus efficient in converting light energy into heat for fast actuation, CNT-LCE nanocomposite is also sensitive to electric fields [15], and silica-coated NPs [16] in LCEs act as nanoheater. Actuation upon the application of magnetic field of Fe_3O_4 NPs in LCEs network [17] is also reported as well. NPs of $PbTiO_3$ [18], molybdenum oxide [19], can be incorporated with LCEs to improve physical properties of nanocomposite. In this chapter, an overview of nanocomposites made of LCEs with various NPs and CNTs with superior properties for their highlighting application in electronics devices will be discussed.

2 Hybridization of 0D and 1D Nanomaterials with LCEs

2.1 Nanoparticles

Nanocomposites of LCEs have been used to further expand the applications beyond the traditional limits [20]. In this context, the design of LC hybrid networks, i.e. containing NPs, NWs, or NTs residues, may also offer original solutions for the development of new generations of stimuli-responsive elastomers [21–23]. Efforts have been made in the design of mechanically adaptive polymer nanocomposites, including LCEs containing carbon particles or NTs with promising electroresponsive performances, and LC networks inserting gold, cadmium selenide (CdSe), and ferrite NPs with interesting optical and magnetic properties [22, 24–28]. Chambers et al. [29] created surface-integrated layer using an existing network and conducting carbon NPs, whereas layer allows the effective resistivity to be reduced from highly insulating to usable values for electrical actuation and withstands large geometrical shape changes. In another work, authors have also developed electrically conductive LCE/carbon black (CB) NPs nanocomposites by passing electric currents, and the temperature of LCE/CB can be increased to initiate the shape change [22].

As compared to other types of NPs, the unique localized surface plasmon resonance (LSPR) [30, 31] and the associated LSPR-enhanced photothermal effect [32, 33] of gold nanoparticles (AuNPs) make them especially interesting in developing multifunctional LCE/AuNP composites with remote triggering capability. Sun et al. presented a composite containing gold nanospheres (AuNSs) of diameter of about 2 nm into micrometer-sized cylindrical actuators using soft lithography and polymerization. These AuNS act as heat transducers that allow for an efficient localized transfer of heat from a focused infrared laser beam. Authors demonstrated that photothermal control of ordering in the liquid crystal elastomer using scanned beams allows for a robust control of colloidal particles, enabling both reversible and irreversible changes of shape [34]. Similar approach was taken by Montazami for enhancing the thermal conductivity of LCEs, and AuNS of about 3 nm in diameter has been shown to be able to significantly reduce the time response of the LCEs shape change when heated directly [28]. Very recently, Lui et al. [35] successfully incorporated AuNS and gold nanorods (AuNR) into polyacrylate-based LCEs elastomer to fabricate LCE/AuNR and LCE/AuNS micropillars or microactuators, as shown in Fig. 1. Inclusion of less than 1 wt% of AuNR, the very high actuation strain (30 %), and rapid response (a few seconds) have been achieved in LCE/AuNR micropillar actuators under 635 nm laser irradiation.

Parallelly, introduction of magnetic NPs into LCEs also provides novel materials [26, 36] with the ability to shape changes in static inhomogeneous fields by means of an applied rotating magnetic field, driving the mesophase-to-isotropic liquid transformation (local temperature rise, i.e., hyperthermia effect) [37, 38]. In addition, due to relaxation processes, magnetic NPs convert electromagnetic irradiation energy into heat. Kaiser et al. [27] prepared of nematic-chain LCEs

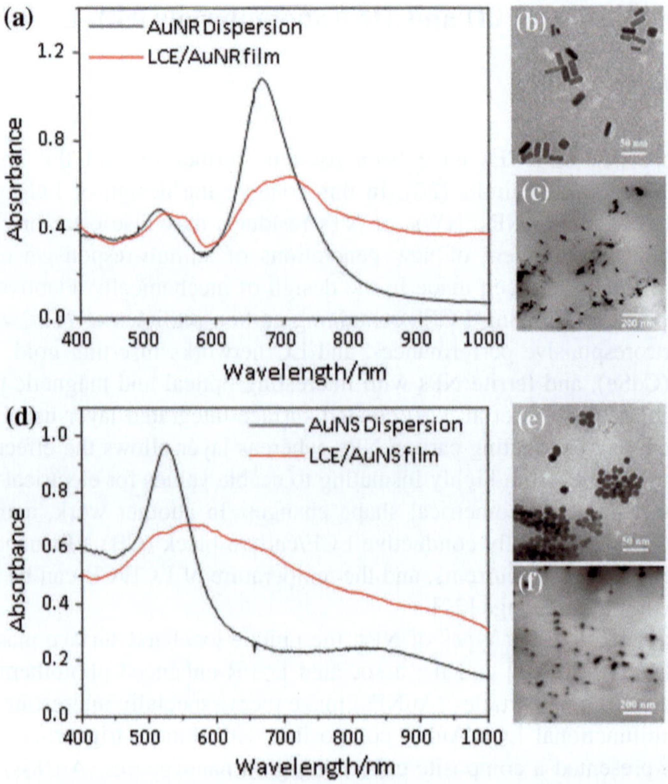

Fig. 1 **a, d** Comparison of the UV–vis–NIR spectra of AuNR and AuNS in aqueous dispersion and composite film, respectively. **b, e** TEM image of both AuNR and AuNS. **c, f** TEM image of AuNR and AuNS aggregation states in ultrathin sectioning (100 nm) of LCE/AuNS(1 wt%) composite thin film, respectively [35]. Copyright 2015. Reprinted with permission from John Wiley & Sons

composites containing different iron oxide NPs contents for use as magnetoactive shape-changing materials by investigating their thermal, thermomechanical, and magnetomechanical behavior. With short response times, elastomers exhibit contractions up to 27 % of the original length due to fast local temperature rise. Winkler et al. [39] reported that magnetoactive LCEs are accessible by the incorporation of superparamagnetic Fe_3O_4 NPs. In addition, nanocomposite offers a contactless activation pathway to the nematic-to-isotrope transition by local magnetic heating in external fields, and around 30 % contraction is observed under field influence with reversibility. In a similar work reported by Garcia-Marquez et al., the hybrid main-chain LCEs were synthesized using the one-step procedure of pre-functionalized 3.3-nm-diameter iron oxide NPs as cross-linkers to reticulate mesogenic linear oligomers [38]. Recently, Haberl et al. [17] have been synthesized core-shell MNs with a 7:3 maghemite:hematite of silica-coated hematite

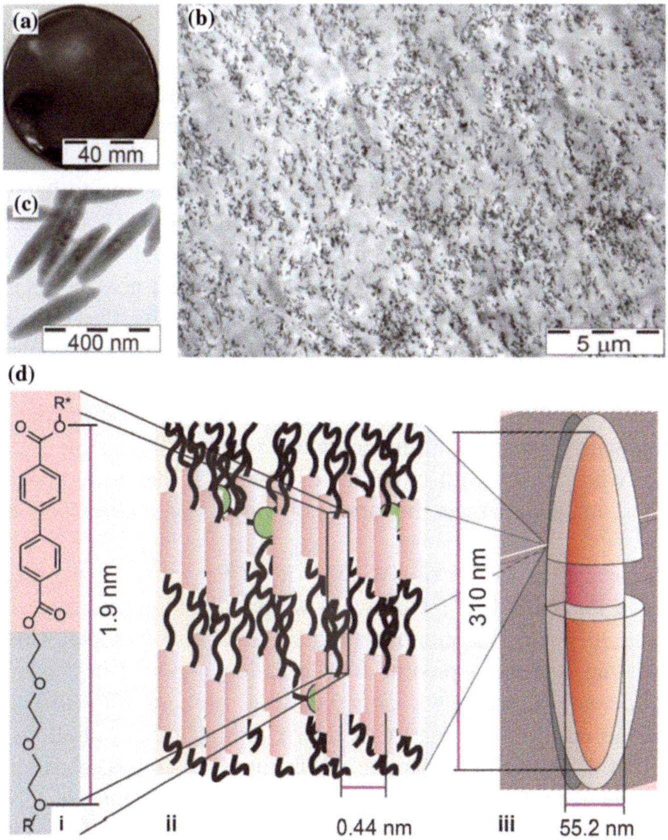

Fig. 2 **a** Photograph of the liquid-crystalline elastomer-magnetic nanocomposite film. **b** TEM micrograph of a stretched sample. **c** TEM micrograph of core-shell ellipsoidal nanoparticles. **d** characteristic features for: (*i*) the calamitic repeating unit in the smectic layering of the liquid-crystalline elastomer, (*ii*) the molecular distance in the layers with the cross-linker molecules (*green spheres*), and (*iii*) the magnetic nanoparticles [17]. Copyright 2013. Reprinted with permission from John Wiley & Sons

MNs with a narrow and monomodal size distribution as illustrated in Fig. 2. A material with magnetic memory can be reversibly stored by mechanical deformation and erased by heating to moderate temperatures. In this case, magnetic particles (MPs) allow manipulation of magnetic properties via the control of parameters and thus greatly expand the scope of magnetic materials. As stored magnetic information is reachable by simple measurements, such as magnetic torque, and can be reversibly tuned, the route presented here increases possibilities in the design of actuators beyond presently achievable imprinted magnetic geometries and shows great promise in strain sensing devices and magnetic information storage applications.

2.2 Nanotubes

CNTs are one of the effective filling materials for nanocomposites due to their one-dimensional structures, nanometer-scale diameters, high aspect ratios, large surface areas, excellent conductivities, and other physical and mechanical properties [40–42]. Recent studies have shown that owing to band gap transitions, both the multiwalled carbon nanotubes (MWCNTs) and single-walled carbon nanotubes (SWCNTs) exhibit a photomechanical response, especially the SWCNTs, show strong absorptions in the visible and near-IR region as well as modified optical and electrical properties [43]. The effective use of CNTs in polymer–composite applications strongly depends on homogenous dispersion of CNTs into polymer matrix. However, the most difficult and complex problem in these approaches is the dispersion capability of CNTs because pristine CNTs are incompatible with most solvents and polymers, which leads to poor dispersion and agglomerate when mixed with most liquids [44, 45]. Therefore, poor dispersion and agglomeration are the main limitations of widespread use of CNTs in composites [46].

Systems for the dispersion of SWCNTs in LCEs systems have recently been developed and studied [44, 47]. However, owing to the high tendency of SWCNTs to aggregate, they are not completely dispersed. For example, Courty et al. reported that the maximum pristine SWCNTs loading-level in a nematic LCEs with a polysiloxane backbone was only 0.02 wt%, because of the difficulty to disperse SWCNTs uniformly in the LCEs matrix at higher concentrations [48]. Due to the inherent difficulty in dispersing CNTs into cross-linked elastomers and in preparing monodomain (well-aligned) LCEs with the presence of CNTs, methods such as shear mixing and ultrasonication have been successfully employed in order to impart sufficient energy to break apart clusters of CNTs [49–51].

A dispersing agent can interact chemically with both the CNT surface and the polymer matrix to avoid reaggregation of CNTs [52]. Chen et al. [61] have reported a dispersing agent of poly(p-phenyleneethynylene) (PPE) used for non-damaging chemistry platform that enables to modify specific CNT surface properties, while preserving the CNT's intrinsic properties. Moreover, dispersion and stabilization of CNTs in common organic solvents can be improved with pyrene end-capping of flexible main-chain liquid-crystalline polymer as shown in Fig. 3. Li et al. [47] have also developed LCEs nanocomposites using SWCNTs as filler in a nematic side-chain LCEs matrix with polysiloxane backbones. Experimental results showed that the addition of SWCNTs into the LCEs matrix lowered the material's nematic–isotropic transition temperature by up to about 19 °C. Microstructures of SWCNT-LCE nanocomposite are represented in Fig. 4. Moreover, reversibly and rapidly actuation of SWCNT-LCE nanocomposites is well demonstrated [47]. However, the mechanism of photo-actuation, in this case, will be very different from the "traditional" photo-isomerizing materials. The disruption of local nematic order will be achieved by local heating when the

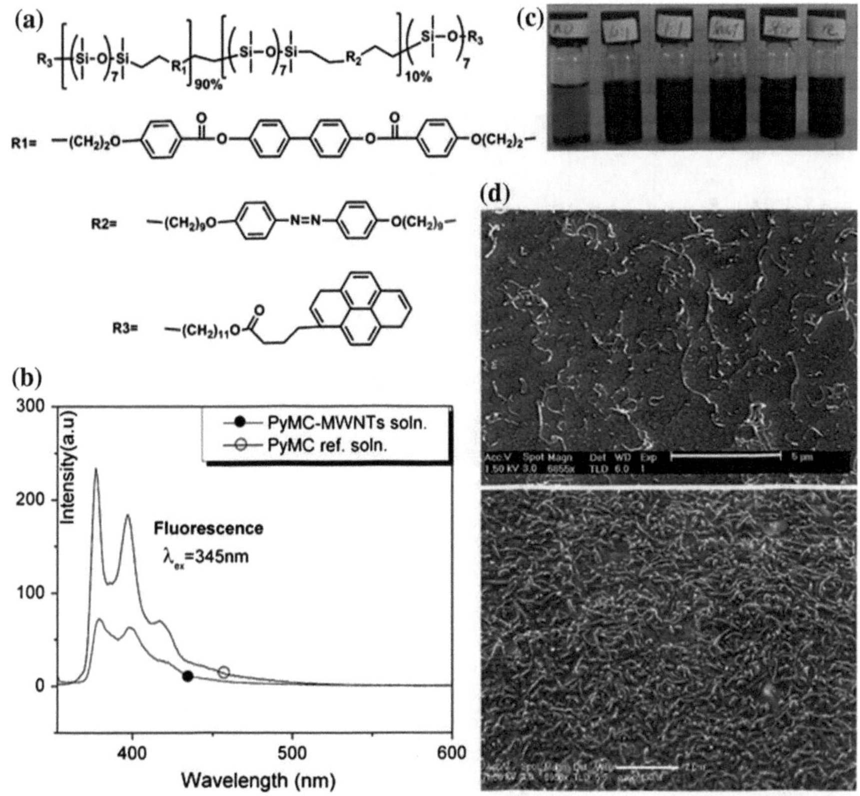

Fig. 3 **a** Chemical structure the pyrene main-chain (PyMC) copolymer. **b** Fluorescence spectra of the PyMC-MWCNT solutions. **c** Appearance of PyMC-MWCNT solutions prepared by different processes. **d** SEM images showing the morphology of the well-separated CNTs [62]. Copyright 2010. Reprinted with permission from John Wiley & Sons

NTs release the energy of absorbed light into heat. As an added benefit, the CNT-loaded matrix has a much better thermal conductivity, which serves to increase the speed of actuator recovery. Basu has recently shown that when a small amount of CNTs is doped in a nematic LC, the LC + CNT matrix shows a faster response to an electric field compared to that of the pure LC. This accelerated response could be used to develop faster LCDs [53].

The CNTs, embedded in the LC matrix, follow the LC reorientation, enhancing the overall dielectric anisotropy of the system. In general, the presence of free ions in liquid crystals would result in an increase in rotational viscosity, slowing down the director's response to an external field. The CNTs in the media act as ion-catchers with a high ion-trapping coefficient. Thus, the CNTs, being suspended in the system, lower the free ion concentration, decreasing the rotational viscosity for the overall system.

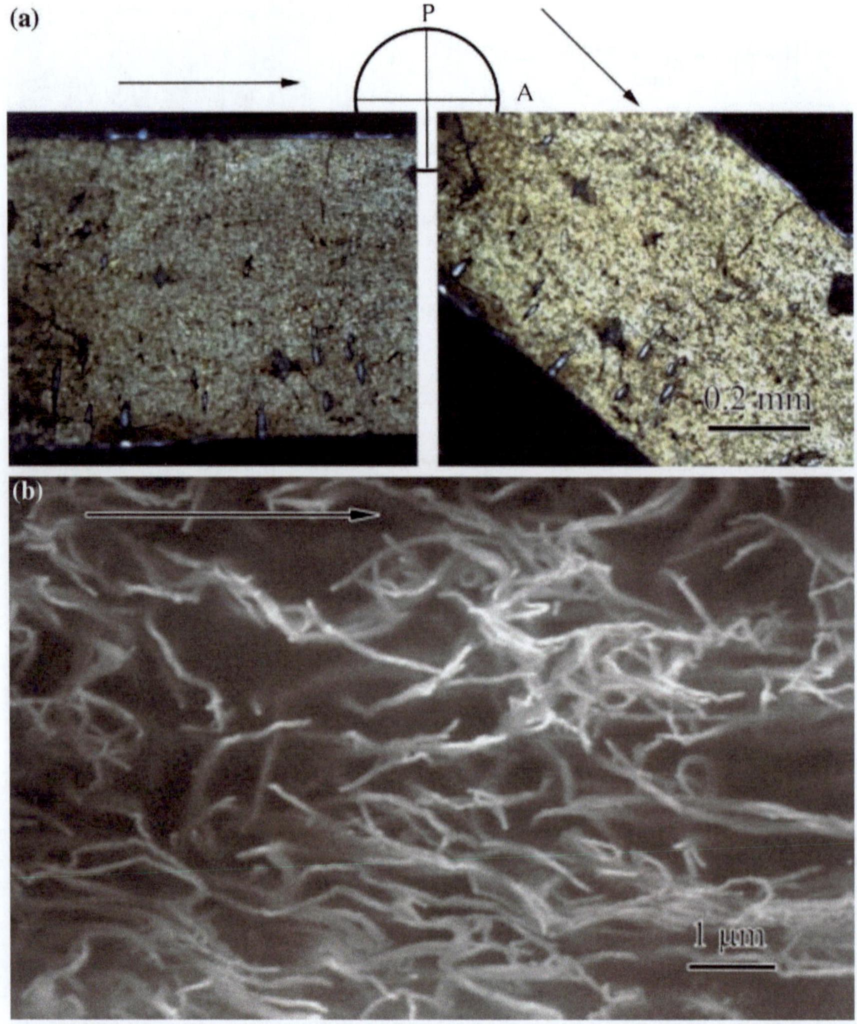

Fig. 4 Microstructures of a 0.1 wt% SWCNT-LCE nanocomposite film of **a** optical microscopy images and **b** SEM image of SWCNTs in the LCEs matrix [44]. Copyright 2008. Reprinted with permission from John Wiley & Sons

3 Characteristics and Devices Application

The incorporation of nanophase materials into LCEs has been of considerable interest since resulting in achieving remote light-controlled actuation of LCEs materials with a faster response time and better control. The nanophase materials selected for LCEs composites possess special optical, thermal, and physicochemical properties. In fact, they can efficiently absorb and convert photon energy into thermal energy and can serve as a nanoscale heat source and thermal conduction pathway to heat the matrices effectively. Hence, those 0D and 1D nanostructures can offer superior opportunities for the photo-actuation of thermal response [47]. The advantage is that the NPs have as little effect on the macroscopic structure and thermal properties of the material. On the other hand, the presence of a low concentration of NPs results in a slight shift of the nematic-isotropic (N-I) transition temperature and glass transition temperature. The transition temperature tends to be few degrees decrease as compared to neat LCEs. This is dependent on the concentration of NPs [16], particle range, or sample preparation [22]. Sometimes, resulting nanocomposites exhibited weakened thermal actuation, which might due to the increase of material stiffness resulting from incorporation of NPs. Moreover, the NPs may act as impurities, leading to a reduction in the LC nematic order by the decrease in the N-I transition temperature.

Montazami et al. [28] developed a thermomechanical nematic LCEs actuator containing AuNP. Embedding a low concentration of AuNP enhances the thermal conductivity of the actuator with minimal effect on the elasticity; therefore, improved thermal conductivity allows more efficient transfer of heat into the elastomer. Since the heat penetration through the elastomer is enhanced in samples doped with metallic nanocolloids, polymeric network surrounding the metallic nanocolloids is subjected to heat in a shorter time domain; thereby, the mechanical actuation occurs at a faster rate with external stimuli. Under fast heating conditions, the LCEs films doped with 0.096 mol% AuNP actuators exhibited more than a 100 % increase in the rate of change of strain with respect to time (Fig. 5b).

Haberl et al. [20] investigated the strain-induced macroscopic magnetic anisotropy from smectic liquid-crystalline elastomer–maghemite nanoparticle hybrid nanocomposites. Several samples were synthesized with different particle contents ranging from 0.5 to 10 wt/wt%, whereas a reference elastomer was prepared without any particle content. Uniaxial stress–strain behavior of samples is represented in Fig. 6a.

The effect of particles on the modulus of the material was evaluated, and three different regions were found, which is shown in Fig. 6b. In the first region (NPs effect in this region is less pronounced), weak interactions in the polymer have to be disrupted, and the relaxed non-oriented liquid-crystalline polymer chains absorb energy to enable the deformation. In the second region (the particle effect is most prominent), enough stress is available and domain reorientation is the dominant process. In this region, particles contribute to elongation via a reorientation process coupled to the reorientation of the LCEs domains. In the third region, the majority of liquid-crystalline domains have already been oriented into a monodomain, and the polymer backbone has to be stretched. It has to be noticed that there is no indication that the NP orientation order is having a dominant effect on the tensile strength, and continuous orientation is assumed in these regions.

The results of actuation induced by infrared (IR) light shown in Fig. 7a indicate key actuation role of CNT alignment in the matrix [62]. However, actuation stresses independent of CNT loading due to unimpeded mechanical motion in different CNT loading. The presence of CNT in LCEs increased the dielectric response as well conductivity (Fig. 7b). In Fig. 7c, absorption of samples

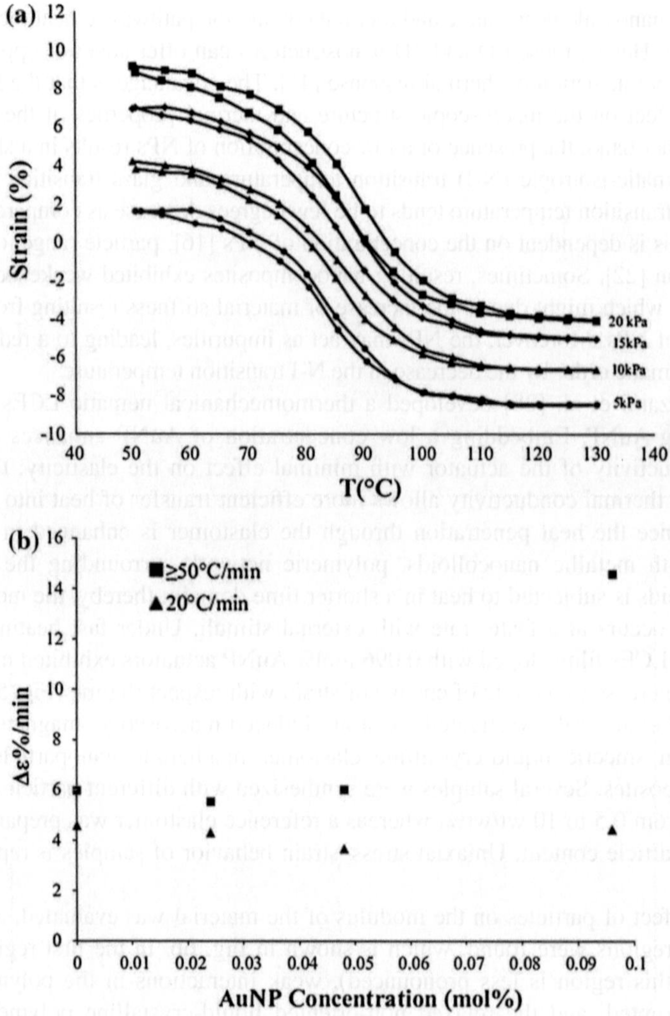

Fig. 5 **a** Thermoelastic measurements of elastomer doped with 0.048 mol% AuNP at 0.5 C/min heating and cooling rate under application of different static forces from 5 to 40 kP. **b** Actuation speeds of LCEs actuators doped with different concentrations of AuNP under slow and fast heating conditions [28]. Copyright 2012. Reprinted with permission from Elsevier

increased with frequency with the presence of CNT, and 3 wt% of CNT containing sample where polarization is parallel to the CNT alignment axis exhibited remarkably higher absorption coefficient than perpendicular polarization. Finally, refractive index and absorption coefficient information indicate the usefulness of samples for optical application in high frequency region (Fig. 7d).

Specific devices application of refreshable braille display, soft robotics, artificial muscles, artificial heliotropism, magnetic actuation, and shape memory of CNT-LCE composite explained in details in the rest part of this chapter.

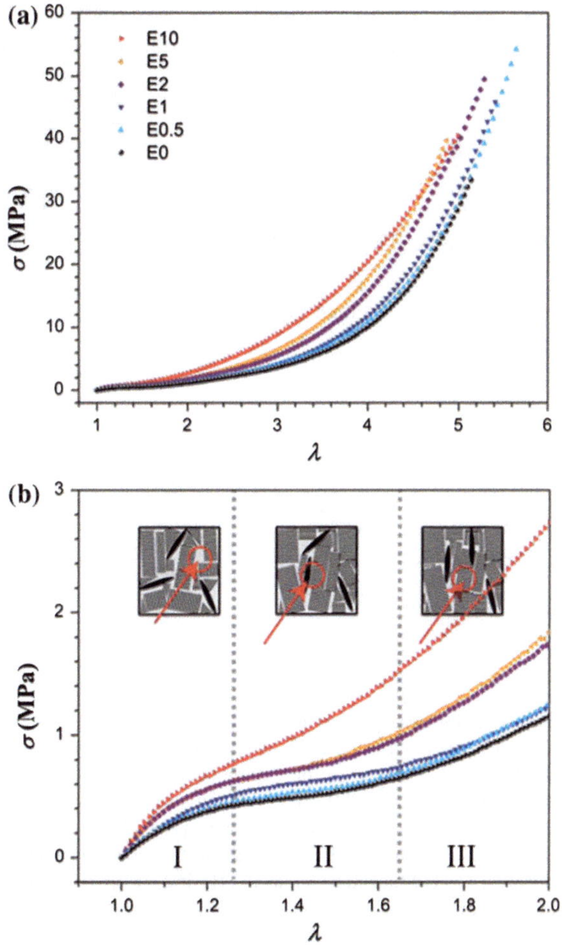

Fig. 6 **a** Uniaxial stress–strain curves for the liquid-crystalline elastomer and nanocomposites. **b** Zoom-in at the initial strain values. The *inset* shows the model for the reorientation processes in the three regions I, II, and III [20]. Copyright 2013. Reprinted with permission from Royal Society of Chemistry

3.1 Refreshable Braille Displays

Refreshable braille displays are electronics device that use vertically moving (up and down) pins to read text tactually [54]. Refreshable braille displays usually assist peoples who cannot use a computer monitor to read text output. Camargo et al. [55] reported composite of MWCNTs with diameter 30–50 nm and LCEs for actuation system to be used in tactile devices by means of blister-shaped actuator shown in Fig. 8b, c. As shown in Fig. 8a, both mechanical stretching and thermal cross-linking demonstrated for creating sufficiently well-aligned liquid crystal units to fabricate blister-shaped patterned LCE-CNT to be used as optomechanical actuators.

Significant reversible change of height (decrease) of the blister has been observed in LCE-CNT for incident light. Shape change of elastomer in Fig. 9a remains almost indifferent with the absence of CNT; however, low concentration

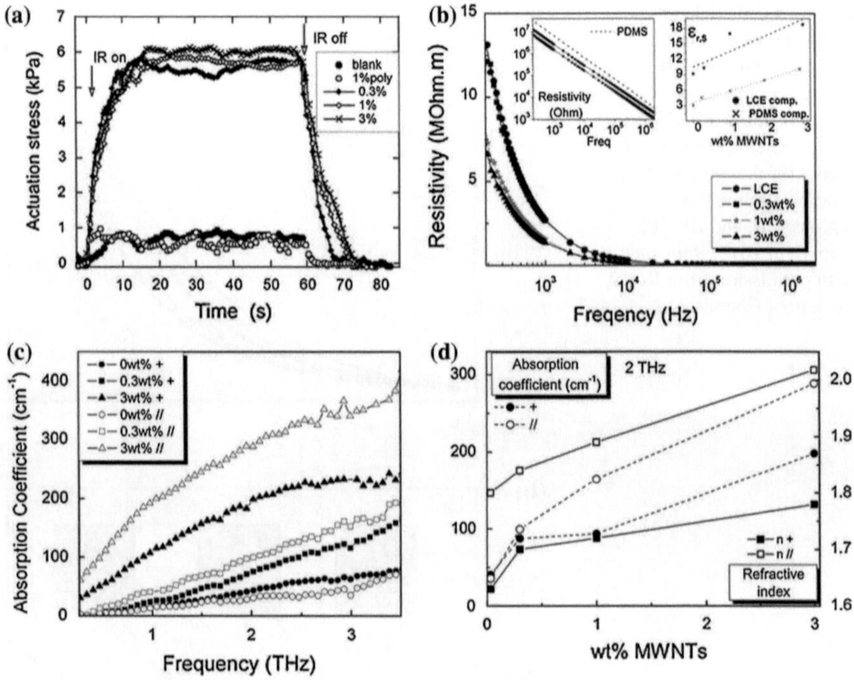

Fig. 7 **a** Infrared photoinduced actuation in constant-strain geometry. **b** Complex resistivity against frequency for the CNT-LCE composites of different CNT loadings. The right *inset* shows the large improvement in the extrapolated effective dielectric constants when CNTs are added to the LCEs matrix. **c** THz time-domain spectroscopy: absorption coefficient spectra of a blank, 0.3 % and 3 wt% CNT composites, when the THz polarization is parallel, //, and perpendicular, +, to the direction of LCEs. **d** The absorption coefficient and the refractive index of samples with different CNT concentrations at 2 THz [62]. Copyright 2010. Reprinted with permission from John Wiley & Sons

(0.1 wt%) of CNT added in elastomer proved optimal for the effective conversion of light into local heat throughout the sample thickness. Contraction of samples follows an exponential growth, whereas relaxation shows an asymptotic log decay. Both relaxation and contraction stabilization time tested at different power input range between 5 and 35 mW (Fig. 9b). It is found that relaxation stabilization time remains almost indifferent with power input, whereas contraction stabilization time of LCE-CNT contraction becomes faster (linearly) at higher powers.

Furthermore, mechanical response of two type's actuators tested with incident light (Fig. 10b). Blister of 1.0 and 1.5 mm diameters with 400-μm wall thickness contract under irradiation by a laser diode with optical power ranging 10–60 mW. Actuation with errors (calculated from 20 actuation cycles) of 1.5-mm blister represented for optical powers between 8.6 and 51.5 mW. It is clear from graph that contraction actuation of 35 μm is saturated with 51.5 mW incident power and lower activation found to be 3 μm at 8.6 mW. Moreover, little variation observed

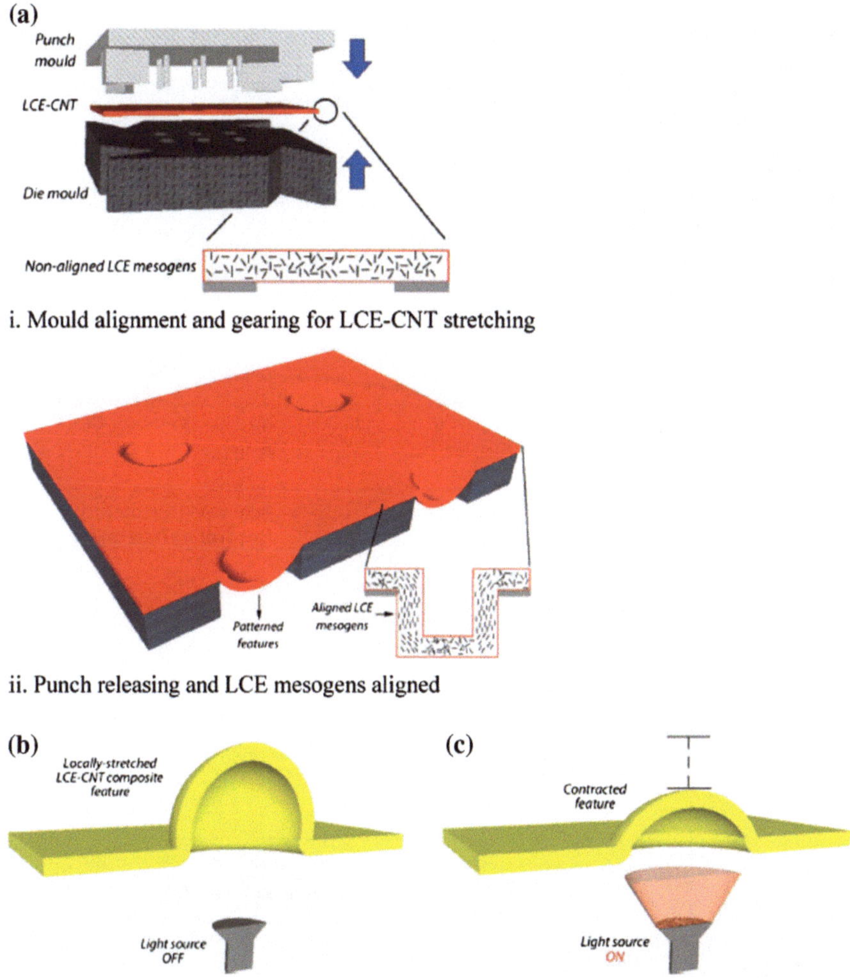

Fig. 8 **a** Aligning LCEs mesogens through stretching process. **b** Normal state of actuator in the absence of light. **c** Contraction of blister height upon incident of light [55]. Copyright 2011. Reprinted with permission from John Wiley & Sons

at actuation cycles. 1.5-mm-diameter blister actuator exhibits higher actuation than 1-mm blister which might happen due to larger light absorption (larger light incident surface), relative higher stretching, mechanical interaction between silicon molds, and the LCE-CNT film during the stamping process for 1.5-mm-diameter blister. 60 on–off actuation cycles carried out at 38.5 mW incident optical power on 1.5-mm dot to test repeatability of the actuation and insignificant performance degradation indicated in last graph. Well-demonstrated research to develop tactile display based nanocomposite of LCEs can also be found here [14, 56, 57].

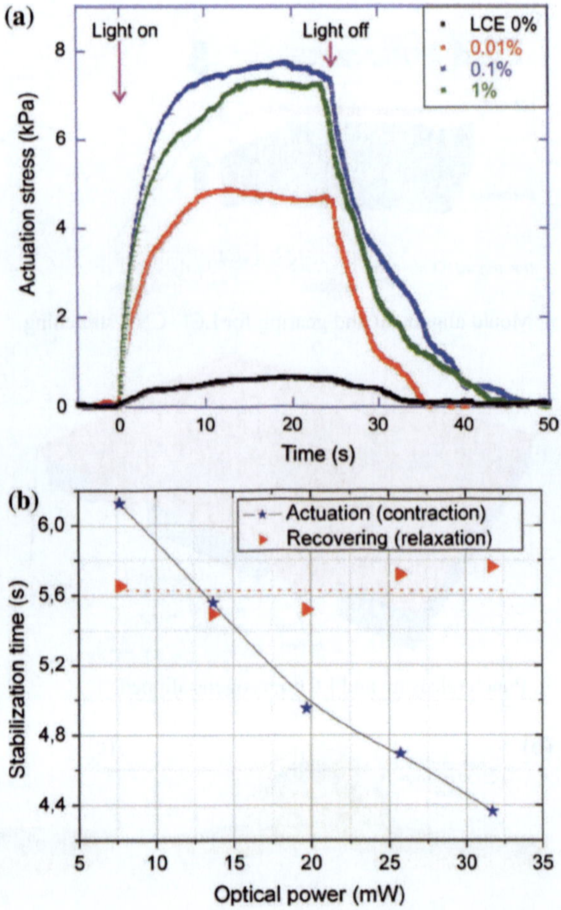

Fig. 9 **a** Actuation performance of LCE-CNT composite with increase percentage of CNT to an IR light source. **b** Stabilization time (contraction and relaxation) of actuation at different power input range between 5 and 35 mW [55]. Copyright 2011. Reprinted with permission from John Wiley & Sons

3.2 *Soft Robotics*

Composite SWCNT-LCE/silicone bilayer hinges actuated by incident wavelength selective, IR light in soft robotics [58] application such as gripper actuator to transfer objects (even under water), inchworm walker crawl up incline plane, reversibly fold and unfold active origami, etc. Advantage of IR, compared to UV and visible light, in most polymeric materials IR can penetrate much deeper with negligible damage. SWCNTs are capable of transferring IR light into thermal energy; thus, sufficient amount of thermal energy converted from IR light leads LCEs N–I phase transition to change the shape of nanocomposite film.

Upon incident of IR light, temperature of 0.1 wt% SWCNTs composite film reaches over 80 °C, thus proving of role of SWCNTs as heat source with the presence of IR. However, in-plane negative strain can be adjusted by IR light intensity, silicon layer thickness, etc. (Fig. 11).

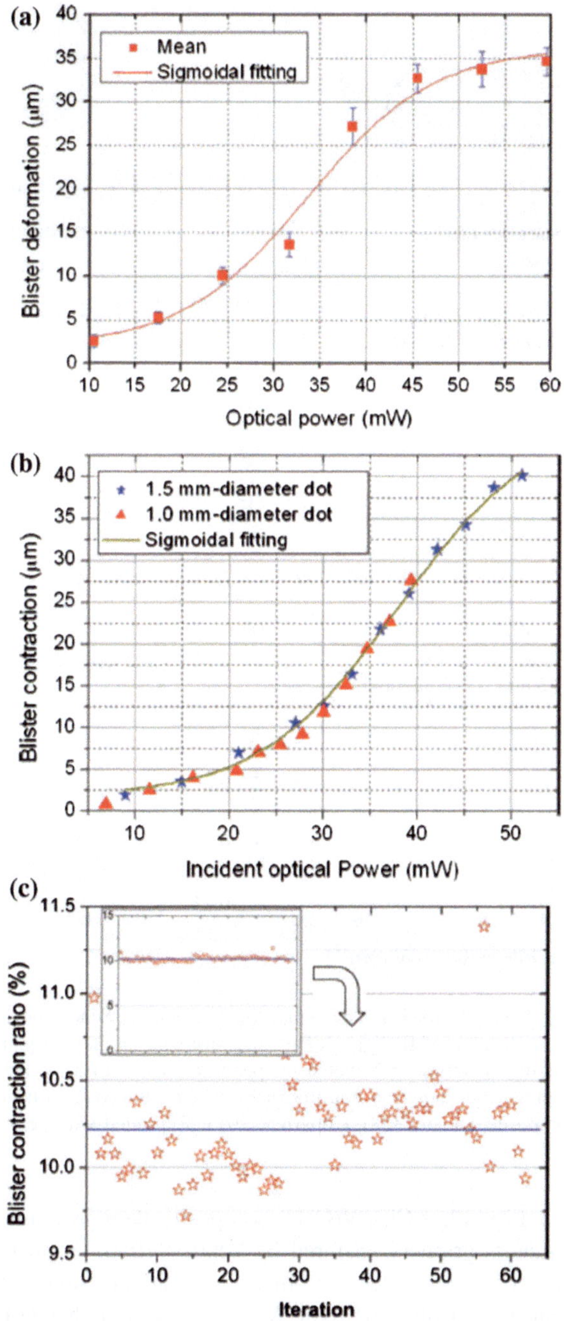

Fig. 10 **a** Optomechanical actuation performance curve of 1.5-mm blister. **b** Optomechanical performance comparison: 1.5- and 1.0-mm blisters. **c** Stable performance of actuator can be concluded from repeatability test during 60 on–off actuation cycles of 1.5-mm blister at 38.5 mW input light power [55]. Copyright 2011. Reprinted with permission from John Wiley & Sons

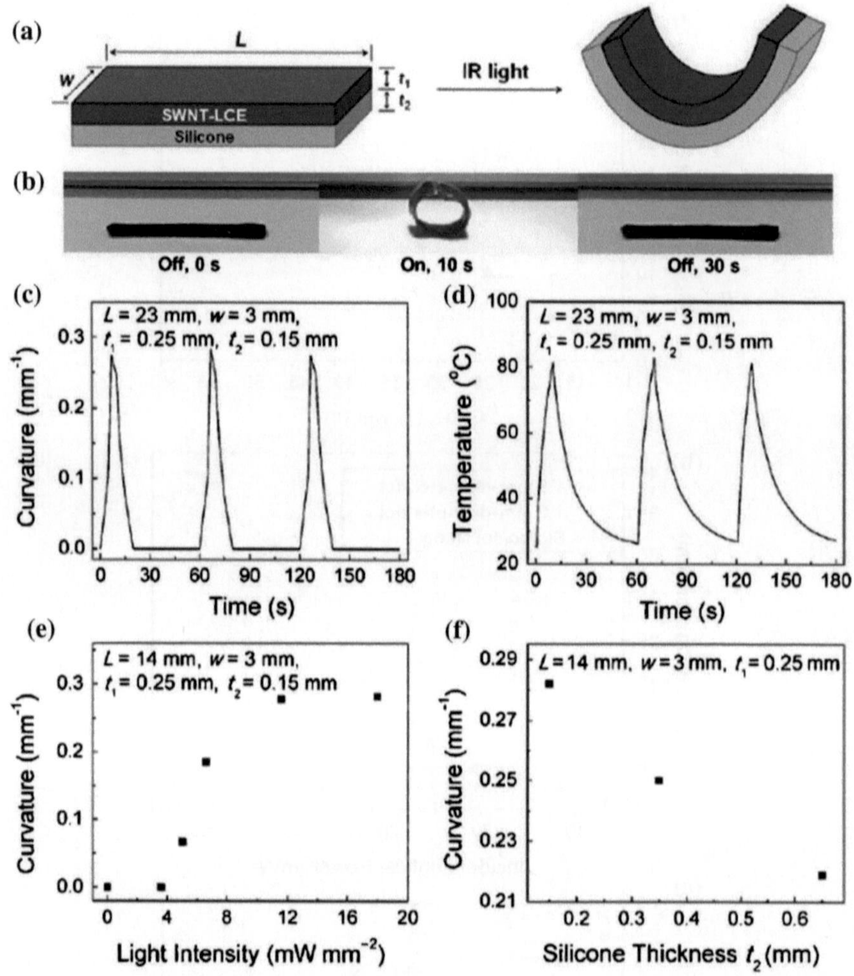

Fig. 11 0.1 wt% SWCNTs-loaded nanocomposite actuator. **a** Schematic bending behavior of bilayer film in application of IR light. **b** Fast and reversible in-plane negative strain demonstration. **c**, **d** Bending curvature and related temperature change of same actuator. **e**, **f** Effect of another two factors (light intensity and thickness of silicone layer) on bending curvature of SWCNT-LCE composite actuator [58]. Copyright 2013. Reprinted with permission from John Wiley & Sons

As shown in Fig. 12, LCE-SWCNT composite demonstrated rapid opening and closing actuators grippers, inspired by Venus flytrap. Grippers were capable of picking, transferring, and releasing objects of relatively large and thin structure with the incident of controlled IR light. Moreover, with a change in design, grippers can pick up objects submerged in water; thus, the strength of a gripper could be enhanced by increasing the number of hinges used in the gripper.

Figure 13 demonstrated inchworm walker made from asymmetric SWCNT-LCE/silicone bilayer film and two PC films with different shapes which can crawl

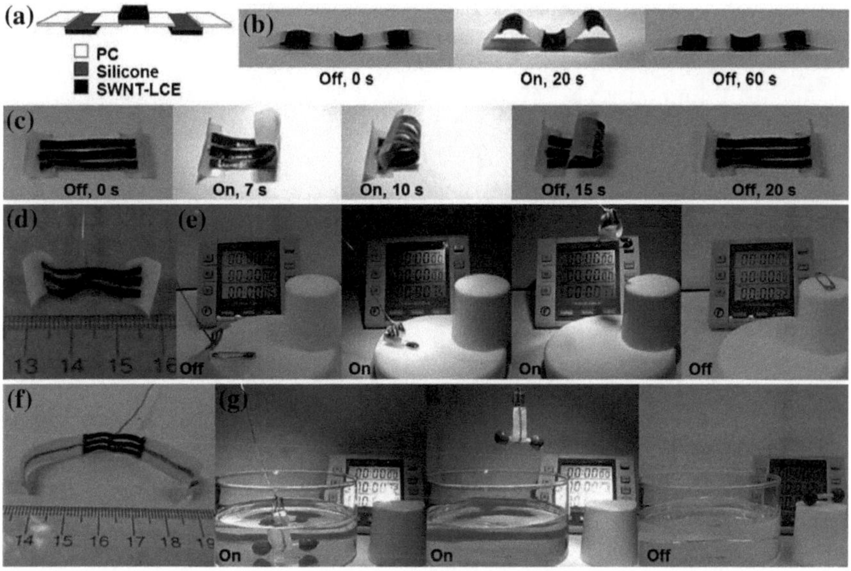

Fig. 12 Developed actuator and capability of actuators represented using NIR light. **a** Sample developed structure of origami made from films of polycarbonate, SWCNT–LCEs composite and silicone. **b** Photomechanical folding and unfolding actuation of origami sample. **c** Quick response of gripper actuator inspired by venus flytrap made from polycarbonate (*white*), SWCNT-LCE composite (*black*) films. **d, e** Sample structure of short gripper actuator and relatively light sample transportation. **f, g** Sample structure of long gripper actuator and under water object pickup and transfer [58]. Copyright 2013. Reprinted with permission from John Wiley & Sons

up a hill at a 50° incline, driven by IR. Inchworm walker's inward and outward bending of front and back part allows it to climb upward. Directional movement of inchworm walker is enabled by the directional movement of IR light from the front to the center of the device. Movement of IR light-driven inchworm walker is approximately 0.1 mm/s at 50° tilt surface which is faster than chemically responsive polymer-gel device on a flat surface [59].

3.3 Artificial Muscles

Artificial muscle are aimed to reproduce characteristics of reversibly contract, expand, or rotate within one component with the respond to various external stimulus such as electric field, magnetic field temperature, and light. Research on artificial muscles are going on based on piezoelectric ceramics, shape memory alloys, polymer, etc., for wide range of applications including but limited to haptic displays, adaptive optics, flat conformal loudspeakers, and, potentially, implantable active medical prosthetics, etc. Electrically controllable monodomain nematic liquid-side chain crystal elastomers–graphite (LSCE-G) composites are a good example of liquid crystal elastomer–graphite (LCE-G) composites in biomimetic

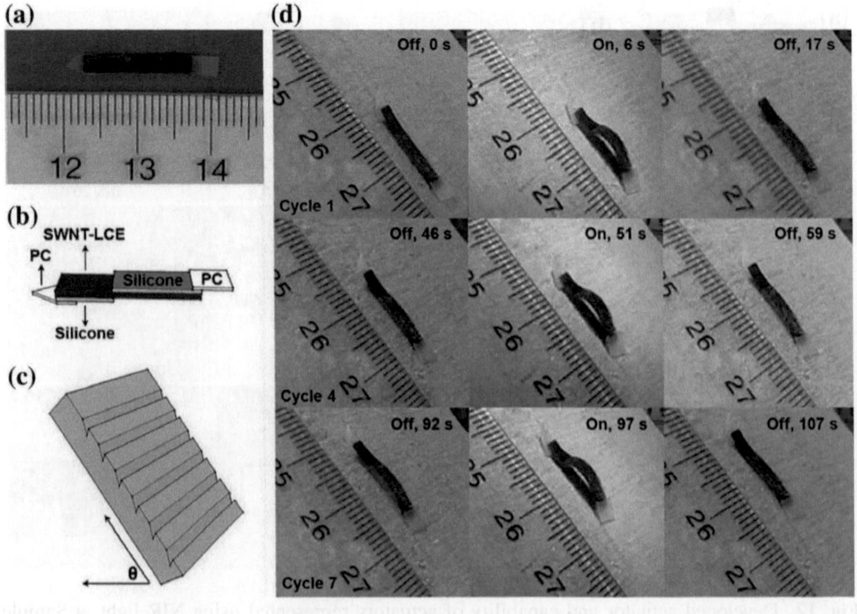

Fig. 13 **a** Illustration of inchworm walker actuator. **b** Layer structure of inchworm walker. **c** Ratcheted wood substrate used for crawling operation of inchworm walker. **d** Step-by-step representation of inchworm walker's hill climbing process [58]. Copyright 2013. Reprinted with permission from John Wiley & Sons

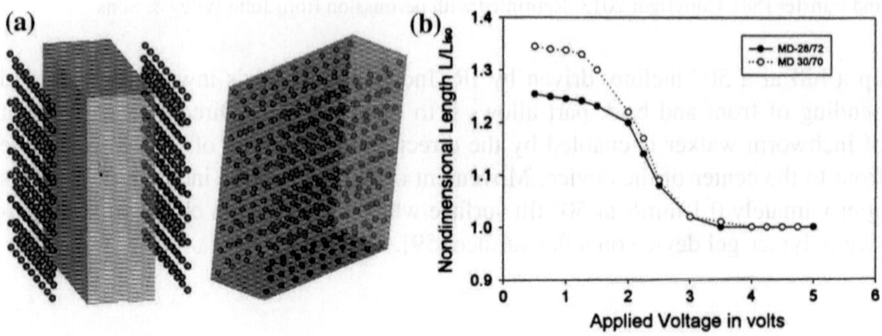

Fig. 14 **a** Schematic of preparation of graphite-loaded LCEs by using pressing method. **b** Contraction of LCE–graphite composite with the application of voltage for two different main-chain to side-chain mesogen ratio [60]. Copyright 2002. Reprinted with permission from SPIE

artificial muscles stated by Finkelmann et al. [60]. Liquid crystalline to isotropic phase transition can cause conformational changes of macromolecules, which are sufficient to obtain a large mechanical response of polymer networks. Principle of electric Joule heating can be applied to manipulate conductive LCEs electrically controlled N-I phase transitions. Conducting particles penetrated inside the elastomer by simple press method (shown in Fig. 14a) to improve conductivity of composite to trigger the elastomer shape-changing effects by rapid Joule heating.

To ensure conductivity of elastomer more than 50 % graphite powder (2 micron in average diameter) was uniformly embedded in LCE during composite construction to reach percolation threshold. Uniform distribution of graphite in LCEs ensures uniform internal joule heating. For composite, effective resistivity of about 1.6 ohms/square and 0.8 ohm/square is achieved on surface and thickness, respectively. Excellent result of contraction is achieved for composite films even at low voltage. Samples exhibit quick length contraction (in around 1 s) strain of 25 % under 10 kPa stress in application of DC −0.5 to 5 V (for 4 s). However, negligible change observed along transverse and thickness direction. In extend, around 5 s were good enough to revert back to initial length during cooling. Such thermal behavior can be explained by network anisotropy of nematic state. Figure 14b depicts the general trend of contraction of graphite fiber-interlaced LSCE sample with Joule heating.

3.4 Artificial Heliotropism

Photo-thermomechanical effect of LCE/SWCNT nanocomposite actuators utilized to develop artificial heliotropic devices for solar cells demonstrated strong actuation capability with the presence of direct natural sunlight [23] (Fig. 15). Artificial heliotropism fabricated using actuators can tilt the solar cells toward the sun by contraction from direct incident sunlight resulting significant improvement in the photocurrent output from the solar cells. Utilizing direct actuation

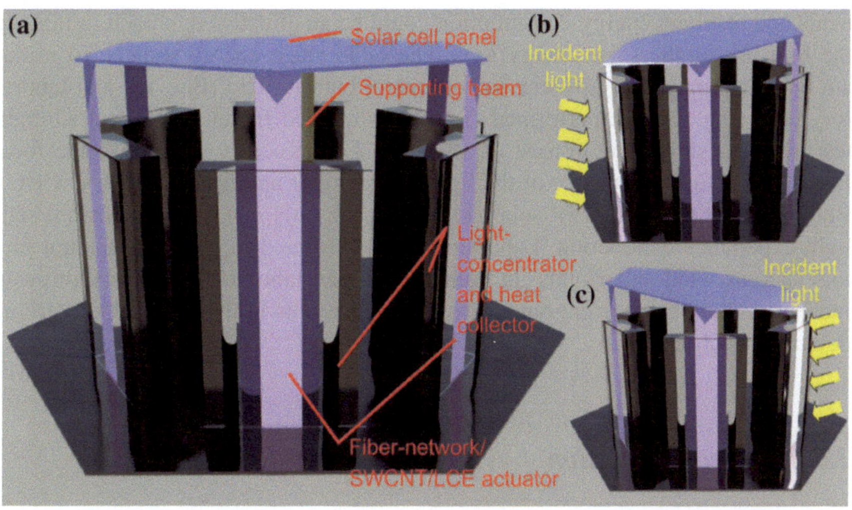

Fig. 15 **a** Design of the artificial heliotropism. **b, c** 3D schematic operation of device and with the incident of sunlight, actuator(s) contracts, and tilt the solar cell toward the sunlight [23]. Copyright 2012. Reprinted with permission from John Wiley & Sons

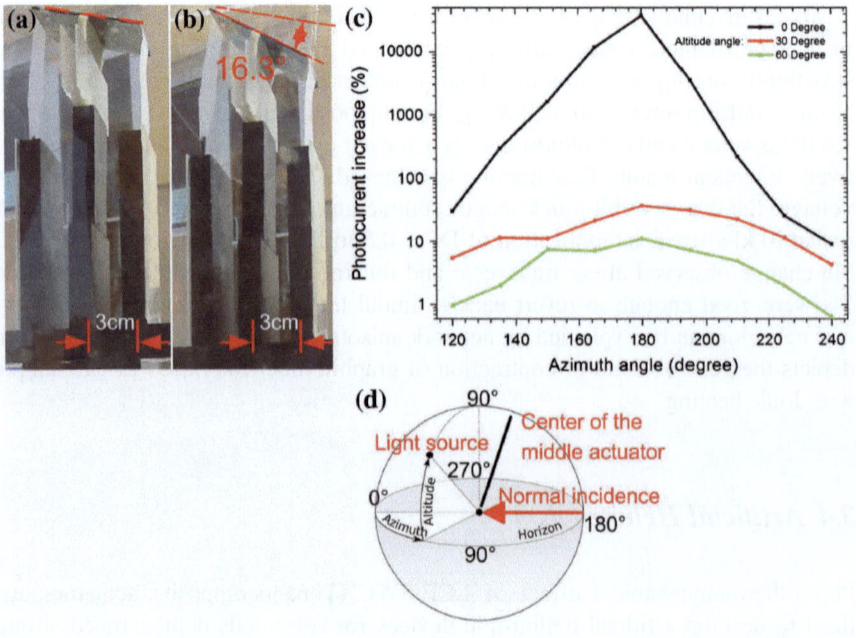

Fig. 16 **a**, **b** In-field testing. **c** Resultant photocurrent increase with azimuth angle. **d** The altitude–azimuth coordinate system used. The origin was the center of the actuator facing the light. The normal incidence direction was 0° altitude and 180° azimuth [23]. Copyright 2012. Reprinted with permission from John Wiley & Sons

by the sunlight, and eliminating the need for additional mechatronic components and resultant energy consumption could be an effective approach to making the best use of the solar energy as a renewable clean energy source. To evaluate the performance of artificial heliotropism, both in-field (Fig. 16) and laboratory tests (Fig. 17) are performed. Two units utilized for in-field direct sunlight test, whereas single unit consist of a CNT-LCE film actuator placed at the focal line of the cylindrical mirror of the light concentrator and heat collector (LCHC). Both units act as support for solar cells; however, only one unit face toward sunlight experience contraction. Laboratory test performed to eliminate natural factors such as wind and clouds. In both in-field and laboratory tests, a significant increase in the photocurrent output from the solar cells in the artificial heliotropic devices was observed.

3.5 Magnetic Actuation

Magnetic iron oxide NPs embedded in LCEs forming nanocomposites presented by Kaiser et al. [27] that can dissipate local heat upon exposed to electromagnetic fields leads to magnetoactive shape change. Subsequently, with the increase

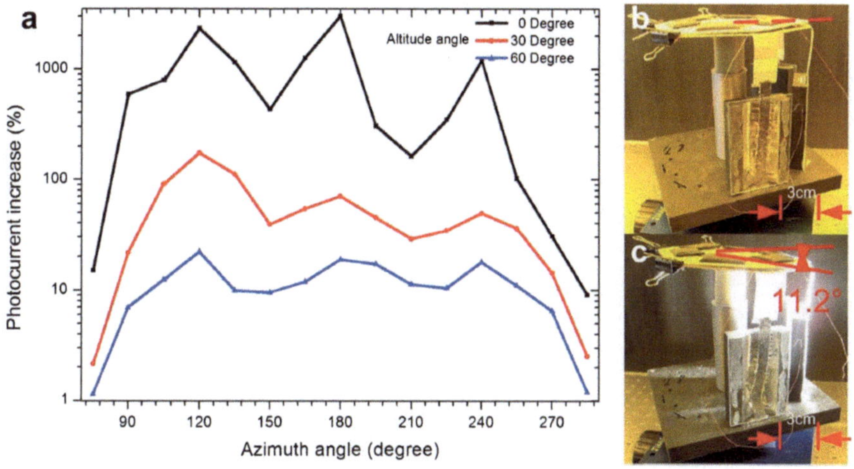

Fig. 17 **a** Increase in photocurrent. **b, c** Artificial heliotropism laboratory testing before irradiation and 30 s after irradiation of light [23]. Copyright 2012. Reprinted with permission from John Wiley & Sons

of temperature, nanocomposite contracted (up to 27 %) at the T_{ni} of the elastomer. Incorporation of NPs acts as a heat sources that lead to a faster response and resulting contactless, remote-controlled shape change. Contraction rate of nanocomposites depends on both heat production rate and the heat transfer coefficient; therefore, thermal, thermomechanical, and magnetomechanical characteristics are investigated with different MNP contents. Two-step cross-linking process performed to prepare magnetoactive side-chain LCEs; however, this section will solely focus actuation characteristics of nanocomposite. Both thermomechanical and magnetomechanical properties are investigated for composite of MNP-LCE with particle contents between 0 and 1.64 vol.%. Sample with 0.31 vol.% of MNP contracts 68.8 % at 80 °C relative to 25 °C; however, contraction was not affected by the particle load.

In case of constant magnetic field (Fig. 18a), immediate contraction of composite is observed and length contraction is saturated in short period of time (couple minutes). Contraction amplitude changes with the amount of MNP particle volume fraction, and relative length (ratio between contracted length and original length) of sample decreases with the increase of particle. Around 74 % of relative sample length observed with highest amount of particle loading. However, particle-free sample exhibits slight contraction due to resultant heat in polymeric material in the presence of constant magnetic field. In Fig. 18b, magnetomechanical behavior of highest particle containing sample (1.64 vol.% of Fe_3O_4) extended applying different field strength (between 8.5 and 42.6 kA m^{-1}) of the applied AC field. Shape changes for AC field are slightly lower than thermomechanical process. It is obvious that higher field strengths cause higher temperatures thus larger contractions in length.

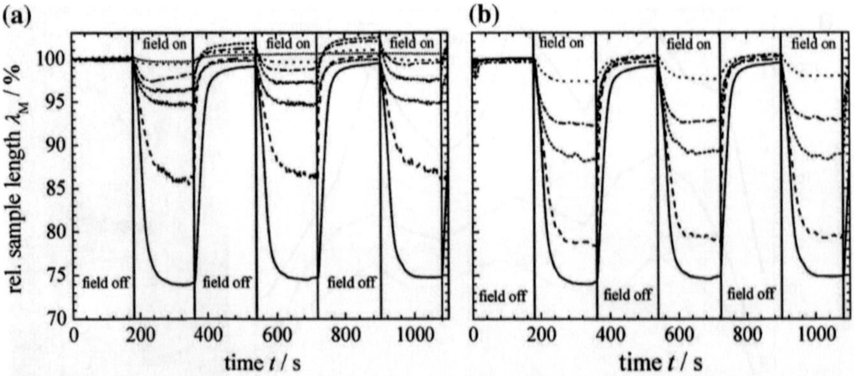

Fig. 18 Change in length (relative) of LCEs with **a** different nanoparticles loading in constant magnetic field amplitude, **b** different field strength amplitude of the applied AC magnetic field [27]. Copyright 2009. Reprinted with permission from Royal Society of Chemistry

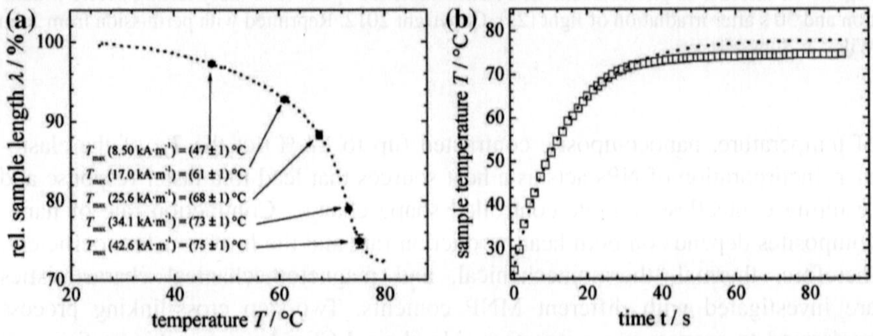

Fig. 19 **a** Temperature estimation from correlated length contraction. **b** Estimated time–temperature relationship at 42.6 k Am^{-1} [27]. Copyright 2009. Reprinted with permission from Royal Society of Chemistry

To estimate the temperatures corresponding to different field amplitudes, saturation contraction in the magnetomechanical experiments correlated to temperature curve of thermoelastic experiment is shown in Fig. 19a. A temperature–time diagram for the magnetomechanical experiment of 1.64 vol.% NPs at 42.5 kA m^{-1} is shown in Fig. 19b. Due to magnetic heating of the NPs under the field influence, a constant heat flux is generated within the sample when the field is switched on.

3.6 Shape Memory

Lower liquid crystalline–isotropic phase transition temperature, higher decomposition temperature, tensile strength, and the elongation can be achieved by nanocomposite of SiO_2 nanoparticles (SNPs)-incorporated liquid-crystalline networks [16].

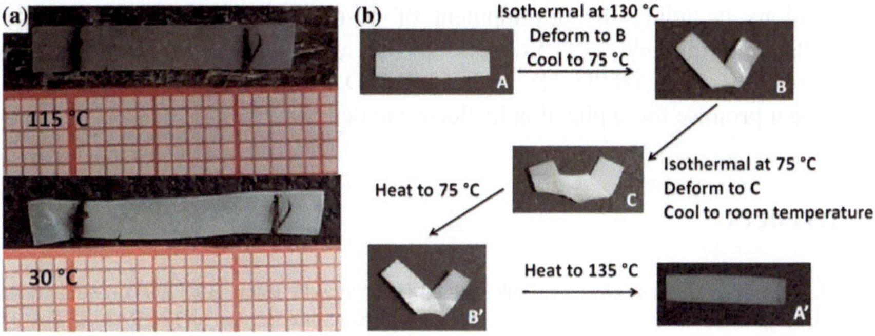

Fig. 20 The change of natural length of a free-standing monodomain LCE/SiO_2 composites (5 wt% SNP content) between the liquid-crystal phase (30 °C) and the isotropic phase (115 °C). **b** Triple shape memory of a nanocomposites (5 wt% SNP content) sample [16]. Copyright 2014. Reprinted with permission from MDPI

As shown in Fig. 20, nanocomposite of SNP changes length with the change of temperature, and isotropic phase of tested sample at 115 °C elongated uniaxially when it cooled down to room temperature (30 °C). Triple shape memory also observed due to at two different temperature stage related to liquid crystalline-isotropic phase transition temperature and glass transition temperature. Flat shape of actuator deformed to a V-shape when temperature increased to 130 °C (same shape retained until cooled down to 75 °C), another shape change found when it cooled down at 75 °C which remain fixed until room temperature. Triple shape memory can be revised to original flat shape during heating process.

4 Conclusions

In this chapter, we reviewed nanomaterials including NPs- and NTs-incorporated LCEs composite materials synthesis, characterization, and specific devices application, which could be a good reference chapter for future research on LCEs nanocomposite. The hybrid of 0D and 1D nanomaterials and LCEs shows tremendous advantages, such as fast response/recovery and low operating temperature, that could be ideal for the development of the next generation of optomechanical devices.

Different ways of synthesis and characterization of LCEs composite have been established to optimize performance of developed devices. Excellent properties of mechanical flexibility and durability can be achieved by incorporation of NPs in LCEs for devices application. Moreover, triggering the phase transition is only possible by the incorporation of nanomaterials in the application of different external stimuli of white, UV, IR light, electric, magnetic field, external heat, etc. However, there are still challenges that need to be overcome for future practical applications including the development of cost-effective, scalable productions methods, and reliability of products. If the aforementioned issues are fully addressed, NPs included 0D, 1D, and recently 2D (graphene) embedded LCEs will hold great promise for application in electronic devices.

References

1. Thomsen DL, Keller P, Naciri J, Pink R, Jeon H, Shenoy D, Ratna BR (2001) Liquid crystal elastomers with mechanical properties of a muscle. Macromolecules 34(17):5868–5875
2. Shenoy DK, Thomsen DL III, Srinivasan A, Keller P, Ratna BR (2002) Carbon coated liquid crystal elastomer film for artificial muscle applications. Sens Actuators A 96(2–3):184–188
3. Li MH, Keller P (2006) Artificial muscles based on liquid crystal elastomers. Philos Trans R Soc A 364:2763–2777
4. Li Y, Pruitt C, Rios O, Wei L, Rock M, Keum JK, McDonald AG, Kessler MR (2015) Controlled shape memory behavior of a smectic main-chain liquid crystalline elastomer. Macromolecules. doi:10.1021/acs.macromol.5b00519
5. Liu C, Qin H, Mather PT (2007) Review of progress in shape-memory polymers. J Mater Chem 17(16):1543–1558
6. Warner M, Bladon P, Terentjev EM (1994) Soft elasticity—deformation without resistance in liquid-crystal elastomers. J Phys II 4(1):93–102
7. Tajbakhsh AR, Terentjev EM (2001) Spontaneous thermal expansion of nematic elastomers. Eur Phys J E 6(2):181–188
8. Li M-H, Keller P, Li B, Wang X, Brunet M (2003) Light-driven side-on nematic elastomer actuators. Adv Mater 15(7–8):569–572
9. Sawa Y, Urayama K, Takigawa T, DeSimone A, Teresi L (2010) Thermally driven giant bending of liquid crystal elastomer films with hybrid alignment. Macromolecules 43(9):4362–4369
10. Yamada M, Kondo M, Mamiya J-I, Yu Y, Kinoshita M, Barrett CJ, Ikeda T (2008) Photomobile polymer materials: towards light-driven plastic motors. AngewandteChemie Int Ed 47(27):4986–4988
11. Sánchez-Ferrer A, Fischl T, Stubenrauch M, Wurmus H, Hoffmann M, Finkelmann H (2009) Photo-crosslinked side-chain liquid-crystalline elastomers for microsystems. Macromol Chem Phys 210(20):1671–1677
12. Sánchez-Ferrer A, Fischl T, Stubenrauch M, Albrecht A, Wurmus H, Hoffmann M, Finkelmann H (2011) Liquid-crystalline elastomer microvalve for microfluidics. Advanced Materials 23(39):4526–4530
13. Finkelmann H, Kim ST, Muñoz S, Palffy-Muhoray P, Taheri B (2001) Tunable mirrorless lasing in cholesteric liquid crystalline elastomers. Adv Mater 13(14):1069–1072
14. Camargo CJ, Campanella H, Marshall JE, Torras N, Zinoviev K, Terentjev EM, Esteve J (2012) Batch fabrication of optical actuators using nanotube–elastomer composites towards refreshable Braille displays. J Micromech Microeng 22(7):075009 (9 pp). doi:10.1088/0960-1317/22/7/075009.15
15. Torras N, Zinoviev KE, Marshall JE, Terentjev EM, Esteve J (2011) Bending kinetics of a photo-actuating nematic elastomer cantilever. Appl Phys Lett 99(25):254102
16. Li Z, Yang Y, Qin B, Zhang X, Tao L, Wei Y, Ji Y (2014) Liquid crystalline network composites reinforced by silica nanoparticles. Materials 7(7):5356–5365
17. Haberl JM, Sánchez-Ferrer A, Mihut AM, Dietsch H, Hirt AM, Mezzenga R (2013) Liquid-crystalline elastomer-nanoparticle hybridswith reversible switch of magnetic memory. Adv Mater 25(12):1787–1791

18. Domenici V, Zupančič B, Laguta VV, Belous AG, V'yunov OI, Remškar M, Zalar B (2010) $PbTiO_3$ nanoparticles embedded in a liquid crystalline elastomer matrix: structural and ordering properties. J Phys Chem C114(24):10782–10789
19. Domenici V, Conradi M, Remškar M, Viršek M, Zupančič B, Mrzel A, Chambers M, Zalar B (2011) New composite films based on MoO_{3-x} nanowires aligned in a liquid single crystal elastomer matrix. J Mater Sci 46(10):3639–3645
20. Haberl JM, Sánchez-Ferrer A, Mihut AM, Dietsch H, Hirtc AM, Mezzenga R (2013) Strain-induced macroscopic magnetic anisotropy from smectic liquid-crystalline elastomer–maghemite nanoparticle hybrid nanocomposites. Nanoscale 5(12):5539–5548
21. Ji Y, Marshall JE, Terentjev EM (2012) Nanoparticle-liquid crystalline elastomer composites. Polymers 4(1):316–340
22. Chambers M, Finkelmann H, Remskar M, Sanchez-Ferrer A, Zalar B, Zumer S (2009) Liquid crystal elastomer–nanoparticle systems for actuation. J Mater Chem 19(11):1524–1531
23. Li C, Liu Y, Huang X, Jiang H (2012) Direct sun-driven artificial heliotropism for solar energy harvesting based on a photo-thermomechanical liquid-crystal elastomer nanocomposite. Adv Funct Mater 22(24):5166–5174
24. Gascon I, Marty JD, Gharsa T, Mingotaud C (2005) Formation of gold nanoparticles in a side-chain liquid crystalline network: influence of the structure and macroscopic order of the material. Chem Mater 17(21):5228–5230
25. Shandryuk GA, Matukhina EV, Vasil'ev RB, Rebrov A, Bondarenko GN, Merekalov AS, Gas'kvo AM Talroze RV (2008) Effect of H-bonded liquid crystal polymers on CdSe quantum dot alignment within nanocomposite. Macromolecules 41(6):2178–2185
26. Song HM, Kim JC, Hong JH, Lee YB, Choi J, Lee JI, Kim WS, Kim JH, Hur NH (2007) Magnetic and transparent composites by linking liquid crystals to ferrite nanoparticles through covalent networks. Adv Funct Mater 17(13):2070–2076
27. Kaiser A, Winkler M, Krause S, Finkelmann H, Schmidt AM (2009) Magnetoactive liquid crystal elastomer nanocomposites. J Mater Chem 19(4):538–543
28. Montazami R, Spillmanno CM, Naciri J, Ratna BR (2012) Enhanced thermomechanical properties of a nematic liquid crystal elastomer doped with gold nanoparticles. Sens Actuators A 178:175–178
29. Chambers M, Zalar B, Remškar M, Žumer S, Finkelmann H (2006) Actuation of liquid crystal elastomers reprocessed with carbon nanoparticles. Appl Phys Lett 89(24):243116
30. Willets KA, Duyne RPV (2007) Localized surface plasmon resonance spectroscopy and sensing. Annu Rev Phys Chem 58:267–297
31. Murray WA, Barnes WL (2007) Plasmonic materials. Adv Mater 19(22):3771–3782
32. Cao L, Barsic DN, Guichard AR, Brongersma ML (2007) Plasmon-assisted local temperature control to pattern individual semiconductor nanowires and carbon nanotubes. Nano Lett 7(11):3523–3527
33. Huang X, El-Sayed IH, Qian W, El-Sayed MA (2006) Cancer cell imaging and photothermal therapy in the near-infrared region by using gold nanorods. J Am Chem Soc 128(6):2115–2120
34. Sun Y, Evans JS, Lee T, Senyuk B, Keller P, He S, Smalyukh II (2012) Optical manipulation of shape-morphing elastomeric liquid crystal microparticles doped with gold nanocrystals. Appl Phys Lett 100(24):241901
35. Liu X, Wei R, Hoang PT, Wang X, Liu T, Keller P (2015) Reversible and rapid laser actuation of liquid crystalline elastomer micropillars with inclusion of gold nanoparticles. Adv Funct Mater. doi:10.1002/adfm.201500443
36. Park J, An K, Hwang Y, Park J-G, Noh H-J, Kim J-Y, Park J-H, Hwang N-M, Hyeon T (2004) Ultra-large-scale syntheses of monodispersenanocrystals. Nat Mater 3:891–895
37. Lu AH, Salabas EL, Schuth F (2007) Magnetic nanoparticles: synthesis, protection, functionalization, and application. AngewandteChemie Int Ed 46(8):1222–1244
38. Garcia-Marquez A, Demortiere A, Heinrich B, Guillon D, Begin-Colin S, Donnio B (2011) Iron oxide nanoparticle-containing main-chain liquid crystalline elastomer: towards soft magnetoactive networks. J Mater Chem 21(25):8994–8996

39. Winkler M, Kaiser A, Krause S, Finkelmann H, Schmidt AM (2010) Liquid crystal elastomers with magnetic actuation. Macromol Symp 291–292(1):186–192
40. Iijima S (1991) Helical microtubules of graphitic carbon. Nature 354:56–58
41. Kim SN, Rusling JF, Papadimitrakopoulos F (2007) Carbon nanotubes for electronic and electrochemical detection of biomolecules. Adv Mater 19(20):3214–3228
42. Zhou W, Bai X, Wang E, Xie S (2009) Synthesis, structure, and properties of single-walled carbon nanotubes. Adv Mater 21(45):4565–4583
43. Hamon MA, Itkis ME, Niyogi S, Alvaraez T, Kuper C, Menon M, Haddon RC (2001) Effect of rehybridization on the electronic structure of single-walled carbon nanotubes. J Am Chem Soc 123(45):11292–11293
44. Yang L, Setyowati K, Li A, Gong S, Chen J (2008) Reversible infrared actuation of carbon nanotube-liquid crystalline elastomer nanocomposites. Adv Mater 20(12):2271–2275
45. Chen J, Hamon MA, Hu H, Chen Y, Rao AM, Eklund PC, Haddon RC (1998) Solution properties of single-walled carbon nanotubes. Science 282(5386):95–98
46. Marshall JE, Ji Y, Torras N, Zinoviev K, Terentjev EM (2012) Carbon-nanotube sensitized nematic elastomer composites for IR-visible photo-actuation. Soft Matter 8(5):1570–1574
47. Li C, Liu Y, Lo C-W, Jiang H (2011) Reversible white-light actuation of carbon nanotube incorporated liquid crystalline elastomer nanocomposites. Soft Matter 7(16):7511–7516
48. Courty S, Mine J, Tajbakhsh AR, Terentjev EM (2003) Nematic elastomers with aligned carbon nanotubes: new electromechanical actuators. Europhys Lett 64(5):654–660
49. Koshio A, Yudasaka Zhang M, Iijima S (2001) A simple way to chemically react single-wall carbon nanotubes with organic materials using ultrasonication. Nano Lett 1(7):361–363
50. Paredes JI, Burghard M (2004) Dispersions of individual single-walled carbon nanotubes of high length. Langmuir 20(12):5149–5152
51. Huang YY, Terentjev EM (2010) Tailoring the electrical properties of carbon nanotube-polymer composites. Adv Funct Mater 20(23):4062–4068
52. JiY Huang YY, Tajbakhsh AR, Terentjev EM (2009) Polysiloxane surfactants for the dispersion of carbon nanotubes in nonpolar organic solvents. Langmuir 25(20):12325–12331
53. Basu R (2013) Effect of carbon nanotubes on the field-induced nematic switching. Appl Phys Lett 103(24):241906
54. Gu W, Zhu X, Futai N, Cho BS, Takayama S (2004) Computerized microfluidic cell culture using elastomeric channels and Braille displays. Proc Natl Acad Sci 101(45):15861–15866
55. Camargo CJ, Campanella H, Marshall JE, Torras N, Zinoviev K, Terentjev EM, Esteve J (2011) Localised actuation in composites containing carbon nanotubes and liquid crystalline elastomers. Macromol Rapid Commun 32(24):1953–1959
56. Camargo CJ, Torras N, Campanella H, Comrie JE, Campo EM, Zinoviev K, Terentjev EM, Esteve J (2011) Light-actuated CNT-doped elastomer blisters towards braille dots. In: 16th international, solid-state sensors, actuators and microsystems conference (TRANSDUCERS), pp 1594—1597. doi:10.1109/TRANSDUCERS.2011.5969797
57. Torrasa N, Zinovieva KE, Camargoa CJ et al (2014) Tactile device based on opto-mechanical actuation of liquid crystal elastomers. Sens Actuators A 208:104–112
58. Kohlmeyer RR, Chen J (2013) Wavelength-selective, IR light-driven hinges based on liquid crystalline elastomer composites. AngewandteChemie 52(35):9234–9237
59. Maeda S, Hara Y, Sakai T, Yoshida R, Hashimoto S (2007) Self-walking gel. Adv Mater 19(21):3480–3484
60. Finkelmann H, Shahinpoor M (2002) Electrically controllable liquid crystal elastomer-graphite composite artificial muscles. In: Proceedings of SPIE 4695, smart structures and materials 2002: electroactive polymer actuators and devices (EAPAD). doi:10.1117/12.475190
61. Chen J, Ramasubramaniam R, Liu H (2004) Carbon nanotube-induced planarization of conjugated polymers in solution. In: Proceedings of materials research society symposium, vol 858. doi:http://dx.doi.org/10.1557/PROC-858-HH12.4
62. Ji Y, Huang YY, Rungsawang R, Terentjev EM (2010) Dispersion and alignment of carbon nanotubes in liquid crystalline polymers and elastomers. Adv Mater 22(31):3436–3440

MIX
Papier aus verantwortungsvollen Quellen
Paper from responsible sources
FSC® C105338

If you have any concerns about our products,
you can contact us on
ProductSafety@springernature.com

In case Publisher is established outside the EU,
the EU authorized representative is:
Springer Nature Customer Service Center GmbH
Europaplatz 3, 69115 Heidelberg, Germany

Printed by Libri Plureos GmbH
in Hamburg, Germany